Human-Environment Interactions

Human-Environment Interactions

VOLUME 1

For further volumes:
http://www.springer.com/series/8599

Eduardo S. Brondízio • Emilio F. Moran
Editors

Human-Environment Interactions

Current and Future Directions

Springer

Editors
Eduardo S. Brondízio
Department of Anthropology
Anthropological Center for Training
and Research on Global Environmental
Change
School of Public and Environmental Affairs
Indiana University
701 E. Kirkwood Avenue
Student Building 130
Bloomington, IN 47405
USA

Emilio F. Moran
Department of Anthropology
Anthropological Center for Training
and Research on Global Environmental
Change
School of Public and Environmental Affairs
Indiana University
701 E. Kirkwood Avenue
Student Building 130
Bloomington, IN 47405
USA

ISBN 978-94-007-4779-1 ISBN 978-94-007-4780-7 (eBook)
DOI 10.1007/978-94-007-4780-7
Springer Dordrecht Heidelberg New York London

Library of Congress Control Number: 2012952490

Printed on acid-free paper

Springer is part of Springer Science+Business Media (www.springer.com)

Eduardo Brondízio and Emilio Moran wish to dedicate this book to their dear colleague and friend **Elinor Ostrom**, *who was kind enough to read the manuscript and write the Foreword despite her illness. She passed away on June 12, 2012. She was an inspiration to us and to everyone who met her. Lin will be profoundly missed. She left a legacy that will be lasting, and her contributions to human-environment interactions are evident in this book and in the work of hundreds of scholars throughout the world.*

Foreword

Eduardo Brondízio and Emilio Moran have both contributed to the study of human-environment interactions (HEI) with their own in-depth studies of these interactions in a variety of settings. They have now edited a path-breaking volume that addresses human-environment interactions in a fascinating and cumulative manner.

In their introduction to the volume, they provide an innovative and useful analysis of three broad themes that help readers gain a useful overview of the variety of relevant theories related to human-environment interactions: environmental determinism, cultural determinism, and human-environment interaction. The first two are unidirectional and pose either the environment or culture as the dominant factor. The third poses human-environment interactions as the result of extensive interactions between the environment and the humans living in it. Reading Brondízio and Moran's overview of these three approaches is very useful for researchers like me, who have read some of the earlier scholars they analyzed without a recognition of these underlying contradictory approaches. Brondízio and Moran's analysis provides a firm foundation for the excellent chapters that follow their introduction.

The 17 chapters in this volume are largely written by younger scholars who come from multiple countries. All of them take complex views of the basic relationships between ecological and social factors affecting human-environment interactions over time. In the first part on "Health and Adaptation Approaches," the seven authors of three chapters focus on over time health affecting the Xavante Indians from Central Brazil, residents living in rural Tibet, and the subtle relationships between humans and wildlife interactions and emerging infectious diseases. In the second part on "Land Change and Land scape Management Approaches," the first of four chapters looks at change in natural resource management in Scotland while the other three compare agroforestry or forestry systems in India, the United States, and Africa.

The third part is devoted to "Institutions and Political Ecology Approaches." As Brondízio and Moran stress in their introduction, institutional approaches are an important current development that make a strong emphasis on human-environment interactions. In this part, scholars are focusing on interactions in the Lower Amazonian Floodplain, in the Amazon, in western Honduras, and in fast-growing communities in the United States and related counties, looking at the urban-rural fringe. Chapter 10 is an experimental chapter that carefully studies whether having a regulation by itself or the size of the penalty imposed is the most important in convincing people to conserve natural resources.

Part IV focuses on "Historical and Archeological Approaches" and addresses some of the key issues that dominated the literature at an earlier juncture, including the adaptable agroforestry systems of Japanese immigrants in Brazil, looking at the intensification of agroforestry systems in Uganda from 1890 to 1995, and whether agriculture was a key productive activity in pre-colonial Amazonia. In the last part on "Future Directions," Eduardo Brondízio and Rinku Roy Chowdhury look at future directions in human-environment interaction research, drawing on the chapters in this volume and amending some of the approaches taken in traditional work by scholars who focus on one and only one discipline to explain human-environment interactions.

The overall quality of this volume is so high that I cannot recommend a particular part to the reader. I think all chapters in this volume are worth a thorough reading. I hope to be able to use many of these chapters in my own graduate seminar on institutional analysis, as there is so much to learn from a careful reading of this book.

By Elinor Ostrom

About the Authors

Sanchayeeta Adhikari is the Berg postdoctoral fellow in the Geography Department at Macalester College and teaches environmental remote sensing and GIS. She was born in India and completed master degrees at Banaras Hindu University and the Indian Institute of Technology, with respective majors in geography, specializing in regional planning, and planning and development. She completed her PhD in geography with a minor in natural resource management at the University of Florida, where her work focused on studying land change science within the broader realm of human-environment interactions using spatial analysis tools such as remote sensing, GIS, and spatial statistics. She continues to work on issues relating to conservation policies, forest ecology and land use, and land-cover change.

Mateus Batistella was born in São Paulo, Brazil, and graduated in biology at the University of São Paulo and in philosophy at the Catholic University of São Paulo. He has a master degree in ecology from the University of São Paulo and a PhD in environmental science from Indiana University (USA). He works with landscape ecology, land-use planning, and geotechnologies coordinating projects in distinct Brazilian sites. Mateus is currently the director of Embrapa Satellite Monitoring, professor at State University of Campinas, and a visiting research associate at the Anthropological Center for Training and Research on Global Environmental Change at Indiana University, USA.

Édson Luis Bolfe received his PhD from the Department of Geography of Campinas State University in 2010. He is currently a researcher at the Brazilian Agricultural Research Corporation—Embrapa in Campinas, Brazil, and assistant director for research and development at Embrapa Satellite Monitoring. His main activities include research in development of geotechnologies for identifying and monitoring agroforestry systems, monitoring mechanisms for greenhouse gas emissions, and WebGIS for technical applications in mapping of tropical agriculture.

Eduardo S. Brondízio is professor of anthropology in the Department of Anthropology, where he served as chair from 2005 to 2012, and adjunct professor of environmental sciences at the School of Public and Environmental Affairs at Indiana University–Bloomington. He is the associate director of the Anthropological Center for Training and Research on Global Environmental Change and affiliated faculty of the Vincent and Elinor Ostrom Workshop in Political Theory and Policy Analysis at Indiana University. For over two decades, his work has been centered in the Brazilian Amazon studying (longitudinally) the formation and transformation of rural families and communities as they interact with government policies and development programs, regional and global markets, demographic dynamics, and environmental change. He has written extensively on environmental anthropology and human-environment interaction, households and land-use change, Amazonian small farmers and global markets, rural development and peasantry identity, and geospatial analysis and integrative multilevel methodologies. He is a member of the Scientific Committee of the International Geosphere Biosphere Program. He is also involved with several collaborative and comparative international research programs examining human dimensions of global environmental change, sociocultural dimensions of ecosystem services, and efforts to develop collaborative frameworks for social-ecological systems research.

Lin Cassidy works at the University of Botswana's Okavango Research Institute focusing on the spatial aspects of governance. She has a PhD in geography from the University of Florida. Her general research interests are in human-environment interactions, social-ecological systems, and integrated conservation and development. Specifically, she studies political ecology, land-use/land-cover change, and community-based natural resources management.

Geoff Childs received a PhD from Indiana University's Departments of Anthropology and Central Eurasian Studies in 1998. Currently associate professor of anthropology at Washington University in St. Louis, he specializes in the study of Tibetan demographic processes, economic changes, and household-level transformations. He has conducted research in the Tibet Autonomous Region of China, Tibetan exile settlements in India, and Nepal's ethnically Tibetan enclaves of Nubri and Tsum. He is the author of *Tibetan Transitions: Historical and Contemporary Perspectives on Fertility and Family Planning* (Brill, 2008) as well as numerous articles dealing with demography, anthropology, development, and culture.

Carlos E. A. Coimbra Jr. received his PhD from the Department of Anthropology at Indiana University in 1990 and postdoctoral training at the University of Massachusetts medical anthropology program. He is currently a professor of medical anthropology and public health at the National School of Public Health, Oswaldo Cruz Foundation, in Rio de Janeiro. His principal research interests include human ecology, health, and nutrition of indigenous peoples. He has received research grants from Wenner Gren and Ford foundations and the Brazilian National Research Council. He has published many papers in anthropological and medical journals and coauthored the book *The Xavante in Transition*, which was awarded the best interdisciplinary book prize by the General Anthropology Division of the American Anthropological Association in 2003.

Fabio de Castro (PhD environmental sciences, Indiana University) is assistant professor in human ecology and Brazilian studies at the Center for Latin American Research and Documentation in Amsterdam, the Netherlands. He is also a collaborating researcher at the Anthropological Center for Training and Research on Global Environmental Change, Indiana University, and at the Centre for Maritime Research, University of Amsterdam.

Stefano Fiorini is an adjunct assistant professor of anthropology and associated faculty in the Anthropological Center for Training and Research on Global Environmental Change (ACT) at Indiana University–Bloomington, where he received his PhD in social/cultural anthropology. As a research scientist with the James Hutton Institute in Aberdeen, Scotland, he developed an interest in collaborative and participatory approaches to the management of natural resources that continues today.

Burnell "Burney" C. Fischer received his PhD from the Department of Forestry and Natural Resources, Purdue University. He is currently clinical professor, School of Public and Environmental Affairs, Indiana University, and adjunct professor of forestry, Department of Forestry and Natural Resources, Purdue University. He is also codirector of the Vincent and Elinor Ostrom Workshop in Political Theory and Policy Analysis. He was assistant professor of forestry at the University of Massachusetts, 1974–1977, professor and extension forestry specialist at Purdue University, 1977–1990, and state forester and director of the Division of Forestry, Indiana Department of Natural Resources, 1990–2005.

Célia Futemma received her MA in cultural anthropology from Tulane University and PhD from the School of Public and Environmental Affairs at Indiana University. She is a professor at the Center for Research and Studies on Environment at the University of Campinas in Brazil. She received her postdoctoral training at the University of São Paulo and at the Anthropological Center for Training and Research on Global Environmental Change at Indiana University. She has conducted fieldwork in the Amazon and Atlantic Rainforests, working with small-scale farmers with diversified sociocultural backgrounds. Her expertise lies in environmental anthropology, land-use and land-cover dynamics, people and parks, and common-pool resources.

Cerian Gibbes is currently an assistant professor in the Department of Geography and Environmental Studies at the University of Colorado at Colorado Springs. She earned a PhD in geography (2011) from the University of Florida. Her research utilizes land-use/land-cover change modeling, remote sensing, and spatial statistics to expand knowledge regarding the interplay among climate, land, and people. Her work has been geographically focused on the tropics in Latin America, the Caribbean, and Africa with an emphasis on semiarid systems. Her research has been supported by funding from the National Aeronautics and Space Administration, National Science Foundation, and the Social Science Research Council.

Melvyn C. Goldstein is John Reynolds Harkness professor of anthropology and codirector of the Center for Research on Tibet at Case Western Reserve University. He is a member of the US National Academy of Sciences. He has conducted extensive research in Tibet over the past 24 years on topics that include pastoral nomadism, monasticism, modern Tibetan history, oral history, and the impact of recent reforms and modernization on life in rural Tibet. He is author or coauthor of more than 100 articles and books on Tibet, his latest books being *A Tibetan Revolutionary: The Political Life and Times of Bapa Phuntso Wangye* (U. of California Press, 2004), *A History of Modern Tibet, Volume Two, 1951–1955* (U. of California Press, 2007), and *On the Cultural Revolution in Tibet* (U. of California Press, 2009).

Joel Hartter received his PhD from the Department of Geography at the University of Florida in 2007. He is currently an assistant professor and Roland H. O'Neil Professor in the Department of Geography and a faculty fellow of the Carsey Institute at the University of New Hampshire. His work focuses on human-environment impacts in protected area landscapes. His primary geographic focus is the Albertine Rift, where he has been working since 2005. While at the University of Florida (2004–2007), he was a US Department of Education Foreign Language and Area Studies (FLAS) fellow.

Maria Claudia Lopez received her PhD in resource economics from the University of Massachusetts–Amherst in 2008 with funding from a Fulbright/Laspau scholarship. Then she completed a 2-year postdoctoral fellowship at the Workshop in Political Theory and Policy Analysis at Indiana University. She is currently an assistant professor at the School of Environmental and Rural Studies of the Pontificia Universidad Javeriana, Bogota, Colombia. For more information go to http://mariaclaudialopez.jimdo.com/.

Pinki Mondal received her PhD from the Department of Geography of the University of Florida in 2011. She is currently a postdoctoral research scientist at the Department of Ecology, Evolution and Environmental Biology of Columbia University. She has also worked as a postdoctoral research associate at the Department of Environmental Conservation of the University of Massachusetts–Amherst. Her research broadly examines nature-society interactions at multiple spatial and temporal scales using geospatial techniques. She is interested in a variety of interdisciplinary topics, including mapping and monitoring forested ecosystems and agroecosystems, climate variability, protected areas, impacts of policies on land-use/land-cover change, and impacts of land-use change on biodiversity. She has been actively involved in research projects funded through the National Aeronautics and Space Administration and the U.S. Forest Service.

Emilio F. Moran is distinguished professor and the James H. Rudy professor of anthropology at Indiana University, professor of environmental sciences, adjunct professor of geography, and director of the Anthropological Center for Training and Research on Global Environmental Change at Indiana University. He is the author of 10 books, 14 edited volumes, and more than 150 journal articles and book chapters. His research has been supported by NSF, NIH, NOAA, and NASA for the past two decades. His three latest books, *Environmental Social Science* (Wiley/Blackwell 2010), *People and Nature* (Blackwell 2006), and *Human Adaptability* 3rd edition (Westview 2007), address broader issues of human interaction with the environment under conditions of change. His most recent book, *Meio Ambiente & Florestas* (Editora SENAC Sao Paulo 2010), addresses the value of forests in Brazil and the world. His book *Developing the Amazon* (Indiana University Press 1981) was the first book-length study of the human and environmental impacts of the Transamazon Highway. He is a fellow of the Linnean Society of London, fellow of the American Anthropological Association and the Society for Applied Anthropology, and fellow of the American Association for the Advancement of Science and was elected to the National Academy of Sciences in 2010.

Michael P. Muehlenbein received MsPH in tropical medicine and biostatistics from Tulane University and an MPhil and a PhD in biological anthropology from Yale University. He is an associate professor of anthropology, core faculty member of the Program in International Studies, and associate faculty of the Anthropological Center for Training and Research on Global Environmental Change and the Center for the Integrative Study of Animal Behavior at Indiana University. He is an associate editor for several journals and the Cambridge Studies in Biological and Evolutionary Anthropology series. His primary research interests include emerging infectious diseases associated with human-wildlife contact, particularly as a result of tourism/travel and land-use change.

Darla K. Munroe received her PhD from the Department of Geography at the University of Illinois in 2000. She is currently associate professor of geography at the Ohio State University. Her research focuses on the impacts of globalization and economic restructuring on land-use change in agricultural and forested environments, particularly at the urban-rural interface. She was a postdoctoral scholar at the Center for the Study of Institutions, Population, and Environmental Change at Indiana University. In 2011–2012, she was a Fulbright scholar at the Geography Institute of Humboldt University in Berlin, Germany.

Harini Nagendra obtained her PhD in ecological sciences at the Indian Institute of Science in 1998. She is currently a Ramanujan fellow at the Ashoka Trust for Research in Ecology and the Environment, Bangalore, India, and Asia research coordinator at the Center for the Study of Institutions, Population, and Environmental Change, Indiana University. She was a Society in Science Branco Weiss fellow at ETH Zurich and has conducted postdoctoral research at the University of California San Diego and Indiana University. Her research examines how human-environment interactions impact conservation, biodiversity, and sustainability in forested and urban contexts in South Asia.

Eduardo Góes Neves has a BA in history from the University of São Paulo and a PhD in archaeology from Indiana University. He is currently a professor in the Graduate Program in Archaeology at the Museum of Archaeology and Ethnology and coordinator of the Laboratory of the Archaeology of the Tropics at the University of São Paulo, Brazil. His focus of research is the understanding of the long-term relationship between people and nature in the humid tropics, especially in the Amazon basin. He is a former president of the Brazilian Archaeological Society and a current member of the Board of Directors of the Society for American Archaeology.

Elinor Ostrom was a university distinguished professor, Arthur F. Bentley professor of political science, and co-founder and senior research director of The Vincent and Elinor Ostrom Workshop in Political Theory and Policy Analysis at Indiana University. She was also founding director of the Center for the Study of Institutional Diversity at Arizona State University. She was a member of the American Academy of Arts and Sciences, the National Academy of Sciences, and the American Philosophical Society and a recipient of the Sveriges Riksbank Prize in Economic Sciences in Memory of Alfred Nobel (2009). Her authored and coauthored books include *Governing the Commons* (1990), *Rules, Games, and Common-Pool Resources* (1994), *Trust and Reciprocity* (2003), *The Commons in the New Millennium* (2003), *Understanding Institutional Diversity* (2005), and *Working Together* (2010). Elinor passed away on June 12, 2012, and will be greatly missed by everyone.

Rinku Roy Chowdhury is assistant professor of geography, Indiana University. She received a PhD in geography from Clark University. Her research focuses on institutional, ecological, and spatial dynamics of forest-agricultural mosaics, urban landscapes, and coastal mangroves in the USA, Mexico, Bangladesh, and parts of Central and South America. Her current funded research includes NASA and NSF projects on land change science, long-term ecological research, and socio ecosystem vulnerability to climate and anthropogenic change. She integrates interdisciplinary theoretical, methodological, and field approaches to investigate the interplay of institutional structures and local agency in shaping landscapes and livelihoods.

Tatyana B. Ruseva received her PhD from the School of Public and Environmental Affairs at Indiana University in 2010. She is an assistant professor of environmental policy and politics at the Department of Government and Justice Studies at Appalachian State University. She was a graduate student and research associate with the Center for the Study of Institutions, Population, and Environmental Change, Indiana University, an interdisciplinary research center focused on long-term, systematic studies of deforestation and afforestation. Her research explores the role of social networks and collaborative management in private forestry and natural resource governance.

Ricardo Ventura Santos received his PhD from the Department of Anthropology, Indiana University, in 1991. He is a professor at the National School of Public Health, Oswaldo Cruz Foundation, and an associate professor of anthropology at the National Museum in Rio de Janeiro. He received his postdoctoral training at MIT and the University of Massachusetts. He has published many papers in anthropological and medical journals and coauthored *The Xavante in Transition* (University of Michigan Press, 2002), *Demografia dos Povos Indígenas no Brasil* (Editora Fiocruz, 2005), *Racial Identities, Genetic Ancestry, and Health in South America* (Palgrave, 2011), and *The Biological Anthropology of Living Human Populations* (special issue of *Current Anthropology*, 2012).

Jane Southworth is a professor in the Department of Geography, Land Use and Environmental Change Institute, and Florida Climate Institute at the University of Florida. Her research interests are based on the study of human-environment interactions within the field of land change science, more specifically remote sensing of land use, land-cover change, and land-change modeling; the implications of scale and scaling in remote sensing and modeling analyses; people and parks; and modeling of the impacts of climate change on human-environment systems.

Catherine M. Tucker is an associate professor of anthropology at Indiana University and affiliated faculty with the School of Public and Environmental Affairs, the Vincent and Elinor Ostrom Workshop in Political Theory and Policy Analysis, and the Anthropological Center for Training and Research on Global Environmental Change. She received her PhD from the University of Arizona in 1996. Her recent publications include *Nature, Science and Religion: Intersections Shaping Society and the Environment* (edited volume, SAR Press, 2012), *Coffee Culture: Local Experiences, Global Connections* (Routledge 2010), and articles in *Global Environmental Change*, *Ecology and Society*, *Human Ecology*, and *Human Organization*.

Nathan D. Vogt is a postdoctoral fellow at the National Institute of Space Research in Brazil investigating the role of institutions and institutional arrangements with land-use and land-cover change on Amazonian frontiers. He also continues to participate in research of the adaptation of livelihoods and resource-use systems of traditional populations in the Amazon estuary to global market demands and change in tidal and rainfall regimes. He received his PhD in environmental science, with a minor in human dimensions of global environmental change from the School of Public and Environmental Affairs at Indiana University in 2005, where he was a fellow at the Center for the Study of Institutions, Population, and Environmental Change.

Puchung Wangdui is a researcher at the Tibet Academy of Social Sciences in Lhasa. He holds an MA degree in anthropology from Case Western Reserve University and has conducted extensive research on change and modernization in rural Tibet.

James R. Welch received his PhD from the Department of Anthropology at Tulane University in 2009 and is currently a researcher at the National School of Public Health, Oswaldo Cruz Foundation, in Rio de Janeiro. He is coeditor of *Ethnobiology Letters* and is author of the upcoming book *Sprouting Valley: Historical Ethnobotany of the Northern Pomo from Potter Valley, California*. His research has been funded by a Fulbright-Hays Fellowship, the National Indian Foundation (Brazil), and the Brazilian National Research Council. His research interests include the interface between culture, the environment, and health among indigenous peoples in South America and California.

Abigail M. York has a joint PhD in public policy from the School of Public and Environmental Affairs and the Department of Political Science at Indiana

University. She is currently an assistant professor of governance and public policy at the School of Human Evolution and Social Change, a senior sustainability scientist in the Global Institute of Sustainability, and an affiliated faculty member in the Center for the Study of Institutional Diversity at Arizona State University. Her research focuses on governance and institutional dynamics within a social-ecological system.

Acknowledgments

This edited volume began as a workshop held at Indiana University in February 2010. The idea was to bring together a set of young to mid-career scholars to share their latest thinking of human-environment interactions research. The discussion was lively and productive—and led to invitations to other scholars who were unable to come to that workshop to join this volume and ensure that the coverage of the book was more complete both geographically and conceptually. The editors thank the authors for their patience and their excellent papers.

The workshop was made possible by financial support from Indiana University's Center for Research on Environmental Science (CRES) and the Anthropological Center for Training and Research on Global Environmental Change (ACT). The arrangements for the workshop were ably handled by ACT's Assistant Director for Administration Linda Day and her assistant Kelsey Scroggins. Without their attention to detail and their responsiveness to the invitees' needs, the intellectual excitement felt at the meeting would have been constrained by concerns with other practical issues. We also wish to recognize the enormous contribution of Joanna Broderick, who worked with individual authors, edited the whole volume for clarity and grammar, and ensured that citations and references were consistent. Editors Fritz Schmuhl and Takeesha Moerland-Torpey at Springer were encouraging from start to end in moving the manuscript forward through the review and production process.

Acknowledgments

Contents

1 **Introduction to Human-Environment Interactions Research** 1
Emilio F. Moran and Eduardo S. Brondízio

Part I Health and Adaptation Approaches

2 **A Half-Century Portrait: Health Transition in the Xavante Indians from Central Brazil** 29
Ricardo Ventura Santos, Carlos E.A. Coimbra, Jr., and James R. Welch

3 **Balancing People, Policies, and Resources in Rural Tibet** 53
Geoff Childs, Melvyn C. Goldstein, and Puchung Wangdui

4 **Human-Wildlife Contact and Emerging Infectious Diseases** 79
Michael P. Muehlenbein

Part II Land Change and Landscape Management Approaches

5 **Change in Natural Resource Management: An Experiment with "Participatory GIS"** 97
Stefano Fiorini

6 **Peopled Parks: Forest Change in India's Protected Landscapes**...... 113
Harini Nagendra, Pinki Mondal, Sanchayeeta Adhikari, and Jane Southworth

7 **Public-Private Interactions in the Conservation of Private Forests in the United States**... 141
Tatyana B. Ruseva and Burnell C. Fischer

8 The Monitoring of Land-Cover Change and Management across Gradient Landscapes in Africa 165
Cerian Gibbes, Lin Cassidy, Joel Hartter, and Jane Southworth

Part III Institutions and Political Ecology Approaches

9 Between Cooperation and Conflict: The Implementation of Agro-Extractive Settlements in the Lower Amazon Floodplain 213
Fabio de Castro

10 Conservation of Natural Resources: Which Matters – Having a Regulation or the Size of the Penalty Imposed? 235
Maria Claudia Lopez

11 Small-Scale Farmers and the Challenges of Environmental Conservation and Rural Development: Case Studies from the State of SãoPaulo and the Amazon Region 251
Célia Futemma

12 Institutional Evolution, Forest Conservation, and Rapid Change in Rural Honduras 275
Catherine M. Tucker

13 Land-Use Institutions and Natural Resources in Fast-Growing Communities at the Urban-Rural Fringe 295
Abigail M. York and Darla K. Munroe

Part IV Historical and Archeological Approaches

14 Agroforestry in Tomé-Açu: An Alternative to Pasture in the Amazon 321
Mateus Batistella, Édson Luis Bolfe, and Emilio F. Moran

15 Changing Driving Forces, Imposed Tenure Regimes, and Tree-Cover Change on Village Landscapes in the West Mengo Region of Uganda, 1890 –2002 343
Nathan D. Vogt

16 Was Agriculture a Key Productive Activity in Pre-Colonial Amazonia? The Stable Productive Basis for Social Equality in the Central Amazon 371
Eduardo Góes Neves

Part V Future Directions

17 Human-Environment Research: Past Trends, Current Challenges, and Future Directions 391
Eduardo S. Brondízio and Rinku Roy Chowdhury

Index 401

List of Figures

Fig. 2.1 Locations of Xavante indigenous reserves, Mato Grosso 32

Fig. 3.1 Linking background variables with fertility outcomes 58
Fig. 3.2 Total fertility rate by year and survey, rural Tibet (*Sources*: Childs 2008 (for 1945–1983); Childs et al. 2005 (for 1984–1997); Goldstein et al. 2006 survey (for 1991–2005)) 64
Fig. 3.3 Tibetan fertility, pre-1959 66
Fig. 3.4 Tibetan fertility, 1960s–1970s 67
Fig. 3.5 Tibetan fertility, mid-1980s to early 1990s 68
Fig. 3.6 Age-specific fertility rates by 3-year periods (*Sources*: Childs et al. 2005 (for 1984–1997 analysis); Goldstein et al. 2006 survey (for 1991–2005 analysis)) 69
Fig. 3.7 Tibetan fertility, 1990s 71

Fig. 5.1 Case study areas and distribution of main land-use objectives (from Irvine et al. 2009) 100
Fig. 5.2 Participatory GIS cycle 101
Fig. 5.3 Reporting structure on an NGO-managed property 106
Fig. 5.4 Decision scheme on an NGO-managed property 107
Fig. 5.5 Examples of the effects of section 7 on a private- and a community-managed estate. Deer cull figures are given as percentage of stags culled on the total cull in each estate and total deer culled by hectare. Estates A and B have been managed by the same organization since year 7. Estate B is currently affected by a section 7 agreement. Estate C is terminating a section 7 agreement that has affected its management since year 5 108

Fig. 6.1 Locations of protected areas in India 116
Fig. 6.2 Forest-cover distribution in Pench Tiger Reserve in 1977, 1989, 2000, and 2007 123

Fig. 6.3 Forest-cover distribution in Tadoba Andhari Tiger Reserve in 1989 and 2001 126
Fig. 6.4 Forest-cover distribution in Mahananda Wildlife Sanctuary and the adjacent Baikunthapur Reserve Forest in 1990 and 2000 130
Fig. 6.5 Forest-cover distribution in Bannerghatta National Park in 1973, 1992, and 2007 133

Fig. 7.1 Challenges and opportunities in the governance of private forests 157

Fig. 8.1 Study area map of Kibale National Park and Corridor region and its location in Uganda 173
Fig. 8.2 Land-cover classifications resultant change trajectory 176
Fig. 8.3 Standardized NDVI images showing variation over space and time for these vegetated forest regions 179
Fig. 8.4 Map showing the location of the Southern Buffalo Fence, visible as a boundary between areas of high and low reflectance in Landsat imagery from 2007. The fence separates two forms of land use—the economically important wildlife management areas where safari tourism is conducted and the socially important tribal grazing areas where rural people raise cattle, goats, and sheep 183
Fig. 8.5 False-color composite Landsat imagery showing pre- and post-fence vegetation conditions along the sandveld tongue. Vegetated areas appear *dark* to *bright red*, water appears *black*, and areas of senescent or no vegetation appear *green-blue* to *white*. (**a**) shows the location of the study area relative to the entire Okavango Delta. (**b**) shows the study area in 1979, prior to the construction of the Southern Buffalo Fence, while (**c**) and (**d**) respectively show 8 and 25 years post-construction 184
Fig. 8.6 Locations of paired points and polygons for analyses. The paired points (*red* and *yellow dots*) follow along the fence, with points 250 m apart across the fence and pairs 2 km apart along the fence. The *polygons* are shaped to avoid the influence of riparian zones and to cover comparable soil units 185
Fig. 8.7 Overlay of sampling *points* and *polygons* on soils map, showing large *brown polygons* in the Matsibe area (eastern pair) covering predominantly arenic soils on the sandveld tongue and *narrow orange polygons* in the Habu area (western area) on calcic to sodic soils 186

Fig. 8.8 Differences in mean Landsat-derived surface temperature values between wildlife and livestock areas for wet and dry seasons from 1990 to 2007, showing one standard deviation. The low standard deviations (compared to data in other figures below) are in part an artifact of the coarser resolution of the thermal band 189
Fig. 8.9 Differences in mean Landsat-derived surface moisture values between wildlife and livestock areas for wet and dry seasons from 1990 to 2007, showing one standard deviation .. 190
Fig. 8.10 Differences in mean Landsat-derived soil brightness values between wildlife and livestock areas for wet and dry seasons from 1990 to 2007, showing one standard deviation .. 191
Fig. 8.11 Study region showing the larger regional context of the study site, the study area as determined by the extent of the IKONOS imagery (outlined in *red*) and the two dominant land management types used in the region 195
Fig. 8.12 Schematic showing the relationships between the actual tree coverage, the idealized object-oriented classification results in *polygon* form, and NDVI for the same area at the Landsat TM scale (Figure modified from Gibbes et al. 2010) 197
Fig. 8.13 Results of the object-based classification. Figures shown are (*left*) the IKONOS imagery for a focus area (color composite RGB = near infrared, red, green where *red* represents vegetation), (*middle*) the corresponding object-based classification for the focus area and (*right*) Landsat NDVI also for the focus area, in which the difference in scale is apparent (modified from Gibbes et al. 2010) 198
Fig. 8.14 Spatial clustering of trees (*left*) results of the Getis-Ord Gi* statistic with Bonferroni correction applied to determine statistically significant tree clusters, where *red dots* indicate clusters of trees with large crown size, while *blue dots* indicate the statistically significant clusters of trees with relatively smaller crown size (*right*) Landsat TM NDVI shown for the same study region 199
Fig. 8.15 Boxplots of the NDVI values for each tree-cover class (1–5) as determined by proportion of tree cover per Landsat TM 30 m × 30 m pixel 200

Fig. 9.1 Spatial distribution of 15 PAEs in the Lower Amazon 226

Fig. 9.2 Social interactions between the four main actors in the implementation of the PAEs in the Lower Amazon (*solid line box*: included actor; *solid line arrow*: collaborative interaction; *dashed line box*: excluded actor; *dashed line arrow*: conflictive interaction) 230

Fig. 10.1 Round average individual contribution levels by treatment 243
Fig. 10.2 Average round individual earnings by treatment 243
Fig. 10.3 Distribution of individual contribution decisions by treatment 247

Fig. 11.1 Distribution of land use and land cover among the Tapajós communities, 1986–1999. Numbers 1–14 refer to riparian communities. Number 15 refers to the upland community 264

Fig. 12.1 View of La Campa landscape 276
Fig. 12.2 Escarpment above the Centro Urbano of La Campa 280
Fig. 12.3 La Campa's forest cover in 1996 (**a**) and 2006 (**b**) 291

Fig. 14.1 Study area in Tomé-Açu, Pará State, eastern Brazilian Amazon... 326
Fig. 14.2 Property based on agroforestry systems in Tomé-Açu in 2005 (Photo: M. Batistella, region of Tomé-Açu, Pará, Brazil, 2005) 330
Fig. 14.3 Sampling strategy for agroforests in Tomé-Açu 332
Fig. 14.4 Classes, parameters, and schematic representation of agroforestry systems in Tomé-Açu 333
Fig. 14.5 Different classes of agroforestry systems in Tomé-Açu (*Source*: Bolfe 2010: Figs. 5a and 5b on p. 137, Fig. 5c on p. 139, and Fig. 5d on p. 142) 334
Fig. 14.6 Schematic view and carbon content estimated for agroforestry systems in Tomé-Açu (*Source*: Bolfe 2010: 149) 336
Fig. 14.7 Extraction of values for carbon content and vegetation indices in Tomé-Açu 336
Fig. 14.8 Carbon content estimates for agroforestry class AFS 3 in 2005 337

Fig. 15.1 Maps of the study area, including those showing the subcounty (*gombolola*) units nested within both district and traditional county (*ssaza*) units. These maps illustrate clear territoriality, fixing boundaries between groups and defining rights to and control over specific sets of natural resources that evolved long before British colonization 344
Fig. 15.2 Soil catena of typical, traditional village unit in West Mengo, Uganda 346

Fig. 15.3 1955 aerial photograph showing village *mitala* (zone B) void of forest and trees, cleared in the early colonial era, and stable forest cover in village valleys 352
Fig. 15.4 1995 aerial photograph showing increased tree cover and valley forest dynamism on village landscapes compared to 1955 aerial photo (Fig. 15.3) 360
Fig. 15.5 Increased tree cover in agroforestry systems and home gardens on village *mitala* since the early colonial era (see Fig. 15.3) (Photos by Nathan Vogt, 2002) 364
Fig. 15.6 Increasing spatial and temporal land-use complexity in village valley forests: one land-use sequence following deforestation (Photos by Nathan Vogt, 2002) 366

List of Tables

Table 3.1	Diminishing per capita landholdings	60
Table 3.2	Percentage of 15- to 49-year-olds in nonfarm work	64
Table 3.3	Landholdings and household sizes, Kyirong taxpayers, 1958	65
Table 3.4	Never-married females (percent) by age	67
Table 3.5	Number of household members by relative size of landholding	74
Table 3.6	Children ever born (CEB) and marriage age of married women aged 30–39 in 1998	74
Table 6.1	Trajectories and drivers of land-cover change in various protected forested areas within India	117
Table 6.2	Trajectories of land-cover change (%) in Pench Tiger Reserve and the surrounding 10-km area	124
Table 6.3	Trajectories of land-cover change (%) in Tadoba Andhari Tiger Reserve and the surrounding 10-km area	127
Table 6.4	Trajectories of land-cover change (%) in Mahananda Wildlife Sanctuary, Baikunthapur Reserve Forest, and the surrounding 5-km area	130
Table 6.5	Trajectories of land-cover change (%) in Bannerghatta National Park and the surrounding 5-km area	133
Table 7.1	Four spheres of public-private interactions in private forestry	146
Table 7.2	Possible governance arrangements for forest goods and services	154
Table 8.1	Forest-cover change over time across the different landscape components	177
Table 8.2	NDVI values and proportion of forest for each image date	177
Table 8.3	Differences in general land-cover characteristics at 20 paired points 250 m apart on either side of the buffalo fence	187

Table 8.4 Density of large trees within each of the two land management types—KCA, government-managed ($\sim$132.1 km^2) versus community-managed lands ($\sim$146.48 km^2) 199

Table 9.1 Features of the four main actors involved in the implementation of the PAE in the Lower Amazon 221
Table 9.2 List of the PAEs in implementation phase in the Lower Amazon 227

Table 10.1 Reminder received if player was inspected at the end of a round 241
Table 10.2 Number of groups in each treatment 242
Table 10.3 Average group contributions and average group earnings for the 15 rounds, the first round, and the last round (standard deviations in parentheses) 244
Table 10.4 Random effects Tobit estimation of individual extraction (standard error in parentheses) 245
Table 10.5 Random effects logit estimation of individual compliance (standard error in parentheses) 247

Table 11.1 Frequency of families from the Tapajós communities ($n = 1{,}438$) and from the Ipanema settlements ($n = 167$) that cultivated annual crops 258
Table 11.2 Absolute (area) and relative (percent) distribution of land-cover changes from 1986 to 1999 according to different types of institutional arrangements: public and community 263
Table 11.3 Distribution of property lots according to size of properties, size of APP (gallery forest) per property at settlements Ipanema 1 and Ipanema 2 (2009) 265

Table 13.1 Comparison of three study sites 304
Table 13.2 Comprehensive plan principles 308

Table 14.1 Phases and characteristics of agroforestry systems' dynamics in Tomé-Açu, PA, Brazil 325
Table 14.2 Allometric equations used to estimate aboveground biomass 335
Table 14.3 Best adjusted models for carbon estimation in Tomé-Açu agroforestry systems 337

Table 15.1 Tree species on village *mitala* ranked by importance value 365

Table 16.1 Climate changes in the Amazon Basin 378

Chapter 1
Introduction to Human-Environment Interactions Research

Emilio F. Moran and Eduardo S. Brondízio

Abstract Contemporary theories about the reciprocal interactions of human beings with the environment are only fully intelligible in the light of the historical roots of such theories. In this introduction we provide an overview of the major Western intellectual currents up to those that are commonly used today. Three main themes help organize this broad array of theories and approaches: environmental determinism, cultural determinism, and human-environment interaction concerned with the processual relationships between people and environment as grounded in historical, social, and ecological contexts. This chapter also provides an overview of the four parts of the book and discusses the coverage, diversity, and parallels in themes and approaches across all chapters.

Contemporary theories about the reciprocal interactions of human beings with the environment are only fully intelligible in the light of the historical roots of such theories. Modern notions of homeostasis reflect ancient concerns and assumptions about the order of nature, just as our current fascination with chaos theory reflects a contemporary jaundiced view of the social order. Every society has philosophical explanations about the natural world and human beings' place in it. It is through such explanations that members of a society articulate their normative rules and the broad outline of how they can best function as societies that depend on natural resources for their survival.

E.F. Moran (✉) • E.S. Brondízio
Department of Anthropology, Anthropological Center for Training and Research on Global Environmental Change, School of Public and Environmental Affairs, Indiana University
Bloomington, IN, USA
e-mail: moran@indiana.edu; ebrondiz@indiana.edu

E.S. Brondízio and E.F. Moran (eds.), *Human-Environment Interactions: Current and Future Directions*, Human-Environment Interactions 1,
DOI 10.1007/978-94-007-4780-7_1, © Springer Science+Business Media Dordrecht 2013

Human-environment interaction (HEI) provides a framework that brings together scholarship sharing both disciplinary depth and interdisciplinary scope to examine past, present, and future social and environmental change in different parts of the world. Key to all of these approaches is that they must be interdisciplinary and cut across the social and the natural sciences. While building upon disciplinary expertise, this type of work asks new questions that purely disciplinary research tends not to ask; it brings new methods and theories to the challenges posed by societal concerns and connects theory and practice in ways that address problems that arise from human interactions with environment. This volume brings senior and junior scholars together and, in so doing, connects these historically influential traditions to new and cutting-edge approaches that give us a glimpse into current and future trends in interdisciplinary science of human-environment interaction.

In this introduction, we provide a broad view of the major Western intellectual currents up to those that are commonly used today. Three main themes help organize this broad array of theories and approaches: environmental determinism—the determining effect of nature upon society, cultural determinism—that sees cultural context as the only way to understand our place in nature, and human-environment interaction—concerned with the *interaction* of people and environment. These themes represent three points on the intellectual spectrum. One view overemphasizes the influence of environment, while the second overemphasizes the role of human culture. The third view bridges the gap between the other two themes, providing a framework to examine HEI as dialectical and diachronic processes rather than unidirectional.[1]

Deterministic explanations in HEI date back a long time in human history. Yet they seem recurrent. In academia, the history of deterministic explanations has promoted inter- and intradisciplinary divisions fueled by theoretical dualisms (e.g., nature-culture, agency-structure, materialist-idealist, rational-moral) rather than having societal problems or crosscutting questions informing the direction of scholarship. Our current environmental and societal issues defy reductionist and deterministic interpretations as well as panacea policies (Ostrom 2007), but we are still learning how to move from segmented to complementary disciplinary knowledge and integrative science. An emphasis on human-environment interaction recognizes the complexity of historical and contemporary factors affecting society and environment at various scales; because of the value put on fieldwork, HEI research has an explicit concern with corroborating deductive and inductive perspectives. As such, it encourages interdisciplinary collaboration constructed around shared questions, common frameworks, and metalanguages across disciplines. A glimpse into the traditions leading to these perspectives may serve as a useful reminder of the challenges and opportunities ahead.

---

[1] Discussions that follow are based in part on a more detailed discussion that can be found in Moran (2006).

1.1 Environmental Determinism

Determinism assigns one factor as a dominating influence over the whole system. From Greco-Roman times through the early part of the twentieth century, scientific theories stressed single-factor explanations to the neglect of the complex interactions of biological systems. At the heart of their argument was the role that their strategic location in the Mediterranean played in the acquisition and maintenance of such power. Writers rose to the task by explaining that the "middle latitudes" (i.e., Greece) were most conducive to favorable cultural developments because in that locale, humans were subject to an ideal proportion of the basic four elements (fire, water, earth, air). A hot tropical climate was believed to foster idleness and resignation (Thomas 1925: 227), while the climate of Greece, with its seasonal changes, balanced the exposure to the elements and thus was the most conducive to progress. These ideas, which were endorsed by Hippocrates, Aristotle, and other major figures of ancient Greece, set a trend that was followed by the Romans. Roman writers cited geoclimatic reasons for the Roman conquest of the rest of the civilized world. Cicero attributed this success to the strategic location of Rome itself. Like Greek authors before him, Vitruvius felt the optimal location was one midway between the two extremes of hot and cold. He pointed to Rome's dominance as proof of the correctness of his judgment. Because of the protection afforded classical learning in Christian monasteries and Arab centers of scholarship, the human-environment theories of classical times survived the turmoil that followed the breakdown of Roman rule (Castaglioni 1958: 258–263).

Arab scholars elaborated on classical theories. The Arab conception of the human-environment relationship was twofold. One part consisted of an astrological explanation that considered humans to be part of the cosmos, resulting in their character and outlook being determined by the ruling stars of their environment. The other part was a purely geographical explanation based on climatic considerations (Alavi 1965: 68). The Arab scholar Al-Mas'udi discussed the importance of the availability of water, natural vegetation, and topography in determining the sites of human settlements. He also correlated the climate to the humors of the body, showing how a certain climate can give rise to humoral imbalances and thus to particular virtues or vices (Alavi 1965: 69–70). Arab scholars preserved and translated the Greco-Roman classics and, in the process, added some of their own interpretations to the texts. As a result, when the classics were read in twelfth- and thirteenth-century Europe, it was with the addition of commentaries by Arab and Jewish scholars from Cordoba, Seville, Toledo, Baghdad, and Damascus.

With the discovery of the East Indies and the New World, Europeans were thrust into contact with cultures and environments that differed considerably from their own. Among the adventurers, missionaries, and merchants of those days were naturalists and curious travelers; their accounts of the strange habitats and ways of native Asians, Africans, and Americans excited intellectual interest in explaining cultural and environmental differences. The role of human culture in buffering the impact of environment upon society began to be appreciated, and the scope of possible explanations for similarities and differences in human populations expanded.

In the late nineteenth century, a general trend toward organizing increasing amounts of archaeological and ethnological data resulted in an attempt to illuminate the processes by which human cultural history changes. A very simple heuristic device was quickly discovered—many cultures with similar artifacts and customs could be grouped by geographic location. Geographers and, later, ethnographers seized upon this notion. They viewed the interrelation of groups with their habitats as producing specific kinds of cultural traits. Friedrich Ratzel (1844–1904), a scholar with broad ethnographic interests who was the founder of anthropogeography, emphasized the primacy of habitat in bringing about cultural diversity (Helm 1962: 630). He explained human cultural evolution as being spurred by the conflicts over territory between migrating peoples. His thesis thus centered on the migration of groups, which promotes the diffusion of cultural traits (Harris 1968: 383). While diffusion may produce divergence in the original traits, Ratzel also believed that migratory peoples usually "hold fast to their natural conditions of existence [that is, culture]" (quoted in Thomas 1925: 140).

The environmental determinist trend continued in the twentieth century with the work of geographers Ellsworth Huntington (1915) and Griffith Taylor (1951). Huntington believed that variations in temperature and humidity were beneficial, provided they were not taken to extremes. He eventually postulated what he came to regard as an ideal climate for maximum human efficiency: one with moderate seasonal changes, average humidity, and abundant storms. Although Huntington was far from naïve, he formulated his generalizations as if climate was the only important factor.

An elementary problem with these deterministic theories is their misuse of inductive reasoning. The inductive approach requires that one observe the facts and then form a generalization that will fit all the observed facts. It has been more common among determinists to formulate a generalization first and then set out to prove it with an unclear methodology and an inadequate sample. Selective sampling led to confirmation of many deterministic generalizations. At a time when so little was known about the workings of the physical world, it is striking how broad the scope of these generalizations was.

1.2 Cultural Determinism

Unlike environmental deterministic theories, which emphasize the influence of nature on human behavior and institutions, cultural deterministic theories viewed nature as a relatively static factor or backdrop, and human history and culture as shaping human communities. Franz Boas (1858–1942) proposed what has come to be called historical possibilism—that is, nature circumscribes the possibilities for humans, but historical and cultural factors explain what is actually chosen. Boas (1896: 901–908) rejected the idea that the environment was a primary molder of culture and sought explanation for cultural differences in the particular cultural history of a people. Boas did not come to this position immediately, but rather after initial acceptance of the environmental deterministic views of his day. When Boas went on his trip to Baffin Island to study the Eskimo, he did so "with a strong presumption in favor of the primacy of geographical factors in the life of the Eskimo" (Boas 1964[1888]; Harris 1968: 265). In *The Mind of Primitive Man*, Boas (1963[1911]) pointed out that the environment furnishes the material out of which people shape and develop the artifacts of daily life as well as their theories, beliefs, and customs (Thomas 1925: 278). Yet while he granted that the environment had a general influence, Boas criticized the one-sided notion that the same type of environment will, in a given stage of culture, produce the same results everywhere. While the followers of Boas insisted that there was no Boasian school, there was a certain common ground that they shared. Regarding the environment, their approach was a reaction to environmental determinism. Lowie (1883–1957), for one, in his *Culture and Ethnology* (1917) set out to disprove the environmental deterministic notions that "culture reaches its highest stages in temperate regions," that the concept of liberty is directly correlated with altitude, and that island inhabitants are accomplished seafarers. Lowie argued that under the same geographical conditions, radically different cultures have developed. Alexander Goldenweiser saw the environment as a static force and culture as the dynamic element that shaped the use of natural resources. He also suggested (cf. Ferndon 1959) that humans change the natural environment (e.g., turning forests into cultivated fields) and, as a result, make their own environment instead of being determined by it (Goldenweiser 1937: 452–453). As we will see later, this view (that people do not adapt to environment but modify their environment to suit them) has returned as historical ecology, emphasizing historical context and agency (Balée 1998; Balée and Erickson 2006).

Alfred Kroeber (1876–1960), like other anthropologists of his day, subscribed to the Boasian credo that the physical environment is there merely to be acted upon by human culture. Kroeber's (1939) approach in *Cultural and Natural Areas* can be likened to that of his contemporary, British geographer/anthropologist C.D. Forde (1902–1973). Both of them emphasized the need for collecting ecological data and viewed such data as potentially valuable in explaining cultural similarities. Forde, for example, after summarizing the history of economic systems in relation to ecology and social organization, concluded that neither an evolutionary sequence of "economic stages" nor the nature of the subsistence base could explain the changes in culture (Forde 1934). Economic and social activities, he concluded, are products of the long, but largely unpredictable, processes of cultural accumulation and integration.

Kroeber's conclusions in *Cultural and Natural Areas* resemble those of Forde's in *Habitat, Economy and Society* (1934). With that effort, Kroeber's fleeting use of technoenvironmental explanation ended, and he turned his idea of culture area increasingly toward notions of diffusion and "areas of culture origins." In regard to his earlier culture/environment explorations, Kroeber (1939: 205) became baffled and wrote, "The interaction of culture and environment become increasingly complex when followed out. And this complexity makes generalization unprofitable, on the whole." He noted that in each situation or area, different natural factors are likely to be "impinging on culture with different intensity" (ibid.). In spite of its merit recognizing cultural differences and the complexity of HEI as one moves from local to higher levels of analysis, an overemphasis on the singularity of local phenomenon continued to plague theoretical approaches that insist on cultural determinism, discouraging attention to comparative work informing robust generalizations.

1.3 Human-Environment Interactions

The eighteenth century was a period when natural historians concerned with human progress formulated evolutionary sequences that attempted to explain human society in terms of increased human control over nature. One of these figures, Anne-Robert-Jacques Turgot (1727–1781), foreshadows the cultural ecological approach of Julian Steward. In his *Universal History* (1750), Turgot interpreted the band organization of hunters as a response to the necessity of pursuing game over vast areas. Such pursuit resulted, in turn, in the dispersal and diffusion of peoples and ideas. On the other hand, where easily domesticated species were present, a pastoral way of life with greater population concentrations and greater control over resources might emerge. During this period, the Scottish School (an intellectual elite in Scotland in the eighteenth century) made efforts to correlate social organization with subsistence. A major figure of this school was historian William Robertson, whose book *The History of America* (1777) is a landmark for its discussion of the conditions for cultural similarities around the world. Robertson believed that cultural similarities were evidence of independent invention, arguing that similarities between the resource bases of two groups would lead to similar adaptive responses. Whenever Robertson encountered seemingly "nonadaptive" traits, he attributed such behaviors to the group's borrowing the trait, despite its nonadaptiveness, from neighbors with whom they had had previous contact. Thus, Robertson dealt with two of the major research questions in cultural ecology: diffusion vs. innovation and explaining adaptive vs. maladaptive cultural behavior.

The Scottish School included many writers, such as Adam Smith (1723–1790), Adam Ferguson (1723–1816), David Hume (1711–1776), and James Millar (1735–1801), all of whom examined the evolution of complex societies and the cultural and materialistic forces that lead to social stratification (Voget 1975: 90). These men looked at the interrelation of cultural units, especially those involved in the economics of a society, rather than the evolution of ideas per se. Adam Smith emphasized the division of labor as basic to understanding the increasing complexities of a modern nation (Voget 1975: 78). Ferguson and Millar attempted to correlate various institutions, such as land tenure, marriage, and slavery, to the subsistence base found in various cultures. In so doing, they tried to correct some of the distorted accounts and explanations of prehistoric life by utilizing a variety of data sources and by avoiding racial and ethnocentric ideas about primitive "nature" or "intellect" (Harris 1968: 29–31). Millar and others also emphasized control over resources and accumulation of an economic surplus as accounting for differing institutions. In the writings of these men, we see a growing awareness that any explanation of cultural diversity must include a consideration of a broad range of factors. They do not use single-factor deterministic explanations, nor do they overemphasize individual choice, cultural determinism, or the purposeful movement of nature toward "progress" and higher civilization. Turgot and the Scottish philosophers emphasized adaptation from one subsistence mode to another.

The nineteenth century was the heyday of the naturalists. The similarities and differences in living organisms impressed them and stimulated their search for explanations. The contributions of Charles Darwin (1809–1882) to ecological theory are particularly notable. Darwin found inspiration for his theory of evolution in the works of Charles Lyell and Thomas Malthus (1766–1834). Darwin took a copy of Lyell's (1830) *Principles of Geology* with him on his HMS Beagle voyage and confided in his diary that it "altered the whole tone of [his] mind." Through Lyell's account of the geological record, Darwin saw an alternative to the narrow Biblical time scale and was impressed by the relationship between environmental change and modifications in biological forms. Malthus's (1798) *An Essay on the Principle of Population* influenced Darwin with its idea that the natural trend of the human population was to increase unless stopped by disease, war, or famine. Darwin extended this notion to plant and animal populations.

Darwin's synthesis appeared in 1859 under the title *On the Origin of Species by Means of Natural Selection*. In this work, Darwin began by assuming that all living things are related and that the diversity of species results from a continual branching out. Such branching is a product of the process known as natural selection. According to the principle of natural selection, those organisms most fit to survive and reproduce in a given environment will outreproduce less well-adapted organisms and that species not adapted to current environmental conditions will be reduced to insignificant numbers and possibly to extinction.

If biological evolution reflected only the process of adaptation to environment, it would be a static, nonevolutionary process. For evolutionary change to occur, there must be random changes in species that are not responses to current needs but that under given circumstances give an advantage to individuals who share the trait in a population. It is easy to misunderstand Darwin's view of natural selection. The Lamarckian idea (Jean-Baptiste Lamarck, 1744–1829) that organisms improve themselves by their own efforts and that they pass on these advantages to their offspring appeals to common sense and to the notion that evolution travels along a progressive path (Gould 1980: 76). However, it is not in this manner that species evolve. The specter of Lamarck in evolutionary theory can be traced to his central notion that organisms respond to felt needs, and indeed, Lamarck's ideas are relevant to the notion of specific evolution as proposed by Marshall Sahlins and Elman Service (1960) and to Steward's (1955) emphasis on adaptive processes in local environments. In contrast, Darwinian theory emphasizes that genetic variation arises randomly and proceeds undirected. Selection acts upon unoriented variation and changes result from reproductive success. Darwinian theory's power derives from its complexity, from its refusal to be a mechanistic theory driven by environmental determinism (Gould 1980: 81) or by purpose. It is less appealing than Lamarck's theory because it presents us with a universe devoid of intrinsic meaning or direction. As Roy Rappaport noted (1984), human beings have had to invent culture and ritual "to give meaning to a world devoid of meaning." Lamarck's theory, while failing to explain how species evolve, suggested instead how human cultural evolution occurs. It is in this realm wherein we can expect rapid acquisition of adaptive traits through ideological and behavioral change and its transmission through socialization. Technological change and cultural change work in Lamarckian ways, and they have unleashed a rate of change inconceivable in the slower, undirected process of natural selection.

Modern evolutionary theory and genetics have put to rest the simplistic notions of determinism. The functions and forms of organisms can be understood only by careful accounting of complex processes of interaction. This is best expressed in the contrast between genotype and phenotype. The genotype refers to the hereditary potential of an organism. The phenotype, on the other hand, is the product of the interaction between the genotype and the environment where the organism is located. Some species tolerate a minimum of environmental change and exhibit a minimum of phenotype variation (i.e., highly specialized species). Bacteria, for one, tolerate only minute differences in habitat temperature. The human species, by contrast, manifests great phenotypic variations and can tolerate a wide range of environmental conditions (i.e., we are a generalist species).

The development of the field of evolutionary ecology has been particularly vigorous since the 1980s (e.g., Boyd and Richerson 1985; Durham 1990; Smith 1991; Winterhalder and Smith 1981). The attention that it gives to the great complexity of the environment distinguishes it from earlier approaches. However, if one is interested in the exceptional plasticity and diversity of behavior within a species, an evolutionary ecological approach is more appropriate. It is more concerned with why diversity of behavioral outcomes occurs and less with how such adaptations effectively address the needs of species or individuals. For the latter, an adaptationist or functional approach is still more appropriate.

Steward's early writings broke with both environmental and cultural determinism by emphasizing the use of the comparative method to test causal connections between social structure and modes of subsistence. Steward's approach was a functionalist one, concerned with the operation of a variable in relation to a limited set of variables, not in relation to an entire social system. The cultural ecological approach proposed by Steward involves both a problem and a method. The problem is to test whether the adjustments of human societies to their environments require specific types of behavior or whether there is considerable latitude in human responses (Steward 1955: 36). The method consists of three procedures: (1) to analyze the relationship between subsistence system and environment, (2) to analyze the behavior patterns associated with a given subsistence technology, and (3) to ascertain the extent to which the behavior pattern entailed in a given subsistence system affects other aspects of culture (Steward 1955: 40–41). In short, the cultural ecological approach postulates a relationship between environmental resources, subsistence technology, and the behavior required to bring technology to bear upon resources.

The crucial element in Steward's approach is neither nature nor culture but, rather, the process of resource utilization, that is, the interaction between people and their resources. The reasons for the priority he gave to subsistence are clear: Obtaining food and shelter is an immediate and urgent problem in all societies, and patterns of work at a given level of technology are limited in their ability to exploit resources. The approach is best illustrated by his study of the Western Shoshone. The Shoshone inhabited the Great Basin of North America, a semiarid land with widely dispersed resources. The Shoshone were hunter/gatherers with simple tools and relied heavily on the collection of grass seeds, roots, and berries. Steward showed how almost every resource could best be exploited by individuals—except rabbits and antelope, which required seasonal group hunting. Each fall, the Shoshone gathered pine nuts that were stored for the long, cold winter. Although in winter they formed larger population concentrations, they did not form stable social units because pine nuts were not available in the same places each year, and groups therefore had to remain fluid to adequately exploit the basin. Thus, the requirements of subsistence produced fluid and fragmentary social units. To Steward, the Shoshone presented an extreme case of the limitations placed by environment on the workable options available to a culture. Steward hypothesized that the immediate impact of environment upon behavior decreased as technological complexity improved the human capacity to modify the environment. He suggested that in complex societies, social factors may be more important in explaining change than subsistence technology or environment (Steward 1938: 262). The research strategy proposed by Steward is all the more striking if one considers its historical backdrop. Until Steward's time, human-environment theories either dealt in broad generalities lacking a firm grounding in empirical research or emphasized lists of cultural traits. Cultural ecology put the emphasis on careful analysis of social interaction, recording of movement, timing of work activity, and so forth. Through such research, it was possible to more effectively delimit the field of study and arrive at cause-and-effect relationships. Another, although less developed and popularized component of his theory, was a concern with levels of cultural and social integration. Steward stressed the importance of understanding complex social systems not as the average of behavioral norms. He focused on understanding levels of social-cultural integration with the idea that in any society there is a succession of horizontal and vertical organizational types along a continuum representing emergent social and political forms.

Steward has been criticized by some scholars because his approach is difficult to operationalize in the field and because it assigns primacy to subsistence behaviors. The focus on subsistence is essential to the cultural ecological approach. There are cases when other factors may have far greater control over a social system, and over the years, Steward (1955: 93) expanded the scope of cultural ecology to include political, religious, military, and aesthetic features of culture. Clifford Geertz (1963) concluded in his study of Indonesian agriculture that historical and political factors are part of the total environment to which populations adapt and must not be dismissed as secondary. A few years later, Rappaport (1968) showed how ritual could play a central role in the maintenance of a society's balance with resources. The contribution of Steward was to delimit, more than anyone before him, the field of human-environment interaction. He did so by emphasizing behavior, subsistence, and technology. The weaknesses of such an approach became apparent within a decade and spawned other research strategies.

Dissatisfaction with the research approach of cultural ecology led some scholars to search for new theories, new data-collecting techniques, and new analytical tools. The major influence on this new research approach came from general, or biological, ecology. The ecosystem concept provided a conceptual framework more satisfactory to some scientists than the behavior/social structure equation stressed by Steward. Michael Little and George Morren (1976: 5) succinctly expressed the strategy: "We are concerned with those cultural and biological responses, factors, processes and cycles that affect or are directly connected with the survival, reproduction, development, longevity, or spatial positions of people. This set of questions rather than the traditional division of scientific labor defines the subject matter."

Rappaport and Andrew Vayda gave the strongest impetus to an ecosystem approach in the field of anthropology. In fact, they preferred the term *ecological anthropology* because they felt that the emphasis on "culture" suggested by the term *cultural ecology* obscures the applicability of principles from biological ecology to the study of human adaptation (Vayda and Rappaport 1976: 20–21). Given that humans are but one species in nature, subject to the same laws as other species, use of the principles, methods, and analytical tools of the ecological sciences would greatly add to our understanding of our own species. Vayda and Rappaport believed that anthropologists should not hesitate to adopt biological units such as population, community, and ecosystem as units of study since it allows a more comprehensive approach to ecological studies. Even the topics of research can be couched in terms that make sense across both disciplines. Vayda and Rappaport pointed out that ecologists have shared various areas of interest with anthropology: ways of defining territorial rights, ways of establishing group identity, and mechanisms for establishing buffer zones. All these can be viewed "ecologically" as regulating behavior or serving a homeostasis function. To test ecological hypotheses properly, a wealth of information is required, and no single researcher can expect to succeed in gathering it all—and it is no surprise that their own experience of field work in Papua New Guinea as part of a large interdisciplinary team studying war in that region led them to see the value of ecosystem as an integrator of work across ecology, geography, and anthropology (Vayda and Rappaport 1976: 23).

Vayda's study of how warfare in New Guinea is related to population fluctuations, changes in man/resource ratios, and the competition of different highland clans for gardens and pigs is a notable example of the ecological approach (Vayda 1974, 1976). Rappaport, working with the Tsembaga Maring in the same region, was more concerned with how ritual serves to regulate: (1) the size of the pig herd, (2) the frequency of warfare, (3) the availability of horticultural land within reasonable walking distance of the village, (4) the length of the fallow cycle, and (5) the military strength and alliances of a tribe and the likelihood that it will hold on to its claimed territory.[2] Rappaport is not really concerned with the individual decisions of the Tsembaga Maring as they see their pig herd increase to the point that they become a threat to the human ecological system. Rather, he finds that the system "senses" the increased burden of having too many pigs. When a system threshold is reached, the elders call for a ritual pig slaughter. The ritual reduces the number of pigs and facilitates the creation of alliances between neighboring groups. Warfare follows, and its occurrence serves to distribute the population over the landscape and to return the system to "initial conditions" or a state of equilibrium.

An important issue raised by Rappaport's study (1968) is the utility of the concept of homeostasis. As used by Rappaport, the concept was equivalent to equilibrium—a view shared by some biological ecologists and reminiscent of the Greco-Roman search for order in nature. In equilibrium models, attention is paid to how cultural practices help maintain human populations in a stable relationship with their environment. This view is the prototype of neofunctionalism, and it has its drawbacks. It views the current state of the system as the norm and overemphasizes the functions of negative feedback to the neglect of the dynamics of change accelerated by positive feedback. This viewpoint tends to preclude the possibility that behaviors might be maladaptive, which they surely are in certain situations (Alland 1975; Eder 1987).

Adaptation to environment is, however, not a simple matter of negative feedback. System correction through negative feedback operates most effectively at lower levels in a system. Higher levels operate at a more general level wherein ambiguity and vagueness permit constant reinterpretation and restructuring of system properties as responses to perturbations. Homeostasis and dynamic equilibrium do not imply changelessness. On the contrary, they require constant adjustment of system parts and even some change in structure in response to perturbations (Rappaport 1977: 169). In other words, while systems have lower-order mechanisms geared to the maintenance of stability, they also have higher-level, less specialized responses that can reorder the system to assure its survival—a view echoed in more sophisticated ways today in terms of emergent properties of systems.

A number of problems must be recognized in how the ecosystem concept was used: a tendency to reify the ecosystem and to give it properties of a biological organism, an overemphasis on energetic flows and measurement of calories, a tendency for models to ignore time and structural change (and to overemphasize homeostasis), a tendency to neglect the role of the individual, a lack of clear criteria for defining boundaries of systems, and level shifting between field study and analysis of data (see review in Moran 1990). Problems of reification have been addressed in recent years by an emphasis on how individuals modify the environment and not simply adapt to a reified nature (Balée 1998; Boster 1983, 1984; Crumley 1994). Today, few scholars would suggest that measurement of energy flow ought to be a central concern of ecosystem studies. Concerns have

[2]More details on the Tsembaga Maring studies by Rappaport can be found in Rappaport's ethnographic account (1968, 1984).

shifted to nutrient cycling, decision making, complexity in systems, and loss of biodiversity (Jordan 1987; Lansing 2003; Levin 1998; NRC 1999; NSB 1989). Studies show increasing attention to historical factors and even whole "schools of thought" on historical ecology, environmental history, environmental geography, and other spinoffs. The role of individuals and households has also blossomed (Lees and Bates 1990; Rindfuss et al. 2003; Roy Chowdhury and Turner 2006; Wilk 1990).

In the future, studies are likely to be most fruitful when they integrate a general systems approach with the study of how actors develop individual strategies. There is no reason why both perspectives cannot be used, and there is evidence that researchers have already begun to balance a concern for the individual with a concern for the population. One way to overcome the tendency toward static equilibrium models might be to study how populations adapt to certain kinds of stress. By studying the response of individuals to hazards, we can answer such questions as the following: Who responds? Does stress lead to changes in the structuring of the population? Are cultural patterns changed? How do people perceive the severity of the stress to which they are responding? How does the human population adjust to termination of the stress? These questions are more likely to be productive in outlining systemic interrelations in populations experiencing changing situations than in those with stable situations (cf. Lees and Bates 1990; McCabe 2004; Vayda 1983).

Historical ecology offers valuable insights to scholars from all disciplines interested in global environmental change, as well as a bridge between the social sciences and the humanities. History represents the recent record of what we know as the longer record of evolution, except that the historical record tends to be more detailed, more nuanced, and closer to contemporary conditions and offers provocative insights into alternatives to our current environmental dilemma. Global models tend to be coarse in scale and lack anything like an adequate representation of human variability and real biotic differences. One of the current and most exciting areas of research is the collaboration of paleoclimatologists, archaeologists, and historians in reconstructing the record of the past 300 years and eventually of the past 6,000 years (Johnson et al. 2005). Landscape history (Crumley 1994: 6) refers to the study of changing landscapes over time and in space. Human beings adapt to and bring about modifications in ecosystems—and have done so for thousands, if not millions, of years. Historically informed environmental analysis is a necessity if for no other reason than to correct the misperception that past environments were "pristine" and that only recently have humans begun to have an impact on the earth (Jacobsen and Firor 1992). Hardly any spot on earth is unaffected by human action, and humans have brought about changes, both positive and negative, in all landscapes. This record of human impact on environment offers a rich menu of choices we have made and their consequences, providing a view of alternatives much richer than a focus on the present would ever provide—choices to avoid and alternatives to be taken. Historical ecology brings together the approaches of ethnography, archaeology, history, and paleoscience to address environmental issues at regional and global scales (Crumley 1994), but the marriage of environmental history with historical ecology has not been consummated (Winterhalder 1994).

The differences between these two approaches come from the former coming from the discipline of history and therefore being reluctant to theorize, while the latter sees itself as a research program that emphasizes agency and historical context. It is not an insurmountable problem and is one that could be resolved by more interaction, given the desire of some environmental historians to ally themselves with ecological anthropological theory (Wooster 1984). To focus together on a given historical problem or landscape is likely to be the way forward.

In a recent and important volume, William Balée and Clark Erickson (2006) present historical ecology as a research program distinct from previous approaches. They suggest that their strategy is distinct from that of landscape ecology because they focus on how human beings bring about changes in landscapes. They take a strong position that there are no pristine environments but, rather, as soon as humans enter into an environment, it is made into a human landscape and modified by human actions for human objectives. They argue that human beings do not adapt to the physical conditions of the environment by adjusting their population size and settlement size to initial environmental conditions. Rather, they propose that humans transform those constraints into negligible analytical phenomena (Balée and Erickson 2006: 4) through transformation of soils, drainage, cropping practices, and so on. Further, they dismiss cultural ecology, ecosystem ecology, adaptationist approaches, and systems ecology because they "ultimately deny human agency" in positively changing the environment over time (ibid.). While there is value in emphasizing how local populations modify an environment to achieve their goals (Balée 1998), it is an overstatement to say that the adaptation approaches deny human agency (Rival 2006).

One current and popular theoretical approach used by environmentally oriented anthropologists and geographers is that of political ecology. Recently, the section on cultural ecology of the Association of American Geographers, for example, was renamed the section on cultural and political ecology (cf. Jarosz 2004). Human ecologists have become increasingly aware that power relations affect human uses of the environment. We are now aware of the role of environmental movements in exerting pressure on political bodies, corporations, and institutions. There is a renewed awareness of the potential value of human ecology in influencing policy and understanding the future of how humans impact the environment (Brosius 1999; Greenberg and Park 1994).

Few if any places in the world today are untouched by global forces such as climate change, capitalism, media, and the reach of the United Nations (Blaikie and Brookfield 1987; Braudel 1973; Rappaport 1993; Wolf 1982, 1999). We cannot ignore in environmental analysis the ways these relationships of local to global systems lead to particular outcomes (Moran 1982).

Political ecology bears great affinity with political economy, as both explore the role of power relations in affecting human uses of the environment, particularly the impact of capitalism on developing societies (Brosius 1997, 1999; Gezon 1999; Kottak 1999). Unlike political economy (with its central interest on class relations), political ecology is centered on the ravages that capitalism brings upon the environment and on human-habitat relations (Johnson 1995; Lansing 1991; Peet and Watts 1994; Rappaport 1993). It has been noted that political ecology has a tendency to privilege the local scale as more desirable than other scales, often viewing larger scales as oppressive of the local, and that this "local trap" can lead to major analytical errors (Brown and Purcell 2005). As a relatively new approach, political ecology still lacks a robust theory or a settled paradigm (Biersack 1999). The scale preference noted above is just one of several philosophical and theoretical traps that remain to be solved. As it matures, one sees some trends developing closer to the concerns of the environmental social sciences (Bates and Lees 1996; Crumley 1994) while others develop closer to so-called critical theory and cultural studies (Biersack 1999; Peet and Watts 1996).

At present, the bulk of political ecological analysis has stayed well within the concerns of the social sciences and distant from the physical and biological sciences in its data collection and methods of research. It has been more concerned with cultural and political critique and has only rarely presented a substantive body of environmental data as part of the analysis of political ecology. In short, it has been stimulating on the politics but less substantive on the environmental side. Vayda and Bradley Walters (1999) take issue with what they consider to be the dominant role claimed for political and political-economic influences in advance of the research (Bryant and Bailey 1997) instead of empirically examining a broader set of factors in which the outcome of what is most important is not known in advance. Lisa Gezon (1997, 1999), among others espousing political ecology, focuses on examining how people engage politically in contesting access to resources but only rarely presents environmental data on the resource being contested. Vayda and Walters (1999: 170) argue that ignoring the biological data can lead to unwarranted conclusions about the primacy of political influences. This may be a sign of political ecology's need to address the valid concerns of environmentalism and other political causes. But if its results cannot be integrated with the enormous efforts at understanding human dimensions of global change, conservation biology, environmental NGOs, and other local and regional agencies engaged in environmental protection, it may grow marginal to the very policy world it wishes to influence. It is important for political ecologists to join biophysical scientists in examining together the complex forces at play. As any other complex adaptive system, human ecosystem outcomes are nonlinear, have emergent properties, and can be remarkably counterintuitive. Political ecology and other ecologies used by environmental social scientists need to seek ever new ways to integrate knowledge and advance understanding of the complexities inherent in ecological systems.

One of the fields that expanded in reaction to environmental problems was institutional analysis. The field of institutional analysis has been interdisciplinary from the beginning, drawing on anthropology, sociology, political sciences, economics, and geography, among other disciplines, but fundamentally concerned with the management of common-pool resources (Acheson 2006; Agrawal 2003). There is significant overlap between political ecology and institutional analysis but enough differences in terms of intellectual community, research framework, and design to treat their development as distinct.

During the 1960s and the early 1970s debates on the catastrophic effects of human behavior, the uncontrolled increase in population and pollution, in addition to the dangers of irreversible environmental damages caused by overexploitation of natural resources at local and global scales, triggered the development of a variety of government regulations and policies to reduce and control the impact of human activities on the earth's resources (McCay and Acheson 1987). The drive to regulate through government intervention was made even more urgent by an extremely influential paper by Garrett Hardin (1968) published in the journal *Science.* Based on this work, "The Tragedy of the Commons," the unsustainable exploitation of natural resources and environmental services, caused by an increase in population and maximization of per capita consumption in the absence of rules of use, could be controlled only through privatization or centralized government. This oversimplification of common-pool resource management (Dietz et al. 2003; Ostrom et al. 1999) stimulated an extremely fertile area of investigation, focused on institutional arrangements to function as mediator between population and natural resources.

Institutional analysis of empirical case studies based on ethnographic work carried out by anthropologists soon uncovered the existence of a variety of successful institutional arrangements for the management of natural resources (McCay and Acheson 1987; Ostrom 1990). This analysis not only revealed that humans were not inherently destructive of their environment but also showed that what can be characterized as rational behavior (i.e., individuals maximizing some objective function) does not necessarily result in the negative outcomes Hardin suggested (Ostrom 1990). Research efforts within the largely interdisciplinary community involved with institutional analysis helped to identify two nodes that could contribute to solve the problem of common-pool resource management: restricting access to resources and creative incentives for responsible use. They also recognized several challenges facing the global commons (e.g., oceans, atmosphere, fisheries), among them the problems involved in expanding local- and national-level arrangements to manage global environmental resources, the effect of cultural differences in defining common rules, the primacy of national political and economic interests, understanding the complexity introduced by the interaction of various resources, and the rapid rate of environmental and social change (Brondízio et al. 2009; Ostrom et al. 1999).

Attention to common-pool resources occurred parallel to and in connection with the rise of indigenous and local social movements and reclamation of access to resources. This occurred as a reaction to expanding agrarian systems into land held under various institutional arrangements, a boom in the creation of parks in previously occupied forests, and a looming crisis for global fisheries. Building upon the now classic work of political scientist Elinor Ostrom (*Governing the Commons*, 1990), this field has developed with a rare combination of theoretical concern (e.g., collective action, game theory) and applied contributions. It arrived at a consensus about several elements significant to the successful management of common-pool resources at least at the local level. Thomas Dietz et al. (2003) discussed some of the key elements for effective management of common-pool resources: (1) effective, clear, and low-cost monitoring of resources; (2) moderate rates of change in resources and social settings; (3) intense communication and dense social networks within communities (also defined as high social capital); (4) the possibility of excluding outsiders from the resource at low cost; and (5) common agreement and support for the institutions in place.

A combination of "actor"- and "community"-centered approaches was particularly suitable for framing and testing the environmental outcomes of specific institutional arrangements. This line of investigation was formalized, for instance, in the institutional analysis and development (IAD) framework (Tucker and Ostrom 2005). In the words of Catherine Tucker and Ostrom (2005: 87), IAD depends on "theories of rational choice, collective action, common property, and social capital," and it focuses on "the action situation, which is composed of participants, positions, actions that respond to information and relate to potential outcomes, and the costs of benefits associated with actions and outcomes." Environmental conditions were important in this approach since they influenced the actors' choices and defined part of the single actor's assets and information.

Research on institutional analysis has also centered attention on scaling up lessons learned at the local level to larger scales (Berkes 2006), which is indicative of the strong presence of institutional perspectives within interdisciplinary research programs concerned with human dimensions of global environmental change (Young et al. 2008). Several examples illustrate the productive engagement of anthropology with institutional research, such as studying the intersection of community management and policy intervention in irrigation (Trawick 2001), fisheries and markets (Acheson 2003; McCay 1998), comanagement systems (Castro and McGrath 2003), and commodity markets (Brondízio 2008; Tucker 2008), among others. This research community continues to grow in an interdisciplinary manner, taking a less dogmatic approach toward the determinants of change in human and environmental systems. Recent efforts in institutional analysis have focused on developing a social-ecological system (SES) framework to support multilevel interdisciplinary research (Ostrom 2009). The basic structure of Ostrom's SES framework is organized in four main domains of analysis (resource systems, resource services and units, governance systems, and actors), each of which has a nested set of tiers of level-specific variables (McGinnis 2011; Ostrom 2009; Nagendra and Ostrom, submitted to *Ecology & Society*). An SES allows for the development of a shared lexicon of variables at different levels, data types and code systems, and integration, and as such it represents a promising tool to support collaborative research and cumulative knowledge on human-environment interaction.

1.4 A Road Map to the Chapters

Chapters in this volume embody the challenges of linking disciplinary expertise and interdisciplinary approaches to the practice of HEI research. In different degrees, chapters contribute ways of overcoming deterministic explanations (cultural, environmental, technological, or otherwise) in favor of historically and politically situated human interactions with the environment. We are aware that this coverage is not exhaustive, but it offers a microcosm of contemporary HEI research in terms of thematic, theory and methodology, level of analysis, and regional coverage. Drawing on research from eleven countries across four continents, chapters bring perspective from various specialties in anthropology and human ecology, institutional analysis, historical and political ecology, geography, archaeology, and land change sciences.

The ensemble of chapters in this volume also aimed at providing, although not all inclusive, a comprehensive sample of theoretical approaches and levels of analysis, regional problems, methodological design, and data collection tools, lending themselves useful to comparative research and to the training of graduate and undergraduate students. The rich array of methods deployed across chapters includes various applications of remote-sensing data (illustrating various forms of fusion of data with different spatial resolution); standard and participatory GIS; statistical, archival, and policy analyses; market surveys; and institutional analysis tools (e.g., IAD), in all cases informed by field research. This is perhaps the hallmark feature of contemporary HEI, that is, the imperative of understanding issues from both a bottom-up perspective informed by the empirical reality of people and localities, analyzed however within a regional framework. Chapters break away from reductionists' disciplinary confines to illustrate cross-sectional, longitudinal, and comparative approaches applied to indigenous, rural, peri-urban, and urban contexts. Field research techniques include a range of ethnographic and standard survey tools used to collect sociodemographic, health, and nutritional data; household- and community-level organization; institutional analysis; experimental economics; vegetation ecology; land-use/cover change (LUCC) inventories; and not least archaeological techniques.

The four parts are organized to reflect approaches to four dimensions of HEI research: health and adaptation approaches, land change and landscape management approaches, institutional and political-ecology approaches, and historical and archaeological approaches. These parts reflect not only attention to different societal problems but illustrate the complementarity of different analytical foci to these problems. While the book is organized in four thematic parts, one will find significant cross-sectional overlap in research approaches and underlying concepts across chapters. This indicates the shared conceptual and methodological basis and shared terminology within the interdisciplinary HEI research community. It is a recognition that the issues at hand cannot (and should not) be approached in isolation, that is, as contained within the domains of either social or physical science. It is this shared understanding of theory, concepts, and methods that offers a metalanguage for collaborative and comparative research addressing problems of societal interest. We see common underlying themes and factors intrinsic to HEI being analyzed across chapters, such as local livelihoods, the impacts of development and policy making, trajectories of urbanization, community and household change, changes in property regimes, conflicts between people and protected areas, and the pressure of globalization on resource systems. One of the underlying crosscutting themes is LUCC. This is not surprising, as it represents an integrative theme in HEI that links external pressures, human behavior and decision making, institutions, and biophysical process from local to global scales. Chapters ground the analyses of these themes in historical and institutional contexts, paying attention to trajectories of change and the interplay between sociodemographic, cultural, environmental, and political-economic variables; in other words, they approach problems by avoiding deterministic interpretations or decontextualized analysis.

The first part of the book brings together three dimensions of research on population health and adaptation, which capture, on one hand, the long-term impact of national development projects and regional transformations on indigenous people, and, on the other hand, emerging infectious diseases. The latter is a growing field of research that opens new interdisciplinary bridges between social, environmental, biological, and medical sciences in HEI research.

These chapters illustrate applications of several theories relevant to HEI research, while placing particular studies within broader trends of fertility, nutritional, and epidemiological transitions. The first and second chapters are unique for their rich longitudinal ethnographic perspectives capturing the ways development programs and major sociocultural transformations around and within indigenous areas have contributed to change the lives and livelihoods of indigenous people in lowland South America and the highlands of Tibet. By placing cases within broader comparative trends, these chapters speak to issues relevant throughout the world. The third chapter, on the other hand, benefits from a case study approach to reveal relationships between environmental and socioeconomic change in the spread of important global diseases such as SARS, Nipah virus, Ebola, malaria, and Lyme disease. Placing infectious diseases within the context of different forms of people-wildlife interaction, the chapter reviews pathways related to wildlife consumption, different forms of LUCC, and primate-based ecotourism. In discussing these pathways, it calls attention to the ways human behaviors act as direct or indirect drivers of change facilitating the spread of infectious diseases. This analysis is particularly relevant to inform our understanding of current and future trends in infectious diseases and possible mitigation pathways.

Chapters in Part II offer a rich array of studies on land change and landscape management across four continents, addressing problems of wildlife management, people and protected areas, and forest conservation across a range of private, public, and common-property regimes. As other chapters in this volume, it contributes to the emerging literature on landscape management and ecosystem services, in particular, by bringing attention to institutional arrangements, policy and economic incentive systems, and sociodemographic and cultural dimensions influencing decision making. The range of research approaches is equally informative to HEI research as it combines participatory GIS and qualitative methodologies, meta-analysis of case studies, institutional analysis, and a variety of remote-sensing techniques coupled with field assessments. Chapter 5 uses a participatory GIS and qualitative approaches to integrate the views of different stakeholders involved with red deer management in Scotland. It focuses on understanding (and finding solutions to) the mismatches between deer ecology and the institutional organization of landscapes bounded by different property and management regimes. Chapter 6, on the other hand, reveals a gradient of sociodemographic and economic conditions surrounding protected areas throughout India. Reviewing the drivers and pressures on 15 protected areas, four of which are studied in detail, it calls attention to the challenges of conservation in areas of high population pressure. As the previous chapter, it calls attention to the importance of considering conservation within a landscape perspective and the limits of conserving islands of resources in an increasingly interconnected world. Chapter 7 provides a comprehensive overview of forest protection in private areas in the United States; it highlights different forms of interactions between private and public decisions regarding the use of forest resources. Using an institutional perspective, it offers a useful approach to examine forests as bundles of property rights and bundles of ecosystem services interacting differently at different scales. Finally, Chap. 8 focuses on the challenges of monitoring landscapes representing a gradient of land-cover types in the African continent. Three case studies (Uganda, Botswana, and Namibia) are examined in detail to review the overlap between vegetation gradients and institutional arrangements representing different types of management areas.

Expanding on approaches to institutional analysis illustrated in the preceding part, Part III brings together five cases discussing the role of local histories, national policies, infrastructure change, and economic pressures and opportunities upon the evolution of institutional arrangements affecting natural resource management and urbanization. Chapters provide analysis representative of different social groups—from farmers and fishery communities to urban residents—living along a gradient of rural-urban settlements across the Americas. They illustrate the sophistication of institutional approaches in incorporating multiple methodologies and research tools, including longitudinal and cross-sectional approaches, experimental economics, remote sensing and GIS, surveys, and institutional analysis frameworks (i.e., IAD). Benefiting from longitudinal ethnography and historical research, Chap. 9 provides a long-term perspective to the evolution of institutional arrangements in the Amazonian floodplains. It examines the roles of federal agencies, NGOs, local fisheries unions, and local residents and reviews the advances and pitfalls of efforts to regularize and regulate land tenure and local resource management systems. Chapter 10 brings an experimental economics perspective to examine the role of incentives and sanctions on rural populations' behaviors toward natural resources in Colombia. It calls attention to the limits and potentially counterproductive results of penalty systems on small farmers' land-use decisions. Chapter 11 takes a comparative perspective to examine small farmers living within and around national forests in two contrasting regional realities of Brazil, that is, the Atlantic Forest in the state of São Paulo and the lower Amazon in the state of Pará. It pays particular attention to the role of national policies affecting small farmers in protected areas, some of which encourage production while others restrict their ability to make land-use decisions in areas of national forests. Building upon long-term ethnographic research, Chap. 12 provides a careful review of phases of institutional evolution in rural Honduras and its implications for local well-being and the forest environment. Integrating remote-sensing analysis and local socioeconomic indicators, it points to the mixed outcomes of economic and institutional changes to the local population.

Closing this part, Chap. 13 focuses on the fast process of urbanization in the United States and the differential roles of federal, state, and local land-use policies in shaping the directions of urbanization and exurbanization. Contextualized within a broader literature review, it examines in detail urbanization cases in the states of Ohio, Indiana, and Arizona and integrates institutional and policy analysis within a spatial framework, which allows the authors to compare the outcomes of different land-use policies and their regional particularities.

Finally, Part IV is dedicated to studies highlighting the value of historical and archaeological approaches to HEI. Ranging from decadal to centennial to millennial, these cases employ a range of methods to study trajectories of land-use intensification resulting from demographic and economic pressures and mediated by institutional arrangements and property systems. Chapters illustrate the influence of colonial policies in Uganda, discuss a unique group of colonists from Japan in post-WWII Brazil, and suggest the expansion of manioc-based agriculture in the Amazon resulting from European conquest. As in previous parts, one finds a diverse set of methodologies and evidence supporting historical approaches to HEI. Together, these chapters illustrate the integration of archival research, institutional analysis, meta-analysis of published studies, remote sensing of various time depths, vegetation ecology, and archaeological field investigation. Chapter 14 describes the fascinating history of Japanese colonization in the Amazon and their trajectories toward a leading position in global black-pepper production following WWII. After a crash in the black-pepper economy, the community pioneered intensive agroforestry using local and exotic fruit crops. The study goes further in examining the consequences of agroforestry intensification for land-cover change, carbon sequestration, and other ecosystem services vis-à-vis a growing tendency toward conversion to pasture in the region. Chapter 15 extends the time scale to the late nineteenth century to examine the impact of English colonial policies on land tenure in the West Mengo region of Uganda. Using aerial photographs, satellite imagery, archival research, and several fieldwork methods to assess land use and institutional arrangements, it reveals that contrary to dominant narratives of degradation by smallholders, the area has experienced increased tree cover since WWII concomitant with increased pressure for natural resources from urban and rural areas. The closing chapter of the volume takes us back to fundamental questions about the intensification of agriculture in pre-Colombian South America and the rise of sociopolitical complexity. It offers a fascinating account of the connection between language dispersal and diversity and plant and agricultural domestication in the region. It flips traditional theories of environmental determinism explaining the structure and distribution of pre-Colombian Amazon societies. It argues for the lack of evidence indicating intensive agriculture as the basis for large pre-Colombian populations in the region's floodplains and proposes a provocative hypothesis suggesting that diverse forms of intensive agroforestry, water resource management, and trade and regional resource acquisition as central to the economy of the region until the arrival of Europeans.

In the concluding chapter, authors review some of the underlying themes of the book and call attention to the recurrent challenges of developing cross-scale analysis and integrative frameworks to overcome deterministic approaches to HEI. We invite the reader to enjoy the richness of each localized chapter while exploring the various thematic, theoretical, and methodological threads connecting them independent of temporal and spatial scale or region of the world. Together, they are representative of a broader "epistemic community" concerned with advancing understanding of the interdependence of social and environmental problems through a network of interdisciplinary collaboration build upon the complementarity with disciplinary knowledge.

References

Acheson, J. M. (2003). *Capturing the commons: Devising institutions to manage the Maine lobster industry*. Hanover: University Press of New England.

Acheson, J. M. (2006). Institutional failure in resource management. *Annual Review of Anthropology, 37*, 117–134.

Agrawal, A. (2003). Sustainable governance of common-pool resources: Context, methods, and politics. *Annual Review of Anthropology, 32*, 243–262.

Alavi, S. M. Z. (1965). *Arab geography in the ninth and tenth centuries*. Aligarh (India): Aligarh Muslim University, Department of Geography.

Alland, A. (1975). Adaptation. *Annual Review of Anthropology, 4*, 59–73.

Balée, W. (Ed.). (1998). *Advances in historical ecology*. New York: Columbia University Press.

Balée, W., & Erickson, C. (Eds.). (2006). *Time and complexity in historical ecology: Studies in the neotropical lowlands*. New York: Columbia University Press.

Bates, D., & Lees, S. (Eds.). (1996). *Case studies in human ecology*. New York: Plenum.

Berkes, F. (2006). From community-based resource management to complex systems: The scale issue and marine commons. *Ecology and Society, 11*(1), 45 [online]. http://www.ecologyandsociety.org/vol11/iss1/art45/

Biersack, A. (1999). From the new ecology to the new ecologies. *American Anthropologist, 101*, 5–18.

Blaikie, P., & Brookfield, H. (1987). *Land degradation and society*. London: Methuen.

Boas, F. (1896). The limitations of the comparative method of anthropology. *Science (New Series), 4*, 901–908.

Boas, F. (1963[1911]). *The mind of primitive man*. New York: Greenwood Press. (Reprint of original published by Macmillan, New York)

Boas, F. (1964[1888]). *The central Eskimo*. Lincoln: University of Nebraska Press. (Reprint of original published by Smithsonian Institution Press, Washington, DC)

Boster, J. (1983). A comparison of the diversity of Jivaroan gardens with that of the tropical forest. *Human Ecology, 11*, 69–84.

Boster, J. (1984). Inferring decision-making from preferences and behavior: An analysis of Aguaruna Jivaro manioc selection. *Human Ecology, 12*(4), 343–358.

Boyd, R., & Richerson, P. (1985). *Culture and the evolutionary process*. Chicago: University of Chicago Press.

Braudel, F. (1973). *The Mediterranean and the Mediterranean world in the age of Philip II*. 2 vols. New York: Harper & Row.

Brondízio, E. S. (2008). *The Amazonian caboclo and the açaí palm: Forest farmers in the global market*. New York: New York Botanical Garden Press.

Brondízio, E. S., Ostrom, E., & Young, O. R. (2009). Connectivity and the governance of multilevel social-ecological systems: The role of social capital. *Annual Review of Environment and Resources, 34*, 253–278.

Brosius, J. P. (1997). Endangered forest, endangered people. *Human Ecology, 25*, 47–69.
Brosius, J. P. (1999). Green dots, pink hearts: Displacing politics from the Malaysian rain forest. *American Anthropologist, 101*, 36–57.
Brown, J. C., & Purcell, M. (2005). There's nothing inherent about scale: Political ecology, the local trap, and the politics of development in the Brazilian Amazon. *Geoforum, 36*(5), 607–624.
Bryant, R. L., & Bailey, S. (1997). *Third world political ecology*. London: Routledge.
Castaglioni, A. (1958). *A history of medicine* (2nd ed.). New York: Knopf.
Castro, F., & McGrath, D. (2003). Community-based management of lakes and sustainability of floodplain resources in the lower Amazon. *Human Organization, 62*(2), 123–133.
Crumley, C. L. (Ed.). (1994). *Historical ecology: Cultural knowledge and changing landscapes*. Santa Fe: School of American Research Press.
Darwin, C. (1859). *On the origin of species by means of natural selection*. London: Murray.
Dietz, T., Ostrom, E., & Stern, P. (2003). The struggle to govern the commons. *Science, 302*(5652), 1907–1912.
Durham, W. (1990). Advances in evolutionary culture theory. *Annual Review of Anthropology, 19*, 187–210.
Eder, J. (1987). *On the road to tribal extinction: Depopulation, deculturation and adaptive well-being*. Berkeley: University of California Press.
Ferndon, E. (1959). The drying oases of central Iran. *Southwestern Journal of Anthropology, 15*, 1–19.
Forde, C. D. (1934). *Habitat, economy and society*. New York: Dutton.
Geertz, C. (1963). *Agricultural involution*. Berkeley: University of California Press.
Gezon, L. (1997). Political ecology and conflict in Ankarana, Madagascar. *Ethnology, 36*, 85–100.
Gezon, L. (1999). Of shrimps and spirit possession: Toward a political ecology of resource management in northern Madagascar. *American Anthropologist, 101*, 58–67.
Goldenweiser, A. (1937). *Anthropology*. New York: F. S. Crofts and Company.
Gould, S. (1980). *The panda's thumb*. New York: Norton.
Greenberg, J., & Park, T. (1994). Political ecology. *Journal of Political Ecology, 1*, 1–12.
Hardin, G. (1968). The tragedy of the commons. *Science, 163*(3857), 1243–1248.
Harris, M. (1968). *The rise of anthropological theory*. New York: Crowell.
Helm, J. (1962). The ecological approach in anthropology. *The American Journal of Sociology, 67*, 630–639.
Huntington, E. (1915). *Civilization and climate*. New Haven: Yale University Press.
Jacobsen, J., & Firor, J. (Eds.). (1992). *Human impact on the environment: Ancient roots, current challenges*. Boulder: Westview.
Jarosz, L. (2004). Political ecology as ethical practice. *Political Geography, 23*(7), 917–927.
Johnson, B. (1995). Human rights and the environment. *Human Ecology, 23*, 111–123.
Johnson, C. D., Kohler, T. A., & Cowan, J. (2005). Modeling historical ecology, thinking about contemporary systems. *American Anthropologist, 107*(1), 96–107.
Jordan, C. F. (Ed.). (1987). *Amazonian rain forest*. New York: Springer.
Kottak, C. (1999). The new ecological anthropology. *American Anthropologist, 101*, 19–35.
Kroeber, A. (1939). *Cultural and natural areas of native North America*. Berkeley: University of California Press.
Lansing, S. (1991). *Priests and programmers*. Chicago: University of Chicago Press.
Lansing, S. (2003). Complex adaptive systems. *Annual Review of Anthropology, 32*, 183–204.
Lees, S., & Bates, D. (1990). The ecology of cumulative change. In E. F. Moran (Ed.), *The ecosystem approach in anthropology*. Ann Arbor: University of Michigan Press.
Levin, S. (1998). Ecosystem and the biosphere as a complex adaptive system. *Ecosystems, 1*(5), 431–436.
Little, M., & Morren, G. (1976). *Ecology, energetics and human variability*. Dubuque: William C. Brown.
Lowie, R. H. (1917). *Culture and ethnology*. New York: Peter Smith.
Lyell, C. (1830). *Principles of geology* (Vol. 1). London: John Murray.

Malthus, T. R. (1798). *An essay on the principle of population*. London: J. Johnson, in St. Paul's Church-yard.
McCabe, J. T. (2004). *Cattle bring us to our enemies: Turkana ecology, politics, and raiding in a disequilibrium system*. Ann Arbor: University of Michigan Press.
McCay, B. (1998). *Oyster wars and the public trust: Property, law, and ecology in New Jersey history*. Tucson: University of Arizona Press.
McCay, B. J., & Acheson, J. M. (1987). *The question of the commons: The culture and ecology of communal resources*. Tucson: University of Arizona Press.
McGinnis, M. (2011). An introduction to IAD and the language of the Ostrom workshop: A simple guide to a complex framework. *Policy Studies Journal, 39*(1), 169–183.
Moran, E. F. (1982). The evolution of Cape Verde's agriculture. *African Economic History, 11*, 63–86.
Moran, E. F. (Ed.). (1990). *The ecosystem approach in anthropology: From concept to practice*. Ann Arbor: University of Michigan Press.
Moran, E. F. (2006). *Human adaptability* (3rd ed.). Boulder: Westview.
NRC (National Research Council). (1999). *Global environmental change: Research pathways for the next decade* (Committee on Global Change Research). Washington, DC: National Academies Press.
NSB (National Science Board). (1989). *Loss of biological diversity* (Committee on International Science). Washington, DC: National Science Foundation.
Ostrom, E. (1990). *Governing the commons: The evolution of institutions for collective action*. Cambridge: Cambridge University Press.
Ostrom, E. (2007). A diagnostic approach for going beyond panaceas. *Proceedings of the National Academy of Sciences of the United States of America, 104*, 15181–15187.
Ostrom, E. (2009). A general framework for analyzing sustainability of social-ecological systems. *Science, 325*(5939), 419–422.
Ostrom, E., Burger, J., Field, C., Norgaard, R., & Policansky, D. (1999). Revisiting the commons: Local lessons, global challenges. *Science, 284*(5412), 278–282.
Peet, R., & Watts, M. (1994). Development theory and environmentalism in an age of market triumphalism. *Economic Geography, 69*, 227–253.
Peet, R., & Watts, M. (Eds.). (1996). *Liberation ecologies: Environment, development, social movements*. London: Routledge.
Rappaport, R. (1968). *Pigs for the ancestors*. New Haven: Yale University Press.
Rappaport, R. (1977). Ecology, adaptation and the ills of functionalism. *Michigan Discussions in Anthropology, 2*, 138–190.
Rappaport, R. (1984). *Epilogue: Pigs for the ancestors* (Rev. ed.). New Haven: Yale University Press.
Rappaport, R. (1993). The anthropology of trouble. *American Anthropologist, 95*, 295–303.
Rindfuss, R., Walsh, S., Mishra, V., Fox, J., & Dolcemascolo, G. (2003). Linking household and remotely sensed data: Methodological and practical problems. In J. Fox, S. Walsh, & V. Mishra (Eds.), *People and the environment: Approaches for linking household and community surveys to remote sensing and GIS* (pp. 1–30). Dordrecht: Kluwer.
Rival, L. (2006). Amazonian historical ecologies. *The Journal of the Royal Anthropological Institute, 12*(1), 79–94.
Robertson, W. (1777). *The history of America*. Dublin: Messrs. Price, Whitestone, W. Watson, Corcoran, R. Cross [and 41 others].
Roy Chowdhury, R., & Turner, B. L., II. (2006). Reconciling agency and structure in empirical analysis: Smallholder land use in southern Yucatan, Mexico. *Annals of the Association of American Geographers, 96*(2), 302–322.
Sahlins, M. D., & Service, E. (Eds.). (1960). *Evolution and culture*. Ann Arbor: University of Michigan Press.
Smith, E. A. (1991). *Inujjuamint foraging strategies: Evolutionary ecology of an Arctic hunting economy*. Chicago: Aldine.

Steward, J. (1938). *Basin-plateau Aboriginal sociopolitical groups* (Bulletin 120, Bureau of American Ethnology). Washington, DC: Smithsonian Institution.
Steward, J. (1955). *The theory of culture change*. Urbana: University of Illinois Press.
Taylor, G. (Ed.). (1951). *Geography in the twentieth century*. London: Methuen.
Thomas, F. (1925). *The environmental basis of society*. New York: The Century Company.
Trawick, P. (2001). Successfully governing the commons: Principles of social organization in an Andean irrigation system. *Human Ecology, 29*(1), 1–25.
Tucker, C. (2008). *Changing forests: Collective action, common property and coffee in Honduras*. New York: Springer.
Tucker, C., & Ostrom, E. (2005). Multidisciplinary research relating institutions and forest transformations. In E. F. Moran & E. Ostrom (Eds.), *Seeing the forest and the trees: Human-environment interactions in forest ecosystems* (pp. 81–103). Cambridge, MA: MIT Press.
Vayda, A. P. (1974). Warfare in ecological perspective. *Annual Review of Ecology and Systematics, 5*, 183–193.
Vayda, A. P. (1976). *Warfare in ecological perspective*. New York: Plenum.
Vayda, A. P. (1983). Progressive contextualization: Methods for research in human ecology. *Human Ecology, 11*(3), 265–281.
Vayda, A. P., & Rapppaport, R. (1976). Ecology, cultural and noncultural. In P. Richerson & J. McEvoy (Eds.), *Human ecology* (pp. 6–25). North Scituate: Duxbury.
Vayda, A. P., & Walters, B. (1999). Against political ecology. *Human Ecology, 27*(1), 167–180.
Voget, F. (1975). *A history of ethnology*. New York: Holt, Rinehart and Winston.
Wilk, R. (1990). Household ecology: Decision-making and resource flows. In E. F. Moran (Ed.), *The ecosystem approach in anthropology* (pp. 323–355). Ann Arbor: University of Michigan Press.
Winterhalder, B. (1994). Concepts in historical ecology: The view from evolutionary ecology. In C. Crumley (Ed.), *Historical ecology* (pp. 17–41). Santa Fe: School of American Research Press.
Winterhalder, B., & Smith, E. A. (Eds.). (1981). *Hunter gatherer foraging strategies: Ethnographic and archaeological analyses*. Chicago: University of Chicago Press.
Wolf, E. (1982). *Europe and the people without history*. Berkeley: University of California Press.
Wolf, E. (1999). *Envisioning power*. Berkeley: University of California Press.
Wooster, D. (Ed.) (1984). History as natural history: An essay on theory and method. *Pacific Historical Review, III*, 1–19.
Young, O. R., King, L., & Schroeder, H. (2008). *Institutions and environmental change: Principal findings, applications, and research frontiers*. Cambridge, MA: MIT Press.

Part I
Health and Adaptation Approaches

From the lowlands of South America to the highlands of the Himalayas and to a global tour of case studies, this section reveals the underlying processes and the local realities associated with broad fertility, nutritional, and epidemiological transitions. The chapters discuss the short- and long-term impacts of national policies and development projects, regional transformations, environmental change, and, broadly speaking, the impacts of globalization on indigenous peoples in particular and human health in general. They illustrate the complementarity of cross-sectional and longitudinal approaches to the understanding of human adaptation and health and the use of different human-environment interaction theories to link local case studies to broader national and global patterns. As also illustrated in other parts of the book, cross-sectional approaches add value to comparative analysis, while longitudinal research reveals the processual nature of human-environment interactions and the consequences, both expected and unexpected, of different drivers of change.

Santos, Coimbra, and Welch open this part of the volume, demonstrating the interrelatedness of health and various ecological, political, economic, gender, and sociocultural transformations affecting indigenous populations at different temporal scales. In Chap. 2, they review in detail the long history of interaction between the Xavante indigenous peoples of Central Brazil and national society within which contemporary changes should be analyzed. They provide an overview of their studies of health transition among Xavante communities since 1990 by examining three dimensions in detail: demographic changes, oral health, and nutritional transition in the context of socioeconomic differentiation. The authors illustrate the way longitudinal research is complementing cross-sectional approaches to understand the consequences of change and implications for population health and community well-being. They examine the long-term relationship between health profiles and emergent forms of socioeconomic differentiation (such as those defining gender relations) internal to specific Xavante communities. The authors give particular attention to the importance of understanding these transitions in the context of local histories, including variability internal to Xavante communities. They do so, however, by contextualizing these trends within broader patterns of health, demographic, and nutritional transitions in this and other parts of the world.

In Chap. 3, Childs, Goldstein, and Wangdui examine interactions between household-level reproductive decisions and available land resources, government policies, and new perceptions of family organization and the role of children on household economic strategies. Based on long-term research in three villages in Shigatse Prefecture of China's Tibet Autonomous Region, the authors integrate survey-based longitudinal demographic data gathered with in-depth interviews with household heads about factors affecting their decisions on family size. Two main theoretical orientations explaining connections between landholding and fertility are reviewed: the land-security and land-labor demand hypotheses. The authors emphasize the role of institutional factors mediating the relationship between landholding and fertility. The chapter provides a rich history of the region in five periods: "the old society," Tibet's commune systems, privatization and the family responsibility system, the implementation of fertility control, and the period of "replacement fertility." This section is followed by attention to the impact of China's birth control policy and the process by which it is contributing to the formation of different notions of family and images of modernity in rural Tibet. These discussions offer the background to examine trends in fertility decline as related to decreasing land per capita, state policies on birth control, and the changing roles that children play in households' long-term economic strategies. The authors highlight that while the connection between land and fertility is still important, it is becoming less relevant because of changes in economic conditions, including China's policy to promote rapid development in Tibet.

In Chap. 4, Muehlenbein takes a case study approach to reveal relationships between environmental and socioeconomic change contributing to emerging infectious diseases such as SARS, Nipah virus, Ebola, Malaria, and Lyme disease in different parts of the world. Muehlenbein places infectious diseases in historical context by discussing the different drivers affecting epidemiological transitions in different societies. Advances in sanitation and health care have shifted attention in some areas from infectious to chronic diseases. At the same time, major changes in population movement due to migration and various forms of tourism, climate change, urbanization, and land-cover change have contributed to the rapid evolution of new or more resistant infectious diseases. This overview provides a context to discuss different forms of human-wildlife interaction and their implications for infectious diseases. The chapter pays particular attention to the ways human actions may impact wildlife and in the process release pathogens previously buffered from human populations. Using data from various case studies, Muehlenbein examines three dimensions of human-wildlife interaction in detail. First, he examines the role of wildlife consumption, not only as food but also in other forms such as medicine. He then examines the complex ways land-use/cover change and climate change interact in affecting wildlife habitat and consequently may trigger the spread and rate of incidence of zoonotic diseases. Finally, Muehlenbein examines the expansion of primate-based ecotourism, including those promoting direct physical contact between human and nonhuman primates. The chapter calls attention to the direct and indirect drivers affecting emerging infectious diseases at different scales, the potential future implications of these changes, and suggests possible mitigation pathways to minimize the vulnerability of human societies and wildlife.

Chapter 2
A Half-Century Portrait: Health Transition in the Xavante Indians from Central Brazil

Ricardo Ventura Santos, Carlos E.A. Coimbra, Jr., and James R. Welch

Abstract This chapter discusses some of the most recent studies that have been conducted among the indigenous Xavante people in Central Brazil by members of the research group Health, Epidemiology and Anthropology of Indigenous Peoples, coordinated by Ricardo Ventura Santos and Carlos E. A. Coimbra Jr., from the Oswaldo Cruz Foundation, Rio de Janeiro. Members of this research group first studied the Xavante in 1990 and have since published on such diverse topics such as nutrition, subsistence, demography, epidemiology of infectious and parasitic diseases, and social organization. Recently, the group has focused on the relationship between health profiles and emergent forms of socioeconomic differentiation internal to specific Xavante communities. Additionally, they have begun publishing the results of longitudinal studies that document phenomena that had not yet been evident in cross-sectional approaches. In this chapter, they illustrate these new directions with examples of recent research on dental health, demography, and nutrition transition. These examples demonstrate the interrelatedness of health and various ecological, political, economic, and sociocultural transformations of different temporal scales.

R.V. Santos (✉)
Escola Nacional de Saúde Pública, Fundação Oswaldo Cruz and Departamento de Antropologia, Museu Nacional, Universidade Federal do Rio de Janeiro, Rio de Janeiro, Brazil
e-mail: santos@ensp.fiocruz.br

C.E.A. Coimbra, Jr. • J.R. Welch
Escola Nacional de Saúde Pública, Fundação Oswaldo Cruz, Rio de Janeiro, Brazil
e-mail: coimbra@ensp.fiocruz.br; welch@ensp.fiocruz.br

E.S. Brondízio and E.F. Moran (eds.), *Human-Environment Interactions: Current and Future Directions*, Human-Environment Interactions 1,
DOI 10.1007/978-94-007-4780-7_2,

2.1 Introduction

The first visit by members of our research group to the Xavante Indians, more specifically to Pimentel Barbosa village, also known as Etênhiritipá, was in July 1990. In Central Brazil, the months in the middle of the year are dry and nights are cold. It is a season with few clouds, making it possible to visually appreciate the vast and varied landscape. At the time we first visited, the Xavante were involved in many ritual activities related to youth initiation rites. That first visit made such an impression on us that we have returned many times. Although at the time we may not have anticipated the trajectory of our relationship with the Xavante of Pimentel Barbosa, our mutual involvement has deepened and diversified profoundly over the years.

That visit marked the beginning of a research program in the areas of health, human ecology, and biological anthropology that has continued to the present. The research team at that initial moment was coordinated by Nancy M. Flowers, Carlos E.A. Coimbra Jr., and Ricardo Ventura Santos. Nancy had conducted her doctoral dissertation research on subsistence, nutrition, and demography for 14 months in 1976–1977 in that same community and under the supervision of Daniel R. Gross at City University of New York (CUNY). At the time, Carlos and Ricardo were novices in studies regarding the Gê peoples of Central Brazil, though both had field experience in other regions of Brazilian Amazonia. Carlos was then a visiting researcher at the National School of Public Health (ENSP) of Fundação Oswaldo Cruz, in Rio de Janeiro, and had concluded his doctorate in anthropology at Indiana University under the supervision of Emilio F. Moran (with a dissertation in 1989 on medical anthropology about the Suruí of Rondônia, Brazil). In 1990, Ricardo was a doctoral student in biological anthropology (under the supervision of Paul L. Jamison and with Moran on his committee, he would defend his dissertation on the Tupi-Mondé of Rondônia in 1991).

In the following decades, many other researchers, the majority of which were master and doctoral students at ENSP, were affiliated with our group on health and indigenous peoples and conducted research at Pimentel Barbosa. Numerous articles, chapters, and books related to our research have been published on topics such as nutrition, subsistence, demography, epidemiology of infectious and parasitic diseases, and social organization, among others (see Coimbra et al. 2002). In 2002, we published a monograph titled "The Xavante in Transition: Health, Ecology and Bioanthropology in Central Brazil" (as part of a series at University of Michigan Press edited by Emilio F. Moran) that synthesized the studies that we and our colleagues had completed by that time. One of the people who joined our research group more recently was anthropologist James R. Welch, who conducted his doctoral dissertation on social organization and life cycles among the Xavante, defending his dissertation at Tulane University under the supervision of William Balée. James is currently a researcher at ENSP and an active participant in the research group on health and indigenous peoples.

Since the beginning of our research with the Xavante, our theoretical interest centered on processes of change and their interfaces with health. As we point out in the introduction of our book *The Xavante in Transition*,

> Our principal aim, in collecting and analyzing demographic, biological, epidemiological, and ecological data, has been to produce a diachronic view of the long and complex interaction between the Xavante people, especially those of the Pimentel Barbosa community, and the surrounding Brazilian national society. On a broader scale, our research may be seen as an attempt to understand how local systems interact with larger social, economic, and political institutions and processes (Coimbra et al. 2002: 1).

Today, we not only continue conducting health research with the Xavante but do so in a dynamic and collaborative manner that has led us to become involved in other research endeavors of interest to the community, including land issues and cultural heritage conservation.

Our objective in this chapter is to reflect upon the most recent studies that have been conducted among the Xavante by our research group. We are especially interested in highlighting theoretical questions and methods that were not emphasized in the first decade of our research (1990–2000) but became prominent in later years. Increasingly, our research addressed questions related to the influence of emergent forms of socioeconomic differentiation, especially on health profiles. Along these lines, our recent research tends to emphasize differentiation in health profiles internal to specific communities in addition to making broader generalizations about the Xavante population as a whole. We consider this an important aspect of our evolving research agenda, since elucidating generalizations may be made through an understanding of local particularities. This is especially true for ethnic segments of national societies, for which assumptions of uniformity often mask important underlying complexities. Additionally, we have begun publishing the results of longitudinal studies that document phenomena that had not yet been evident in cross-sectional approaches. In this chapter, we illustrate these new directions with examples of recent research on dental health, demography, and nutrition transition.

2.2 The Xavante: Ecology, Territory, and History

Xavante territory is located in the central region of Brazil, characterized by tropical scrub vegetation known as *cerrado*. At present, the Xavante number approximately 14,500 distributed among nine separate indigenous reserves (Fig. 2.1). In 2004, the villages in these territories totaled 160. Most of them (about 80%) had populations under 100.

According to contemporary elders at Pimentel Barbosa, the Xavante were previously highly mobile, once residents of the Atlantic coast, and more recently occupants of the wide expanse of the Central Brazilian Plateau to the east of the Araguaia River. They explain that the increasing presence of nonindigenous colonists forced them to migrate to their present territory relatively recently, an

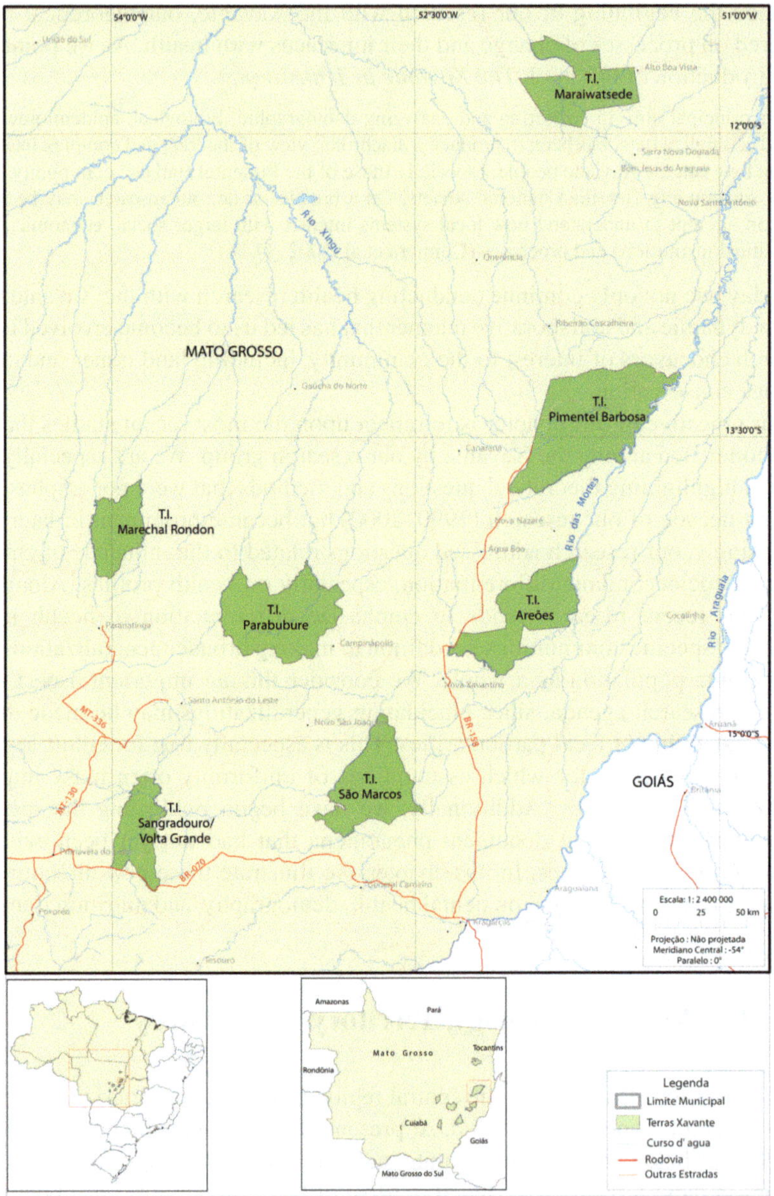

Fig. 2.1 Locations of Xavante indigenous reserves, Mato Grosso

event scholars estimate occurred in the mid-nineteenth century. These Xavante oral historians identify the banks of the Rio das Mortes in Mato Grosso as the location of the first permanent village, which differed from previous settlements because it was maintained for multiple years in succession (Coimbra et al. 2002).

According to historical documents, in the recent precontact era (pre-1940s), the Xavante were seminomadic, occupying vast territories in eastern and central Mato Grosso (Lopes da Silva 1992). Relatively large villages numbering hundreds of individuals were built near gallery forests along rivers, where a handful of crops (including maize, beans, and squash) were cultivated in modest quantities with slash-and-burn horticulture. These villages dispersed periodically into smaller trekking groups that subsisted for weeks to months by hunting and gathering. Foraging provided a significantly greater portion of dietary resources than horticulture at all times of the year, including periods of residence in permanent villages (Coimbra et al. 2002; Maybury-Lewis 1967; Santos et al. 1997).

In 1941, one year after a Brazilian Air Force airplane carrying Getulio Vargas, the president of Brazil, flew over the region to observe Xavante villages, an Indian Protection Service team entered the territory by land with the intent of "pacifying" them (Coimbra et al. 2002; Garfield 2001). At this time, the Brazilian government was engaged in the effort to occupy its central region, seen as "wilderness," and open it to settlement and economic integration with the rest of the country. Direct state intervention consisted in large part of road building and carrying out the enormous enterprise of building the new national capital Brasilia. As a result of the thrust of settlement and development during the 1950s and 1960s, a number of Xavante groups were dislocated from their lands. By the end of the 1950s, they were reduced to ten small patches of population (Maybury-Lewis 1967). In the words of Lopes da Silva (1992: 369), "pressure on their land and their villages had increased so much that, defeated by the indirect contact that decimated them, they saw no alternative than to deliberately seek peaceful coexistence with the whites."

During the same decades, the Xavante suffered from epidemic diseases that severely reduced their population. The large number of deaths resulted in constant movement of villages, forcing enemy factions to live in close proximity, which led to intense internal conflict. According to Lopes da Silva (1992: 372), "it was a period of great epidemics; it was also the time when people came in daily contact with the catechizing of the missions and with industrial goods. The fission and fusion of political factions and villages led to migration, but from this time it was only within a limited area, disputed piece by piece with non-Indians."

After being confined to reservations, the Xavante became increasingly sedentary and increased their dependence on agriculture. Trekking was gradually discontinued and is now rare. At this time, when the Xavante population was in steep decline from epidemic diseases and persecution, the military government of Brazil promoted a number of programs specifically aimed at the occupation and economic development of the Amazon region, emphasizing mechanized agriculture and cattle-raising. Migration from the south of the country to Mato Grosso intensified. As a consequence, the population of Mato Grosso increased rapidly, transforming its economy and its urban structure (Coimbra et al. 2002).

Toward the end of the 1970s, the National Indian Foundation (FUNAI), a government agency, introduced a project aimed at turning the Xavante into large-scale rice producers (Coimbra et al. 2002). The expectation was that the project would integrate them into the local market economy as sedentary and economically self-sufficient farmers. The scheme failed, but it left a trail of dependence on rice monoculture and changes in food habits. Rice is now the indispensable staple of Xavante diet, and at times, it is the only food available. In 1996, the Brazilian government, as part of a social project known as "United Communities" began to distribute a ration of basic foods on the reservations. This led to a decline in agriculture, an increase in village divisions, and greater external economic dependence.

Also in the 1970s, the Xavante began to reassert their land claims with vigor, seeking to recover possession of at least a portion of their historical territory. When their reservations were finally officially recognized, mainly in the 1980s and 1990s, the boundaries failed to include much of the land they considered their own. That struggle continues today, and the authors of this chapter have recently completed a land demarcation study for FUNAI at the request of the Xavante communities of the Pimentel Barbosa Reserve. Xavante lives continue to be conditioned by interaction with Brazilian society and political and economic events that occur beyond reservation boundaries (Lopes da Silva 1992).

During the 1990s, the question of how to provide health care for indigenous groups in Brazil came under intense discussion in the national political arena. As a result, the Indigenous Health Care Subsystem was instituted in 1999. At that time, the responsibility of providing indigenous health care was transferred from FUNAI to the National Health Foundation (FUNASA), which provides services under the Ministry of Health (Santos et al. 2008). A special health service for indigenous peoples was instituted and structured according to geographical districts called Special Indigenous Health Districts, or DSEI, which are linked to the Unified Health System (SUS). The headquarters of the Xavante district, one of the 34 DSEIs in Brazil, is located in Barra das Garças, Mato Grosso.

At present, salaries earned through remunerated public service positions (e.g., FUNAI employees, municipal school teachers, healthcare assistants), coupled with government retirement pensions and freelance work, constitute the major sources of income that are largely used by Xavante families to buy commercially available consumer goods, including food, at stores in neighboring towns outside the reservation. Those economic changes directly affect the quality of the food system. Today, rice accounts for most of the caloric bulk of the diet. Sugar, coffee, salt, pasta, soda drinks, crackers, and cooking oil are now regular items in Xavante meals (Coimbra et al. 2002; Gugelmin and Santos 2001; Welch et al. 2009). According to our observations, these and other foods circulate within and between households according to a strong morality of sharing and complex norms of kinship-based reciprocity, such that the nutritional impacts of food shortages and excesses may potentially be buffered. As we will see in this chapter, these dietary changes are closely associated with the ongoing health transition taking place among the Xavante.

It is important to mention that the Xavante people are among the most studied indigenous groups in Brazil, as evidenced in a variety of publications on diverse themes such as ecology, health, demography, ethnohistory, and social structure and organization (for a review, see Coimbra et al. 2002). The Xavante have factored importantly in public discourse since the mid-twentieth century for, among other things, being pioneers in the field of Brazilian indigenous politics (Garfield 2001) and being pivotal in the development of anthropological structuralism (see Maybury-Lewis 1979).

2.3 Health Transition and the Xavante

We now present a series of three case studies in health transition from our recent research among the Xavante. The first, focusing on demography, addresses the entire Xavante population resident on officially recognized reserves. The other two, addressing dental health and nutritional status of adults, focus on the population at Pimentel Barbosa, among which we have worked since the 1990s.

2.3.1 Recent Demographic Change

In this first section we present an overview of Xavante demography according to recent research, including a doctoral dissertation defended by Luciene Souza in 2008 at ENSP (Souza 2008; Souza et al. 2011). Whereas previous demographic studies focused on individual Xavante communities, this study addressed the population of all nine Xavante reserves. Demographic data covered the period from 1999 to 2004. They derived from household censuses and records of vital statistics collected by FUNASA local health teams, which were systematized and analyzed by Souza.

One of the difficulties of doing demographic research among indigenous peoples in Brazil is obtaining reliable and complete data. Populations are often small in number and geographically dispersed. It is therefore not surprising that many studies of anthropological demography carried out in Amazonia are based on data from specific communities, seldom considering whether they are representative of the entire ethnic group. Among the few exceptions are a study by Early and Peters (1990) of the Yanomama of Mucajai in the state of Roraima and a recent participatory sociodemographic investigation among the Sateré-Mawé of the state of Amazonas (Teixeira and Brasil 2005). These studies analyzed the demographic dynamics of relatively large and geographically dispersed populations. The Xavante study we present here is similar to these in scope.

A first aspect that deserves attention is that the Xavante population is not evenly distributed among the reservations. Pimentel Barbosa, with more than 300,000 ha, is the largest reservation and has the lowest population density, with 0.44 persons per square kilometer. Parabubure and São Marcos reserves, among others, have higher densities with 2.14 and 1.6 persons per square kilometer, respectively.

The overall results show that from 1999 to 2004 the mean annual growth of the Xavante population was 4.4%. The highest annual growth rate was at Pimentel

Barbosa (5.2%) and the smallest at São Marcos and Sangradouro-Volta Grande (3.9%). The high population growth rate of the Xavante is similar to those reported for a number of other indigenous peoples in Brazil (Pagliaro et al. 2005) and elsewhere in Latin America (McSweeney and Arps 2005).

As Azevedo and Ricardo (2002) note, at present, one of the great challenges for demography of indigenous peoples in Brazil is to identify the factors that may explain this increase. In the Xavante case, the figures clearly indicate that natural increase, i.e., the difference between births and deaths, is the principal factor responsible for the striking increase in population. Migration is an irrelevant component. Fertility analysis indicates a total fertility rate of 8.2 children per woman, from 1999 to 2004. In a society where marriage and maternity are practically universal, women's reproductive years start in their early teens, peaking at the ages of 20–29. As a result, the population is very young, with 54% under 15 years of age. Xavante population structure and their specific demographic indices are quite similar to those reported for other indigenous peoples in Brazil with high population growth (Pagliaro et al. 2005).

A very important aspect of the Xavante population in both demographic and epidemiological terms is that along with the high growth rate, death rates remain high. The crude death rate from 1999 to 2004 was 12.1 per 1,000. Infant mortality was 96.7 per 1,000 live births. In comparison, the infant mortality rate for all Brazil in 2000 was 29.6 per 1,000, and for the Central-West (Centro-Oeste) region, it was 21.2 per 1,000 (Simões 2002), pointing to major health disparities.

Demographic analyses of the entire Xavante population also demonstrated considerable variation in epidemiological and demographic indicators between Xavante groups. Notable differences are evident in practically all measurements analyzed. The total fertility rate varied from 6.7 at Marechal Rondon to 10.9 at Pimentel Barbosa; the infant mortality rate varied from 74.4 per 1,000 live births at Sangradouro-Volta Grande to 129.5 at São Marcos. Some differences between Xavante groups have been described in the ethnological literature. For example, Lopes da Silva (1986) divided Xavante groups into three major blocks according to the following criteria: (1) intensity and characteristics of contact between villages, (2) whether or not the group still occupied its traditional territory, and (3) whether agents of contact were representatives of the state or of the church. According to this model, the three blocks consist of (1) Pimentel Barbosa and Areões; (2) Batovi (Marechal Rondon), Culuene (Parabubure), and part of Couto Magalhães (Parabubure); and (3) Sangradouro-Volta Grande, São Marcos, and most of Couto Magalhães. Our demographic results do not permit in-depth analysis of the demographic characteristics of different Xavante groups according to an interpretive system such as that which Lopes da Silva (1986) suggested. However, recognizing their different historical trajectories is important in order to avoid unjustified generalizations about the Xavante as a whole.

The demographic data reveal some relevant trends when comparisons are made between the first half of the study period (1999–2001) and the second half (2002–2004). Despite being a relatively short period of time, by taking into consideration all the reservations, we can observe important changes from the first 3-year period to the second: a decline in crude birth rate from 61.2 to 57.3 per 1,000, a decline in total fertility rate from 8.5 to 7.1 children per woman, a decline in crude death rate from 12.7 to 11.5 per 1,000, and a marked decline in infant mortality rate from 103.3 to 89.8 per 1,000 live births. Similar patterns are observed for all seven individual reserves, although there is variation among them, probably due in part to fluctuation in small numbers.

Even though death rates are still high among the Xavante, there has been some decline since the implementation of the Indigenous Health Care Subsystem and the beginning of the demographic study period (both in 1999). There are indications that the system, while suffering from major deficiencies, has nevertheless afforded indigenous peoples greater access to health services than they had before 1999.[1] In an analysis of the first 3 years of the subsystem, Garnelo et al. (2003: 77) note that national health coverage provided by FUNASA to indigenous peoples became more inclusive and more health professionals entered the system to provide basic care. Lunardi et al. (2007) found that there was an increase in the use of hospital services by the Xavante population after 1999. They observed that rates of hospitalization increased for all four Xavante reserves analyzed. Including all reserves, the number of hospitalizations per 100 inhabitants almost doubled, from 5.1 to 9.5. It is important to emphasize that, despite increased access to hospital services, the major causes of hospitalization in Xavante children were diarrhea and acute respiratory infections, indicating major deficiencies in primary health care in villages (Lunardi et al. 2007).

While the reduction in Xavante mortality might be associated with greater investments made in indigenous health at the national level, the relationship between better health care and the observed decline in natality and fertility is neither obvious nor direct. A possible cause for this reduction might be more access to contraceptive technologies, although in the Xavante case this is probably not the explanation, since FUNASA did not provide routinely family planning or contraception services. There are no specific studies available on this politically delicate topic in Brazil, although the public position of some activists in the indigenous movement is opposition to contraceptives (Coimbra and Garnelo 2004).

Perhaps the most important finding that emerges from our demographic analysis of the Xavante population is that the decline in fertility is not the result of directed family planning but appears to be rather a consequence of the reduction in infant mortality through better health services. In other words, it may be due to mortality-fertility interaction. According to Wood (1990), fertility levels may be affected more by mortality than by reproduction itself, especially in populations with high mortality. As he explains, "when a nursing child dies, lactation is terminated and the mother resumes ovulating sooner than she otherwise would

[1] As we write this chapter, responsibility for indigenous health in Brazil has once again undergone a major reorganization, being transferred in late 2010 from FUNASA to the newly created Special Secretariat of Indigenous Health. The implications of this most recent change will not become apparent for some years to come.

have" (Wood 1990: 233). Therefore, paradoxically, elevated infant mortality may result in increased total fertility of a group of women. Since ovulation resumes when the death of an infant interrupts lactation, the interval between births is reduced, a phenomenon Wood calls "reproductive compensation." Alternatively, decreased infant mortality, as observed in the Xavante, may result in lower total fertility. This argument parallels the discussion of the "proximate determinants" of fertility by Bongaarts and Potter (1983), which cites the duration of lactation as an important factor in regulating fertility.

Survival analysis of Xavante interbirth intervals showed them to be consistent with the idea that the reduction of fertility from 1999–2001 to 2002–2004 could be due to fertility-mortality interaction. In the first place, there was an increase in average length of time between births of around 6%. Moreover, the results suggest a close relationship between the survival of infants and longer interbirth intervals. It is also notable that the decline in fertility was spread across all age cohorts, which suggests that the process was general and unfocused. If it were only in the older age brackets, we might suspect the use of contraceptives for family limitation. In the short length of time analyzed, it is improbable that fertility decline among the Xavante was due to extensive sociocultural change leading to new normative values regarding family size.

To conclude this section, it is appropriate to mention a major recent transformation in the demographic outlook for Brazil's indigenous population. During almost all of the twentieth century, it was believed that indigenous peoples were undergoing biological and cultural extinction and would not survive as socioculturally distinct populations in the twenty-first century. Only very recently, in the 1980s and 1990s, did it become evident that these populations were not only coming to occupy important political spaces (especially through the indigenous and indigenist movements in the aftermath of the 1988 constitutional revision), but also their population growth rates indicated rapid demographic recovery and growth. The demographic data from the Xavante point to an additional factor, whereby improved health services not only decrease mortality but also indirectly influence the indigenous population's fertility levels. These findings suggest that recent public health policies regarding indigenous peoples may have unintended effects, a subject that deserves research attention at the national level.

2.3.2 The Case of Oral Health

Research regarding health transition among the Xavante of Pimentel Barbosa benefits from an exceptional element, the existence of a detailed genetic and medical study conducted in the same population in 1962 by James V. Neel and Francisco M. Salzano. One aspect of their published analyses (Neel et al. 1964) addressed dental health conditions, especially the occurrence of caries. They observed a low prevalence of caries, which they explained as the result of a diet based on hunting and collecting, with limited dependence on agricultural products.

As in several other parts of the world (Grim et al. 1994; Jamieson et al. 2006; Niendorff and Jones 2000), the oral health of indigenous peoples in Brazil, exemplified by several cases showing a trend toward increased rates of caries and other pathological conditions, attests the interplay of dietary changes, insufficient access to preventive resources, and inadequate availability of health care. Notwithstanding those limited data, all too little is known about the human biological and epidemiological conditions of indigenous peoples related to changes in oral health conditions in Brazil (Arantes 2003; Arantes et al. 2001, 2010).

Very few studies have focused on the dental health of indigenous peoples in Brazil (Arantes et al. 2010). Those that were conducted were cross-sectional; none were prospective. It is especially productive to follow the trajectories of changing dental health conditions in indigenous peoples because the epidemiological transition may take different tracks, depending on ethnic group and socioeconomic and environmental contexts.

While national surveys have documented general tendencies toward reduction in the prevalence of dental caries in the Brazilian schoolchildren population (Narvai et al. 2006), studies in specific indigenous communities have noted an increase, which may be explained by dietary changes and limited access to health services (see review in Arantes et al. 2010). However, the notion that these peoples formerly had low prevalences of dental disease and are uniformly moving to higher rates should be examined with care. Some case studies fail to corroborate this tendency, indicating that consumption of sugar and length of contact are not the only determining factors of caries epidemiology in indigenous peoples (Arantes 2003).

We recently published the results of the first longitudinal study on oral health carried out in an indigenous group in Brazil (Arantes et al. 2009). The research was based on data collected in two surveys (1999 and 2004) and was carried out as part of a doctoral research project by Rui Arantes at ENSP. Following WHO methodology (WHO 1997), the study included 128 individuals, 63 (49.2%) males and 65 (50.8%) females, divided into four age groups (6–12, 13–19, 20–34, 35–60 years of age). The decayed, missing, and filled teeth index and incidences (difference between 1999 and 2004) were calculated for each individual.

Arantes et al. (2009) observed that exposure to risk of developing caries varied significantly according to age group and gender. Higher incidence of caries was found in females compared to males and in adults over 20 years of age compared to the 6–12 and the 13–19 age groups. Our interpretation is that these risk differentials may be related to a protective factor active in the younger age group and to a vulnerability factor that intensifies the incidence of caries in adults over 20 years of age. The protective factor may involve improved access to preventive measures, which were implemented at Pimentel Barbosa by Arantes in the mid-1990s and continued until 2009. These measures included regular monitoring of caries lesions, use of fluoride topical application and varnishes, supervised tooth brushing, and distribution of dental hygiene materials (toothbrushes, toothpaste, and dental floss).

We also observed that the high incidence of caries among women observed in the Xavante community at Pimentel Barbosa was an important factor in raising the incidence of caries in adults over 20 years of age. Men in the 20–34 age bracket showed an increase of only 1.08 teeth with caries, while women showed an increase of 4.37 teeth with caries. The mean increment in women was four times greater than that in men, thereby raising the increment of the age bracket as a whole.

Our interpretation is that certain socioeconomic and cultural factors may determine different oral health profiles between Xavante males and females, such as those that emerge from gendered production and consumption patterns and differential access to information, health services, and education. In a comparative study of the health determinants of indigenous women of the Americas based on data from seven countries (Canada, Ecuador, Guatemala, Mexico, Nicaragua, Peru, and the United States), researchers called attention to the centrality of gender to the understanding of health differences in indigenous communities (PAHO 2004; see also Verbrugge 1985). That study found that indigenous females show higher rates of dropping out of school and lower education levels, lower rates of bilingualism, and limited access to health services. It seems that a similar pattern may be present among the Xavante, whereby men and women have different roles in society, causing unequal exposure to factors responsible for the development of dental decay. For example, males and females not only have different reproductive roles but may also have unequal access to education and health preventive measures.

As we mentioned above, demographic data show that many of Brazil's indigenous populations experienced population recoveries after initial reductions caused by epidemics and sociocultural upheaval following contact with non-Indians. In the Xavante case, conditions improved somewhat in the 1960s, and, in the late 1970s, the population at Pimentel Barbosa began a process of recovery and demographic growth through an increase in women's fertility and a decline in infant mortality (Coimbra et al. 2002). The analyses carried out by Coimbra et al. (2002) showed that in the period from 1970 to 1990, the fertility rate of Xavante women increased relative to earlier periods. The rate for women between the ages of 15 and 40 was 7.9 births, with the highest age-specific fertility being that of women between the ages of 20 and 30. As we showed in the previous section, the total fertility rate at Pimentel Barbosa reached the impressively high rate of 10.2 in the period 1999–2004.

Clinical studies have shown that oral tissues may be affected by pregnancy (Laine 2002). Hormonal changes during pregnancy increase the susceptibility of gum tissues to local irritating factors and inflammatory processes. An increased concentration of estrogen in the saliva produces more epithelial shedding, creating an environment propitious to bacterial growth. Levels of the bacteria *Streptococcus mutans* and *Lactobacillus* spp. rise in the buccal milieu, and while there is no significant change in the flow of saliva, the saliva has a lowered pH and buffering capacity. The concentration of calcium and phosphate is slightly lowered during pregnancy, which may affect the remineralization of initial caries lesions. It appears that the effects of pregnancy are more closely related to the buccal milieu than to teeth themselves. These changes alter unfavorably the organism's capacity to resist the various risk factors for caries and gingivitis (Laine 2002; Lukas and Largaespada 2006).

Due to their high fertility, Xavante women spend most of their reproductive years pregnant or lactating, on average experiencing eight pregnancies. Consequently, alterations in the buccal environment due to pregnancy may be both frequent and long-lasting. This situation, related to the high value attributed by Xavante society to large families (Coimbra et al. 2002; Maybury-Lewis 1967), may make women more susceptible than men to mouth diseases like caries and gingivitis.

The Xavante economy was historically based on seasonal cycles of wild food procurement and horticulture, with marked gender roles in productive activities. In recent decades, reservation life has led to profound changes in the group's economy and the emergence of new configurations of gender roles. In the past, Xavante families spent much of the year engaged in hunting and gathering activities during treks throughout a large territory (Maybury-Lewis 1967). At that time, women provided the caloric basis of the dietary through gathering activities and men concentrated their attentions on hunting. Despite their distinct work activities, both women and men were very mobile. At present, Xavante women and men retain primary responsibility for gathering and hunting, respectively, although both now invest a great deal more effort in gardening. Furthermore, women and men have undergone other contrastive transformations in their modes of economic production. Today, women spend the greater proportion of their time in the village or at their gardens, where they attend to child care, handcraft production, and food preparation. Men, on the other hand, now tend to be more involved in economic activities that require interactions beyond the village. At Pimentel Barbosa, nearly all salaried jobs associated with the village school and the health post are occupied by men. Men are also more mobile than women, tending to visit neighboring towns with greater frequency than women. Due to this combination of factors, Xavante women have limited fluency in Portuguese as compared to men, have fewer years of schooling, and therefore may be less exposed to new habits that might impact oral health, including brushing and dental flossing.

The arguments we have outlined here suggest that changing sociocultural, economic, and dietary circumstances have caused Xavante women to face greater exposure to caries than men. The influence of gender upon oral health operates at several levels in this population. The social expectations for large families, perhaps the result of regional demographic changes and federal policies regarding indigenous settlement, cause female adulthood to be closely associated with pregnancy and breast-feeding, thereby causing specific adverse alterations in the buccal environment. At the same time, additional socioeconomic determinants of oral health, highly influenced by gender roles, place women at a disadvantage in accessing public health measures that may prevent caries.

2.3.3 Nutrition Transition and Socioeconomic Differentiation

The records of early visitors to Xavante communities often mention witnessing impressive physiques. A journalist who visited the Xavante village of São Domingos in the 1940s wrote, "Although they are not really giants, as was rumored at first, the Xavante physical presence is admirable. Bronzed, of medium height, among them we can find some individuals who are tall, others who are short. However, what seems to be most characteristic of the tribe are their fine, at times even Olympian, physical proportions. They are strong, vigorous, athletic" (Souza 1953: 99).

Fit Xavante bodies impressed not only journalists and visitors but also physicians and biologists who did research among them. When reporting on health conditions among the Xavante in 1954, Amaury Sadock de Freitas-Filho (1955: 155) referred to their good health and "fine physical appearance." The doctor even suggested that the Xavante physique represented "the physical standard that should be the Brazilian ideal" (Freitas-Filho 1955: 165). In the early 1960s, when James Neel, Francisco Salzano, and their collaborators made a detailed biomedical study of the Xavante at São Domingos, almost 20 years after first contact, they were hardly less emphatic. They wrote, "The general impression of the men was of exuberant health and vitality. They were erect in carriage, deep-chested, and very well muscled, with a notable absence of adiposity" (Neel et al. 1964: 110).

Subsequent reports began to depict a very different scenario. They suggest that the Xavante, especially those belonging to certain communities, are experiencing an accelerated nutritional and epidemiological transition. Endocrinologist João Paulo Botelho Vieira-Filho, from the Escola Paulista de Medicina, wrote, based on his many years of experience in providing medical care in the Xavante reservations of Sangradouro and São Marcos:

> The Xavante, who were slim before contact ... have become overweight or obese (some weighing more than 100 kg) because of dietary changes that came with the government rice-growing project ... [and] high consumption of rapidly absorbed carbohydrates. Their consumption of vegetable fiber is drastically reduced. Every time I visit the western Xavante I see more cases of diabetes. Twenty years ago there was no diabetes among them. (Vieira-Filho 2000: 2)

As it is widely recognized, excess weight is a major health concern throughout the world, both in industrialized countries and in developing regions undergoing rapid urbanization (Popkin 2001, 2006; WHO 1998). Some authors refer to obesity as a worldwide epidemic and emphasize its greater prevalence in low socioeconomic status populations in developing countries (Caballero 2007). High-fat and low-fiber diets, combined with low levels of physical activity, play an important role in the increase of overweight and obesity. Obesity may have a number of different impacts on health: it is associated with diabetes, cardiovascular disease, cancer, hypertension, gallbladder disease, and dyslipidemia (Popkin 2001, 2006; WHO 1998).

Recent case studies carried out among indigenous peoples in Brazil show an emerging pattern of high prevalence rates of overweight and obesity in adults (see review in Coimbra and Santos 2004). Several of those studies were carried out in the Amazon region. Comorbidities associated with obesity (e.g., hypertension and non-insulin-dependent diabetes mellitus) also emerged recently as important health challenges to indigenous peoples in the region. These changes are related to swift and profound shifts in nutritional profiles, associated with modifications in dietary intake and nutrient expenditure patterns, indicative of an ongoing nutrition transition.

The adoption of Western diets (i.e., high in saturated fats, salt, and sugar), reduced levels of physical activity, and increased use of alcohol are considered important aspects of the nutrition transition in developing regions (Batista-Filho and Rissin 2003; Kain et al. 2003; Popkin 2001, 2006). In Amazonia, such transition takes place in contexts of rapid sociocultural and environmental transformations accompanying increased involvement in the market economy. Recent studies show that this process can be internally heterogeneous, with intracommunity variation in socioeconomic status correlating with body composition measures. One study carried out among the Suruí Indians from the southwestern Brazilian Amazon showed that both adult males and females in wealthier households, as reflected in household infrastructure and presence of consumer durable goods, had significantly higher mean values of weight, body mass index (BMI), and skinfold thickness (Santos and Coimbra 1996). More recently, Dangour (2003) demonstrated that greater BMI among the Wapishana from Guyana was strongly associated with differential wealth between households, approximated by the value of household consumer durables. Godoy et al. (2005a, b) found among the Tsimane from lowland Bolivia a stronger correlation between nutritional status and wealth, defined in terms of physical assets, than between nutritional status and income. Lourenço et al. (2008) restudied the Suruí Indians previously investigated by Santos and Coimbra (1996) and characterized the nutritional transition taking place in the population in relation to socioeconomic and gender issues. Together, these studies demonstrate that nutrition transition among indigenous Amazonian populations varies within societies and point to household wealth measured in terms of durable goods as a useful proxy for internal socioeconomic differentiation as it relates to health discrepancies.

In our 2002 book *The Xavante in Transition*, we analyzed in detail the epidemiological transition that the Xavante have been experiencing. We argued that although infectious and parasitic diseases were still the leading causes of sickness and death, it was clear that chronic noncontagious diseases, including hypertension and diabetes, were becoming increasingly important. Among the Xavante, changes that have taken place in settlement patterns, levels of physical activity, and diet are affecting health conditions. We compared anthropometric and other health-related data collected in the 1990s with data collected in the 1960s. The trend was toward changes in body composition (weight gain) and emergence of hypertension. We argued that in order to understand the trend toward weight gain at Pimentel Barbosa, altered patterns of physical activity might be of even greater importance than nutritional shifts.

We recently concluded a study that aimed to readdress the issue of nutritional transition among the Xavante (Welch et al. 2009). In this investigation, our first goal was to compare contemporary data with historical information collected in 1962 and 1990 in the same community, permitting a diachronic perspective of anthropometric changes associated with ecological and socioeconomic transformations. With a time depth of 45 years, this diachronic view has no parallel in human ecological and bioanthropological studies in the Brazilian Amazon. Our second objective was to compare two sets of original field data collected in 2006, one that assessed household socioeconomic status and a second that recorded anthropometric measurements of individual participants. Those two data sets were compared statistically to ascertain how internal socioeconomic differentiation may be associated with differences in body morphology.

In order to assess within-community socioeconomic variation, we identified two apparent and measurable socioeconomic variables (monetary income and industrial durable consumer goods) to serve as proxies for household socioeconomic status. We opted to work with household monetary income and industrial goods for several reasons. First, because we sought to assess aspects of socioeconomic differentiation related to recent nutritional change, we emphasized types of income and wealth that reflected market integration. This strategy differs from studies that include subsistence production in income calculations and locally produced goods in wealth calculations (e.g., Godoy et al. 2005a, b). Second, the particulars of the Xavante situation facilitated our assessment of both monetary income and wealth, two commonly used indices of socioeconomic differentiation. Some other studies of socioeconomic differentiation and anthropometric variation in Amazonian societies assessed wealth but not monetary income due to income unpredictability or participants' reluctance to disclose incomes (e.g., Dangour 2003; Lourenço et al. 2008). In contrast, at Pimentel Barbosa, we found monetary income to be relatively predictable and income sources to be willingly disclosed.

Income was assessed by tabulating the quantity and type of routine personal income sources for all members of each household. Quantification of consumer goods owned by members of each household was based on a comprehensive list of privately owned durable industrial consumer goods present in the community at the time of the study. The items were motorcycles, satellite dishes, televisions, VCR players, DVD players, audio recorders, portable stereos, cellular telephones, film and digital cameras, gas stoves, sewing machines, firearms, handguns, bicycles, and fishing nets. Local market value of each consumer item included in the sample was later estimated by informal survey of local sources and consumers.

In comparing our results with data from studies undertaken in the Pimentel Barbosa population at earlier points in time, we observed that the adult Xavante population has experienced rapid and substantial increase in adiposity, as indicated by significant differences in mean values for weight and BMI between 1962 and 2006, while, at the same time, stature remained stable. A dramatic finding is that, on average, adult Xavante men and women are 8.9 and 13.8 kg heavier, respectively, than in 1962 when Neel and his collaborators conducted their original study (see Table 3 in Welch et al. 2009).

As is the case among other recently studied indigenous Amazonian peoples (Lourenço et al. 2008), the substantial weight gains observed through time among Xavante adults, which reached a frequency of overweight or obesity of 68.6% for individuals $\geq$20 years of age, resulting from a combination of changes in diet and physical activity. Recent research conducted in other indigenous communities in Brazil has also found very high prevalence of overweight or obesity in adults, surpassing 70% in many cases (see review in Coimbra and Santos 2004).

The few studies that have also analyzed the anthropometric profile of the same indigenous population at different moments have confirmed the rapid pace and pervasiveness of the ongoing nutrition transition among Brazilian indigenous peoples (see, for instance, the studies carried out by Lourenço et al. (2008) and Santos and Coimbra (1996) among the Suruí from Rondônia and Sampei et al. (2007) among the Kamayurá of Mato Grosso). As in the Xavante case, those studies emphasize the association between the emergence of overweight and obesity with the commoditization of indigenous economies, bringing about the Westernization of indigenous diets and the reduction of physical activity levels. Not surprisingly, chronic metabolic disorders and cardiovascular disease are also rapidly becoming part of the health-disease profile of indigenous peoples in Brazil (Coimbra and Santos 2004).

As we have mentioned, in recent years, there has been an increase in the number of studies pointing to associations between body composition measures and socioeconomic conditions in indigenous peoples in Lowland South America. Common to those studies is the finding that the processes of socioeconomic change, including increased market involvement, create internal differences within communities that come to be expressed in health differentiation. Our findings show that socioeconomic differentiation is associated with anthropometric measurements in the study population. Notably, the direction of the relationship is the opposite of that often observed in developing countries. Whereas increased weight is often associated with poverty (Caballero 2007), in Xavante society, it is more prevalent among adults from households with greater socioeconomic status. Although our cross-sectional data preclude direct assessment of the historical relationship between socioeconomic change and changes in the nutritional profile, they suggest that socioeconomic status is a factor in the internal contours of the nutrition transition that is taking place among the Xavante. The results also suggest that the process of differentiation and its relation to anthropometric variables is a recent one, as we did not observe differences in stature among adults, which reflects long-term linear growth.

It is relevant to recall that in the Xavante case we are dealing with a society that, in the recent past, did not show significant internal economic differentiation, given that the mode of production was largely identical for all households. Furthermore, even today the Xavante maintain a strong morality of equality and sharing. Pervasive social-leveling mechanisms, such as kinship-based reciprocity, multiple intersecting social affiliations that structure sharing, and the social expectation that those with greater resources should give freely to those with less, may serve to minimize the dietary effects of household income and wealth disparities among male and female individuals. Accordingly, sharing behaviors may mitigate the health effects of socioeconomic differentiation among members of higher-income households and perhaps transfer and dilute them as those benefits are passed to others. Yet our data corroborate the findings of other studies which show that despite reciprocity and other sharing practices, a clear relationship remains between socioeconomic status and adiposity (e.g., Godoy et al. 2005a, b).

Among the most striking findings of our recent study is the very clear role of gender in internal adiposity differentiation. Socioeconomic indicators were strongly associated with adiposity measures for females, considered separately, but not for males. That pattern has been documented elsewhere in the Amazon region, such as the case reported by Lourenço et al. (2008). As in that case, our ethnographic data suggest that the gender discrepancy observed in the relationships between adiposity and income and wealth measures may be explained by gender roles and food sharing.

Among the Xavante, females and males may have different access to food resources, such that female diets reflect more closely the income and wealth characteristics of their households. At Pimentel Barbosa, males tend to eat away from home and share food resources more frequently and according to more diverse social configurations than females. This occurs for several reasons. Males tend to go to town more often than females and during those trips eat in groups in restaurants and purchase food in supermarkets. Males also eat from their yields during hunting and fishing excursions before taking them back to their households. Participation in those subsistence activities was not limited by income-related factors, since 75% of incomes were retirement pensions and only 11 men had employment obligations that partially reduced their availability to hunt or fish. According to our observations, men with high personal and household incomes were among the most avid hunters in the community. Furthermore, males often share food resources according to male-only principles involving extended kinship ties, relative age, and ceremonial group affiliations (other sharing principles apply to women and to both sexes). For example, male members of younger age sets often furnish food to male members of older allied age sets. Also, sons-in-law are generally permitted to eat freely at their natal households throughout life, whereas females frequently do not have an analogous alternative venue within the village. Taken together, these male food consumption and sharing practices may mitigate against residence-based income and wealth disparities manifesting in fatness differentials among men.

Besides the epidemiological component of nutrition transition in our recent research at Pimentel Barbosa, we have also sought to address in greater depth aspects related to Xavante perceptions of the transition in process. Important in this respect have been sociocultural analyses carried out by James R. Welch. As we will see below (and detailed in Welch et al. 2009), these analyses show that Xavante perspectives of health and body weight are highly compatible with, if not in technical agreement with, our finding of an association between socioeconomic indicators and adiposity measures. Taking those compatibilities as a point of departure, we have sought new avenues for interchange between Xavante perspectives and our research in order to stimulate critical and productive attention within the Xavante community to the problem of nutrition transition.

Xavante adults unanimously agree that contemporary life at Pimentel Barbosa is characterized by poorer diets and greatly reduced levels of physical activity than in the past, and that those changes are important factors (among others) in what they perceive to be an overall deterioration in health and nutritional status throughout the population. However, they also frame those changes in terms of a distinction between "strengthening" foods and behaviors (*danhiptetezé*), which promote health, and "weakening" ones (*danhip'uwazé*), which diminish it. Although opinions vary, important strengthening foods include wild tubers, traditional maize, wild game meat, and other collected, gathered, and traditional horticultural foods. Examples of weakening foods are cooking oil, salt, pasta, candy, alcoholic beverages, and other introduced and industrial foods. Certain lifestyle and behavioral habits are similarly believed to promote strength and health. For example, traveling in the forest by foot rather than by vehicle is considered to strengthen the body. Also, for men, participating in spiritual rituals and other demanding ceremonial rites is also thought to be a vital source of strength, whereas abandoning them is thought to cause weakness. A theme common to each of these distinctions is an association between foods and activities perceived to be traditional and healthiness. Importantly, according to the Xavante point of view, excess fatness (*dahöi'pi*—"fat skin") is a common result of nontraditional diets and lifestyles.

Those Xavante interpretations of nutritional transition and adiposity are compatible with epidemiological explanations even if they are not in absolute technical agreement. Specifically, Xavante individuals tend to attribute excessive weight gain not only to specific foods ("weakening" foods) and types of behaviors (avoidance of resistance-building activities) but also more generally to a lack of traditionalist values. Furthermore, many Xavante see money as a potentially corrupting influence that can diminish cultural continuity and compromise traditionalist values. Although we do not make such value judgments, our finding that adiposity is positively associated with income is compatible with the Xavante perspective. More generally, we perceive an emergent awareness at Pimentel Barbosa of the broad deleterious health effects of weight gain, many of which are not accounted for by Xavante ethnomedicine, and a strong disposition toward reducing its prevalence through reconsideration of dietary practices. There is also keen awareness of the practical difficulties in motivating and sustaining behavioral changes to the effect that community leaders explicitly seek solutions that are congruent with contemporary lifestyles and values.

A product of conversations between members of our research group and the community at Pimentel Barbosa regarding diet and health is an ongoing collaborative initiative sponsored by the Museu do Índio, Rio de Janeiro, in which Xavante youth, trained in audiovisual documentation, record traditional subsistence techniques and foodways from their own culture. One of our shared goals for this endeavor is to stimulate interest in traditional foodways among youth and provide tools that facilitate intergenerational communication. It is their hope, and ours, that these new efforts may improve youth appreciation of traditional Xavante dietary culture and thereby better equip young people to make lifestyle choices that implicate good health (Welch 2010).

2.4 Final Considerations

In *Disease and Social Diversity: The European Impact on the Health of Non-Europeans*, Stephen Kunitz (1994) makes a comparative analysis of what he calls the "changing epidemiological regimes" of indigenous peoples in different parts of the world, drawing his examples from North America, Australia, New Zealand, and Polynesia. Kunitz emphasizes the ways in which processes of health and disease overlap with colonial policies, with political institutions, and with indigenous cultures and patterns of social organization, in order to demonstrate that "diseases rarely act as independent forces but instead are shaped by the different contexts in which they occur" (1994: 5).

Stressing the importance of recognizing diversity and giving attention to local contexts, Kunitz argues that although it is possible to find parallels in the situations of indigenous peoples in different parts of the world—when placed in national context invariably they have lower life expectancies, higher morbidity and mortality, and higher rates of violent death compared to nonindigenous peoples—this inequality is produced and reproduced in multiple and diverse forms (see also Stephens et al. 2009). Kunitz does not believe that a few explanatory schemes can be made to render, in any satisfactory way, the complexities faced by human groups living in different historical, social, economic, and political situations. This is even truer of indigenous peoples, who have such diverse histories of interaction with Western expansion and colonialism:

> At our present stage of knowledge and in the wake of the recent collapse of many old certainties, it is more useful to understand in detail the myriad ways in which different causes of morbidity and mortality in populations are affected by social processes, rather than strive to build grand theories. (Kunitz 1994: 4–5)

In our research on the Xavante, we have tried to adhere to an analytic focus that recognizes the influence of local history, or, as Kunitz writes, adopt a "particularistic approach to the study of diseases in populations" (1994: 4). This approach is especially pertinent to our discussion of the changing health and disease patterns of the Xavante. The subject of gender, mentioned in the introduction of this chapter, is an important example. In the case of demography, certain female social and biological processes, such as fertility, are shown to be closely linked to changing national health policies. Similarly, oral health data demonstrate a link between gendered health disparities and Xavante sociocultural values that favor large families, which may in turn be associated with historical political processes at regional and national levels. Nutrition data also point to a relationship between a more elevated weight gain among women, as compared to men, and emergent socioeconomic differentiation accompanying the sociopolitical process of increasing market integration. When considering the various forms that the "epidemiological transition" may assume, we find that they are made up of elements that include political history, demography, human ecology, and epidemiology—in short, the themes we have been addressing in our research over the years.

A benefit of reflecting on one's past research is that it offers an opportunity to recognize patterns that are not otherwise obvious. An important pattern that emerges from the three studies we have highlighted in this chapter is the interrelatedness of health and various ecological, political, economic, and sociocultural transformations of different temporal scales. In the Xavante case, this dynamic may be framed as involving interaction between both biology and culture and internal and external sociocultural factors. For example, although gender was not a primary focus of our research, each of the three studies we discuss in this chapter reveals gender to be an important component of health. Gender, considered in relation to biology, demography, social behavior, and political policy, proves to be an important factor in diverse aspects of health transition among the Xavante. Such interconnections are central to our ongoing research and motivate us to pay special attention to relationships between health and environmental context, which we consider in both its physical/biological and sociopolitical dimensions.

Acknowledgment This chapter was originally presented in the "Human-Environment Interactions Workshop" organized by Emilio Moran and Eduardo Brondizio at the Anthropological Center for Training and Research on Global Environmental Change (ACT), Indiana University, Bloomington, Indiana, February 25–27, 2010. We thank the Xavante of Pimentel Barbosa for their continual support and interest in our research activities over the years. We also thank Luciene Souza and Rui Arantes, who came to ENSP in the 1990s to carry out graduate training in indigenous health and became close collaborators in our research group and collaborated in the case studies described in this chapter. Over the years, our research with the Xavante has been supported by the Wenner Gren Foundation, the Brazilian National Research Council (CNPq), Fundação Oswaldo Cruz (Fiocruz), and the Fulbright-Hays Doctoral Dissertation Abroad (DDRA) Program.

References

Arantes, R. (2003). Saúde bucal dos povos indígenas do Brasil: panorama atual e perspectivas. In C. E. A. Coimbra Jr., R. V. Santos, & A. L. Escobar (Eds.), *Epidemiologia e saúde do povos indígenas no Brasil* (pp. 49–72). Rio de Janeiro: Editora Fiocruz/ABRASCO.

Arantes, R., Santos, R. V., & Coimbra, C. E. A., Jr. (2001). Saúde bucal na população indígena Xavante de Pimentel Barbosa, Mato Grosso, Brasil. *Cadernos de Saúde Pública, 17*, 375–384.

Arantes, R., Santos, R. V., Frazão, P., & Coimbra, C. E. A., Jr. (2009). Caries, gender and socio-economic change in the Xavante Indians from Central Brazil. *Annals of Human Biology, 36*, 162–175.

Arantes, R., Santos, R. V., & Frazão, P. (2010). Oral health in transition: The case of indigenous peoples from Brazil. *International Dental Journal, 60*, 235–240.

Azevedo, M. M., & Ricardo, F. P. (2002). *Censo 2000 do IBGE revela contingente "Indígena" pouco conhecido*. São Paulo: Instituto Socioambiental.

Batista-Filho, M., & Rissin, A. (2003). A transição nutricional no Brasil: tendências regionais e temporais. *Cadernos de Saúde Pública, 19*(Suppl. 1), S181–S191.

Bongaarts, J., & Potter, R. G. (1983). *Fertility, biology, and behavior: An analysis of the proximate determinants*. New York: Academic Press.

Caballero, B. (2007). The global epidemic of obesity: An overview. *Epidemiologic Reviews, 29*, 11–15.

Coimbra, C. E. A., Jr., & Garnelo, L. (2004). Questões de saúde reprodutiva da mulher indígena no Brasil. In S. Monteiro & L. Sansone (Eds.), *Etnicidade na América Latina: um debate sobre raça, saúde e direitos reprodutivos* (pp. 153–173). Rio de Janeiro: Editora Fiocruz.

Coimbra, C. E. A., Jr., & Santos, R. V. (2004). Emerging health needs and epidemiologic research in indigenous peoples in Brazil. In F. M. Salzano & M. Hurtado (Eds.), *Lost paradise and the ethics of research and publication* (pp. 89–109). Oxford: Oxford University Press.

Coimbra, C. E. A., Jr., Flowers, N. M., Salzano, F. M., & Santos, R. V. (2002). *The Xavante in transition: Health, ecology, and bioanthropology in Central Brasil*. Ann Arbor: The University of Michigan Press.

Dangour, A. D. (2003). Cross-sectional changes in anthropometric variables among Wapishana and Patamona Amerindian adults. *Human Biology, 75*(2), 227–240.

Early, J. D., & Peter, J. F. (1990). *The population dynamics of the Mucajai Yanomama*. San Diego: Academic Press.

Freitas-Filho, A. S. (1955). Inquérito médico sanitário entre os Índios Xavante. In M. F. Simões (Ed.), *Relatório de atividades do Serviço de Proteção aos índios durante o ano de 1954* (pp. 145–172). Rio de Janeiro: Serviço de Proteção aos Índios.

Garfield, S. (2001). *Indigenous struggle at the heart of Brazil: State policy, frontier expansion, and the Xavante Indians, 1937–1988*. Durham: Duke University Press.

Garnelo, L., Macedo, G., & Brandão, L. C. (2003). *Os povos indígenas e a construção da política de saúde no Brasil*. Brasília: Organização Pan-Americana da Saúde-OPAS.

Godoy, R., Byron, E., Reyes-García, V., Vadez, V., Leonard, W. R., Apaza, L., et al. (2005a). Income inequality and adult nutritional status: Anthropometric evidence from a pre-industrial society in the Bolivian Amazon. *Social Science & Medicine, 61*(5), 907–919.

Godoy, R., Reyes-García, V., Vadez, V., Leonard, W. R., Huanca, T., & Bauchet, J. (2005b). Human capital, wealth, and nutrition in the Bolivian Amazon. *Economics and Human Biology, 3*(1), 139–162.

Grim, C. W., Broderick, E. B., Jasper, B., & Phipps, K. R. (1994). A comparison of dental caries experience in Native American and Caucasian children in Oklahoma. *Journal of Public Health Dentistry, 54*, 220–227.

Gugelmin, A. S., & Santos, R. V. (2001). Ecologia humana e antropometria nutricional de adultos Xavante, Mato Grosso, Brasil. *Cadernos de Saúde Pública, 17*, 313–322.

Jamieson, L. M., Armfield, J. M., & Roberts-Thomson, K. F. (2006). Oral health inequalities among indigenous and nonindigenous children in the Northern Territory of Australia. *Community Dentistry and Oral Epidemiology, 34*, 267–276.

Kain, J., Vio, F., & Albala, C. (2003). Obesity trends and determinant factors in Latin America. *Cadernos de Saúde Pública, 19*(Suppl. 1), S77–S86.

Kunitz, S. J. (1994). *Disease and social diversity: The European impact on the health of non-Europeans*. New York: Oxford University Press.

Laine, M. A. (2002). Effect of pregnancy on periodontal and dental health. *Acta Odontologica Scandinavica, 60*, 257–264.

Lopes da Silva, A. (1986). *Nomes e amigos: da prática Xavante a uma reflexão sobre os Jê*. São Paulo: FFLCH/USP.

Lopes da Silva, A. (1992). Dois séculos e meio de história Xavante. In M. C. Cunha (Ed.), *História dos Índios no Brasil* (pp. 357–378). São Paulo: Companhia das Letras.

Lourenço, A. E., Santos, R. V., Orellana, J. D. Y., & Coimbra, C. E. A., Jr. (2008). Nutrition transition in Amazonia: Obesity and socioeconomic change in the Suruí Indians from Brazil. *American Journal of Human Biology, 20*(5), 564–571.

Lukas, J. R., & Largaespada, L. L. (2006). Explaining sex differences in dental caries prevalence: Saliva, hormones and "life-history" etiologies. *American Journal of Human Biology, 18*, 540–555.

Lunardi, R., Santos, R. V., & Coimbra, C. E. A., Jr. (2007). Morbidade hospitalar de indígenas Xavante, Mato Grosso, Brasil (2000–2002). *Revista Brasileira de Epidemiologia, 10*(4), 441–452.

Maybury-Lewis, D. (1967). *Akwê Xavante society*. Oxford: Clarendon.
Maybury-Lewis, D. (Ed.). (1979). *Dialectical societies: The Gê and Bororo of Central Brazil*. Cambridge: Harvard University Press.
McSweeney, K., & Arps, S. (2005). A "demographic turnaround": The rapid growth of indigenous populations in lowland Latin America. *Latin American Research Review, 40*, 3–29.
Narvai, P. C., Frazão, P., Roncalli, A. G., & Antunes, J. L. (2006). Dental caries in Brazil: Decline, polarization, inequality and social exclusion. *Revista Panamericana de Salud Pública, 19*, 385–393.
Neel, J. V., Salzano, F. M., Junqueira, P. C., Keiter, F., & Maybury-Lewis, D. (1964). Studies on the Xavante Indians of the Brasilian Mato Grosso. *American Journal of Human Genetics, 19*, 463–489.
Niendorff, W. J., & Jones, C. M. (2000). Prevalence and severity of dental caries among American Indians and Alaska natives. *Journal of Public Health Dentistry, 1*(60 Suppl), 243–249.
Pagliaro, H., Azevedo, M., & Santos, R. V. (Eds.). (2005). *Demografia dos povos indígenas no Brasil*. Rio de Janeiro: Editora Fiocruz.
PAHO (Pan-American Health Organization). (2004). *Gender, equity, and indigenous women's health in Americas*. Washington, DC: PAHO-Gender and Health Unit.
Popkin, B. M. (2001). The nutrition transition and obesity in the developing world. *Journal of Nutrition, 131*(3), 871S–873S.
Popkin, B. M. (2006). Global nutrition dynamics: The world is shifting rapidly toward a diet linked with noncommunicable diseases. *American Journal of Clinical Nutrition, 84*(2), 289–298.
Sampei, M. A., Cano, E. N., Fagundes, U., Lima, E. E. S., Rodrigues, D., Sigulem, E., et al. (2007). Avaliação antropométrica de adolescentes Kamayurá, povo indígena do Alto Xingu, Brasil Central (2000–2001). *Cadernos de Saúde Pública, 23*, 1443–1453.
Santos, R. V., & Coimbra, C. E. A., Jr. (1996). Socioeconomic differentiation and body morphology in the Suruí of the southwestern Amazonia. *Current Anthropology, 37*(5), 853–856.
Santos, R. V., Flowers, N. M., Coimbra, C. E. A., Jr., & Gugelmin, S. A. (1997). Tapirs, tractors, and tapes: Economy and ecology of the Xavante Indians of Central Brazil. *Human Ecology, 25*, 545–566.
Santos, R. V., Cardoso, A. M., Garnelo, L., Coimbra, C. E. A., Jr., & Chaves, M. B. G. (2008). Saúde dos povos indígenas e políticas públicas no Brasil. In L. Giovanella, S. Escorel, L. Lobato, J. C. Noronha, & A. I. Carvalho (Eds.), *Políticas e sistema de saúde no Brasil* (pp. 1035–1056). Rio de Janeiro: Editora Fiocruz & Cebes.
Simões, C. C. S. (2002). *Perfis de saúde e de mortalidade no Brasil: uma análise de seus condicionantes em grupos populacionais específicos*. Brasília: Organização Pan-Americana da Saúde.
Souza, L. (1953). *Entre os Xavante do Roncador*. Rio de Janeiro: Serviço de Documentação, Ministério da Educação e Saúde.
Souza, L. G. (2008). *Demografia e saúde dos índios Xavante de Mato Grosso*. Doctoral thesis in public health, Escola Nacional de Saúde Pública, Rio de Janeiro.
Souza, L. G., Santos, R. V., Pagliaro, H., Flowers, N. M., & Coimbra, C. E. A., Jr. (2011). Demography and health of the Xavante Indians from Central Brazil. *Cadernos de Saúde Pública, 27*(10), 1891–1905.
Stephens, C., Porter, J., Nettleton, C., & Wilis, R. (2009). Disappearing, displaced, and undervalued: A call to action for indigenous health worldwide. *Lancet, 367*, 2019–2028.
Teixeira, P., & Brasil, M. (2005). Estudo demográfico dos Sateré-Mawé: um exemplo de censo participativo. In H. Pagliaro, M. M. Azevedo, & R. V. Santos (Eds.), *Demografia dos povos indígenas no Brasil* (pp. 135–154). Rio de Janeiro: Editora Fiocruz/ABEP.
Verbrugge, L. M. (1985). Gender and health: An update on hypotheses and evidence. *Journal of Health and Social Behavior, 26*, 156–182.
Vieira-Filho, J. P. B. (2000). Genótipo econômico e produção de insulina. In *Diabetes, saúde, & cia* (p. 2). São Paulo: Aventis Pharma.

Welch, J. R. (2010). O Projeto Danhiptetezé. Newsletter of the Project Danhiptetezé: inciativa de Cultura Alimentar Xavante, Indigenous Language and Culture Documentation Project. *Museu do Índio/FUNAI, 1*(1), 3.

Welch, J. R., Ferreira, A. A., Santos, R. V., Gugelmin, S. A., Werneck, G., & Coimbra, C. E. A., Jr. (2009). Nutrition transition, socioeconomic differentiation, and gender among adult Xavante Indians, Brazilian Amazon. *Human Ecology, 37*, 13–26.

WHO (World Health Organization). (1997). *Oral health surveys – Basic methods* (4th ed.). Geneva: WHO.

WHO (World Health Organization). (1998). *Obesity: Preventing and managing the global epidemic*. Geneva: WHO.

Wood, J. W. (1990). Fertility in anthropological populations. *Annual Review of Anthropology, 19*, 211–242.

Chapter 3
Balancing People, Policies, and Resources in Rural Tibet

Geoff Childs, Melvyn C. Goldstein, and Puchung Wangdui

Abstract This chapter examines ways that household-level decisions about reproduction are shaped by available land resources, political discourse on population and impoverishment, and changing perceptions of children's utility to long-term economic strategies. The research centers on three villages in Shigatse Prefecture of China's Tibet Autonomous Region and is based on a combination of longitudinal demographic data gathered through surveys and in-depth interviews with parents about social, economic, and political factors that influenced their decisions to limit family size. The first part of the chapter discusses policy changes in the 1980s that dismantled Tibet's commune system and gave families control over set amounts of arable land, and ensuing processes that led to a sharp reduction in per capita land holdings. The second part of the chapter discusses China's birth control policy in terms of how it is rooted in a vision to create a modern society and how this policy applies in Tibet. The third part of the chapter documents the timing and magnitude of the recent fertility decline in rural Tibet and links it with (1) the reduction in per capita land holdings, (2) China's birth control policy, and (3) the changing roles that children play in households' long-term economic strategies. The concluding section discusses how human-environment interactions are one among several variables involved in the complex reproductive decision-making process in rural Tibet.

G. Childs (✉)
Department of Anthropology, Washington University, St. Louis, MO 63130, USA
e-mail: gchilds@wustl.edu

M.C. Goldstein
Department of Anthropology, Case Western Reserve University, Cleveland, OH 44106, USA
e-mail: mcg2@case.edu

P. Wangdui
Tibet Academy of Social Sciences, Lhasa, Tibet Autonomous Region, China
e-mail: puchungwangdui@yahoo.com

E.S. Brondízio and E.F. Moran (eds.), *Human-Environment Interactions: Current and Future Directions*, Human-Environment Interactions 1,
DOI 10.1007/978-94-007-4780-7_3,

3.1 Introduction

Academic inquiries into the relationship between land, fertility, and population growth have a long history. Two centuries ago, Thomas Malthus (1989[1803]: 293) hypothesized that checks on population growth arose first and foremost from an "insufficiency of subsistence," and implied a direct relationship between fertility and land by stating, "Plenty of rich land, to be had for little or nothing, is so powerful a cause of population as generally to overcome all obstacles." Fertility entered the equation either as a static variable in societies where population is regulated by "positive checks" (i.e., via mortality) or as a fluctuating variable in societies where population is regulated by "preventive checks" that operate through changes in the frequency and timing of marriage.

A more contemporary debate emerged in the 1980s when scholars postulated the existence of a direct and discernible relationship between landholdings and fertility (Cain 1985, 1986; Schujter and Stokes 1984; Stokes et al. 1986). The debate centered on the roles that landholding size and land tenure security play in reproductive decision-making. Since that debate was waged, however, high fertility is rapidly becoming a relic of the past as people gain access to reliable means for birth control. Nowadays, a more pertinent question seems in order: How do landholdings and land tenure influence reproductive decision-making in the context of a fertility decline?

This chapter addresses the above question by examining a recent and rapid demographic transition that occurred in rural Tibet.[1] From the mid-1980s to 2000, the total fertility rate[2] fell from six births per woman to the replacement level of two births per woman (Childs et al. 2005). The transition occurred at a time when population growth and land losses were eroding per capita landholdings, tempting the conclusion that people were modifying their reproductive behavior in an effort to maintain a balance between family size and landholdings. However, such an explanation is complicated by the fact that the fertility decline coincided with three societal-level transformations in rural Tibet: the replacement of rural communes with a land tenure system whereby households became the primary units of production, the implementation of China's birth control policy, and a campaign to rapidly develop Tibet. The purpose of this chapter is to explore the relationship between land and fertility in a more inclusive context of social, political, and economic factors that influence reproductive decision-making.

---

[1]In this chapter, Tibet refers exclusively to China's Tibet Autonomous Region.

[2]The total fertility rate is a synthetic cohort estimate of the average number of children who would be born to each woman in a population if current age-specific fertility rates remain constant. It is one of the most widely used barometers of childbearing used by demographers.

3.2 Landholdings and the Proximate Determinants of Fertility

Research on the relationship between landholding and fertility has centered on two hypotheses: the land-security hypothesis and the land-labor demand hypothesis. The land-security hypothesis predicts an inverse relationship between fertility and the level of confidence that people have in their long-term ability to access agricultural land. Contrary to the proposition that old-age security is a motivating factor for having children (Nugent 1985), the land-security hypothesis posits that the value of children (and consequently the demand for children) as a form of social security declines for those who have secure access to land (Cain 1985; Jensen 1990). For example, in a Nigerian setting where written land deeds are rare, having many children is a strategy to stake claims to land (Renne 1995) which implies that high fertility is related to insecure land tenure. A more recent study in a frontier region of the Ecuadorian Amazon also found evidence to support the hypothesis: women living in households with legal land titles had considerably fewer children than women living in households without legal land titles (Carr et al. 2006).

The land-labor demand hypothesis posits that the size of one's landholding affects the demand for labor, which can be satisfied at the household level through reproduction. If this hypothesis holds true, then landholding size is an important determinant of fertility (Schujter and Stokes 1984; Stokes et al. 1986). However, attempts to test the land-labor demand hypothesis produced mixed results. Several studies found a positive correlation between landholdings and household sizes (summarized in Netting 1993: 85–87), between landholdings and fertility (Mueller and Short 1983; Schujter and Stokes 1984; Schujter et al. 1983), and even between first-birth timing and the proportion of land under agricultural use (Ghimire and Hoelter 2007). Yet Cain (1985: 12–13) challenged the hypothesis by pointing out that any statistical association between landholdings and fertility must be explained in relation to fertility-related motivations, which can be difficult to discern in studies that rely exclusively on survey data. Also, because institutional factors have an undeniable bearing on the relationship between landholdings and fertility, the mere size of a farmer's landholding cannot be a reliable predictor of a reproductive outcome. Other factors demand consideration, including the terms and security of land tenure, mechanisms of property right enforcement, and opportunities for family members to engage in nonfarm labor (Cain 1985, 1986). Cain (1985) thereby argued that a positive relationship between landholdings and fertility could very well be spurious or an "unintended by-product of other behavioral patterns that are associated with landholding status."

Other research indicates that the causal relationship between landholdings and fertility may stem from land scarcity rather than abundance. Clay and Johnson (1992) supported this position by citing a study from historical France (Goldscheider 1971) that found bequeathing equal parcels of land to all children provided a powerful incentive to limit marital fertility, and a study of frontier regions of the United States (Easterlin et al. 1978) where fertility declined when increasing population density transformed land abundance into scarcity. In a more recent study, Shreffler and Dodoo (2009) found that rural Kenyans considered children beneficial when landholdings were large, but large families became undesirable as the population grew, and land was parceled through inheritance. People responded by using contraception to limit reproduction and by seeking new educational and off-farm employment opportunities for children.

Although Netting and colleagues did not directly engage the land-labor demand hypothesis, their research on household size, population density, land usage, and labor demands is relevant. Netting (1993: 87) used comparative evidence to conclude that, "smallholders everywhere strike some kind of economic balance between household members and land size," noting that the way land is used has demographic consequences. Drawing on Boserup's thesis that population increase can impel technological change, Netting argued that agricultural intensification requires rising labor input. For Netting (1965, 1993), the relationship between landholdings and household size was mediated through the demand for and quality of labor. Specifically, on the densely populated Jos Plateau, Kofyar household sizes were small, in part because of diminishing returns achieved by adding people to the household's labor force. In contrast, households were larger in a sparsely populated frontier region where production was limited by labor rather than land (Netting 1968). Household sizes increased as more time elapsed since settling the frontier, increases that did not necessarily result from rising fertility but from adding new members through polygynous marriages and forming stem rather than nuclear households (Netting et al. 1989; Stone et al. 1984). In both the Kofyar research and Netting's analysis of smallholders, landholdings have a strong and predictable effect on household size when that relationship is mediated principally by labor demands. Both fertility and household norms change quickly when the marginal utility of adding household workers changes.

Much of the landholding and fertility debate has been waged in the absence of theories developed to explain peoples' motivations for having children (one exception is Carr et al. 2006). In particular, the land-labor demand hypothesis reflects Easterlin and Crimmins' (1985) supply-demand model of reproductive decision-making that considers three variables: the demand for children, the supply of children, and the costs of regulating reproduction. The demand for children is influenced by a family's income, consumer preferences, and cost of living. Meanwhile, the supply of children under natural fertility[3] conditions is shaped by the proximate determinants of fertility (see below), as well as infant and childhood mortality. The motivation to regulate fertility derives from the balance between

[3]The concept of natural fertility was initially defined by Louis Henry (1961: 81) as "fertility which exists or has existed in the absence of deliberate birth control." Demographers generally understand natural fertility to imply that couples can influence the number of children born, for example, through periodic abstinence and other cultural measures that affect spacing between births, age at marriage, and norms of widow and divorcee remarriage, but that any such action is independent of the number of children already born and therefore not meant to control the ultimate number of children born.

supply and demand. If demand exceeds supply, motivation is absent; if supply exceeds demand, motivation is present. Finally, the cost of regulating fertility involves social costs (e.g., the acceptability of using birth control) as well as economic costs (availability and affordability).

The social cost of fertility regulation varies considerably from one society to the next and is related to cultural and religious perspectives on birth control and political agendas that may be either pro- or anti-natalist. The cost is not static; it can vary through ideational changes that originate within a society or spread from one setting to another. According to Cleland and Wilson (1987), a major determinant of fertility decline is the diffusion of new attitudes about, and technologies for, controlling reproduction. Although their findings focused mainly on ideational changes that occurred during the early phase of fertility transition in Europe, in a contemporary state like China, ideational changes can disseminate very quickly through official policies that are designed to change the social and economic costs of childbearing and fertility regulation.

Caldwell's (1982) wealth flow hypothesis links theories emphasizing economic factors with those highlighting ideational changes. According to Caldwell, attitudes toward the value of children can change in association with rising affluence and the availability of secular education, factors that prompt parents to invest more heavily in their offspring. When wealth flows from children to parents, large families make good economic sense. However, when a wealth flow inversion occurs so that parents invest more heavily in children and can expect fewer returns, a motivation arises to control fertility. The motivation is therefore related to an ideational shift in the way parents value the social and economic roles of their children.

The most effective tool for analyzing fertility is the proximate determinants of fertility model (Bongaarts and Potter 1983; Davis and Blake 1956). As Davis and Blake (1956) point out, background variables such as wealth, education, or religion affect fertility through the proximate determinants, which they group into three categories: factors affecting exposure to intercourse ("intercourse variables," e.g., age at entry into sexual unions, voluntary and involuntary abstinence, coital frequency), factors affecting exposure to conception ("conception variables," e.g., lactational amenorrhea, use of contraception), and factors affecting gestation and successful parturition ("gestation variables," e.g., miscarriage and abortion). To illustrate how background variables work through the proximate determinants, consider the oft-repeated aphorism "education reduces fertility." Staying in school longer and subsequently attempting to establish a career may lead women to delay marriage (an exposure to intercourse variable), and literacy may provide them more access to information on birth control methods (an exposure to conception variable). Education per se (the background variable) does not cause women to have fewer children; fewer children result from delayed marriage and contraceptive usage.

Figure 3.1 illustrates the analytical model we use for linking background variables to fertility outcomes. The background variables we focus on in this chapter are the land tenure system, landholdings, household-level labor demands, and China's birth control policy. These variables present people with a range

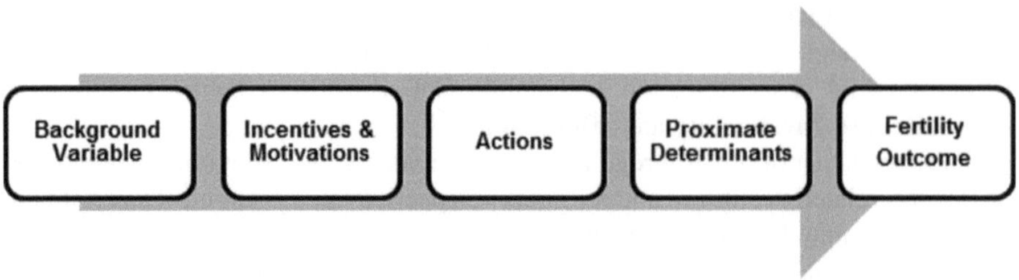

Fig. 3.1 Linking background variables with fertility outcomes

of incentives, motivations, and desires, which lead to actions that shape fertility outcomes via the proximate determinants of fertility. Figure 3.1 should not be interpreted as a deterministic model of how specific background variables invariably lead to specific actions and outcomes. Although a background variable can constrain people's options and provide them with certain incentives, one must never ignore the agency that allows different people to make different choices when presented with similar circumstances. With this important caveat in mind, the model is designed to represent what happens when a background variable influences a significant proportion of a population to select a specific course of action that, in the aggregate, has a discernible effect on fertility.

3.3 Data and Methods

From 2006 to 2009 we conducted four stints of fieldwork for a total of 9 months in three villages in Tibet's Shigatse Prefecture to investigate the impact of modernization on rural families and the elderly.[4] The three villages, while not selected to represent all of Tibet, lie within a major agricultural corridor running between Tibet's two largest cities: Lhasa and Shigatse. This corridor contains about 30% of Tibet's population. Sogang, the least affected by development, is located in Panam County in the upper part of a tributary river valley, while Norgyong, the intermediate site, is situated below Sogang on the main river. Betsag, the third site, is located only 10 km from Shigatse City and was included in the study to represent a wealthy farming village that is more heavily affected by mechanized agriculture and government development programs. Despite these economic differences, the three villages are geographically close, within a two-hour drive of one another, and are part of the same Tibetan subethnic cultural and linguistic zone.

The demography presented in this chapter combines a previous analysis of survey data from Goldstein and Beall's 1997–1998 study of rural Tibet with the analysis

[4]This NSF-sponsored research project (#0527500) was conducted in collaboration with the Tibet Academy of Social Sciences in Lhasa.

of data from Goldstein and Childs' 2006 household survey of Sogang, Norgyong, and Betsag villages. We use the own-children method, a reverse-survival technique designed to estimate age-specific fertility rates and total fertility rates in the absence of detailed data on reproduction (Cho et al. 1986).

We also present descriptive statistics from surveys and qualitative data from in-depth interviews to shed light on various factors that shape reproductive motivations and outcomes. During interviews with people who are currently having children or recently ceased doing so ($n = 73$), we asked a series of questions on ideal family size, contraceptive usage, reproductive decision-making, and the birth control policy. We also interviewed village leaders about how and when the birth control policy was implemented in the area. Before presenting the emic perspectives that emerged, we detail the societal-level changes that have shaped the environment in which people make decisions that have demographic consequences.

3.4 Land Tenure, Birth Control, and Economic Development in Rural Tibet

In this section, we outline three transformative policies that have changed rural Tibetans' relationships with agricultural land and the way they marry and form households: the distribution of land on a per capita basis, the implementation of a birth control policy, and the initiation of a massive development scheme. This contextual information provides a backdrop for analyzing the changing relationship between land and fertility in rural Tibet.

3.4.1 Land Tenure and Per Capital Landholdings

Following a period from the 1960s to 1982, when Tibet's rural population was organized into communes, China implemented the Household Responsibility System (Tibetan: *genzang*) that transferred land tenure from village collectives to individual households (Goldstein et al. 2003). The government allocated land on a per capita basis so that every person alive at the date of decollectivization received one equal share regardless of age, sex, social status, or any other factor. A household consisting of seven members thereby received seven shares of land, whereas a household consisting of three members received three shares.

Technically, all agricultural land still belongs to the state, so it cannot be bought and sold and is basically held as a long-term lease. In our research area a trio of factors has steadily eroded landholdings both in absolute and per capita terms: the government's use of eminent domain for development projects, the loss of land through flooding and other forces of nature, and the natural increase of the

Table 3.1 Diminishing per capita landholdings

Village	Year	mu per household	Percent change	mu per capita	Percent change
Sogang	1982	12.7		2.5	
	1997	11.8	−7.1	1.7	−32.0
	2006	9.3	−21.2	1.3	−23.5
	1982–2006		−26.8		−48.0
Norgyong	1982	22.6		4.2	
	1997	20.9	−7.5	2.8	−33.3
	2006	16.3	−22.0	2.4	−14.3
	1982–2006		−27.9		−42.9
Betsag	1982	26.4		3.8	
	2006	25.8	−2.3	3.4	−10.5

Sources: Goldstein et al. 1998 survey, Goldstein et al. 2006 survey

population due to births outnumbering deaths.[5] The first of these, eminent domain, heightens local concerns over long-term land security. This is especially true in Norgyong where officials have appropriated large tracts of land for development projects. The expansion of Panam from a small county seat of government into a burgeoning regional town necessitated that land be taken to construct streets, markets, a hospital, and various buildings for government and private use. Villagers are paid 7,100RMB ($950) for each mu[6] of appropriated land, which many consider inadequate compensation. Similarly, Betsag lost some land to the expansion of a thoroughfare connecting the region's two urban areas, Shigatse and Gyantse, whereas Sogang lost a few fields to the construction of a health post and other government buildings. More significantly, Sogang was struck by a flood in 2002 that rendered much land useless. Between 1982 and 2006, Betsag lost 2.8% (71 mu), Norgyong 14.2% (350 mu), and Sogang 22.1% (256 mu) of their agricultural land. Whereas development accounts for most of Norgyong's land loss (93.6%), the majority of land lost in Sogang (84.9%) resulted from flooding.

At the same time when farmers were losing land, rural Tibet was undergoing a period of rapid population growth. From 1982 to 2006 Betsag's population increased from 586 to 712 (17.7%), Norgyong's from 575 to 853 (32.6%), and Sogang's from 503 to 665 (24.6%). Table 3.1 shows the extent to which land losses and population increase diminished landholdings. Sogang and Norgyong were especially hard-hit: 48.0% and 42.9% reductions in per capita landholdings, respectively.

[5]Several initiatives have partially counteracted the diminishing landholding trend. For example, below Sogang the government established a "poverty alleviation village" by building a dam and irrigation system on marginal land. This temporarily increased per capita landholdings in Sogang by moving several poor families to the new village.

[6]Mu, the basic land measurement in China, is equal to 1/15 of a hectare.

3.4.2 Birth Control Policy and Implementation

China's initial push toward birth control started in the 1970s with a policy designed around the principles of "later" (commencement of childbearing), "longer" (intervals between births), and "fewer" (total children). By the late 1970s, birth control had become a cornerstone of the nation's drive to achieve rapid economic development and evolved into a massive effort to demographically engineer China toward a smaller population consisting of higher-quality individuals (Greenhalgh 2008; Greenhalgh and Winckler 2005). Since the 1978 inception of China's birth control policy, the government technically has permitted each couple to have a single child. However, in reality, a patchwork of policies evolved catering to regional conditions (Gu et al. 2007). Ethnic minorities like Tibetans have been partially or fully exempt. On one hand, Tibet's population is a miniscule proportion of China's population, so allowing Tibetans a higher birth rate hardly affects national population growth. On the other hand, leaders in Beijing did not want to antagonize potentially restless Tibetans by subjecting them to an unpopular policy. Consequently, Tibet's rural residents are officially permitted to have three children, but this has not been strictly enforced (Goldstein et al. 2002). Most Tibetans living in the neighboring provinces of Qinghai and Sichuan can have only two (Gu et al. 2007; Schrempf 2008).

The government introduced the birth control policy to Tibet during the mid-1980s. In Lhasa, the capital, Han cadres could have only one child and Tibetan employees of the government only two (Goldstein and Beall 1991). Stated penalties for noncompliance included salary deductions, barriers to promotion, and withholding residence cards for excess children, but these were not vigorously enforced (Goldstein et al. 2002).

In the countryside, the government began in the 1980s to propagate the idea that small families are economically advantageous, and announced a limit of three births for rural women in 1984 but did not seriously enforce it (Goldstein et al. 2002). The county in our research area began to disseminate information about the national policy only in 1988 and started implementation in 1989 by having the Maternal Care Office under the Bureau of Health, and the townships' health clinics organize campaigns to encourage people to have fewer children. During these sessions, government representatives emphasized the state's message that "fewer children equal more wealth." Every household was required to send one woman of reproductive age to the meetings, and the task of ensuring attendance was delegated to the local representative of the Women's Association. Households that failed to send a representative had points deducted from their tally of obligatory community labor, which in effect was a small fine. At this early date of the campaign, however, propaganda was not accompanied by birth control services. Most villagers still lacked the means to control their reproduction.

The situation began to change during the mid-1990s when the government started prioritizing agricultural and pastoral communities. The official 1996 document on family planning policy in Tibet said, "In poor areas it is a common situation that 'the more poor someone is, the higher the number of births, and the higher the number of births, the more poor' .... If we do not pay attention to the matter of population and do not carry out family planning ... it will definitely affect the goal of achieving progress and prosperity for the Tibetan nationality."[7] In weekly meetings, local leaders emphasized that one child is best, two is good, three is the limit, and poor families on welfare should have fewer children for their own economic well-being. In addition to promoting the advantages of smaller families, the government instituted disincentives, in the form of fines and penalties, for exceeding the three-child limit.

Villagers reported feeling anxious that the limit would be strictly enforced, although it was the poorer families who were targeted most; leaders of the Women's Association repeatedly visited their homes and encouraged them to use birth control. The rural poor were much more susceptible to government pressure because local officials acted as conduits for benefits such as welfare. Wealthier families felt less compelled to comply with the policy and were subjected to less pressure.

Similar to other parts of China, each village in Tibet was issued a birth quota. To ensure that the quota was met, county and township officials asked village leaders to make certain that women with more than three children were using contraception, particularly more effective methods such as sterilization or IUDs.[8] Such services were provided free at scheduled intervals by traveling medical teams. However, consistent with Greenhalgh's (1994) findings elsewhere in China, local officials in our research area seemed reluctant to compel their relatives and neighbors to comply with the unpopular mandate. Enforcement mechanisms proved to be somewhat of a façade because officials in a township that exceeded its quota could "loan" births to a neighboring township that had fallen short of its quota (Goldstein et al. 2002). Other local leaders engaged in the deliberate manipulation of population figures. One explained that his village's initial quota was ten births, but each year double that number of children was born. Because some people passed away, at the end of the year, he would report a net increase in the village of roughly ten people, thereby implying (without directly stating) that the village had kept within the birth quota. Apparently higher-level officials turned a blind eye to this ruse.

To summarize, China's birth control policy was first announced in rural Tibet during the mid-1980s. Officially, each rural household could have three children, although poorer families were strongly encouraged to have fewer. In the mid-1990s penalties were announced for exceeding the limit, yet it quickly became obvious that enforcement would be lax so families continued to have more than three children when they felt it would be advantageous to do so. For example, of the 113 births that occurred to women in a rural study in 1997, 31.4% were fourth order or higher, and 55.9% of women who had ever given birth ($n = 1{,}100$) had given birth to four or

[7] Document No. 5, Party Committee of Tibet, 1996, as cited in Goldstein et al. (2002).

[8] Individual counties and prefectures had considerable autonomy regarding how to enforce this limit, if at all, and in some areas the limit was changed to allow four children for rich families, three for middle income families, and two for poor families (Goldstein et al. 2002).

more children. Moreover, village leaders also exceeded the limit even though many were party members. The average number of living children for the 20 village heads was 5.1; 70% had four or more, and 60% had five or more (Goldstein et al. 2002).

3.4.3 Rapid Economic Development

With the launch of China's "Develop the West Campaign" in 2000, the government has devoted unprecedented sums of money to a regional development project aimed at rectifying economic disparities between the nation's wealthier eastern provinces and the poorer western provinces, including Tibet. The program includes a series of policies for rapid development such as more investment, preferential tax rates, and huge expenditures for infrastructure. In 2003 alone China invested about 200 billion yuan ($24.3 billion) in large projects in the western regions. In Tibet, most of the initial money was used for large infrastructure projects: highways, buildings, and the new railway to Lhasa. The central government also implemented projects to bolster the quality of life of rural Tibetans by rebuilding townships and schools, expanding electrification, and improving the health care system (Goldstein et al. 2008). More recently, the government initiated a "People First" policy to bring income-generating opportunities into villages (Goldstein et al. 2010).

China's push to develop Tibet has created new demands for nonfarm labor. People in our research area responded by diversifying their traditional agropastoral household economy. Nowadays, the vast majority sends members off farm to earn cash income in the burgeoning labor market, and the overwhelming proportion of nonfarm laborers (84%) and nonfarm income (79%) derives from going outside the village for work.

Villagers generally engage in three broad types of off-farm activities. The largest and least lucrative category is unskilled manual labor, mainly construction workers who carry loads, mix cement, and so forth. A second category is skilled labor, for example, carpenters, masons, and drivers. A third and growing category includes entrepreneurs who purchase vehicles that can be used to generate income, as well as contractors and subcontractors on construction projects. Because we have described the village-level impacts of developments projects elsewhere (Goldstein et al. 2008, 2010), we will not provide details here beyond showing the dramatic, post-2000 rise in the number of people—males and females alike—who engage in off-farm income-generating activities in the two villages for which we have longitudinal data (Table 3.2).

The new land tenure system introduced in the 1980s, the implementation of a birth control policy, and the rapid push to develop Tibet have all affected families' household management strategies and relationship with the land. The following section charts the fertility decline in rural Tibet and then links the societal-level transformations described above with decision-making and its demographic consequences.

Table 3.2 Percentage of 15- to 49-year-olds in nonfarm work

		Males		Females	
Village	Ages	1997	2005	1997	2005
Sogang	15–19	4.8	30.2	8.3	23.7
	20–29	18.0	69.2	7.4	31.9
	30–39	25.5	71.7	2.1	16.7
	40–49	40.9	38.3	7.7	6.7
Norgyong	15–19	14.9	23.9	2.2	3.9
	20–29	28.4	63.7	2.5	28.0
	30–39	29.3	63.5	6.8	28.3
	40–49	35.5	59.2	0.0	16.0

Source: Goldstein et al. (2008)

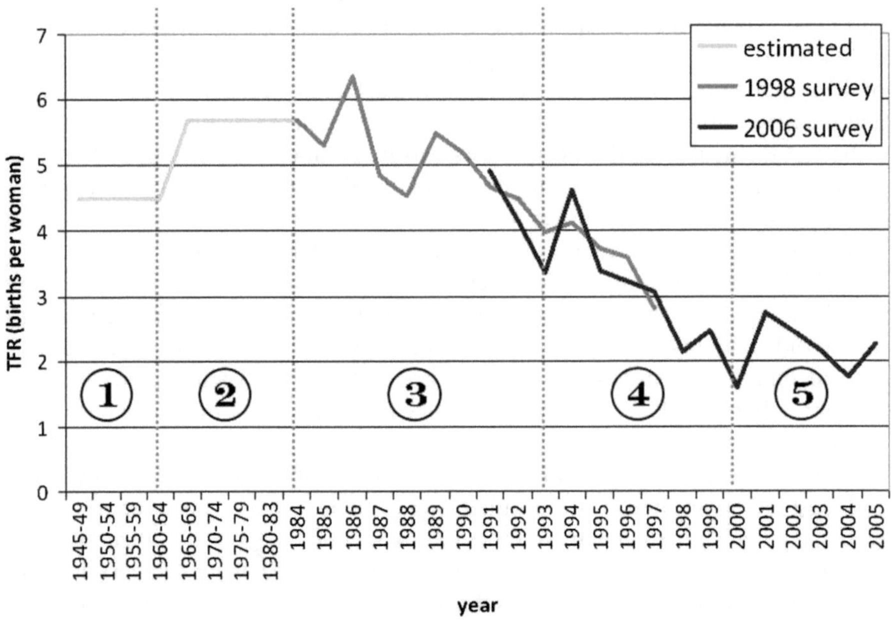

Fig. 3.2 Total fertility rate by year and survey, rural Tibet (*Sources*: Childs 2008 (for 1945–1983); Childs et al. 2005 (for 1984–1997); Goldstein et al. 2006 survey (for 1991–2005))

3.5 The Rural Tibetan Fertility Transition

Figure 3.2 illustrates the total fertility rate in rural Tibet from the 1940s to 2005. The figure is divided into five phases, each of which is associated with a specific set of conditions.

Table 3.3 Landholdings and household sizes, Kyirong taxpayers, 1958

Taxation units (*gang*)[a]	Number of households	Mean number of household members
0–0.9	49	4.5
1–1.9	228	6.4
2–2.9	73	8.5
3+	11	9.5

Source: 1958 Kyirong Household Register (see Childs 2008)
[a]*Gang* was a land unit calculated on the basis of several factors, including the amount of seed that could be sown in a certain area, the fertility of the soil, and local climatic conditions (Goldstein 1971)

3.5.1 Phase 1: The "Old Society"

Prior to China asserting control over Tibet in 1951, the majority of rural farmers in the "old society"[9] were legally bound subjects (*miser*) of a government, monastic, or aristocratic estate. One class of subjects, referred to as "taxpayer" (*trelpa*), held usufruct rights to till fields, the extent of which were specified in a contract with the estate-holding institution. Although it was difficult for a household to expand its landholding, it could pass the usufruct right to succeeding generations in perpetuity providing it met a range of tax obligations that included payments of grain and corvée labor (Goldstein 1971).

From at least the eighteenth century until 1959, the land tenure policy gave taxpayers a strong incentive to practice fraternal polyandry.[10] By doing so, they avoided the need to partition landholdings through inheritance. Retaining multiple males in the family also gave them a better chance to fulfill corvée tax obligations to their lords while engaging in a complex adaptive strategy that included farming, herding, and trade. In Kyirong during the 1950s, a strong correlation existed between the size of a household's usufruct landholding and its membership (Table 3.3). Whether this is evidence of higher fertility among large landholders (Schujter and Stokes 1984; Stokes et al. 1986) or a result of households adding members through other means to increase their labor force (Netting et al. 1989; Stone et al. 1984) is impossible to determine with the data on hand.

Polyandry moderates aggregate fertility through its corollary: a high frequency of female non-marriage. This was first demonstrated by Goldstein (1981) who found that the high proportion of female non-marriage (31%) resulted in a completed fertility rate for all women (6.3 births) that was considerably lower than the completed fertility rate for married women (7.4 births). A subsequent study linking fertility outcomes with the Tibetan manorial estate system found that nearly half of all women aged 25–34 were not formally married into taxpayer households.

---

[9]In contemporary political discourse and in rural areas where we work, pre-1959 Tibet is commonly referred to as the "old society" (*chitsok nyingba*).

[10]Although China asserted control over Tibet in 1951, the traditional manorial estate system continued until 1959.

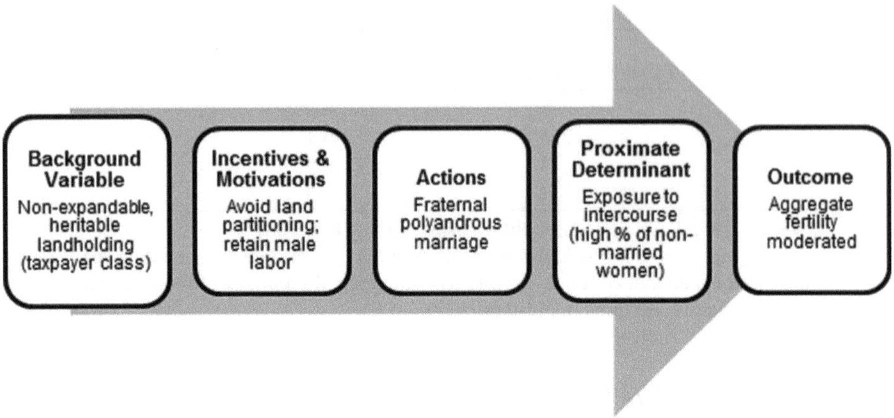

Fig. 3.3 Tibetan fertility, pre-1959

Their relatively low level of fertility (2.2 births per woman) counteracted the marital fertility rate of 6.2 births per woman to result in a total fertility rate of 4.4 births per woman (Childs 2008). Figure 3.3 illustrates how the background variable of limited landholdings provided an incentive to practice fraternal polyandry, which moderated fertility by curtailing marriage—and, by extension, exposure to intercourse—for a significant proportion of the female population.

3.5.2 Phase 2: The Commune Period

Starting in the 1950s, China embarked on a massive campaign to reorganize rural society into communes. Lee and Wang (1999) attribute China's unprecedented high fertility during the 1960s to the commune system in which marriage and reproduction came under the purview of the state rather than the family. This caused a decline in traditional Chinese mechanisms for controlling family size such as coital restraint within marriage and female infanticide. Equally important, living in communes provided incentives to have many children because food and other resources were allocated in part on a per capita basis, so every child entitled a family to additional rations (Lee and Wang 1999: 119–122).

In Tibet, most rural communities were reorganized into communes starting in the mid-1960s. Collectivization of agriculture reduced the incentive to marry polyandrously, at least among the former taxpayer class, because households no longer had their own land to preserve. The decline of polyandry was accelerated by a government mandate that outlawed all forms of marriage except monogamy. It is no coincidence that, at a time when marriage became more universal for women, aggregate fertility rose by approximately one birth per woman. Figure 3.4 illustrates how collectivization changed the marital incentive structure and thereby affected fertility by increasing many women's exposure to intercourse through marriage.

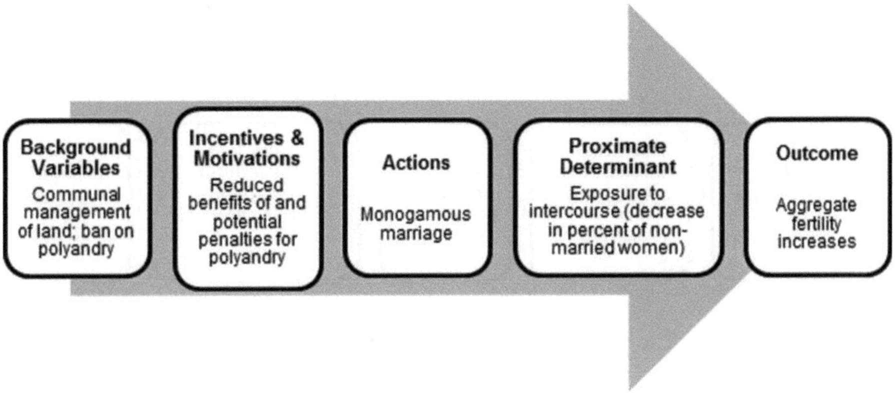

Fig. 3.4 Tibetan fertility, 1960s–1970s

Table 3.4 Never-married females (percent) by age

Age	Rural Tibet, 1998	Sogang, Norgyong, and Betsag, 2006
25–29	39.9	39.1
30–34	25.6	22.5
35–39	24.0	19.2
Total	31.0	27.3

Sources: Goldstein et al. (1998) household survey (rural Tibet), Goldstein et al. (2006) household survey (Sogang, Norgyong, and Betsag)

3.5.3 Phase 3: The Family Responsibility System

In the early 1980s, China dismantled Tibet's communes and instituted the "Household Responsibility System," making households the primary decision-making units for agricultural production. Overnight the government created a direct relationship between landholdings and household size by allocating one share of land per family member. Tibetans in our research area responded by reinstituting polyandry (Fjeld 2006; Goldstein et al. 2002; Jiao 2001).

Unlike pre-1959 society, when only a certain class of rural peasants (the taxpayers) held heritable usufruct rights to land, after 1982, all rural households held this privilege. It is important to note that fertility began to decline steadily in the late 1980s—before the birth control policy had been implemented and before contraceptive methods were widely available in rural areas. Therefore, the rise in polyandrous marriages and associated exclusion of many women from marriage is the most likely instigator of the fertility decline. The percentage of never-married females in their prime reproductive age remains high decades after the Household Responsibility System was implemented (Table 3.4).

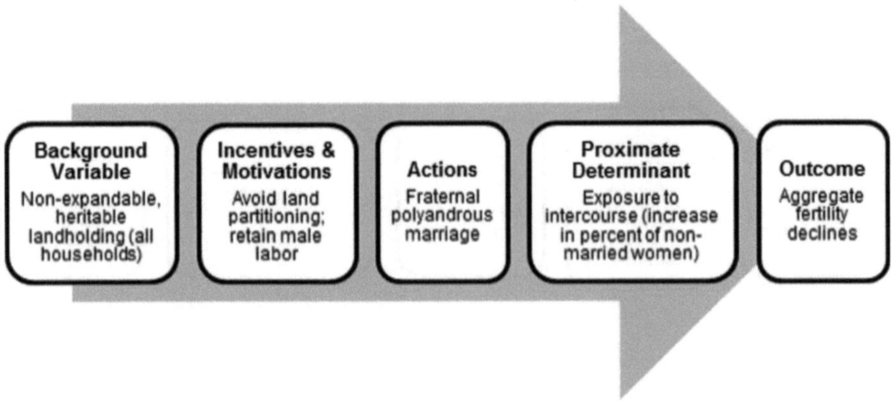

Fig. 3.5 Tibetan fertility, mid-1980s to early 1990s

Figure 3.5 illustrates how the Household Responsibility System changed a key background variable and thereby renewed peoples' incentive to practice polyandry. The result was the resumption of a traditional demographic system in rural Tibet; once again, fertility was moderated by a high frequency of female non-marriage.

3.5.4 Phase 4: Motives and Means for Effective Fertility Control

Fertility continued to decline throughout the 1990s when population growth, eminent domain, and natural disasters were steadily eroding per capital landholdings. Because diminishing landholdings were partially a product of population growth, it is easy to see how—in Easterlin and Crimmins' (1985) economic calculus—the supply of children eclipsed the demand for children. However, people could not act effectively on a desire to limit childbearing unless they had the means to control reproduction.

By the late 1990s, people were fully aware of the birth control policy and had access to a variety of contraceptive methods. In Goldstein and colleagues' 1998 rural fertility survey, 92% of all village women aged 30–44 knew of at least one method and 76.1% knew of four or more ($n = 489$). Regarding policy, 96.7% of married women said that there was a limit to the number of children they could have; 95.3% said the limit was three; and 92.9% believed there was a fine for exceeding the limit. But only a small proportion of married women, 16.0%, said they were currently using contraception because of the birth control policy or that they wanted to avoid fines and difficulties (presumably incurred by acting contrary to the policy) (Goldstein et al. 2002).

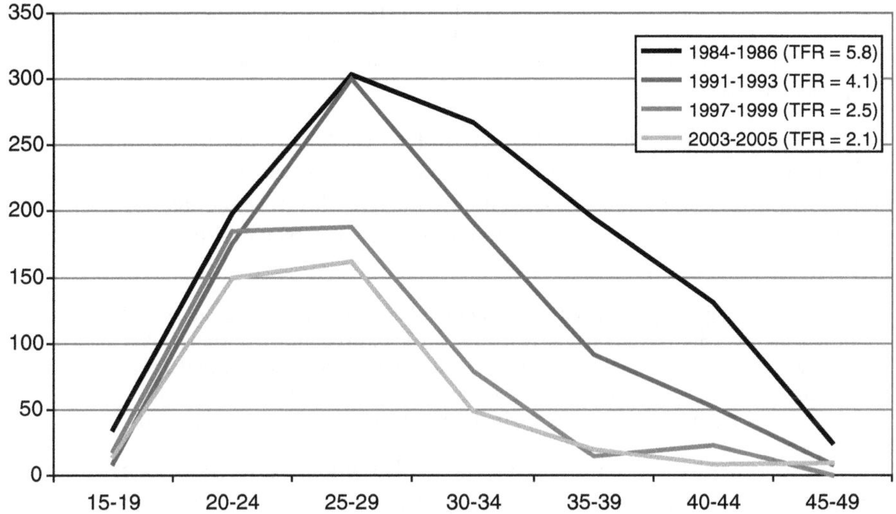

Fig. 3.6 Age-specific fertility rates by 3-year periods (*Sources*: Childs et al. 2005 (for 1984–1997 analysis); Goldstein et al. 2006 survey (for 1991–2005 analysis))

Figure 3.6 demonstrates that a shift took place in rural Tibet from a "natural" to a "controlled" fertility pattern between 1991–1993 and 1997–1999, indicating that a large number of women began to stop reproducing altogether once a certain number of children had been born. This shift coincides with the implementation of China's birth control policy in rural Tibet.

In their 1998 study, Goldstein and colleagues found that many people were motivated to limit childbearing because they were concerned that their landholdings were insufficient to support a large family (Goldstein et al. 2002). A similar finding emerged from our more recent interviews. Several people expressed concerns that their family size had exceeded the number of land shares they held. For example, one farmer told us:

> A while ago it [the birth control policy] seemed serious during community meetings. But nobody came to my home to talk one-on-one about it. I voluntarily had only two children because we have limited land. We have only three shares of land, but now we have six household members. I worried about raising our kids.

Another woman who already had three children said:

> I want to stop, but my in-laws and husband want one more child. I tell them that we have to think of each child's future. There are 14 people registered to our household, but we only have six shares of land. We always have to buy barley.

On the other hand some families were influenced by the government's position, backed by the threat of penalties, that having many kids was detrimental for poor families. For example, a woman with two children, born in 1989 and 1991, stated:

> We are very poor, and the government said that the poor must do the operation. They told us in village meetings that the poor can only have two kids, while wealthier households can have three. I thought, "We are poor with little land, so two children are enough and the operation is free."

In a few cases the threat of repercussions—especially during the mid- to late 1990s when some feared the policy would be rigorously enforced—was a powerful motivation to use birth control. One man with five children, the last born in 1993, reported that his wife was one of the first to be sterilized by a mobile team of doctors. He said:

> At the time many people wanted more children but there was a government birth control movement. My wife was among the first group to do birth control. All women were gathered at a meeting where they announced the names of those who should have the operation. At the time they said we couldn't have more than three kids. Those with more than three did birth control. At the time they didn't talk openly about a fine, but it was implied.

Some people regretted the choice they felt compelled to make. For example, a man with two sons and a daughter born between 1991 and 1995 reported, "The government told us there is a limit. If not for that I'd like to have had four or five kids. They are needed to herd the sheep, farm, and go for income." It is also possible that some women were sterilized against their will during the initial years of the birth control campaign. One person recalled the time when the first mobile operating unit came to her village in the early 1990s. She said that many who were targeted by local officials for the operation were crying, while others were even resisting. She reported that this group of women included many from poorer households who had been pressured into undergoing sterilization. Nevertheless, as the data on high-parity births presented earlier reveal, most women continued to exceed the three-child limit. One man with more than three children said:

> In community meetings they [village officials] announced a birth limit, and said they would fine those who had too many kids. I knew we were acting contrary to that. But later nobody mentioned fines and we kept having children.

These days, officials continue to announce the three-child limit, yet one man stated, "They don't enforce it. Villagers have their own desires for how many kids they want to have." When asked if anybody had told him to stop having children, a man from a wealthy household bluntly stated, "Even if they say that, we won't listen."

The evidence points to the conclusion that some people, notably those living in poor households, may have been pressured to limit their childbearing during the mid-1990s, while others decided to have fewer children in response to rising concerns over their ability to support a large family on limited land. Figure 3.7 represents how the background variables of diminishing per capita landholdings and the implementation of China's birth control policy created an incentive structure for using birth control, thereby reducing aggregate fertility by impacting many women's exposure to conception. Because poorer families were encouraged to use birth control more than wealthier families, the effect on them would presumably be greater.

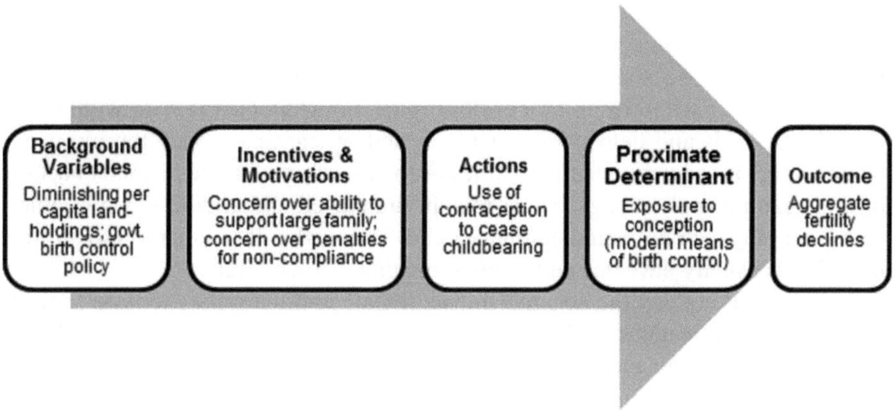

Fig. 3.7 Tibetan fertility, 1990s

3.5.5 Phase 5: Is Replacement Fertility Here to Stay?

Since 2000, fertility in our research villages has fluctuated between two and three births per woman, indicating that the rate has stabilized slightly above replacement level. Local leaders are now convinced that their constituents use birth control willingly and out of economic self-interest, which is consistent with findings from Qinghai Province where Tibetans had been subjected to much more stringent birth control regulations (Schrempf 2008). Leaders in our research area believe that the government's message—a small family is a route to affluence—has taken root. One official explained:

> Even though a household may have three husbands and one wife, they only want to have two or three children. They think this is good enough. They think that the more kids they have, the more hardships they will face as parents. They think that the government's statement "fewer children equal more wealth"[11] is correct. If the first two or three children are girls, then they try for a boy [i.e., larger families result from trying to have a son].

A leader from another village concurred:

> [Nowadays] you don't need to say anything. Women go for birth control on their own volition. I still mention the policy in meetings, but nobody takes it seriously. More people are voluntarily using birth control.

There is certainly merit to these village leaders' assessments. The concern some people have over insufficient landholdings provides a motive to use birth control after having two or three children. Furthermore, in today's expanding economy, many people continue to value polyandrous marriage as a strategy to improve

[11]The government propagates this message through radio, television, newspapers, and in village meetings via members of the Maternal Care Office.

the family's standard of living—a strategy that lies at the heart of a strong son preference found among farmers. The combination of polyandry and a desire to have more than one son is affecting fertility in conflicting ways. On one hand, polyandry continues to result in a bride surplus. Some women remain within their natal households and help with agriculture and by earning money through wage labor. Others move permanently to cities or towns, often with the financial backing of their parents and siblings. Some remain unmarried and childless, while others marry out of love or have an illegitimate child. Those who do not marry into farming households usually have one and rarely more than two children.

Labor needs counter the negative impact female non-marriage exerts on fertility. Goldstein and colleagues (2002) found that wealthy families with large landholdings expressed a desire for more children to fulfill labor demands. This continues to be the case. For example, in 2006 a woman with two sons and two daughters explained, "We have a lot of land and a shortage of labor so we thought having more kids would make life easier." Her husband concurred, "Until now I'm the only one who can do the farming. More children mean more labor." Many families feel strongly that they need at least two sons to thrive in today's economic environment. This was already evident from Goldstein and colleague's 1998 survey. When women were asked to state their ideal number of sons and daughters, the average response for women aged 20–39 (i.e., those in their reproductive years) was 2.0 sons and 1.2 daughters. A strong gender preference is also evident from more recent interviews. For example, while discussing labor needs, an elderly man commented:

> Some rich households feel that even four or five sons are not enough. They have many tractors and farm machinery to operate. If they hire outside labor then they must pay and cannot control the quality of the work or the care of the machinery. For middle-class families at least two sons are good: one to drive the small tractor for income and the other to stay home and farm.

Sons are critical for several reasons, including their higher capacity to do agricultural labor, their social and cultural value as household successors, and their ability to garner higher wages in today's economy. Simply put, a household with two adult sons has one available to fulfill farming tasks and one to send outside to earn income. It therefore comes as no surprise that we found numerous examples of parents whose intense desire for sons prompted them to exceed the government-mandated three-child limit. For example, when we asked a man why he had seven children, he explained:

> The first five daughters are from our first wife.... Our first wife kept having daughters, so it was my idea to bring in her younger sister [as a second wife]. I think that, for a farming family, sons are critical. If you don't have several sons you cannot compete with other households. We waited and waited, but our first wife bore no sons. I thought if we changed wives we could get sons.

Another man with one son and four daughters said:

> We felt that our family size was small, and that we couldn't finish [farming] jobs on time. If we had more children, once they grew up, then we could finish jobs on time. At first we couldn't get a son. After our son was born we wanted a helper for him but only got daughters.

Our research revealed numerous cases whereby families continued having children in the hope of having one, or preferably more than one, son. Conversely, some forces of modernization are counteracting the desire for farming families to have many children. In particular, people feel that the cost of raising children is increasing in the wake of China's recent policy of mandatory ninth grade education. In the words of one man,

> If you have many children and limited money, you have to pay school fees which are high. You will not be able to effectively use the money to educate your children in the best way.... If you have a small family then you can use your money more effectively. All our family members discussed it and agreed that three children are best.... We openly discussed that having fewer children means you can give them better treatment. There are households that have undergone great improvements with only two or three children. You can offer them a good education.

Many parents now stress the importance of educating those children who will not remain in the household, typically daughters and youngest sons, and see this as a strategy for providing externally resident children with independent livelihoods. Caldwell's (1982) thesis on the wealth flow inversion as an incentive to limit childbearing seems to be playing out in rural Tibet as parents invest more heavily in education and thereby diminish their children's roles in the household's labor force.

In summary, aggregate fertility in Tibet is now being moderated by concerns over balancing a household's population with its landholdings, by the increasing cost of raising and educating children, and by a high frequency of female non-marriage generated by polyandry. Through the government's birth control policy, Tibetans now have the ability to control fertility by means of modern contraception. However, some families—especially those with large landholdings—continue to feel that a viable household labor force necessitates having more than three children. In addition, the preference to have two sons so they can marry polyandrously and work jointly to improve the family's standard of living often results in the birth of three or more children. As a result of these conflicting forces, fertility in rural Tibet has apparently stabilized between two and three births per woman or slightly above replacement level.

3.6 Conclusion: Landholdings and Fertility Revisited

In societies that had not yet undergone demographic transitions, scholars hypothesized a direct and predictable relationship between fertility and landholdings (Schujter and Stokes 1984; Stokes et al. 1986) or between household size and landholdings (Netting 1993). In contemporary rural Tibet, even in the wake of a fertility decline, there still exists a strong correlation between landholdings and household size. Table 3.5 divides households according to whether their landholding size is among the top 25%, middle 50%, or lowest 25% in their respective villages.

The correlation between size of landholdings and mean number of residents is consistent: households with the largest landholdings also have the most members.

Table 3.5 Number of household members by relative size of landholding

Village	Lowest 25%	Middle 50%	Highest 25%
Betsag	5.4	8.1	10.3
Norgyong	5.5	7.8	10.0
Sogang	4.7	7.9	10.2
Total	5.2	7.9	10.2

Source: Goldstein et al. (2006) socioeconomic survey

Table 3.6 Children ever born (CEB) and marriage age of married women aged 30–39 in 1998

Land shares	CEB by 1998	Mean age at marriage
2–3	3.0	25.3
4–5	3.7	22.9
6–7	3.8	22.9
8–9	3.9	23.0
10+	4.0	22.3

Source: Goldstein et al. (1998) survey

To a great extent this is a legacy of how the land tenure policy was instituted in 1982: families that were already large received more land because land was parceled on a per capita basis. Demographic fluctuations over time could have leveled the differences, but they did not. Eighty-five percent of households that are now in the top 25% in landholdings started in the top quarter. In other words, most large households in 1982 remain large today, while most small households remain small. This supports Netting's (1993) observation that smallholders tend to strike a balance between their landholdings and household membership.

Table 3.6 shows mean ages at marriage and children ever born according to the number of land shares a household received in 1982. We focus exclusively on the 30–39 cohort of married women because they were aged 14–23 in 1982 and were therefore the first to reproduce in the wake of the Household Responsibility System.

Note that women living in households with the least land shares (2–3) gave birth to fewer children than women living in households with four or more shares. This observation suggests that people with more land want more children than people with less land and end up having more (Schujter and Stokes 1984; Stokes et al. 1986). Our research also found that people living in the poorest households expressed trepidation about having more children than their land could support. Note, however, that those with the least land married significantly later which can partially account for the difference in children ever born. However, poorer families were subjected to more government pressure than wealthier ones to limit reproduction, so fertility outcomes are not merely a function of landholding sizes.

What is clear from our research is that China's policies regarding land tenure have had a considerable impact on the relationship between landholdings and fertility in rural Tibet. Tibetans have adapted their marital and reproductive strategies in response to government policies that first brought agricultural production into a collectivized system, and then redistributed land in such a way that set the size of a household's landholdings according to the size of its membership. The ability to adjust the size of one's landholdings in response to household-level demographic fluctuations is largely absent because land cannot be bought or sold. Polyandry proved to be a resilient strategy for dealing with land as an inelastic resource, and polyandry affected fertility via an intercourse variable: proportion of women in marital unions. Shortly thereafter, diminishing per capita landholdings provided some people with a strong incentive to regulate their reproduction, which affected fertility via a conception variable: use of contraception. However, the ability to act on any desire to limit childbearing only became possible during the 1990s through government programs that made contraception widely available. The government played a further role in shaping fertility outcomes by convincing poorer families with small landholdings, either through persuasion or coercion, to use birth control.

Although landholdings have played a role in shaping fertility outcomes in rural Tibet, the connection between landholdings and fertility is becoming less important over time because of changing economic conditions. Some families with large landholdings still express a desire to have several children, especially sons. Doing so makes sense in the context of emerging economic opportunities stemming from China's policy to rapidly develop Tibet. Households now have the ability to improve their standard of living by deploying members into the nonfarm labor force. Rural families still want two or more sons not just for farm labor, but so they can earn cash as well. This incentive structure exists for all households, regardless of landholding size, and leads many to overshoot the government's official limit of three children. If the current structure of childbearing incentives remains in place, and in the absence of coercive measure to enforce a birth limit, fertility in rural Tibet will likely hover above replacement level but below three births per woman—a range that the government, according to its own policy, would probably consider acceptable.

References

Bongaarts, J., & Potter, R. G. (1983). *Fertility, biology, and behavior: An analysis of the proximate determinants*. New York: Academic.

Cain, M. (1985). On the relationship between landholding and fertility. *Population Studies, 39*, 5–15.

Cain, M. (1986). Landholding and fertility: A rejoinder. *Population Studies, 40*, 313–317.

Caldwell, J. C. (1982). *Theory of fertility decline*. London: Academic.

Carr, D. L., Pan, W. K. Y., & Bilsborrow, R. E. (2006). Declining fertility on the frontier: The Ecuadorian Amazon. *Population and Environment, 28*, 17–39.

Childs, G. (2008). *Tibetan transitions: Historical and contemporary perspectives on fertility, family planning, and demographic change*. Leiden: Brill.

Childs, G., Goldstein, M. C., Jiao, B., & Beall, C. M. (2005). Tibetan fertility transitions in China and South Asia. *Population and Development Review, 31*(2), 337–349.

Cho, L.-J., Retherford, R. D., & Choe, M. K. (1986). *The own-children method of fertility estimation*. Honolulu: East-west Center.

Clay, D. C., & Johnson, N. E. (1992). Size of farm or size of family: Which comes first? *Population Studies, 46*(3), 491–505.

Cleland, J., & Wilson, C. (1987). Demand theories of the fertility transition: an iconoclastic view. *Population Studies, 41*, 5–30.
Davis, K., & Blake, J. (1956). Social structure and fertility: An analytical framework. *Economic Development and Culture Change, 4*(4), 211–235.
Easterlin, R. A., & Crimmins, E. M. (1985). *The fertility revolution: A supply-demand analysis*. Chicago: University of Chicago Press.
Easterlin, R. A., Alter, G., & Condran, G. (1978). Farms and farm families in new and old areas: The northern states in 1860. In T. Hareven & M. Vinovskis (Eds.), *Family and population in nineteenth century America* (pp. 22–84). Princeton: Princeton University Press.
Fjeld, H. (2006). *The rise of the polyandrous house: Marriage, kinship and social mobility in rural Tsang, Tibet*. Ph.D. dissertation, Faculty of Social Sciences, University of Oslo.
Ghimire, D. J., & Hoelter, L. F. (2007). Land use and first birth timing in an agricultural setting. *Population and Environment, 28*, 289–320.
Goldscheider, C. (1971). *Population, modernization, and social structure*. Boston: Little, Brown and Co.
Goldstein, M. C. (1971). Stratification, polyandry, and family structure in central Tibet. *Southwest Journal of Anthropology, 27*, 64–74.
Goldstein, M. C. (1981). New perspectives on Tibetan fertility and population decline. *American Ethnologist, 8*(4), 721–738.
Goldstein, M. C., & Beall, C. M. (1991). China's birth control policy in the Tibet Autonomous Region: Myths and realities. *Asian Survey, 31*(3), 285–303.
Goldstein, M. C., Jiao, B., Beall, C. M., & Tsering, P. (2002). Fertility and family planning in rural Tibet. *The China Journal, 47*, 19–39.
Goldstein, M. C., Jiao, B., Beall, C. M., & Tsering, P. (2003). Development and change in rural Tibet: Problems and adaptations. *Asian Survey, 43*(5), 758–779.
Goldstein, M. C., Childs, G., & Wangdui, P. (2008). "Going for income": A longitudinal analysis of change in farming Tibet, 1997–98 to 2006–07. *Asian Survey, 48*(3), 514–534.
Goldstein, M. C., Childs, G., & Wangdui, P. (2010). Beijing's "people first" development initiative for the Tibet Autonomous Region's rural sector—A case study from the Shigatse area. *The China Journal, 63*, 58–75.
Greenhalgh, S. (1994). Controlling births and bodies in village China. *American Ethnologist, 21*(1), 3–30.
Greenhalgh, S. (2008). *Just one child: Science and policy in Deng's China*. Berkeley: University of California Press.
Greenhalgh, S., & Winckler, E. A. (2005). *Governing China's population: From Leninist to neoliberal biopolitics*. Stanford: Stanford University Press.
Gu, B., Wang, F., & Guo, Z. (2007). China's local and national fertility policies at the end of the twentieth century. *Population and Development Review, 33*(1), 129–147.
Henry, L. (1961). Some data on natural fertility. *Eugenics Quarterly, 8*, 81–91.
Jensen, E. R. (1990). An econometric analysis of the old-age security motive for childbearing. *International Economic Review, 31*(4), 953–968.
Jiao, B. (2001). *Socio-economic and cultural factors underlying the contemporary revival of fraternal polyandry in Tibet*. Ph.D. dissertation, Department of Anthropology, Case Western Reserve University.
Lee, J., & Wang, F. (1999). *One quarter of humanity: Malthusian mythology and Chinese realities*. Cambridge: Harvard University Press.
Malthus, T. R. (1989[1803]). *An essay on the principle of population*. Cambridge: Cambridge University Press.
Mueller, E., & Short, K. (1983). Income and wealth as they affect the demand for children in developing countries. In R. Bulatao & R. D. Lee (Eds.), *Determinants of fertility in developing countries* (pp. 474–517). New York: Academic.
Netting, R. M. C. (1965). Household organization and intensive agriculture: The Kofyar case. *Africa, 35*, 422–429.

Netting, R. M. C. (1968). *Hill farmers of Nigeria: Cultural ecology of the Kofyar of the Jos Plateau.* Seattle: University of Washington Press.
Netting, R. M. C. (1993). *Smallholders, householders: Farm families and the ecology of intensive, sustainable agriculture.* Stanford: Stanford University Press.
Netting, R. M. C., Stone, M. P., & Stone, G. D. (1989). Kofyar cash cropping: Choice and change in indigenous agricultural development. *Human Ecology, 17*, 299–319.
Nugent, J. B. (1985). The old-age security motive for fertility. *Population and Development Review, 11*(1), 75–97.
Renne, E. P. (1995). Houses, fertility, and the Nigerian land use act. *Population and Development Review, 21*(1), 113–126.
Schrempf, M. (2008). Planning the modern Tibetan family in China. In V. Houben & M. Schrempf (Eds.), *Figurations of modernity: Global and local representations in comparative perspective* (pp. 121–151). Frankfurt: Campus Verlag.
Schujter, W. A., & Stokes, C. S. (Eds.). (1984). *Rural development and human fertility.* New York: Macmillan.
Schujter, W. A., Stokes, C. S., & Poindexter, J. R. (1983). Farm size, land ownership, and fertility in rural Egypt. *Land Economics, 59*, 393–403.
Shreffler, K. M., & Dodoo, F. N.-A. (2009). The role of intergenerational transfers, land, and education in fertility transition in rural Kenya: The case of Nyeri District. *Population and Environment, 30*, 75–92.
Stokes, C. S., Schujter, W. A., & Bulatao, R. A. (1986). Is the relationship between landholding and fertility spurious? A response to Cain. *Population Studies, 40*, 305–311.
Stone, G. D., Johnson-Stone, M. P., & Netting, R. M. C. (1984). Household variability and inequality in Kofyar subsistence and cash-cropping economies. *Journal of Anthropological Research, 40*, 90–108.

Chapter 4
Human-Wildlife Contact and Emerging Infectious Diseases

Michael P. Muehlenbein

Abstract The majority of all emerging pathogens in humans are zoonotic (nonhuman animal) in origin. Population, ecological, and behavioral changes that increase contact with wildlife exacerbate emergence of these pathogens. Anthropogenic modification of the physical environment has altered not only our risk of zoonotic infection from wildlife but also the likelihood of pathogen transmission from human to nonhuman animal populations. This is particularly the case for primates that share a number of common infections with humans. In this chapter, I use a series of case studies involving SARS, HIV, Nipah virus, Lyme disease, malaria, and Ebola to exemplify how various anthropogenic factors have facilitated pathogen transmission between human and nonhuman animal populations. The costs and benefits of primate-based ecotourism are also reviewed to better illustrate how human-wildlife contact can affect both populations. Responsible health monitoring of human-wildlife interactions is a necessary prerequisite for prevention of the transmission of future emerging infectious diseases.

4.1 Emerging Infectious Diseases

The patterns of morbidity and mortality in human populations have changed significantly over the past few thousand years. In most populations to date, pestilence and famine decreased as the result of advances in sanitation, medicine, and nutrition. Increases in life expectancy paralleled socioeconomic development and technological changes, and pandemics were replaced with chronic degenerative disorders. Of course, different populations progress through such epidemiologic transitions at different rates and in different times. To use the United States of America as an

M.P. Muehlenbein (✉)
Department of Anthropology, Indiana University, 701 E. Kirkwood Ave.,
Student Building 216, Bloomington, IN 47405, USA
e-mail: mpm1@indiana.edu

E.S. Brondízio and E.F. Moran (eds.), *Human-Environment Interactions: Current and Future Directions*, Human-Environment Interactions 1,
DOI 10.1007/978-94-007-4780-7_4,

example, rapid development throughout the agricultural and industrial revolutions followed by advances in science and public health drove the US Surgeon General William H. Stewart to reportedly tell the Congress in 1969 that it was "time to close the book on infectious diseases, declare the war against pestilence won, and shift national resources to such chronic problems as cancer and heart disease." He did not foresee the massive impact that emerging and reemerging infectious diseases would have on human health and economic prosperity worldwide.

Emerging infectious diseases (EIDs) are those that have recently increased in incidence (number of new cases in a given period of time), expanded in geographic range, moved into a new host population, or are caused by newly evolved or discovered pathogens (Lederberg et al. 1992; Morens et al. 2004; Weiss and McMichael 2004). Examples include dengue hemorrhagic fever, West Nile virus, and H5N1 avian influenza. The primary causes of EIDs in human populations have been the anthropogenic modifications of physical and social environments (Cohen 2000; Daily and Ehrlich 1996; McMichael 2004; Morens et al. 2004; Patz et al. 2000, 2004; Weiss and McMichael 2004). Human and livestock populations continue to grow rapidly, increasing the number of hosts potentially susceptible to novel infections. Mass transportation of people, products, livestock, and vectors of disease brings each of these closer to one another, and more quickly (Kimball et al. 2005; Wilson 2003). Population movements due to war, social disruption, and rural-to-urban migration, in addition to general urbanization, increase the densities of nonimmune human hosts and pose significant sanitation problems. Natural disasters and bioterrorism may destroy public health infrastructure and other resources (Watson et al. 2007). Changes in water usage, such as during the construction of dams, culverts, and irrigation systems, can increase the potential breeding sites of vector species like mosquitoes and snails (Keiser et al. 2005; Steinmann et al. 2006). Sex tourism, intravenous drug abuse, reuse of injectable medical equipment, and improper disinfection or ineffective protective measures in hospitals all contribute to the rapid evolution of resistant and deadly pathogens. The misgovernance of epidemics with decentralized testing, misdiagnoses, and underreporting of cases has resulted in slow public health responses to several pandemics (Cyranoski 2005; Normile 2005). Interactions between human and wildlife populations have also been a source of EIDs.

4.2 Wildlife and Emerging Infectious Diseases

Over half of all human infections are zoonotic (nonhuman animal) in origin (Cleaveland et al. 2001; Woolhouse and Gaunt 2007), and the majority of all emerging pathogens in humans are zoonotic in origin (Jones et al. 2008; Taylor et al. 2001). Population, ecological, and behavioral changes that increase contact with wildlife exacerbate emergence of these pathogens (Daszak et al. 2000). Human encroachment into previously undisturbed areas increases remote area accessibility and introduces more vectors and reservoirs of infection to new hosts. Encroachment, extensification of agricultural land, and urban sprawl all alter population densities and distributions of wildlife, which change disease dynamics (Patz et al. 2004). Forest fragmentation can produce an "edge effect," increasing the flow of organisms across ecotones, novel species contact, and the likelihood of infection transmission between populations.

Wildlife populations can serve as sinks for human pathogens. That is, increased biodiversity can function as a buffer of hosts in an environment, decreasing the likelihood that vectors of infection would feed off of humans. Areas with increased host diversity could be characterized by a reduced probability of pathogen transmission to vectors as well as a reduced encounter rate between vectors and hosts. "Zooprophylaxis," or the purposeful use of wildlife and livestock to protect against zoonotic infection, has been utilized to try to guard humans against a number of infections, including encephalitis and malaria (Hess and Hayes 1970). Biodiversity loss due to global climate change, deforestation, the spread of invasive species, overexploitation, and other causes decreases this important buffer, increasing the likelihood of cross species transmission (Maillard and Gonzalez 2006; Pongsiri et al. 2009). Increased inbreeding and decreased genetic diversity in remaining wildlife populations could even facilitate further outbreaks due to impaired immune functions in host animals (ibid.).

Considering all wildlife as a source of human infections would obviously be a grave mistake; protecting wildlife actually serves a selfish purpose of protecting our own health. For example, species-poor areas in the northeastern United States are characterized as having more white-footed mice (*Peromyscus leucopus*), and thus increased incidence of the Lyme spirochete *Borrelia burgdorferi* (LoGiudice et al. 2003). Squirrels and other "dilution hosts" can protect areas from Lyme disease (ibid.). With more hosts in general, and less competent hosts (those that do not support optimal growth and transmission of the pathogen) in particular, *Ixodes* ticks are less likely to feed on humans and successfully transmit Lyme disease. Predatory vertebrates also help protect humans from Lyme disease and many other infections by keeping the abundance of rodent hosts at bay (Ostfeld and Holt 2004).

Maintaining preferential hosts in an environment places a barrier against vector-borne infections. For example, counties in the eastern United States with greater avian diversity exhibit lower incidence of human West Nile virus cases (Swaddle and Calos 2008), thanks in part to the ornithophilic nature of *Culex* and other mosquito species. Similarly, increased small-mammal diversity protects sympatric rodent populations from the spread of Hantavirus infection (Suzán et al. 2009). Such protection represents a largely overlooked ecosystem service. Twenty-five percent (25%) of all extant mammals are now endangered, and approximately 30% of these will become extinct by the year 2050 (Schipper et al. 2008). Overexploitation of animals and environments has and will continue to produce "spillover" of zoonotic infections into human populations.

4.3 Consumption of Wildlife

Responsible human-wildlife contact requires that we carefully consider how to manage wildlife as sources and sinks of human infections. One way in which wildlife serves as a source of infection is via bushmeat, or the use of wild animals for food, medicinal, or traditional cultural purposes, usually involving ingestion. Bushmeat is an import source of protein and income for millions of people, and the illegal bushmeat trade has been facilitated by the use of modern weapons and communication, logging operations that provide access to forests and transportation of products, lack of economic alternatives, and minimal capacity to enforce laws (Karesh and Noble 2009). Current usage rates of bushmeat are unsustainable, well exceeding 100 kg/km^2/year in many locations (Robinson and Bennett 2000).

Bushmeat hunting, preparation, and consumption are linked with several pandemics and epidemics, most notably human immunodeficiency virus (HIV), Ebola, and severe acute respiratory syndrome (SARS). The current HIV pandemic, with an estimated global prevalence of 33 million people (UNAIDS 2008), appears to have originated from nonhuman primate simian immunodeficiency viruses (SIV). Through the hunting and butchering of nonhuman primates (resulting in cutaneous or mucous membrane exposure) in West Africa, SIV likely entered into the human population several times and became established as HIV around 1900 in what is now the Democratic Republic of Congo (Worobey et al. 2008). Based on overwhelming genetic similarities (genome structure and protein homology), HIV-1 groups M and N likely originated from chimpanzees (*Pan troglodytes troglodytes*) in Cameroon, group O from western lowland gorillas (*Gorilla gorilla gorilla*) in Cameroon, and HIV-2 from sooty mangabeys (*Cercocebus atys*) in Côte d'Ivoire (Apetrei et al. 2005; Bailes et al. 2003; Gao et al. 1999; Hahn et al. 2000; Keele et al. 2006; Peeters et al. 2002; Santiago et al. 2005; Van Heuverswyn et al. 2006). Direct exposure to nonhuman primate blood through hunting and butchering is common in several populations in West Africa (Wolfe et al. 2004), and SIV has been identified in nonhuman primate bushmeat, pet animals, and bushmeat hunters in West Africa (Apetrei et al. 2005; Kalish et al. 2005; Peeters et al. 2002).

In contrast to HIV, *Ebola virus* (subtypes Zaire, Sudan, Côte d'Ivoire, Bundibugyo, and Reston) is an RNA filovirus that has wiped out several nonhuman primate populations over the past 20 years (Bermejo et al. 2006; Leroy et al. 2004; Walsh et al. 2003). The virus appears to be restricted to the rainforests of central and western Africa and Southeast Asia (Monath 1999; Peterson et al. 2004). Marburg, a related virus, appears to be restricted to dry, open areas of central and eastern Africa, and the distributions of Ebola and Marburg likely reflect natural host distribution (ibid.). Nearly all cases of Ebola in humans can be traced back to the handling or consumption of infected wildlife carcasses, particularly that of apes (Leroy et al. 2004; Pourrut et al. 2005). Although not definitively known at this time, several fruit bat species are suspected to be the natural reservoir host for Ebola (subtype Zaire) and possibly other subtypes (Biek et al. 2006; Leroy et al. 2005).

Bats (genus *Rhinolophus*) may also be the natural reservoir for SARS (Lau et al. 2005; Li et al. 2005; Vijaykrishna et al. 2007; Wang et al. 2006). SARS is a coronavirus (Drosten et al. 2003; Rota et al. 2003) that produced a human epidemic between 2002 and 2004, claiming almost 775 lives (Hughes 2004). The epidemic originated in the Foshan Municipality of the Guangdong Province of China, with the earliest infections confirmed in animal traders (Yu et al. 2003). Several species of animals in the wet market tested positive for SARS, including Himalayan palm civets (*Paguna larvata*), raccoon dogs (*Nyctereutes procuyoinboides*), and Chinese ferret badgers (*Melogale moschata*) (Guan et al. 2003). Horseshoe bats were likely fed to farmed civets or came into contact with civet farms, and these infected civets were later sold in wet markets, introducing SARS into the general human population. The main February 2003 outbreak can be traced to a single individual who stayed at a hotel in Hong Kong where the pathogen was then spread via aerosolization to ten hotel guests who then traveled to seven different countries, subsequently spreading the pathogen to many individuals, including hundreds of healthcare workers (Hughes 2004). Fear of decreased tourism, travel, and trade was the primary reason for Chinese government officials' underreporting the actual number of casualties, delayed reporting to the World Health Organization, and initial noncooperation with the US Centers for Disease Control and Prevention (Barry 2004; Bloom 2003; Heymann 2004; Kahn 2003; Parry 2003).

4.4 Land-Use/Land-Cover Change and Climate Change

Consumption of wildlife is obviously not the only mechanism by which zoonoses have been introduced into human populations. Vector-borne and waterborne EIDs have been linked in complex ways with climate change (Epstein 2001; Patz et al. 2008). Examples include outbreaks of hantavirus pulmonary syndrome in the southwest United States (Engelthaler et al. 1999), West Nile virus in the United States (Epstein 2001), Rift Valley fever in Kenya (Linthicum et al. 1999), and cholera in Bangladesh (Colwell 1996). Degradation and fragmentation of habitats also forces the overlap of wildlife, domestic animal, and human ecologies. In fact, an important, but often overlooked externality of land-use/land-cover change is the spread of infection. Examples of EIDs linked with climate and land-use/land-cover changes include Nipah virus, Lyme disease, Chikungunya virus, and malaria.

Nipah virus is a single-stranded RNA virus of the family Paramyxoviridae that causes severe acute febrile encephalitis in humans. Like SARS and many other viruses (Calisher et al. 2006; Halpin et al. 2007; Wong et al. 2007), bats (flying fox, genus *Pteropus*) are the natural reservoir of Nipah virus. Nipah virus resulted in over 100 human fatalities in the Kinta district of northern peninsular Malaysia in 1998–1999. The cause of the outbreak is now attributed to a complex interaction of human-induced environmental changes. Fire-mediated deforestation for the expansion of oil palm plantations produced significant respirable suspended-particulate matter in the region. Combined with a drought produced by the El Niño

Southern Oscillation, the availability of flowering and fruiting forest trees was reduced (Chua et al. 2002). Malayan flying foxes (*Pteropus vampyrus*) and island flying foxes (*Pteropus hypomelanus*) began feeding in durian (*Durio zibethinus*) and rambutan (*Nephelium lappaceum*) orchards planted near pig farms. Bat urine as well as partially eaten fruit entered the pigsties, and the virus amplified within the swine. Despite producing severe respiratory disease, the infected pigs were distributed throughout Malaysia. Transmission between bats, pigs, and humans probably happened several times, beginning even before the El Niño event of 1998. It may also have simply been the case that a rapidly expanding pig population increased the probability of contact between bats and pig farms, or that bats were forced to feed in these human orchards because of hunting in other areas (Field 2009). Habitat loss and climate change may also have facilitated viral transmission via nutritional stress, which alters bat physiology, increasing viral seroprevalence and persistence in these animals, increasing the likelihood of horizontal transmission between the flying foxes, and possibly direct transmission between flying foxes and humans (Plowright et al. 2008). Subsequent human outbreaks have been reported in Bangladesh, with transmission largely attributed to *Pteropus giganteus* (Epstein et al. 2006).

Climate is an important determinant of the distribution of arthropod vectors of disease. This is particularly the case of the deer tick, *Ixodes scapularis*, the primary vector for Lyme disease (Brownstein et al. 2003). Suitable habitat for various tick species is expected to increase dramatically in North America and Europe with global warming (Brownstein et al. 2005; Lindgren et al. 2000; Ogden et al. 2005). Already the *Aedes aegypti* mosquito has expanded its geographic range in response to increased global temperatures, resulting in increased risk of Dengue virus in subtropical and temperate climates (Hales et al. 2002). Similarly, the outbreak of hantavirus pulmonary syndrome in the southwestern United States in the early 1990s was linked to heavy summer rains associated with the El Niño Southern Oscillation effect and the subsequent proliferation of pine nuts and deer mice (*Peromyscus maniculatus*), the natural reservoir of the Hantaan virus (family Bunyaviridae) (Engelthaler et al. 1999).

Land-use and climate changes are associated in complicated ways with altered rates of malaria incidence. Malaria is a mosquito-borne disease caused by protozoa of the genus *Plasmodium* (phylum Apicomplexa, order Haemosporidia, family Plasmodiidae), with 172 named species that parasitize reptiles, birds, and mammals (Coatney et al. 1971; Garnham 1966; Levine 1988). Of these, only 4 usually infect humans (*P. falciparum*, *vivax*, *ovale*, and *malariae*), 19 infect nonhuman primates, and 19 infect various other mammals. The parasites are transmitted by over 50 species of female *Anopheles* mosquitoes (Kiszewski et al. 2004). Members of the genus *Plasmodium* likely diverged from the order Haemosporidia around 500 million years ago, perhaps around the time of the Cambrian explosion (Escalante and Ayala 1994). Worldwide climatic changes throughout the last glaciation and the advent of agriculture would have facilitated the spread of the *Anopheles* vector (Coluzzi 1999). During the agrarian revolution, higher-density, sedentary human populations provided mosquitoes with necessary bloodmeals and potential breeding sites (Hartl 2004; Livingstone 1958). Malaria is now endemic in most tropical regions of the world (Guerra et al. 2008). The global incidence of human malaria is estimated to be more than 300 million new clinical cases each year (Trigg and Kondrachine 1998; WHO 1999). The economic and social impacts of this disease are enormous (Gallup and Sachs 2001).

Deforestation and increased global temperature have led to alterations in mosquito breeding habitats. Forest habitat alterations can create new breeding sites by changing surface-water availability (J. Walsh et al. 1993) and increasing water temperatures (Tuno et al. 2005). The combined effects include more mosquitoes with shortened parasite development time (Afrane et al. 2005, 2006) and increased vectorial capacity (number of infective bites a host receives in a given time period) (Afrane et al. 2008; Vittor et al. 2006). Cultivated areas with elevated water temperatures are also characterized by higher mosquito production rates and increased malaria transmission risk compared to undisturbed areas (Lindblade et al. 2000; Munga et al. 2006). Further shifting of geographic and temporal patterns of incidence of malaria is expected in response to continued climate change (Bhattacharya et al. 2006). This will likely include increased duration of transmission windows, increased mosquito abundance (Pascual et al. 2006), and increased altitudinal distribution of mosquitoes (Ebi et al. 2005; Martens et al. 1995).

The incidence of zoonotic malaria is also expected to increase with continued encroachment of humans into forested areas in Southeast Asia. Zoonotic malaria has probably infected humans for thousands of years, but not until recently has *Plasmodium knowlesi* become recognized as a potential major cause of human morbidity and mortality. *Plasmodium knowlesi* commonly infects long-tailed (*Macaca fascicularis*) and pig-tailed (*Macaca nemestrina*) macaques and is typically transmitted by forest-dwelling *Anopheles hackeri* and *Anopheles latens* in Sarawak, Malaysia, and *Anopheles cracens* in peninsular Malaysia (Vythilingam et al. 2008), and other *Anopheles* mosquitoes of the leucosphyrus group in other countries (Galinski and Barnwell 2009). Many cases of fatal malaria once attributed to *P. malariae* were only recently recognized to have been caused by *P. knowlesi* (Jongwutiwes et al. 2004; Singh et al. 2004). Encroachment into forested areas will likely facilitate continued host switching of malarial parasites (Cox-Singh and Singh 2008). Conflict and natural disasters combined with normal rural-to-urban migration will result in population movements that increase the likelihood of malaria transmission (Martens and Hall 2000).

Other malaria parasites, like *P. cynomolgi* and *P. inui*, that naturally infect some nonhuman primates in Asia may also pose a future zoonotic threat to humans (Galinski and Barnwell 2009). The same could be said for Chikungunya virus, an *Alphavirus* transmitted by *Aedes* mosquitoes and maintained naturally in nonhuman primate and other wildlife populations (Gould and Higgs 2009; Powers and Logue 2007). Infection causes severe joint pain in humans. However, humans will not need to enter forests inhabited by monkeys in order to contract this disease. Rather, this virus is already rapidly spreading to urban areas as well as higher latitudes (ibid.).

Thanks in part to global climate change, EIDs are now affecting residents of even very high latitudes (Greer et al. 2008; Parkinson and Evengård 2009).

Anthropogenic ecosystem disruptions that increase contact between humans, wildlife, and vectors of disease increase the likelihood of not only zoonoses but also anthropozoonoses (pathogens transmitted from human to nonhuman animal populations). Nonhuman animals are very susceptible to our pathogens (Epstein and Price 2009); infectious diseases can cause high levels of morbidity and mortality with subsequent population declines in wildlife (Pedersen et al. 2007; Smith et al. 2009). We must be conscious about how our behaviors, including habitat alteration, meat consumption, and even recreational use of natural areas, may inadvertently affect the risks of zoonotic and anthropozoonotic pathogen transmission.

4.5 Primate-Based Ecotourism

One overlooked aspect of human-wildlife contact (and conflict) includes nature-based tourism. Ecotourism accounts for a significant proportion of all international tourism and contributes billions of dollars to the national income of various countries (Filion et al. 1994). Such revenue can enhance economic opportunities for local residents, support environmental education, and protect the natural and cultural heritage of the area, including the conservation of biodiversity and improvement of local facilities. Wildlife is certainly more valuable alive than dead (Thresher 1981). Unfortunately, rapid, unmonitored development of ecotourism projects can lead to degradation of habitats and deleterious effects on animal well-being (Kuss et al. 1990; Speight 1973). Habituation of animals to human presence can increase the likelihood that animals will actively seek out contact with humans, particularly in the form of crop raiding and invasion of garbage pits and latrines. Habituation also makes wildlife more vulnerable to poaching because of their loss of fear of humans. Other risks associated with ecotourism in protected areas include habitat degradation, pollution, crowding, introduction of invasive species, increased tax burdens for local communities, inappropriate development of infrastructure, and disruption of community activities, particularly when communities do not have adequate control over their involvement in tourism.

Increased human-wildlife contact through ecotourism can contribute to the transmission of bacteria, protozoa, viruses, and helminths through direct and indirect infection routes. Zoonotic and anthropozoonotic infection transmission are of vital consideration given the increasing demand from tourists to experience direct encounters with wildlife. This is particularly the case for primates that are genetically closely related to humans (and thus share susceptibility to many common pathogens) and are threatened by habitat destruction, exploitation and fragmentation, hunting and bushmeat consumption, illegal pet trade, and infectious diseases. Transmission of infections between human and nonhuman primate populations is exacerbated by expansion of local communities, refugee populations, park personnel, researchers, filmmakers and photographers, and tourists. This is despite the fact that primate ecotourism is increasingly perceived as a venue for promoting awareness about conservation issues. Over the past few decades, primate ecotourism activities have been flourishing throughout Asia, Middle East, Africa, and Central and South Americas.

Wild primates function as reservoirs for a number of human infections, including filariasis, yellow fever, and Chikungunya virus (Mak et al. 1982; McIntosh 1970; Monath 2001). More important, though, nonhuman primates are very susceptible to many human infections (Brack 1987), and transmission of these anthropozoonoses poses a significant threat to wildlife (Daszak et al. 2000). These animals are usually immunologically naïve to our pathogens, and ape populations can be quickly decimated because of their slow reproductive rates. To date, several infection transmission events from human to nonhuman primate populations have been either suspected or confirmed, including human respiratory syncytial virus and metapneumovirus in chimpanzees in Côte d'Ivoire (Köndgen et al. 2008) and intestinal pathogens *Giardia* and *E. coli* in mountain gorillas and chimpanzees in western Uganda (Goldberg et al. 2007; Graczyk et al. 2002; Rwego et al. 2008). Although local populations or researchers have been the sources of these infections, tourists pose an uncalculated risk to wildlife, which in turn has the potential of producing devastating health and economic outcomes (Muehlenbein and Ancrenaz 2009).

To better understand the risks of anthropozoonotic infection between tourists and wild nonhuman primates, our research team has surveyed ecotourists at the Sepilok Orangutan Rehabilitation Centre in Sabah, Malaysia. Sepilok is one of the largest primate-based ecotourism destinations in the world, averaging over 100,000 visitors annually. Of 633 individuals surveyed in 2007, over half reported being currently vaccinated against tuberculosis, hepatitis A, hepatitis B, polio, and measles. Fewer participants reported current vaccination status for influenza, rabies, and chicken pox. Despite the fact that the majority of visitors to Sepilok are from temperate regions where influenza is relatively more prevalent, 67.1% of those surveyed with medical-related occupations reported not being currently vaccinated for influenza (Muehlenbein et al. 2008). Respiratory infections, like influenza, would arguably be the easiest to transmit while visiting these primates (primarily orangutan and macaques) at Sepilok. Visitors to other wildlife sanctuaries are likely undervaccinated as well.

Fifteen percent of the 633 tourists self-reported at least one of the following current symptoms: cough, sore throat, congestion, fever, diarrhea, and vomiting (Muehlenbein et al. 2010). Of the sample of tourists, 10.8% self-reported at least one symptom associated with respiratory tract infection (cough, sore throat, or congestion). The participants with recent animal contact (e.g., livestock, wildlife at other sanctuaries, unfamiliar domestic pets) were more likely to report current respiratory symptoms compared to individuals with no such animal contact. Similarly, participants with a medical-related occupation were more likely to report current respiratory symptoms while at Sepilok compared to participants with nonmedical occupations. That is, currently ill and potentially infectious tourists were still visiting a wildlife sanctuary to view endangered species, despite having at least some basic knowledge about infection transmission (i.e., medical-related occupation), or having had animal contact immediately prior to arriving. While participants in nature-based tourism are generally concerned about environmental protection, analyses suggest that a significant proportion of ecotourists are either uninformed of the risks they may pose to nonhuman animal health or chose to ignore such risks. These tourists underestimate their own risk of infection as well as their potential contribution to the spread of diseases.

Despite a plethora of best-practice guidelines produced by various tourism and conservation organizations (Ceballos-Lascuráin 1996; Christ et al. 2003; Eagles et al. 2002; Higginbottom 2004; UNEP/WTO 2005; Wood 2002), tourists are simply not being adequately informed about the risks of zoonotic and anthropozoonotic diseases. Tourist ignorance over infection risks, particularly the risks they pose to the very same wildlife they are interested in protecting, cannot be justified, regardless of the large sums of money people spend to visit these exotic destinations. It is therefore the combined responsibility of the tourism and medical communities to more accurately communicate the risks of zoonotic and anthropozoonotic infections in ways that best support the needs of humans and wildlife alike. The global management of zoonotic and anthropozoonotic epidemics is an obligation that transcends any one discipline. Unregulated ecotourism, climate and land-use/land-cover changes, and bushmeat consumption can all exacerbate human-wildlife contact and the subsequent emergence and reemergence of infectious diseases. Yet arguably these human-wildlife interactions are manageable to some degree, given the right resolve and resources. Responsible health monitoring of human-wildlife interactions is a necessary prerequisite for prevention of future emerging infectious diseases.

References

Afrane, Y. A., Lawson, B. W., Githeko, A. K., & Yan, G. Y. (2005). Effects of microclimatic changes caused by land use and land cover on duration of gonotrophic cycles of *Anopheles gambiae* (Diptera: culicidae) in western Kenya highlands. *Journal of Medical Entomology, 42*(6), 974–980.

Afrane, Y. A., Zhou, G., Lawson, B. W., Githeko, A. K., & Yan, G. (2006). Effects of microclimate changes caused by deforestation on the survivorship and reproductive fitness of *Anopheles gambiae* in western Kenya highlands. *The American Journal of Tropical Medicine and Hygiene, 74*, 772–778.

Afrane, Y. A., Little, T. J., Lawson, B. W., Githeko, A. K., & Yan, G. (2008). Deforestation and vectorial capacity of *Anopheles gambiae* Giles mosquitoes in malaria transmission, Kenya. *Emerging Infectious Diseases, 14*, 1533–1538.

Apetrei, C., Metzger, M. J., Richardson, D., Ling, B., Telfer, P. T., Reed, P., et al. (2005). Detection and partial characterization of simian immunodeficiency virus SIVsm strains from bush meat samples from rural Sierra Leone. *Journal of Virology, 79*, 2631–2636.

Bailes, E., Gao, F., Bibollet-Ruche, F., Courgnaud, V., Peeters, M., Marx, P. A., et al. (2003). Hybrid origin of SIV in chimpanzees. *Science, 300*, 1713.

Barry, J. M. (2004). *The great influenza: The epic story of the deadliest plague in history*. New York: Viking Penguin.

Bermejo, M., Rodríguez-Teijeiro, J. D., Illera, G., Barroso, A., Vilà, C., & Walsh, P. D. (2006). Ebola outbreak killed 5000 gorillas. *Science, 314*, 1564.

Bhattacharya, S., Sharma, C., Dhiman, R. C., & Mitra, A. P. (2006). Climate change and malaria in India. *Current Science, 90*, 369–375.

Biek, R., Walsh, P. R., Leroy, E. M., & Real, L. A. (2006). Recent common ancestry of Ebola Zaire virus found in a bat reservoir. *PLoS Pathogens, 2*(10), e90. doi:10.1371/journal.ppat.0020090.

Bloom, B. R. (2003). Lessons from SARS. *Science, 300*, 701.

Brack, M. (1987). *Agents transmissible from simians to man*. Berlin: Springer.

Brownstein, J. S., Holford, T. R., & Fish, D. (2003). A climate-based model predicts the spatial distribution of the Lyme disease vector *Ixodes scapularis* in the United States. *Environmental Health Perspectives, 111*, 1152–1157.

Brownstein, J. S., Holford, T. R., & Fish, D. (2005). Effect of climate change on Lyme disease risk in North America. *EcoHealth, 2*, 38–46.

Calisher, C. H., Childs, J. E., Field, H. E., Holmes, K. V., & Schountz, T. (2006). Bats: Important reservoir hosts of emerging viruses. *Clinical Microbiology Reviews, 19*, 531–545.

Ceballos-Lascuráin, H. (1996). *Tourism, ecotourism, and protected areas: The state of nature-based tourism around the world and guidelines for its development*. Gland: World Conservation Union.

Christ, C., Hillel, O., Matus, S., & Sweeting, J. (2003). *Tourism and biodiversity: Mapping tourism's global footprint*. Washington, DC: Conservation International.

Chua, K. B., Chua, B. H., & Wang, C. W. (2002). Anthropogenic deforestation, El Niño and the emergence of Nipah virus in Malaysia. *Malaysian Journal of Pathology, 24*, 15–21.

Cleaveland, S., Laurenson, M. K., & Taylor, L. H. (2001). Diseases of humans and their domestic mammals: Pathogen characteristics, host range and the risk of emergence. *Philosophical Transactions of the Royal Society of London Series B, 356*, 991–999.

Coatney, R. G., Collins, W. E., Warren, M., & Contacos, P. G. (1971). *The primate malarias*. Bethesda: National Institutes of Health.

Cohen, M. L. (2000). Changing patterns of infectious disease. *Nature, 406*, 762–767.

Coluzzi, M. (1999). The clay feet of the malaria giant and its African roots: Hypotheses and inferences about origin, spread and control of *Plasmodium falciparum*. *Parassitologia, 41*, 277–283.

Colwell, R. R. (1996). Global climate and infectious disease: The cholera paradigm. *Science, 274*, 2025–2031.

Cox-Singh, J., & Singh, B. (2008). Knowlesi malaria: Newly emergent and of public health importance? *Trends in Parasitology, 24*, 406–410.

Cyranoski, D. (2005). Tests in Tokyo reveal flaws in Vietnam's bird flu surveillance. *Nature, 433*, 787.

Daily, G. C., & Ehrlich, P. R. (1996). Global change and human susceptibility to disease. *Annual Review of Energy and the Environment, 21*, 125–144.

Daszak, P., Cunningham, A. A., & Hyatt, A. D. (2000). Emerging infectious diseases of wildlife – Threats to biodiversity and human health. *Science, 287*, 443–449.

Drosten, C., Günther, S., Preiser, W., van der Werf, S., Brodt, H.-R., Becker, S., et al. (2003). Identification of a novel coronavirus in patients with severe acute respiratory syndrome. *The New England Journal of Medicine, 348*, 1967–1976.

Eagles, P. F. J., McCool, S. F., & Haynes, C. D. (2002). *Sustainable tourism in protected areas: Guidelines for planning and management*. Gland: World Conservation Union.

Ebi, K. L., Hartman, J., Chan, N., McConnell, K. J., Schlesinger, M., & Weyant, J. (2005). Climate suitability for stable malaria transmission in Zimbabwe under different climate change scenarios. *Climate Change, 73*, 375–393.

Engelthaler, D. M., Mosley, D. G., Cheek, J. E., Levy, C. E., Komatsu, K. K., Ettestad, P., et al. (1999). Climatic and environmental patterns associated with hantavirus pulmonary syndrome, Four Corners region, United States. *Emerging Infectious Diseases, 5*, 87–94.

Epstein, P. R. (2001). Climate change and emerging infectious diseases. *Microbes and Infection, 3*, 747–754.

Epstein, J. H., & Price, J. T. (2009). The significant but understudied impact of pathogen transmission from humans to animals. *The Mount Sinai Journal of Medicine, 76*, 448–455.
Epstein, J. H., Field, H. E., Luby, S., Pulliam, J. R. C., & Daszak, P. (2006). Nipah virus: Impact, origins, and causes of emergence. *Current Infectious Disease Reports, 8*, 59–65.
Escalante, A. A., & Ayala, F. J. (1994). Phylogeny of the malarial genus *Plasmodium*, derived from rRNA gene sequences. *Proceedings of the National Academy of Sciences of the United States of America, 91*, 11373–11377.
Field, H. E. (2009). Bats and emerging zoonoses: Henipaviruses and SARS. *Zoonoses and Public Health, 56*, 278–284.
Filion, F. L., Foley, J. P., & Jacqemot, A. J. (1994). The economics of global ecotourism. In M. Munasinghe & J. McNealy (Eds.), *Protected area economics and policy: Linking conservation and sustainable development* (pp. 235–252). Washington, DC: The World Bank.
Galinski, M. R., & Barnwell, J. W. (2009). Monkey malaria kills four humans. *Trends in Parasitology, 25*, 200–204.
Gallup, J. L., & Sachs, J. D. (2001). The economic burden of malaria. *The American Journal of Tropical Medicine and Hygiene, 64*, 85–96.
Gao, F., Bailes, E., Robertson, D. L., Chen, Y., Rodenburg, C. M., Michael, S. F., et al. (1999). Origin of HIV-1 in the chimpanzee *Pan troglodytes troglodytes*. *Nature, 397*, 436–441.
Garnham, P. C. C. (1966). *Malaria parasites and other haemosporidia*. Oxford: Blackwell Scientific Publications.
Goldberg, T. L., Gillespie, T. R., Rwego, I. B., Wheeler, E., Estoff, E. L., & Chapman, C. A. (2007). Patterns of gastrointestinal bacterial exchange between chimpanzees and humans involved in research and tourism in western Uganda. *Biological Conservation, 135*, 511–517.
Gould, E. A., & Higgs, S. (2009). Impact of climate change and other factors on emerging arbovirus diseases. *Transactions of the Royal Society of Tropical Medicine and Hygiene, 103*, 109–121.
Graczyk, T. K., Nizeyi, J. B., Ssebide, B., Thompson, R. C. A., Read, C., & Cranfield, M. R. (2002). Anthropozoonotic *Giardia duodenalis* genotype (assemblage) A infections in habitats of free-ranging human-habituated gorillas, Uganda. *Journal of Parasitology, 88*, 905–909.
Greer, A., Ng, V., & Fisman, D. (2008). Climate change and infectious diseases in North America: The road ahead. *Canadian Medical Association Journal, 178*, 715–722.
Guan, Y., Zheng, B. J., He, Y. Q., Liu, X. L., Zhuang, Z. X., Cheung, C. L., et al. (2003). Isolation and characterization of viruses related to the SARS coronavirus from animals in southern China. *Science, 302*, 276–278.
Guerra, C. A., Giandi, P. W., Tatem, A. J., Noor, A. M., Smith, D. L., Hay, S. I., et al. (2008). The limits and intensity of *Plasmodium falciparum* transmission: Implications for malaria control and elimination worldwide. *PLoS Medicine, 5*, e38.
Hahn, B. H., Shaw, G. M., De Cock, K. M., & Sharp, P. M. (2000). AIDS as a zoonosis: Scientific and public health implications. *Science, 287*, 607–614.
Hales, S., de Wet, N., Maindonald, J., & Woodward, A. (2002). Potential effect of population and climate changes on global distribution of dengue fever: An empirical model. *The Lancet, 360*, 830–834.
Halpin, K., Hyatt, A. D., Plowright, R. K., Epstein, J. H., Daszak, P., Field, H. E., et al. (2007). Emerging viruses: Coming in a wrinkled wing and a prayer. *Clinical Infectious Diseases, 44*, 711–717.
Hartl, D. L. (2004). The origin of malaria: Mixed messages from genetic diversity. *Nature Reviews Microbiology, 2*, 15–22.
Hess, A. D., & Hayes, R. O. (1970). Relative potentials of domestic animals for zooprophylaxis against mosquito vectors of encephalitis. *The American Journal of Tropical Medicine and Hygiene, 19*, 327–334.
Heymann, D. L. (2004). The international response to the outbreak of SARS in 2003. *Philosophical Transactions of the Royal Society of London Series B, 1447*, 1127–1129.
Higginbottom, K. (Ed.). (2004). *Wildlife tourism: Impacts, management and planning*. Altona: Cooperative Research Centre for Sustainable Tourism.

Hughes, J. M. (2004). SARS: An emerging global microbial threat. *Transactions of the American Clinical and Climatological Association, 115*, 361–374.

Jones, K. E., Patel, N. G., Levy, M. A., Storeygard, A., Balk, D., Gittleman, J. L., et al. (2008). Global trends in emerging infectious diseases. *Nature, 451*, 990–994.

Jongwutiwes, S., Putaporntip, C., Iwasaki, T., Sata, T., & Kanbara, H. (2004). Naturally acquired *Plasmodium knowlesi* malaria in human, Thailand. *Emerging Infectious Diseases, 10*, 2211–2213.

Kahn, J. (2003). It's a small world after all: Ethics and the response to SARS. *The Hastings Center Report, 33*(3), 6.

Kalish, M. L., Wolfe, N. D., Ndongmo, C. B., McNicholl, J., Robbins, K. E., Aidoo, M., et al. (2005). Central African hunters exposed to simian immunodeficiency virus. *Emerging Infectious Diseases, 11*, 1928–1930.

Karesh, W. B., & Noble, E. (2009). The bushmeat trade: Increased opportunities for transmission of zoonotic disease. *The Mount Sinai Journal of Medicine, 76*, 429–434.

Keele, B. F., Van Heuverswyn, F., Li, Y., Bailes, E., Takehisa, J., Santiago, M. L., et al. (2006). Chimpanzee reservoirs of pandemic and nonpandemic HIV-1. *Science, 313*, 523–526.

Keiser, J., De Castro, M. C., Maltese, M. F., Bos, R., Tanner, M., Singer, B. H., et al. (2005). Effect of irrigation and large dams on the burden of malaria on a global and regional scale. *The American Journal of Tropical Medicine and Hygiene, 72*, 392–406.

Kimball, A. M., Arima, Y., & Hodges, J. R. (2005). Trade related infections: Farther, faster, quieter. *Globalization and Health, 1*, 3.

Kiszewski, A., Mellinger, A., Spielman, A., Malaney, P., Sachs, S. E., & Sachs, J. (2004). A global index representing the stability of malaria transmission. *The American Journal of Tropical Medicine and Hygiene, 70*, 486–498.

Köndgen, S., Kühl, H., N'Goran, P. K., Walsh, P. D., Schenk, S., Ernst, N., et al. (2008). Pandemic human viruses cause decline of endangered great apes. *Current Biology, 18*, 1–5.

Kuss, F. R., Graefe, A. R., & Vaske, J. J. (1990). *Recreation impacts and carrying capacity* (Vols. I and II). Washington, DC: National Parks and Conservation Association.

Lau, S. K. P., Woo, P. C. Y., Li, K. S. M., Huang, Y., Tsoi, H.-W., Wong, B. H. L., et al. (2005). Severe acute respiratory syndrome coronavirus-like virus in Chinese horseshoe bats. *Proceedings of the National Academy of Sciences of the United States of America, 102*, 14040–14045.

Lederberg, J., Shope, R. E., & Oakes, S. C., Jr. (Eds.). (1992). *Emerging infections: Microbial threats to health in the United States*. Washington, DC: Institute of Medicine of the National Academies.

Leroy, E. M., Rouquet, P., Formenty, P., Souquière, S., Kilbourne, A., Froment, J.-M., et al. (2004). Multiple ebola virus transmission events and rapid decline of central African wildlife. *Science, 303*, 387–390.

Leroy, E. M., Kumulungui, B., Pourrut, X., Rouquet, P., Hassanin, A., Yaba, P., et al. (2005). Fruit bats as reservoirs of Ebola virus. *Nature, 438*, 575–576.

Levine, N. D. (1988). *The protozoan phylum apicomplexa* (Vols. 1 and 2). Boca Raton: CRC Press.

Li, W., Shi, Z., Yu, M., Ren, W., Smith, C., Epstein, J. H., et al. (2005). Bats are natural reservoirs of SARS-like coronaviruses. *Science, 310*, 676–679.

Lindblade, K. A., Walker, E. D., Onapa, A. W., Katungu, J., & Wilson, M. L. (2000). Land use change alters malaria transmission parameters by modifying temperature in a highland area of Uganda. *Tropical Medicine & International Health, 5*, 263–274.

Lindgren, E., Tälleklint, L., & Polfeldt, T. (2000). Impact of climatic change on the northern latitude limit and population density of the disease-transmitting European *Ixodes ricinus*. *Environmental Health Perspectives, 108*(2), 119–123.

Linthicum, K. J., Anyamba, A., Tucker, C. J., Kelley, P. W., Myers, M. F., & Peters, C. J. (1999). Climate and satellite indicators to forecast rift valley fever epidemics in Kenya. *Science, 285*, 397–400.

Livingstone, F. B. (1958). Anthropological implications of sickle cell gene distribution in West Africa. *American Anthropologist, 60*, 533–562.

LoGiudice, K., Ostfeld, R. S., Schmidt, K. A., & Keesing, F. (2003). The ecology of infectious disease: Effects of host diversity and community composition on Lyme disease risk. *Proceedings of the National Academy of Sciences of the United States of America, 100*, 567–571.
Maillard, J. C., & Gonzalez, J. P. (2006). Biodiversity and emerging diseases. *Annals of the New York Academy of Sciences, 1081*, 1–16.
Mak, J. W., Cheong, W. H., Yen, P. K., Lim, P. K., & Chan, W. C. (1982). Studies on the epidemiology of subperiodic *Brugia malayi* in Malaysia: Problems in its control. *Acta Tropica, 39*, 237–245.
Martens, P., & Hall, L. (2000). Malaria on the move: Human population movement and malaria transmission. *Emerging Infectious Diseases, 6*, 103–109.
Martens, W. J. M., Niessen, L. W., Rotmans, J., Jetten, T. H., & McMichael, A. J. (1995). Potential impact of global climate change on malaria risk. *Environmental Health Perspectives, 103*, 458–464.
McIntosh, B. M. (1970). Antibody against Chikungunya virus in wild primates in Southern Africa. *South African Journal of Medical Sciences, 35*, 65–74.
McMichael, A. J. (2004). Environmental and social influences on emerging infectious diseases: Past, present and future. *Philosophical Transactions of the Royal Society of London, Series B, 359*, 1049–1058.
Monath, T. P. (1999). Ecology of Marburg and Ebola viruses: Speculations and directions for future research. *Journal of Infectious Diseases, 179*, S127–S138.
Monath, T. P. (2001). Yellow fever: An update. *The Lancet Infectious Diseases, 1*, 11–20.
Morens, D. M., Folkers, G. K., & Fauci, A. S. (2004). The challenge of emerging and re-emerging infectious diseases. *Nature, 430*, 242–249.
Muehlenbein, M. P., & Ancrenaz, M. (2009). Minimizing pathogen transmission at primate ecotourism destinations: The need for input from travel medicine. *Journal of Travel Medicine, 16*, 229–232.
Muehlenbein, M. P., Martinez, L. A., Lemke, A. A., Andau, P., Ambu, L., Nathan, S., et al. (2008). Perceived vaccination status in ecotourists and risks of anthropozoonoses. *EcoHealth, 5*, 371–378.
Muehlenbein, M. P., Martinez, L. A., Lemke, A. A., Ambu, L., Nathan, S., Alsisto, S., et al. (2010). Unhealthy travelers present challenges to sustainable ecotourism. *Travel Medicine and Infectious Disease, 8*, 169–175.
Munga, S., Minakawa, N., Zhou, G., Mushinzimana, E., Barrack, O. O. J., Githeko, A. K., et al. (2006). Association between land cover and habitat productivity of malaria vectors in western Kenyan highlands. *The American Journal of Tropical Medicine and Hygiene, 74*, 69–75.
Normile, D. (2005). WHO faults China for lax outbreak response. *Science, 309*, 684.
Ogden, N. H., Maarouf, A., Barker, I. K., Bigras-Poulin, M., Lindsay, L. R., Morshed, M. G., et al. (2005). *International Journal of Parasitology, 36*, 63–70.
Ostfeld, R. S., & Holt, R. D. (2004). Are predators good for your health? Evaluating evidence for top-down regulation of zoonotic disease reservoirs. *Frontiers in Ecology and the Environment, 2*, 13–20.
Parkinson, A. J., & Evengård, B. (2009). Climate change, its impact on human health in the Arctic and the public health response to threats of emerging infectious diseases. *Global Health Action*. doi:10.3402/gha.v2i0.2075.
Parry, J. (2003). WHO is worried that China is under-reporting SARS. *British Medical Journal, 326*, 1110.
Pascual, M., Ahumada, J. A., Chaves, L. F., Rodo, X., & Bouma, M. (2006). Malaria resurgence in the East African highlands: Temperature trends revisited. *Proceedings of the National Academy of Sciences of the United States of America, 103*, 5829–5834.
Patz, J. A., Graczyk, T. K., Geller, N., & Vittor, A. Y. (2000). Effects of environmental change on emerging parasitic diseases. *International Journal of Parasitology, 30*, 1395–1405.
Patz, J. A., Daszak, P., Tabor, G. M., Aguirre, A. A., Pearl, M., Epstein, J., et al. (2004). Unhealthy landscapes: Policy recommendations on land use change and infectious disease emergence. *Environmental Health Perspectives, 112*, 1092–1098.

Patz, J. A., Olson, S. H., Uejio, C. K., & Gibbs, H. K. (2008). Disease emergence from global climate and land use change. *Medical Clinics of North America, 92*, 1472–1491.

Pedersen, A. B., Jones, K. E., Nunn, C. L., & Altizer, S. A. (2007). Infectious disease and mammalian extinction risk. *Conservation Biology, 21*, 1269–1279.

Peeters, M., Courgnaud, V., Abela, B., Auzel, P., Pourrut, X., Bibollet-Ruche, F., et al. (2002). Risk to human health from a plethora of simian immunodeficiency viruses in primate bushmeat. *Emerging Infectious Diseases, 8*, 451–457.

Peterson, A. T., Carroll, D. S., Mills, J. N., & Johnson, K. M. (2004). Potential mammalian filovirus reservoirs. *Emerging Infectious Diseases, 10*, 2073–2081.

Plowright, R. K., Field, H. E., Smith, C., Divljan, A., Palmer, C., Tabor, G., et al. (2008). Reproduction and nutritional stress are risk factors for Hendra virus infection in little red flying foxes (*Pteropus scapulatus*). *Proceedings of the Royal Society of London, Series B, 275*, 861–869.

Pongsiri, M. J., Roman, J., Ezenwa, V. O., Goldberg, T. L., Koren, H. S., Newbold, S. C., et al. (2009). Biodiversity loss affects global disease ecology. *Bioscience, 59*, 945–954.

Pourrut, X., Kumulungui, B., Wittmann, T., Moussavou, G., Délicat, A., Yaba, P., et al. (2005). The natural history of Ebola virus in Africa. *Microbes and Infection, 7*, 1005–1014.

Powers, A. M., & Logue, C. H. (2007). Changing patterns of chikungunya virus: Re-emergence of a zoonotic arbovirus. *Journal of General Virology, 88*, 2363–2377.

Robinson, J. G., & Bennett, E. L. (Eds.). (2000). *Hunting for sustainability in tropical forests*. New York: Columbia University Press.

Rota, P. A., Oberste, M. S., Monroe, S. S., Nix, W. A., Campagnoli, R., Icenogle, J. P., et al. (2003). Characterization of a novel coronavirus associated with severe acute respiratory syndrome. *Science, 300*, 1394–1399.

Rwego, I. B., Isabirye-Basuta, G., Gillespie, T. R., & Goldberg, T. L. (2008). Gastrointestinal bacterial transmission among humans, mountain gorillas, and livestock in Bwindi Impenetrable National Park, Uganda. *Conservation Biology, 22*, 1600–1607.

Santiago, M. L., Range, F., Keele, B. F., Li, Y., Bailes, E., Bibollet-Ruche, F., et al. (2005). Simian immunodeficiency virus infection in free-ranging sooty mangabeys (*Cercocebus atys atys*) from the Tai Forest, Cote d'Ivoire: implications for the origin of epidemic human immunodeficiency virus type 2. *Journal of Virology, 79*, 12515–12527.

Schipper, J., Chanson, J. S., Chiozza, F., Cox, N. A., Hoffmann, M., Katariya, V., et al. (2008). The status of the world's land and marine mammals: Diversity, threat, and knowledge. *Science, 322*, 225–230.

Singh, B., Kim Sung, L., Matusop, A., Radhakrishnan, A., Shamsul, S., Cox-Singh, J., et al. (2004). A large focus of naturally acquired *Plasmodium knowlesi* infections in human beings. *The Lancet, 363*, 1017–1024.

Smith, K. F., Acevedo-Whitehouse, K., & Pedersen, A. B. (2009). The role of infectious diseases in biological conservation. *Animal Conservation, 12*, 1–12.

Speight, M. C. D. (1973). *Outdoor recreation and its ecological effects*. London: Department of Botany, Westfield College.

Steinmann, P., Keiser, J., Bos, R., Tanner, M., & Utzinger, J. (2006). Schistosomiasis and water resources development: systematic review, meta-analysis, and estimates of people at risk. *The Lancet Infectious Diseases, 6*, 411–425.

Suzán, G., Marcé, E., Giermakowski, J. T., Mills, J. N., Ceballos, G., Ostfeld, R. S., et al. (2009). Experimental evidence for reduced rodent diversity causing increased hantavirus prevalence. *PLoS One, 4*, e5461.

Swaddle, J. P., & Calos, S. E. (2008). Increased avian diversity is associated with lower incidence of human West Nile infection: Observation of the dilution effect. *PLoS One, 3*, e2488.

Taylor, L. H., Latham, S. M., & Woolhouse, M. E. (2001). Risk factors for human disease emergence. *Philosophical Transactions of the Royal Society of London, Series B, 356*, 983–989.

Thresher, P. (1981). The economics of a lion. *Unasylva, 33*, 34–35.

Trigg, P. I., & Kondrachine, A. V. (1998). The current global malaria situation. In I. W. Sherman (Ed.), *Malaria: Parasite biology, pathogenesis and protection* (pp. 11–22). Washington, DC: ASM Press.

Tuno, N., Wilberforce, O., Minakawa, N., Takagi, M., & Yan, G. (2005). Survivorship of *Anopheles gambiae* sensu stricto (Diptera: Culicidae) larvae in western Kenya highland forest. *Journal of Medical Entomology, 42*, 270–277.

UNAIDS (Joint United National Programme on HIV/AIDS). (2008). *Report on the global AIDS epidemic*. Geneva: World Health Organization.

UNEP/WTO (United Nations Environment Programme and the World Tourism Organization). (2005). *Making tourism more sustainable: A guide for policy makers*. Madrid: World Tourism Organization.

Van Heuverswyn, F., Li, Y., Neel, C., Bailes, E., Keele, B. F., Lie, W., et al. (2006). SIV infection in wild gorillas. *Nature, 444*, 164.

Vijaykrishna, D., Smith, G. J. D., Zhang, J. X., Peiris, J. S. M., Chen, H., & Guan, Y. (2007). Evolutionary insights into the ecology of coronaviruses. *Journal of Virology, 81*, 4012–4020.

Vittor, A. Y., Gilman, R. H., Tielsch, J., Glass, G., Shields, T. I. M., Lozano, W. S., et al. (2006). The effect of deforestation on the human-biting rate of *Anopheles darlingi*, the primary vector of falciparum malaria in the Peruvian Amazon. *The American Journal of Tropical Medicine and Hygiene, 74*, 3–11.

Vythilingam, I., NoorAzian, Y. M., Huat, T. C., Jiram, A. I., Yusri, Y. M., Azahari, A. H., et al. (2008). *Plasmodium knowlesi* in humans, macaques and mosquitoes in peninsular Malaysia. *Parasites and Vectors, 1*, 26.

Walsh, J. F., Molyneux, D. H., & Birley, M. H. (1993). Deforestation: Effects on vector-borne disease. *Parasitology, 106*, 55–75.

Walsh, P. D., Abernethy, K. A., Bermejo, M., Beyers, R., De Wachter, P., Ella Akou, M., et al. (2003). Catastrophic ape decline in western equatorial Africa. *Nature, 422*, 611–614.

Wang, L.-F., Shi, Z., Shang, S., Field, H., Daszak, P., & Eaton, B. T. (2006). Review of bats and SARS. *Emerging Infectious Diseases, 12*, 1834–1840.

Watson, J. T., Gayer, M., & Connolly, M. A. (2007). Epidemics after natural disasters. *Emerging Infectious Diseases, 13*, 1–5.

Weiss, R. A., & McMichael, A. J. (2004). Social and environmental risk factors in the emergence of infectious diseases. *Nature Medicine, 10*, S70–S76.

WHO (World Health Organization). (1999). *World health report*. Geneva: World Health Organization.

Wilson, M. E. (2003). The traveler and emerging infections: Sentinel, courier, transmitter. *Journal of Applied Microbiology, 94*, 1S–11S.

Wolfe, N. D., Switzer, W. M., Carr, J. K., Bhullar, V. B., Shanmugam, V., Tamoufe, U., et al. (2004). Naturally acquired simian retrovirus infections in central African hunters. *The Lancet, 363*, 932–937.

Wong, S., Lau, S., Woo, P., & Yuen, K.-Y. (2007). Bats as a continuing source of emerging infections in humans. *Reviews in Medical Virology, 17*, 67–91.

Wood, M. E. (2002). *Ecotourism: Principles, practices and policies for sustainability*. Paris: United Nations Environment Programme.

Woolhouse, M., & Gaunt, E. (2007). Ecological origins of novel human pathogens. *Critical Reviews in Microbiology, 33*, 1–12.

Worobey, M., Gemmel, M., Teuwen, D. E., Haselkorn, T., Kunstman, K., Bunce, M., et al. (2008). Direct evidence of extensive diversity of HIV-1 in Kinshasa by 1960. *Nature, 455*, 661–664.

Yu, D., Li, H., Xu, R., He, J., Lin, J., Li, L., et al. (2003). Prevalence of IgG antibody to SARS-associated coronavirus in animal traders – Guangdong Province, China, 2003. *Morbidity and Mortality Weekly Report, 52*, 986–987.

Part II
Land Change and Landscape Management Approaches

The chapters that follow take a landscape-level approach to a range of themes central to human-environment interactions. The chapters go across gradients of institutional arrangement, demographic pressure, resource ownership systems, and ecosystem types. Chapters complement each other in terms of cultural and historical contexts and methodological approaches to research and to applied situations.

In Chap. 5, Fiorini uses a combination of participatory approaches to examine the case of red deer management in Scotland, an iconic and keystone species that poses challenges to the organization of property systems and different views of land use. Fiorini describes this case of natural resource management first by framing it within European and Scottish regulatory contexts, then across different groups of stakeholders and landscapes bounded by overlapping units of institutional arrangements. The research process illustrates the use of different interview techniques and qualitative data analysis, sketch maps, institutional analysis, and population modeling integrated within a participatory geographic information system. The analysis highlights the importance of understanding the views and perspectives of different stakeholders, the need for flexible and adaptive management plans, and the need for a landscape approach to resource management.

In Chap. 6, Nagendra, Mondal, Adhikari, and Southworth focus on the challenge of understanding forest change in the densely populated, forested landscapes in 15 protected areas in India. Selecting 15 case studies of protected areas from the peer-reviewed literature, four of which they examine in detail, the authors use a combination of meta-analysis and remote sensing approaches to examine their effectiveness amid different drivers of change. They evaluate the role of different forms of population pressure—infrastructure, consumption needs, markets, and settlement—by examining the larger landscapes within which protected areas are embedded. They find that protected areas are an important influence on the rural Indian landscape, in some cases isolating users from the resources they depend on. In general, they find that protected areas contribute to maintaining forest cover at least within the core areas of these reserves. They call attention to growing pressures in the rate of change around the borders and peripheries of these areas, even those once distant from denser human settlements. The authors raise questions about the future of protected areas as proximate and distant pressures are projected to increase in this second most populous nation in the world.

In Chap. 7, Ruseva and Fischer overview forest protection in private areas in the United States, a category that encompasses 56 % of US forests of which 92 % are owned by families, mostly in small properties. They highlight different forms of interactions between private and public decisions regarding the use of forest resources. Using an institutional approach, they start by reviewing the interactions between private and public stakeholders involved in private forestry and propose a typology that characterizes four spheres of public and private interaction concerning investments in forestland management. Their focus is on the challenges and opportunities of governing private forests as a bundle of property rights and bundles of ecosystem services. They highlight the importance of considering the spatial and temporal scales used to manage different aspects of forest resources and their implications for the provisioning of ecosystem services to different groups of stakeholders.

Finally, Chap. 8 focuses on the challenges of monitoring landscapes representing a gradient of land-cover types in the African continent. Gibbes, Cassidy, Hartter, and Southworth carry out analyses across three African countries—Uganda, Botswana, and Namibia—representing different gradients of vegetation, institutional arrangements, and types of resource management. This chapter offers critical perspectives to the limitations of categorical land-cover classification approaches to gradient landscapes. Their cases illustrate remote sensing methodologies to overcome these problems and their limitations. Their contributions go far beyond remote sensing methodologies. Their cases illustrate common themes in human-environment interactions in Africa and elsewhere, such as revealing factors affecting land-use/cover change, the impact and effectiveness of protected areas, and management of livestock-wildlife interaction.

Chapter 5
Change in Natural Resource Management: An Experiment with "Participatory GIS"

Stefano Fiorini

Abstract Natural resource management is a dynamic and adaptive science that responds to changes in the social, economic, and ecological factors pertaining to the managed resource(s). It requires and involves interactions and mediations at different spatial scales and of different knowledge pools and stakes. Participatory approaches can facilitate these interactions and mediations if structured so that collaboration can exist, conflicts are resolved, and knowledge exchange is enhanced. In this chapter, I discuss the application of a spatially explicit participatory GIS (PGIS) to bring together communities of managers, ecologists, and government and NGO representatives in two high-conflict areas of Scotland to discuss conflicts and develop a common knowledge and understanding of red deer and their management. I will first discuss the approach taken during a project that consisted of map-based interviews, secondary data collection, analysis, modeling, and two workshops that engaged stakeholders. Next I will discuss the results of the analysis and illustrate how comanagement and adaptation are currently taking place in deer management in

This research is part of the United Kingdom Research Councils' Rural Economy and Land Use (RELU) program's project "Collaborative Frameworks in Land Management" (Project: RES 227-025-0014). RELU is funded jointly by the Economic and Social Research Council, the Biotechnology and Biological Sciences Research Council, and the Natural Environment Research Council, with additional funding from the Scottish Government's Department for Environment, Food, and Rural Affairs. Funding for this project was also provided by the Forestry Commission, United Kingdom. I would like to thank the above for their financial support for this research and other team members who worked with me in this project: Justin Irvine, Steve Yearley, Helen Armstrong, Jim McLeod, and Amy Turner. I also extend my sincere gratitude to the members of the Deer Management Groups and other stakeholders involved with the project who made our work possible.

S. Fiorini (✉)
Department of Anthropology, Indiana University, 701 E. Kirkwood Avenue, Bloomington, IN 47405-7100, USA
e-mail: sfiorini@indiana.edu

E.S. Brondízio and E.F. Moran (eds.), *Human-Environment Interactions: Current and Future Directions*, Human-Environment Interactions 1,
DOI 10.1007/978-94-007-4780-7_5,

Scotland. This discussion will contain an overview of the reaction of the participants to the PGIS approach. In the end, I will outline some critical consideration for discussion of the role participation can and should have in informing and addressing natural resource management.

5.1 Introduction

Natural resource management is a dynamic and adaptive science that responds to changes in the social, economic, and ecological factors pertaining to the managed resource(s). Among the driving forces behind changes in management objectives in recent decades, one can cite the need to preserve biodiversity and ecosystem functionalities in the effort of preserving the ecological foundations of human livelihoods (MA 2005a, b).

Until recently, management of natural resources with the objective of preserving both biodiversity and the functioning of ecosystems has been concentrated in bounded, designated areas such as national parks and national nature reserves. It is evident now that bounding resources is not enough to guarantee their long-term functionality (e.g., Brondizio et al. 2009; Ostrom and Nagendra 2006), and that there is a need to optimize management to respond to social, economic, and environmental needs over an unbounded landscape. Unbounding the managed landscape breaks those boundaries that separated different practices and objectives for land and resource management, requiring, more often than not, the mediation of conflicting cultural-economic understandings of the managed space and transforming management into a social process.

In various European countries, ecosystem-scale management has jumped to the forefront of the agenda of statutory bodies due to the need of meeting international obligations incorporated into European and national legislations. This has been the case in recent years in Scotland, the site of the case studies discussed in this chapter. Here, governmental organizations have recognized the importance of developing partnerships with private landowners, NGOs, communities, and research providers, building on preexisting forms of comanagement arrangements, like the Deer Management Groups, or building new ones, for example, management groups tasked to address the requirements of the Water Framework Directive. These partnerships would ideally work as "bridging organizations," arenas for developing the needed elements of adaptive governance and comanagement, that is, "knowledge production, trust building, sense making, learning, vertical and horizontal collaboration, and conflict resolution" (Berkes 2009: 1695; see also Folke et al. 2005; Hahn et al. 2006; Olsson et al. 2007, 2008). Participatory approaches, by being processes inclusive of those who hold a stake in the managed resource(s), have the potential for meeting the needs of comanagement partnerships and bridging organizations if structured in a way that conflicts are addressed, collaboration and knowledge exchange is enhanced, and power and resources are shared.

The Scottish case studies this chapter is based on are ideal for investigating the elements affecting contemporary management of natural resources for several reasons. Changes in landownership and increased land fragmentation, combined with the vagility of wild deer in Scotland that knows no boundary (unless, of course a fence, often very costly, obstructs its passage), is increasingly requiring wild deer to be managed over an unbounded landscape. Increasing deer populations are affecting the capacity of ecosystems in various areas to meet the standards set by the legislation. This forces government intervention through the establishment of a variety of partnerships or enforcing changes in deer management. In Scotland, a form of comanagement/a network of bridging organizations, called Deer Management Groups (DMGs), has been in place for nearly three decades.

Various forms of managers' participation in comanagement arrangements and in the working of deer management organizations will provide examples of their diversity. Next to these examples I will illustrate the application of a participatory geographic information system (PGIS) as a platform for knowledge sharing and production, sense making, learning, and conflict resolution. This will highlight the advantages and shortcomings of the application of PGIS as a tool in natural resources management. Among the elements that made the choice of a PGIS relevant is the fact that we are discussing a case of a resource that is mobile and that conflicts over the management of deer derive from the diversity of management objectives existing over the landscape. Various case studies have discussed the value that GIS-supported participatory approaches have for collecting information, supporting analysis, and integrating knowledge (Bacic et al. 2006; Irvine et al. 2009; Ramsey 2009). It has also been shown that PGIS has the potential for enhancing participation, mediation, and planning (Fedra 1995; Janssen et al. 2006; Sandström et al. 2003; Smith 2002; Wang et al. 2008). PGIS was originally chosen as a method in this project as a tool for integrating scientists and practitioners' knowledge (see Irvine et al. 2009). During the project design, this original objective was expanded to include analysis of conflicts and participants' evaluation of PGIS effectiveness for landscape-level deer management, incorporating emergent discussions on the importance and opportunity that PGIS offers in bringing together different forms of knowledge and enhancing power sharing (e.g., Ramsey 2009; Wright et al. 2009).

5.2 Data and Methods

The case studies discussed here were selected based on the advice of a panel of representatives of deer management NGOs, environmental NGOs, and governmental institutions. The selection aimed at maximizing the diversity of settings in which deer management occurs in Scotland, in terms of properties' sizes, characteristics of the landscape, and vicinity to urban centers, as well as investigating areas in which changes in management objectives, due to shifts in public policies or changes in landownership, increased the intra-DMGs conflicts. The work was conducted between 2006 and 2009.

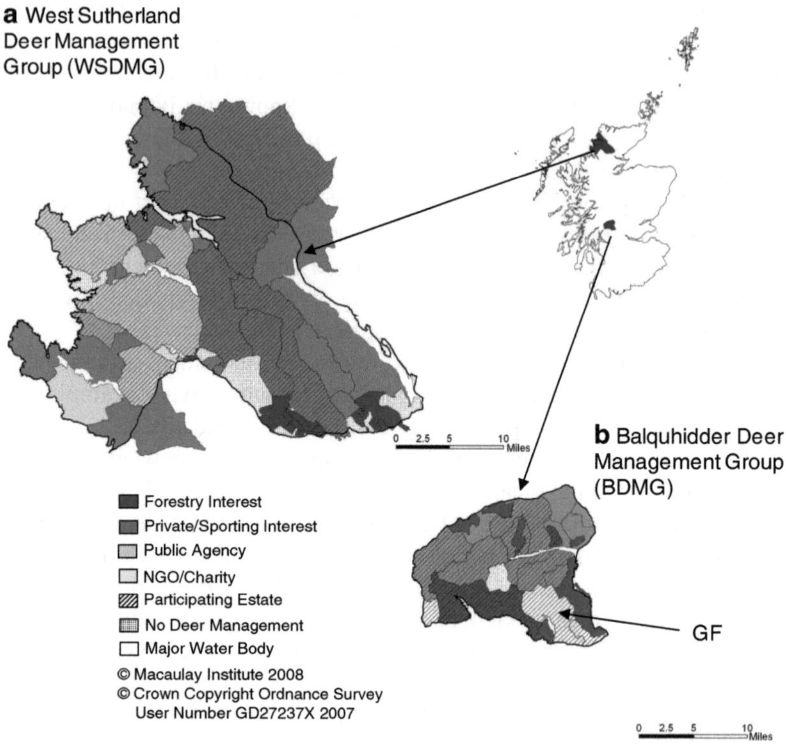

Fig. 5.1 Case study areas and distribution of main land-use objectives (from Irvine et al. 2009)

Figure 5.1 shows the location of the two case studies and maps the diversity of land-use objectives/ownerships. The selected areas coincide with the boundaries of two DMGs. DMGs are networks of voluntary organizations with the objective of bringing together interested parties to facilitate discussion of and coordinate deer management in units of a landscape that are thought to contain a limited number of deer populations. Even if examples of DMGs existed from the late 1960s (Finnie 2004), it is with the support of a statutory body, the Red Deer Commission, the precursor of the Deer Commission for Scotland (DCS), which was merged in August of 2010 with the Scottish National Heritage (SNH), that the number of DMGs grew from a total of 10 in 1979 to about 45 in 2001 (Nolan et al. 2001). Today there are more than 70 DMGs in Scotland, including groups and subgroups (ADMG 2008). Two groups—Balquidder and West Sutherland—are discussed in this chapter.

Balquidder DMG (BDMG) is located in central Scotland and covers an area of about 44,000 ha. At the time of the survey, the average size of the subgroups for which the research team had data was about 1,700 ha, with a maximum of about 7,000 ha and a minimum of about 200 ha. West Sutherland DMG (WSDMG) is located in northwest Scotland and has an approximate total area of 149,900 ha.

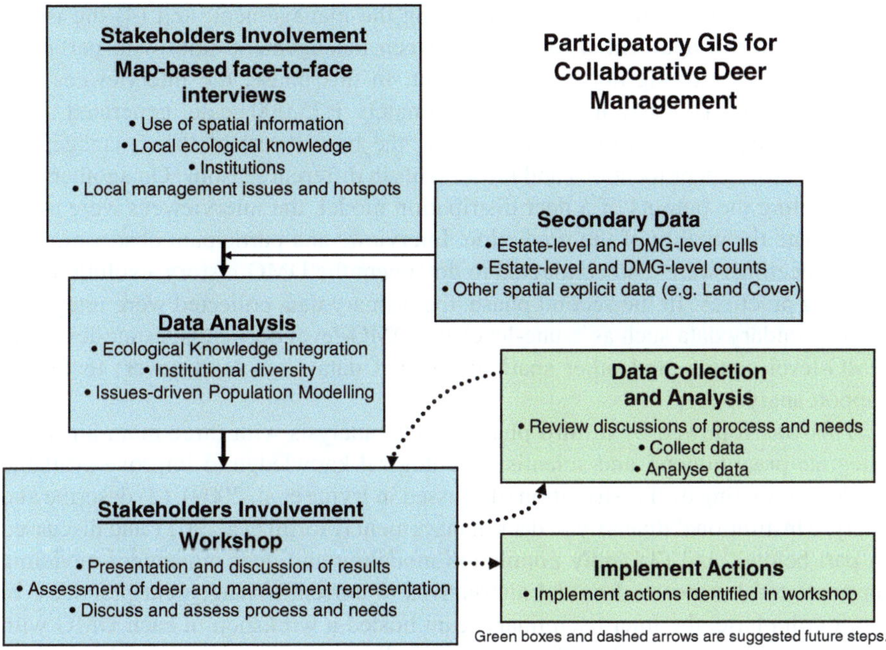

Fig. 5.2 Participatory GIS cycle

The average size of its subgroups was 4,300 ha, with a maximum of 17,700 ha and a minimum of 200 ha. The selection of the interviewees was aimed at providing wider documentation of deer management practices and objectives and covering a representative portion of the landscape included in each DMG. A total of 12 management units actively participated in the survey in BDMG, giving a representation of 74% of the area. In WSDMG, 11 management units took active part in the project, giving a representation of 67% of the DMG territory (see Irvine et al. 2009 for further discussion of the land cover in the DMGs).

Data collection consisted of an interactive process between team members and the chairmen and members of the two DMGs. Participants included both private and NGO landowners and officers of government organizations such as the Deer Commission for Scotland, which had the mandate to oversee matters related to deer management, and the Scottish Natural Heritage, which is in charge of monitoring and advising on the management of areas under various sorts of natural conservation status. The process followed the scheme illustrated in Fig. 5.2.

The first phase of stakeholders' involvement consisted of a series of map-based, face-to-face interviews. The interviews covered four main areas: (1) the use of spatial information to document and inform management planning; (2) the local ecological knowledge on deer use of the landscape that was considered important for deer management; (3) the process followed to decide objectives for deer management, planning, and conducting the management; and (4) the issues that affected management within and between management units with particular attention to identifying hotspots of conflict. In this phase, the interviewees also annotated maps (A1-size maps of approximately 1:25,000 scale, generated using ESRI ArcMap 9.1 with OS MasterMap as the base layer) of their management units, recording deer locations and movements in different seasons. On another map representing the outputs of a deer distribution model, the interviewees were asked to evaluate the quality of the prediction. Interviews and participant observations at DMG meetings were also conducted to document the DMG history, evolution, and working practices. In the second phase, the primary data collected were integrated with secondary data such as estate-level and DMG-level deer culls, estate-level and DMG-level counts, and other spatially explicit data (e.g., land cover) to further support analysis.

This was followed by a third phase of data analysis with three main aims: (1) integrate practitioners' and scientists' ecological knowledge to improve available models predicting deer distribution (discussed in Irvine et al. 2009), (2) describe and analyze institutional diversity in deer management (Fiorini et al. 2011 and discussed in part below), and (3) apply population modeling and deer movement modeling on contested hotspot and model alternatives to current management practices. In the fourth phase, the interdisciplinary team hosted a workshop in each DMG with the stakeholders involved in the research to discuss the outcomes of the project. During the workshops, presentations of the results were followed by discussions of their implications for management, and participants were asked to assess the deer and management representation provided by the research team. They were also asked to discuss and assess the process applied and its contribution to DMG-level management planning and implementation.

Two of the phases suggested in the proposed PGIS cycle, that is, implementation of actions and further data collection (see green boxes in Fig. 5.2), could not be conducted due to limitation of funding and the "experimental" nature of the process that did not require the participants to change management practices in response to the outcomes of the study. Data analysis applied both quantitative and qualitative techniques. Qualitative data was processed using the analysis software NVivo 8 and consisted of theme identification to allow the descriptions of the management process in decision making, identification of constraints, practices, and resources applied in management, summarizing DMG histories, and listing practitioners' ecological knowledge.

The outcome of this analysis is in part described in Sect. 5.3. I will focus particularly on various forms of local-level comanagement arrangements that developed as a response to conservation needs and pressures. Conservation and natural habitat regeneration in Scotland often requires an increase in the number of deer culled. This is considered to be damaging to the interests of those managers who rely on deer hunting as a sporting activity to supplement an estate's income or as an estate's reason to exist.

5.3 Results and Discussion

5.3.1 Deer Commission for Scotland and DMGs: Earlier Attempts at Landscape-Level Deer Management

Deer management in Scotland is regulated by the Deer (Scotland) Act 1996 that established the Deer Commission for Scotland (DCS). DCS had duties to "further the conservation, control and sustainable management of deer, and keep under review all matters, including their welfare, relating to deer" (Scottish Office of the UK Government 1996, part 1, paragraph 1, section a). With the Public Service Reform (Scotland) Act 2010, the power, functions, and staff of the DCS were transferred to SNH. In this chapter, I will refer to DCS because that was the organization in charge at the time of our research.

In some instances, in Scotland, legislation can very closely define deer management objectives and practice. For example, where deer are causing damage to agriculture or natural heritage, or becoming a threat to public safety on the roads, the DCS can use legislation to set control areas and define voluntary "control agreements" with land managers to reduce the impact deer are having (under section 7 of the Deer (Scotland) Act 1996). In cases where a control agreement is "not possible" or "is not being carried out," the DCS can undertake a "control scheme" and intervene directly in culling, or otherwise removing, deer (section 8) and can even demand the expenses incurred to be refunded by the landowner (section 9). Voluntary (section 7) agreements have been the preferred approach. Direct interventions of DCS in culling deer on private estates in a couple of cases resulted in severe backlashes to its image and reputation (e.g., ERAD 2004).

As we mentioned above, DCS was instrumental in establishing the DMGs, bridging organizations whose members participate on a voluntary basis and that have represented the preferred avenue for DCS to influence local-level management. DMGs tend to have about two general meetings every year in which information relevant to the sector is shared or reported by representatives of relevant organizations, including, for example, the DCS and the Association of Deer Management Groups. Cull numbers, deer counts, and other management methods are also discussed. During these meetings, minutes are usually taken; they are distributed to the members before the following meeting.

Interviews in the two case studies reveal that the voluntary nature of DMGs promotes participation and sharing of information. At the same time, DMGs do not have the power of generating binding and enforceable agreements. This contributes to the generally shared belief that DMGs are not effective, are just an occasion for chatting without any practical consequence for management, and that what happens in the pub after the meetings is more useful than the meeting discussions. Despite this, the DMGs in the two case studies enjoy wide participation and have been active for about two decades. It is also believed that the greater the diversity of objectives for deer management among members of the group, the less effective the DMG is. Our evidence suggests otherwise. For example, BDMG enjoyed an increase in participation after one of the large estates was acquired by an environmental NGO. In both DMGs, changes in landownership resulted in increased environmental and deer population monitoring, producing an overall increase in the knowledge available for management decisions.

Despite being a useful instrument for supporting adaptation of management systems to changing ecological and social contexts, the fact that DMGs cannot establish binding and enforceable agreements among their members has induced both governmental organizations and conservation NGOs to seek/follow other paths for achieving environmental conservation objectives. However, these strategies are often limited to estates or management units, often resulting in conflicts with management of neighboring areas. These strategies impair the development of unbounded landscape-level management of the resource deer by focusing on delimited areas and increasing conflicts at the DMG level. In Sect. 5.3.2, I will provide two examples of how conservation is implemented in two "voluntary" forms of comanagement arrangements at the management unit level.

5.3.2 Estate-Level Management

In the two case studies, priorities and objectives for deer management are generally decided among owners and their managers (i.e., estate managers, NGO officers, gamekeepers, stalkers [deer hunters], and rangers). Unless more formal (i.e., legally binding) obligations are in place with government agencies such as SNH and DCS, land managers (including NGOs) appear to be quite independent regarding decisions over what to do on their estates. Even when binding obligations are in place, the government agencies need to accept certain flexibility because resources are not sufficient to either monitor or to provide incentives nor to regulate management in the entire countryside.

There are at least two examples of how biodiversity conservation policies directly impact estates' management. The first example consists of obligations under conservation legislation imposed through the establishment of a voluntary agreement with DCS under section 7 of the Deer (Scotland) Act 1996 and in collaboration with SNH. Biodiversity conservation policies apply to the study areas due to the presence of areas that are part of the Natura 2000 network—special areas of conservation for natural habitats and wild fauna and flora designated by European Union member states. The second example consists of conservation NGOs acquiring significant parcels of land and the associated rights to manage deer. Given the environmental objective of the NGO, it develops a "natural" alignment between government objectives and management in these areas. We will start by discussing the latter example.

5.3.2.1 Change in Landownership

In the study areas, since 1993 NGOs have been acquiring estates. This process is driven by both the opportunities offered by land reforms in Scotland and the increased buying power of NGOs for achieving conservation objectives. Acquisitions might have objectives for local economic development, as is the case in two crofters (small-scale farmers) and community buyouts we investigated under this survey, but in general, management is subordinated to the objective of the conservation organization that has control of the land. Hence, in one of the case studies, an alliance between two NGOs and a government agency aims at regenerating a large area of native woodland, and deer management plans have been developed to achieve this goal in the long term. This process, in practice, translates the guiding values and principles of the conservationists into practical deer management. In a similar way, a community buyout was particularly conditioned to achieve regeneration objectives due to the strength of conservation NGO and government representatives among its membership. These shifts in land management are perceived to have an impact on the neighbors who view the change as having a negative impact on the deer resource because of differences in how deer and natural heritage are valued.

Figures 5.3 and 5.4 show the reporting/decision structure that governs the workings of an NGO in one of the case studies and translates NGO objectives into applied management. The estate manager is the individual responsible for management. He works in close collaboration with the stalker, the shepherd, and the cattleman. The manager reports to the top-governing entity of the NGO (the board of trustees) through the Scottish operations manager, who reports to the Scottish operations director, who reports to the NGO or trust operations director, who finally reports to the board of trustees—quite a hierarchical structure. The manager and other people who work on the estate execute the mandates from the board of trustees, who, on its side, fulfills the mandates of the organization's members. Based on the regeneration objectives of the organization and the estimated grazing impact of the deer, the managers and stalkers set the target number of deer that need to be culled each season (Fig. 5.4). This is a more simple arrangement compared to what drives culls on sporting estates.

5.3.2.2 Application of Statutory Powers

As we have briefly discussed above, the Deer (Scotland) Act 1996 gives power to DCS to enforce particular deer management objectives and practice on a private estate. Voluntary (section 7) agreements have been the preferred approach so far. Section 7 agreements are an issue in one of the case studies in which three agreements are currently in place, and others are potentially under development or may be required in the future. Some estate managers try to avoid such agreements by taking proactive steps; however, conservation areas might include more than one estate or receive pressure from deer moving in from neighboring areas. In these cases, section 7 agreements are seen by DCS as ways for developing collaborations.

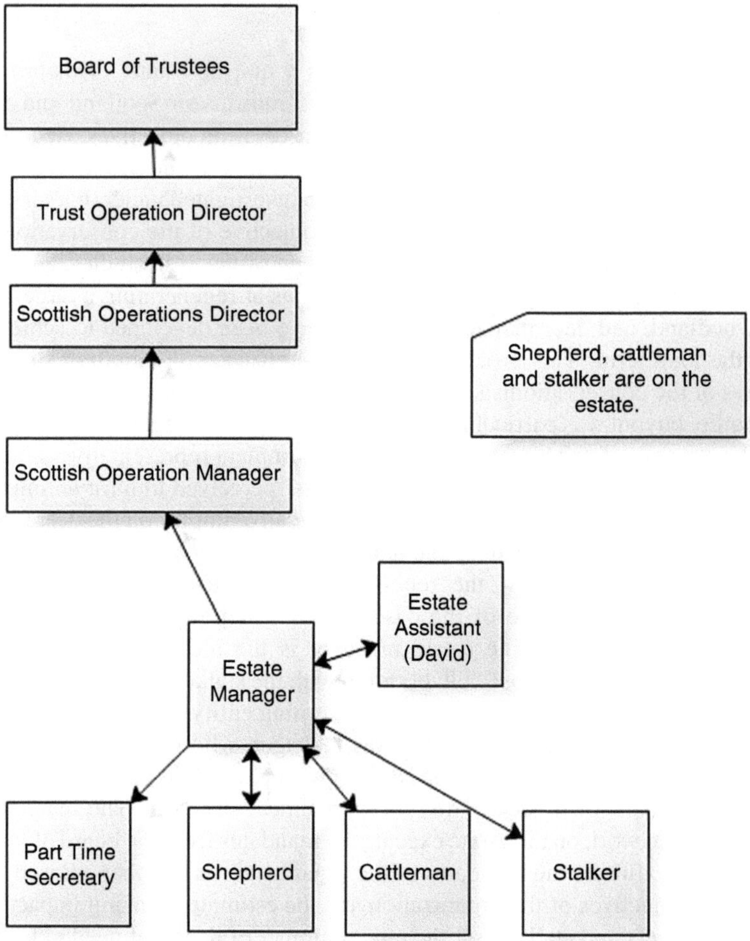

Fig. 5.3 Reporting structure on an NGO-managed property

Managers and gamekeepers traditionally set stocking rates and associated hunting/culling targets based on recruitment levels, winter mortality, and the characteristics of the individual animals they shoot. The agreements force them to adopt different targets based on habitat condition set by SNH and DCS guidelines. Disagreement over what the "right quantity" of deer is can affect the long-term effectiveness of a section 7 agreement, which is usually terminated after the habitats are deemed to be in "favorable conditions." The causes of disagreements are likely to reside in the managers' own experience of what grazing sustainability means, a different way of assessing environmental quality.

The views of traditional estate managers and gamekeepers can be summarized in the following quotes:

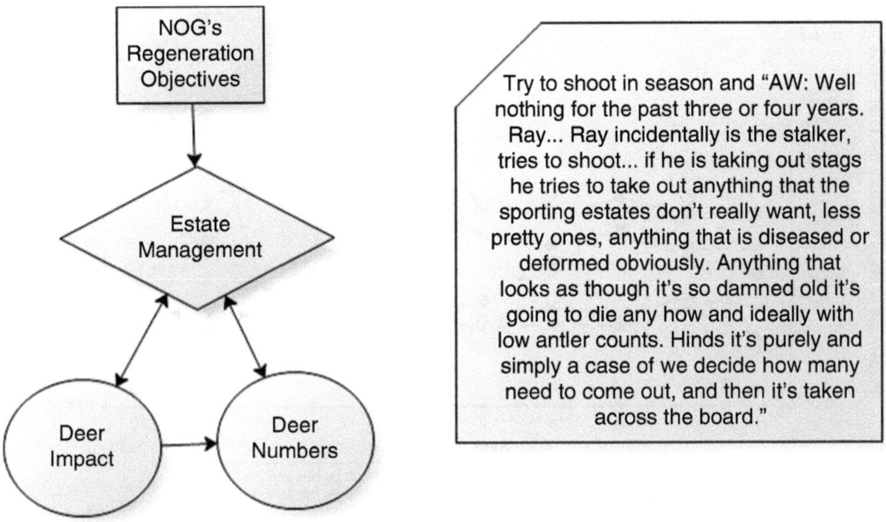

Fig. 5.4 Decision scheme on an NGO-managed property

> The way I would manage them? Well as far as I am concerned the management of deer is controlling them, culling them so you have the right quantity of deer on the ground, so the habitat doesn't get destroyed .... As you know, if you've got too many deer, which they said [x] years ago, you do get a deterioration in the habitat .... At the moment, we have a reduced number of deer on the ground and we know that the habitat is improving.
>
> I would say at the moment, maybe the deer are too low in numbers but that's [because] the SNH and DCS wanted a reduced number of deer on the ground.
>
> We had more deer in the past so we've got a reduced number now. Personally, we had a nice number [x] years ago but of course, they said there was [sic] too many but I didn't think so. Previous to that [ten years earlier], as well as deer we had two dozen head of sheep on it and there was no complaining then. [Laughs] And you know, all these plants were still there. I'd better not say too much into there [refers to the digital recorder].

A clear demonstration of the transient impact of section 7 agreements can be seen in Fig. 5.5. Estates A and B are managed by a new organization beginning in year 7. They manage estate A for sporting income, while estate B is under a section 7 agreement. The data show that for A, there was a constant culling effort per hectare, but an increase in the proportion of males (stags) culled following the section 7 agreement in year 7. After year 7, B focused on woodland regeneration leading to increased cull efforts per hectare and an increased stag cull. C entered a section 7 agreement in year 4 and the observed increase in numbers culled per hectare is mainly due to an increase in the female (hind) cull: the result of a conservative approach to the valuable stag cull. The achievement of favorable conditions in year 9 on estate C has subsequently resulted in a decrease of the total cull mainly due to a decreased hind cull.

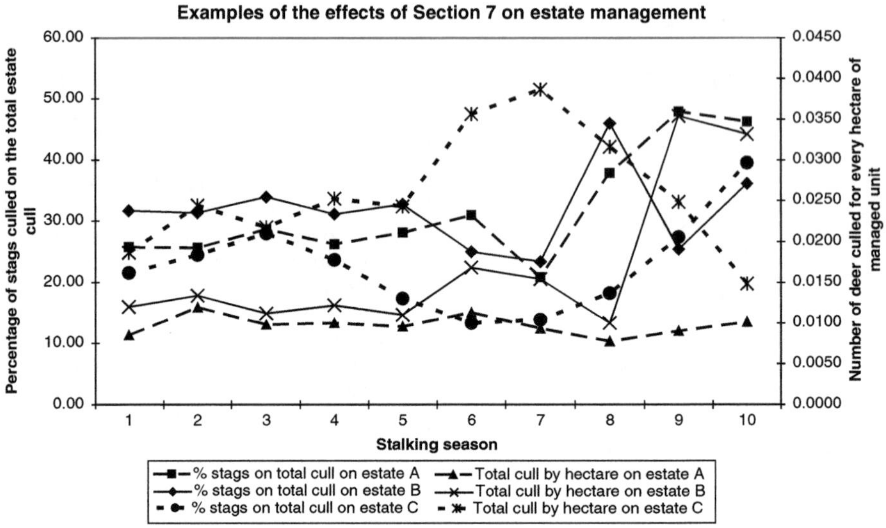

Fig. 5.5 Examples of the effects of section 7 on a private- and a community-managed estate. Deer cull figures are given as percentage of stags culled on the total cull in each estate and total deer culled by hectare. Estates A and B have been managed by the same organization since year 7. Estate B is currently affected by a section 7 agreement. Estate C is terminating a section 7 agreement that has affected its management since year 5

The impact policies have is not limited to a simple matter of deer numbers; they have wider implications for the way people relate to the land. Section 7 on estate C was imposing a different "reading" of the environment on the manager that contrasted with the manager's own cultural perspective. Others expressed the anxiety produced by these changes in driving forces behind land management as feeling like they had been "regulated out of existence." In the words of a crofter:

> The Highlands here today, with all the stock going off the land, the deer are the only natural stock left and what worries me about the countryside today is regulating everything. It almost regulates people out of existence. You can't do anything, you can't be carefree today, or anything compared to my youth.

This regulating for conservation and from a distance pushes people away from the land, this same crofter told us:

> Well, it is abandoning the land. The conservationists say it is putting the land back to wilderness and presumably that's what they'd like but I don't believe that's a good thing. No, I don't.

Regulation also introduces or reinforces an opposition between the local and the global. This emerged, for example, in the comments of a manager who works closely with SNH:

> Our difficulties seem to occur with people higher up the chain of command, people who are less frequently out of the office and on the estate, and are much more aware and influenced by central SNH policies so they end up being less flexible.

The remoteness of the public objectives for deer management, from those involved with practical management, is perceived to go all the way to mainland Europe. This was pointed out by a couple of landowners we interviewed who complained that government organizations such as SNH and DCS were just implementing what "Brussels" wanted them to do.

5.3.3 *Experimenting with PGIS*

The above subsections highlighted a series of lacunae in the way adaptation of management to new needs occurs. While the DMG provides a forum for discussion, information exchange, and communication among different interest groups, its inability to produce binding and enforceable agreements, as well as the lack of effective research and analytical capabilities often results in circular discussions and conflicts based on perceptions held about the managed ecosystem. Estate/management unit-level changes are potentially associated with increases in conflicts at the DMG level or are linked to short-term and unsustainable changes in management practices.

The system currently in place is reactive to changes in policies' objectives; when adaptation is required at the local level, it is often in order to implement or respond to external goals for deer management. In practice, there is no inclusion of local managers in the decision regarding whether change is needed or not. In other words, the current pressure for change is driven by a perceived deterioration of the environment and/or to recreate landscapes of the past (e.g., in Scotland wooded landscapes). Hence, change derives from a conception of environment and of what should drive its management, which is often foreign to those who have done it until now. Our interviewees showed us that managers have a deep knowledge and connection with the environment they manage and set their objectives and value the quality of their environment in the light of their experience and opportunity for achieving objectives defined by themselves and the landowners. Past practices and personal experience and perception of the environment play a great role in setting natural resource management objectives to produce a long-term, sustainable use of the managed resources. This is why the need to change in order to preserve an environment that has sustained certain practices for decades comes as a surprise (see the comments of the gamekeeper quoted in Sect. 5.3.2.2).

An inclusive participatory approach, aimed at effectively integrating diversity of knowledge, what Ramsey (2009) calls "diversity of problem understanding," can provide a way forward for achieving effective adaptive management. This was in part achieved in our PGIS project. I must say "in part," because funding and time constraints prevented the research team from taking the PGIS cycle beyond the final workshops.

After completing the first three phases of the PGIS cycle, the research team held a workshop for participants in each of the two case study areas, one in the second half of 2008 and the other at the beginning of 2009. Hosting the workshops after the one-to-one interviews and analysis was fundamental for preventing the development of new conflicts and providing material for discussions that could bring resolutions to the then-current conflicts. During the workshops, the results of analysis, such as the one illustrated in this section, quantitative analysis of DMGs' historical culls and counts, and deer population and spatial modeling were presented. The participants—local estate managers, landowners, and representatives of DCS, SNH, and NGOs—discussed and questioned the results of the analysis. In this discussion, we noticed that conversations were less based on perceived effects of different management practices and more based on ways to improve the monitoring and analytical systems in place so that better information could be used to devise management practices that would not damage the interest of the various members of the DMG. In a sense, *the analysis grounded the discussion and social interaction on specific, practical issues related to deer and ecosystem management, moving it away from discussion driven by perceived diversity of management due to membership in a particular interest group*; discussions moved away from an interaction based on "us vs. them" to an interaction based on "how to solve or overcome differences."

Participants were overall satisfied with the process and the outcomes and asked us for more of this kind of work. Among the points they raised and what emerged from the discussions during the workshops we find:

- The modeling tool was considered useful for predicting effects of changes in management.
- This approach could allow a meaningful discussion regarding habitat condition, moving the management away from a focus on deer number to a focus on habitat quality.
- They could see how less deer pressure can result in better hunting stocks (a change in knowledge of the ecology underlying management).
- PGIS provided the stakeholders opportunity to give their input in the process of decision making and knowledge production regarding the management of an area.
- PGIS enhanced dialogue between interest groups, for example, private estate owners/managers and government representatives, gamekeepers, and scientists.
- PGIS allowed the stakeholders to challenge the model and quality of the data, prompting discussions on monitoring needs.
- The approach to monitoring and analyzing management was considered valuable if protracted in time.
- PGIS eliminated/reduced support of perceived conflicts between management practices.

Participants also pointed out that:

- The quantitative analysis and models were based on too many assumptions.
- They would have preferred the use of a simpler language and have more explanation of concepts or representations used in the analysis.

5.4 Conclusions and Final Considerations

The analysis described above revealed how the current Scottish deer management system can potentially adapt to changes in deer management needs. However, the governance systems in place for promoting such changes can fail in their effectiveness. For example, DMGs are effective in promoting information exchange, but tend to be lacking in the means for achieving effective knowledge production, trust building, and sense making. Changes in management at the estate level tend to be short term or prone to increase conflicts between management units, increasing both social and management fragmentation at the landscape level and moving the system away from the needed, unbounded management approach discussed at the beginning of this chapter. On the other hand, the addition of PGIS to the tools available to a DMG showed that it can build and produce a collective knowledge that can be applied to enhance horizontal collaborations and support conflict resolution, contributing to the building of a long-term process of management, monitoring, and adaptation.

PGIS is a knowledge- and resource-intensive tool and that might impair the possibility of its long-term implantation in Scottish DMGs. Moreover, in this experiment, researchers assumed the role of an unaligned party, making sure that the visions, ideas, and interests of all the stakeholders involved were represented. This might not be possible when the process is coordinated by someone with, or perceived to have, a vested interest in the outcome of the process.

As the final consideration, PGIS enhanced the horizontal social and cultural interactions needed to expand management adaptation at the landscape level but did not enhance a similar interaction along the vertical axis (i.e., from the local to the regional, national, and international levels). In the end, the participants in this study were asked to incorporate (or had to deal with) objectives for deer management that were guided by models of conservation foreign to their culture and emerging from a different relationship with nature, the environment, and the local context. PGIS, in a sense, was able to facilitate the identification of the "how to change" management, rather than enhance collaboration on deciding the "why change is needed." In short, what kind of participatory approach would allow a meaningful vertical participation in the management of our natural resources?

References

ADMG (Association of Deer Management Groups). (2008). *Deer Management in Scotland.* Retrieved April 23, 2008, from http://www.deer-management.co.uk/dmg2.php

Bacic, I. L. Z., Rossiter, D. G., & Bregt, A. K. (2006). Using spatial information to improve collective understanding of shared environmental problems at watershed level. *Landscape and Urban Planning, 77*, 54–66.

Berkes, F. (2009). Evolution of co-management: Role of knowledge generation, bridging organizations and social learning. *Journal of Environmental Management, 90*, 1692–1702.

Brondizio, E. S., Ostrom, E., & Young, O. (2009). Connectivity and the governance of multilevel socio-ecological systems: The role of social capital. *Annual Review of Environment and Resources, 34*, 3.1–3.26.

ERAD (Land Use and Rural Policy Division). (2004). *Report to ministers: Operations involving the Deer Commission for Scotland in Glenfeshie and Strathglass*. Retrieved April 23, 2008, from http://www.scotland.gov.uk/Publications/2004/06/19474/38560

Fedra, K. (1995). Decision support for natural resources management: Models, GIS, and expert systems. *AI Applications, 9*, 3–19.

Finnie, R. (2004). Written answer to the Scottish executive question #S2W-9212. *Scottish Executive Written Answers*. Retrieved June 20, 2012, from http://archive.scottish.parliament.uk/business/pqa/wa-04/wa0810.htm

Fiorini, S., Yearley, S., & Dandy, N. (2011). Wild deer, multivalence and institutional adaptation: The "deer management group" in Britain. *Human Organization, 70*(2), 179–188.

Folke, C., Hahn, T., Olsson, P., & Norberg, J. (2005). Adaptive governance of social-ecological systems. *Annual Review of Environment and Resources, 30*, 441–473.

Hahn, T., Olsson, P., Folke, C., & Johansson, K. (2006). Trust-building, knowledge generation and organizational innovations: The role of bridging organization for adaptive comanagement of a wetland landscape around Kristianstad, Sweden. *Human Ecology, 34*, 573–592.

Irvine, R. J., Fiorini, S., Yearley, S., McLeod, J. E., Turner, A., Armstrong, H., et al. (2009). Can managers inform models? Integrating local knowledge into models of red deer habitat use. *Journal of Applied Ecology, 46*, 344–352.

Janssen, M. A., Goosen, H., & Omtzigt, N. (2006). A simple mediation and negotiation support tool for water management in the Netherlands. *Landscape and Urban Planning, 78*, 71–84.

MA (Millennium Ecosystem Assessment). (2005a). *Ecosystems and human well-being: Synthesis*. Washington, DC: Island Press.

MA (Millennium Ecosystem Assessment). (2005b). *Ecosystems and human well-being: Policy responses* (Vol. 3). Washington, DC: Island Press.

Nolan, A. J., Hewison, R. L., & Maxwell, T. J. (2001). *Deer management groups: Operation and good practice: A report for the Deer Commission for Scotland*. Aberdeen: The Macaulay Land Use Research Institute.

Olsson, P., Folke, C., Galaz, V., Hahn, T., & Schultz, L. (2007). Enhancing the fit through adaptive co-management: Creating and maintaining bridging functions for matching scales in the Kristianstads Vattenrike Biosphere Reserve, Sweden. *Ecology and Society 12*(1), Article 28. Retrieved June 20, 2012, from http://www.ecologyandsociety.org/vol12/iss1/art28/

Olsson, P., Folke, C., & Hughes, T. P. (2008). Navigating the transition to ecosystem-based management of the Great Barrier Reef, Australia. *Proceedings of the National Academy of Sciences of the United States of America, 105*(28), 9489–9494.

Ostrom, E., & Nagendra, H. (2006). Insights on linking forests, trees, and people from the air, on the ground, and in the laboratory. *Proceedings of the National Academy of Sciences of the United States of America, 103*(51), 19224–19231.

Ramsey, K. (2009). GIS, modeling, and politics: On the tensions of collaborative decision support. *Journal of Environmental Management, 90*, 1972–1980.

Sandström, P., Pahlen, T. G., Edenius, L., Tommervik, H., Hagner, O., Hemberg, L., et al. (2003). Conflict resolution by participatory management: Remote sensing and GIS as tools for communicating land-use needs for reindeer herding in northern Sweden. *Ambio, 32*, 557–567.

Scottish Office of the UK Government. (1996). *Deer (Scotland) Act 1996*. Retrieved April 23, 2008, from http://www.legislation.gov.uk/ukpga/1996/58/contents/enacted

Smith, J. (2002). The clearinghouse approach to enhancing informed public participation in watershed management utilizing gis and internet technology. *Water International, 27*(4), 558–567.

Wang, X., Yu, Z., Cinderby, S., & Forrester, J. (2008). Enhancing participation: Experiences of participatory geographic information systems in Shanxi province, China. *Applied Geography, 28*, 96–109.

Wright, D., Duncan, S. L., & Lach, D. (2009). Social power and GIS technology: A review and assessment of approaches for natural resource management. *Annals of the Association of American Geographers, 99*(2), 254–272.

Chapter 6
Peopled Parks: Forest Change in India's Protected Landscapes

Harini Nagendra, Pinki Mondal, Sanchayeeta Adhikari, and Jane Southworth

Abstract Understanding forest change in the densely populated, forested landscapes of India poses a challenge. Although India has an established approach toward conservation that is focused on protected areas, the long history of human settlement, high population density, and human use of forests and the diversity of biophysical, ecological, social, and institutional conditions create a challenging context for forest protection. In this chapter, we synthesize information from a

---

We thank the forest departments of the states where this fieldwork was conducted for permissions to conduct our research and all the local communities who assisted us with our enquiries. We also thank Rucha Ghate, Sajid Pareeth, Somajita Paul, Sugato Dutt, Ranjith BN, Shankara KP, and Michael Binford for their assistance with field research, data collection, and satellite image processing. Financial support was provided to Harini Nagendra from the Ramanujan Fellowship from the Department of Science and Technology, Government of India, and to Pinki Mondal from a field research grant from the Tropical Conservation and Development Program at the University of Florida.

H. Nagendra (✉)
Ashoka Trust for Research in Ecology and the Environment (ATREE), Royal Enclave, Srirampura, Jakkur Post, Bangalore 560064, India

Center for the Study of Institutions, Population, and Environmental Change (CIPEC), Indiana University, 408 N. Indiana Avenue, Bloomington, IN 47408-3742, USA
e-mail: nagendra@atree.org

P. Mondal
Department of Ecology, Evolution and Environmental Biology, Columbia University, 1200 Amsterdam Avenue, New York, NY 10027, USA
e-mail: pm2658@columbia.edu

S. Adhikari
Geography Department, Macalester College, 1600 Grand Avenue, Saint Paul, MN 55105, USA
e-mail: sadhikar@macalester.edu

J. Southworth
Department of Geography, University of Florida, Gainesville, FL 32611, USA
e-mail: jsouthwo@geog.ufl.edu

E.S. Brondízio and E.F. Moran (eds.), *Human-Environment Interactions: Current and Future Directions*, Human-Environment Interactions 1,
DOI 10.1007/978-94-007-4780-7_6, © Springer Science+Business Media Dordrecht 2013

number of case studies of land-use/land-cover change in India, and examine four case studies in detail to provide insights into the drivers of forest clearing and regrowth in peopled parks. We find that protected areas are an important influence on the rural Indian landscape and have been reasonably successful in maintaining forest cover within their core areas. There is, however, considerable forest change taking place at the periphery of protected areas, with substantial clearing near settlements and accessible locations. In urbanizing landscapes, protected areas continue to play a dominant role, but forest availability has a major influence on vegetation change, with the human footprint expanding to cover distant, hitherto inaccessible areas. This has extremely worrisome implications for the future of forests in an increasingly urbanizing country that represents the second most populous nation in the world.

6.1 Introduction

Protected areas have become an established approach for conservation worldwide, now existing in almost all parts of the world and comprising more than 13% of the world's land cover (Jenkins and Joppa 2009). Yet in many parts of the developing world such as India, these parks coexist with high population densities, leading to conflicted people-park relationships (Nagendra 2008). These are exacerbated by regulations aiming to limit forest clearing by restricting forest access and harvest of resources by local communities, many of which have existed in close proximity to the forests they are now excluded from for decades or even for centuries. In a number of peopled parks across the world, it has become practically impossible to achieve conservation through the protectionist paradigm. This has led to constant conflicts between park authorities and local communities, as well as among academics, conservationists, and practitioners, on the best way to achieve conservation in these peopled parks (Chapin 2004; Schwartzman et al. 2000).

Such conflicts seem to be particularly acute in countries like India, where there is a close coexistence of biodiversity and human population density (Cincotta and Engelman 2000). This large country provides a particularly compelling instance of the need to balance conservation with human requirements, in a species-rich landscape with centuries-long history of settlement. The awareness of the social costs of creating exclusionary parks is increasing (Bawa et al. 2004). At the same time, other and equally valid concerns are raised about rapid habitat loss and fragmentation in parks where local communities continue to depend on access to forest resources (Shahabuddin 2010; Shahabuddin and Rangarajan 2007). Thus, tensions between conservation and development have increased in recent years across the Indian landscape.

These tensions are exacerbated by patterns of development that make their presence felt in the landscapes surrounding protected areas (DeFries et al. 2010). Urbanization, industrial development, population growth, and migration lead to shifts in land use that impact land-cover change and shape demands for forest resources both within protected areas and outside their boundaries (Nagendra et al. 2009). Consequently, there is a need to move the focus of examination beyond looking at specific protected areas to examining larger landscapes within which protected areas are embedded and to examine when and why deforestation and regrowth occur in specific regions within these larger landscapes (Ostrom and Nagendra 2006; Robbins et al. 2007).

Satellite remote sensing and geographic information systems (GIS) provide effective tools for such analyses of forest change in protected-area, peopled landscapes (Ostrom and Nagendra 2006; Southworth et al. 2006). A host of studies in park-embedded landscapes are making it increasingly clear that most parks are embedded in and responsive to activities in the surrounding landscapes within which they are situated. Thus, the patterns and drivers of land-use/land-cover change within parks are closely related to the larger regional drivers of change, such as urbanization, that impact the areas surrounding parks (DeFries et al. 2010; Southworth et al. 2006).

This chapter studies the trajectories and drivers of forest change in India's peopled parks using a combination of a broad meta-analysis and in-depth case studies. First, a meta-analysis of 15 papers from India is conducted to assess the major trajectories of forest change, identify significant patterns of land-use change within and outside parks, and look at common drivers of change. Subsequently, four case studies of specific parks are discussed in detail, enabling the description of the human drivers of forest change to a level of detail not possible in the meta-analysis. These case studies provide explicit comparisons of patterns and processes of forest change within and outside protected areas, and enable us to assess how differences in the locations of these parks, from rural to urban areas, impact the nature of forest change. Integrating findings from these different approaches provides an enriched understanding of the factors impacting land-use/land-cover change in the densely populated, biodiversity-rich protected-area landscapes of India.

6.2 Meta-analysis

6.2.1 Methods

Academic databases such as Google Scholar and Web of Science were used to identify peer-reviewed journal articles that assessed forest change in protected-area landscapes in India using satellite remote sensing. The selected papers used satellite images to map forest cover at one or multiple points in time, and provided discussions of trajectories of land-cover change, as well as information on major drivers of forest change in the protected area. A total of 15 case studies were located, a number much smaller than expected, indicating the relative paucity of studies on land-cover change in India's protected areas. Of these, 11 studies discussed changes

Fig. 6.1 Locations of protected areas in India

in land cover using data from more than one point in time. Four studies reflected single-point satellite image assessments but provided supplementary textual assessments of major trajectories and drivers of land-cover change and were therefore included in the meta-analysis. From these case studies, information was gathered on the distribution of human settlements within and adjacent to the parks, major trajectories of forest change, drivers of change—proximate and underlying—and the actors involved in forest change, from local communities to governmental agencies.

6.2.2 Results

The distribution of the 15 case studies across peninsular India is illustrated in Fig. 6.1. A majority of the parks (80%) indicate the presence of settlements within or outside their boundaries that access forest products (Table 6.1). Three of the

Table 6.1 Trajectories and drivers of land-cover change in various protected forested areas within India

Protected area	Reference	Period of study	Settlements	Plantations within park or surrounding area	Major trajectories of change	Proximate drivers	Underlying drivers	Extraction of forest products
Bhadra Wildlife Sanctuary	Somashekar et al. (2008)	2000–2004	Not mentioned	Teak, coffee, betel nut	Forest degradation, regrowth	Fire, livestock grazing, timber extraction	Population growth	Timber, wildlife hunting
Chandoli National Park	Imam et al. (2009)	2005	32 Villages inside (28 later resettled), 78 outside	None mentioned	Regeneration of grasslands, regrowth of forests	Resettlement of villages outside park	None mentioned	Fuelwood, timber, fodder, house construction material
Corbett Tiger Reserve	Singh et al. (2009)	2005	250 Villages outside	Teak and eucalyptus	None mentioned	Roads and villages	Population growth	Not mentioned
Fambong Lho Wildlife Sanctuary	Sudhakar et al. (2008)	2002	Not mentioned	*Cryptomeria* species	Forest degradation	Not mentioned	Not mentioned	Not mentioned
Indira Gandhi Wildlife Sanctuary	Joseph et al. (2009)	1973–2006	36 villages and 2 tea plantations inside, unspecified number of villages with over 100,000 people outside	Teak, tea, coffee, cardamom	Forest degradation, regrowth, grassland conversion to plantation	Livelihood dependence of locals, infrastructure development, agriculture expansion, plantation forestry	Population growth, changes in social customs, government development policies	Timber, fuelwood, grass for thatching, grazing
Kalakkad Mundanthurai Tiger Reserve	Giriraj et al. (2010)	1973–2004	28 villages inside, 180 outside	Not mentioned	Evergreen forest to semievergreen	Fire, invasion of exotics, erosion	Not mentioned	Not mentioned
Keoladeo National Park	Palria et al. (2005)	1999	14 villages outside	Not mentioned	Wetland degradation	Encroachment of water weeds	Increased agriculture and settlements leading to demand for water	Not mentioned

(continued)

Table 6.1 (continued)

Protected area	Reference	Period of study	Settlements	Plantations within park or surrounding area	Major trajectories of change	Proximate drivers	Underlying drivers	Extraction of forest products
Kumbhalgarh Wildlife Sanctuary	Robbins et al. (2007)	1986–1999	33 villages outside	Not mentioned	Forest degradation, regrowth	Livelihood dependence of locals, livestock grazing, protection by foresters	Economic growth promoting increase in livestock	Fuelwood, timber, lopped fodder, leaf collection, grazing
Mahananda National Park	Nagendra et al. (2009)	1990–2000	9 villages, one small town, and one large town outside	Tea	Deforestation, degradation, reforestation, regrowth	Livelihood dependence of locals, timber extraction for furniture, landslides, ban on commercial timber felling	Population growth, economic growth promoting furniture manufacture, change in national policies	Timber, fuelwood, grazing
Manas National Park	Sarma et al. (2008)	1977–2006	Has villages outside, number not mentioned	Not mentioned	Deforestation, conversion of grassland	Livelihood dependence of locals, agriculture expansion, weed encroachment, poaching	Economic growth leading to expansion in agriculture, ethnic conflicts	Timber, fuelwood, grass for grazing, wildlife hunting
Marine National Park	Nayak and Bahuguna (2001)	1975–1988	Not mentioned	Not mentioned	Initial forest degradation, followed by regrowth	Livelihood dependence of locals, industrial and infrastructure development, conservation measures by park authorities	Economic growth leading to infrastructure expansion	Fuelwood, fodder, grazing
Nanda Devi Biosphere Reserve	Nautiyal and Kaechele (2007)	1979–2002	Has villages outside, number not mentioned	Not mentioned	Degradation of alpine pastures with encroachment of weeds and bushes	Ban on grazing	Park management policies	Grazing

(continued)

Table 6.1 (continued)

Protected area	Reference	Period of study	Settlements	Plantations within park or surrounding area	Major trajectories of change	Proximate drivers	Underlying drivers	Extraction of forest products
Pench Tiger Reserve, Maharashtra	Mondal and Southworth (2010a)	1977–2007	Approx. 40 villages outside	Teak and other types (not mentioned)	Initial deforestation, followed by regrowth	Livelihood dependence of locals, initial commercial tree felling by park managers, later ban on commercial felling	Change in national policies	Fuelwood and grazing
Rajaji National Park	Nandy et al. (2007)	1972–2005	Villages, army cantonment, guesthouse and shopping area within park; major towns nearby	Not mentioned	Deforestation and degradation	Livelihood dependence of locals, demand for forest products from army cantonment, agriculture expansion, increase in grazing area, construction of power canal	Infrastructure expansion, economic growth, presence of army	Timber, fuelwood, lopped fodder, tall grass for rope making
Tadoba Andhari Tiger Reserve	Nagendra et al. (2006)	1989–2001	6 villages within, 53 outside, major town nearby	Not mentioned	Degradation and reforestation	Livelihood dependence of locals, collection of fuelwood and bamboo for sale	Park management policies, economic growth leading to demand for fuelwood and bamboo	Fuelwood, grazing, leaf litter, bamboo

case studies do not mention the presence of settlements, and this may be because they do not assess the drivers of forest change, as none of the three studies clearly mentions that there are no settlements impacting the park. The local populations in these settlements extract a variety of products from the parks. Apart from the extraction of timber, fuel wood, and bamboo, biomass is extracted from the park through grazing, fire, lopping for fodder, removal of leaf litter, and extraction of thatch grass. In addition, minor forest produce is extracted from these parks.

Several parks (40%) indicate the presence of plantations within or adjacent to their boundaries. They included a diversity of types, with plantations of tea, coffee, cardamom, teak, eucalyptus, and *Cryptomeria* sp being mentioned (Table 6.1). A vast majority of parks (87%) indicated some deforestation and degradation within their boundaries, although 53% of the parks indicated that reforestation and regrowth were also taking place in the same parks. Thus, most protected areas appear to be dynamic locations with multiple trajectories of deforestation and regrowth taking place simultaneously in different parts of the landscape—indeed, such spatial variation in trajectories of change appears to be critical for the effectiveness of park management (Nagendra et al. 2009; Robbins et al. 2007). Two out of three studies indicated that local communities' livelihood dependence on the parks—manifesting itself through grazing and extraction of forest products for local consumption—was an important proximate driver of forest change. One out of two studies also mentioned the role played by external factors including the expansion of road and railway networks; the presence of markets for the sale of timber, fuel wood, and bamboo; infrastructure projects like the construction of dams, power lines, and ports; and economic trends that favored increase in livestock holdings or agricultural expansion. Finally, three parks (Rajaji National Park, Mahananda Wildlife Sanctuary, and Tadoba Andhari Tiger Reserve) have major nearby towns that fuel the demand for timber and other forest products like bamboo. Increasing road and railway connectivity in and around these parks appears to be an important factor creating more pressure on the forests within.

One in two studies also cited national and local management policies as important factors responsible for the protection or degradation of the parks. Since Indian-protected areas tend to follow exclusionary strategies of park management (Nagendra et al. 2006), villages located within parks are often relocated outside the park, and this has been mentioned by some studies as having had positive impacts on land cover (Imam et al. 2009), as well as having negative impacts in increasing people-park conflicts (Nagendra et al. 2010). Several studies mentioned the efforts taken by park managers to minimize forest degradation (Mondal and Southworth 2010a; Nagendra et al. 2006, 2009; Nayak and Bahuguna 2001; Robbins et al. 2007), and two studies discuss the positive impacts of national-scale policy changes banning commercial forestry in national parks and wildlife sanctuaries (Mondal and Southworth 2010a, b; Nagendra et al. 2009). Three studies also discuss the negative impacts of government strategies. Joseph et al. (2009) point out the negative impacts of uncoordinated development activities taken up near the Indira Gandhi Wildlife Sanctuary by other government agencies, while Nagendra et al. (2006) discuss the extensive park-people conflicts arising out of park management strategies. Nautiyal and Kaechele (2007) describe how a well-intentioned ban on grazing within the Nanda Devi Biosphere Reserve has led to severe degradation of grassland habitat and encroachment by weeds and thorny bushes.

Thus, we see that population pressure makes its presence felt on the Indian protected-area landscape in a number of ways, from infrastructure projects to markets for timber and forest products to local villagers' subsistence requirements from parks. Yet, most government management efforts focus on relocating forest villages and reducing extraction for subsistence needs (Ostrom and Nagendra 2006; Robbins et al. 2007). Although national-scale challenges like urbanization, extraction for commercial purposes, infrastructure development, and tourism represent increasing dangers to Indian parks (DeFries et al. 2010; Shahabuddin 2010), they appear to be rarely discussed or dealt with at a local or national scale. Focusing all of one's efforts on banning subsistence harvest by local communities only addresses part of the issue, and clearly cannot provide a long-term, sustainable approach to forest management, as the detailed case studies discussed in the following section demonstrate.

6.3 Detailed Case Studies

While the overview of the literature review provides us with interesting information, in-depth local studies are required to develop a more detailed understanding of the drivers of forest change in protected-area landscapes. These drivers must be evaluated across a range of locations, from rural and isolated landscapes to areas near major urban centers. The following four detailed case studies help highlight some of the drivers and changes along this gradient of development and urbanization.

6.3.1 Landscape 1: Pench Tiger Reserve, Maharashtra

6.3.1.1 Study Area

Pench Tiger Reserve is located in the Nagpur district of Maharashtra state in central India (Fig. 6.1). This park shares its name and northern boundary with another tiger reserve in Madhya Pradesh. The vegetation type is southern tropical dry deciduous, dominated by teak (*Tectona grandis*) with frequent presence of other species such as Indian laurel (*Terminalia tomentosa*), white marudah (*Terminalia arjuna*), and bamboo (*Bambusa arundinacea*). Land cover within the park ranges from very dense to moderately open forest with good ground cover, sparse forest, meadows, riverine forest, and aquatic habitat. Pench was declared a national park in 1975 by the Government of Maharashtra, and declared the 25th tiger reserve of India in 1999. This park constitutes one of the smallest tiger reserves in India with an area of only 257 km^2.

Two forest villages were located inside Pench at the time of notification in 1975. One of the villages was relocated after the submergence of the land during the construction of a hydroelectric dam. The other village is still located within the park, and relocation is currently in progress. Until the late 1980s, the forests in the park were subjected to heavy thinning and clear-felling in good-quality forested tracts to meet the requirements of infrastructural developments, as also observed in other parks and forested areas all over India (Pandey 1992; Prasad 2000). In 1988, following a ban on tree felling in national parks, this practice was discontinued.

The forests outside the park are commercially managed by the Nagpur Forest Division under the state government and the Forest Development Corporation of Maharashtra, a public sector unit. The focus of the commercial management has been on replacing natural mixed forests with monoculture plantations of financially profitable species, such as teak. This landscape also witnessed the implementation of forestry projects during the 1990s that encouraged tree plantations with benefit sharing by local communities (World Bank 1991).

The surrounding forests host over 40 villages with significant human and cattle populations. These communities depend on the forests for resource extraction, mostly of nontimber forest produce such as *tendu*, gum, *mahua*, *lac*, fuel wood, and cattle grazing. In addition, the park forests experience human pressure through grazing, collection of *tendu* leaves, fishing, and setting of fires in times of park-people conflict. The park also suffers from excessive tourism pressure, as well as human pressure due to the presence of residential colonies of the Maharashtra Irrigation Department and the Madhya Pradesh Electricity Board within the core forest. To better protect the Pench landscape, a proposal for a buffer area has recently been approved. In the past, efforts were made to minimize human dependence on the forest through several ecodevelopment projects, but as with most of these initiatives across India, these efforts have been largely categorized as "failures" by scholars studying participatory management regimes in Indian inhabited landscapes (Shahabuddin 2010). The park management is currently trying to reduce anthropogenic pressure in the park core by creating employment opportunities for the surrounding villages through park development- and protection-related work.

6.3.1.2 Methods

A multitemporal satellite dataset was utilized for this study, with a 1977 Landsat MSS image, a 1989 Landsat TM image, a 2000 Landsat ETM+ image, and a 2007 ASTER image selected from the post-monsoon season, with minimal cloud cover and phenological variation. The park boundary as well as a 10-km buffer was created in a GIS environment. Pench Tiger Reserve in Madhya Pradesh was excluded from the analysis, as this park is managed by another state administration with a different set of approaches.

A mosaic of two ASTER images from 2007 was utilized as the base map to geometrically rectify the remaining satellite images. An image-to-image registration method was used with a nearest-neighbor resampling algorithm resulting in a root

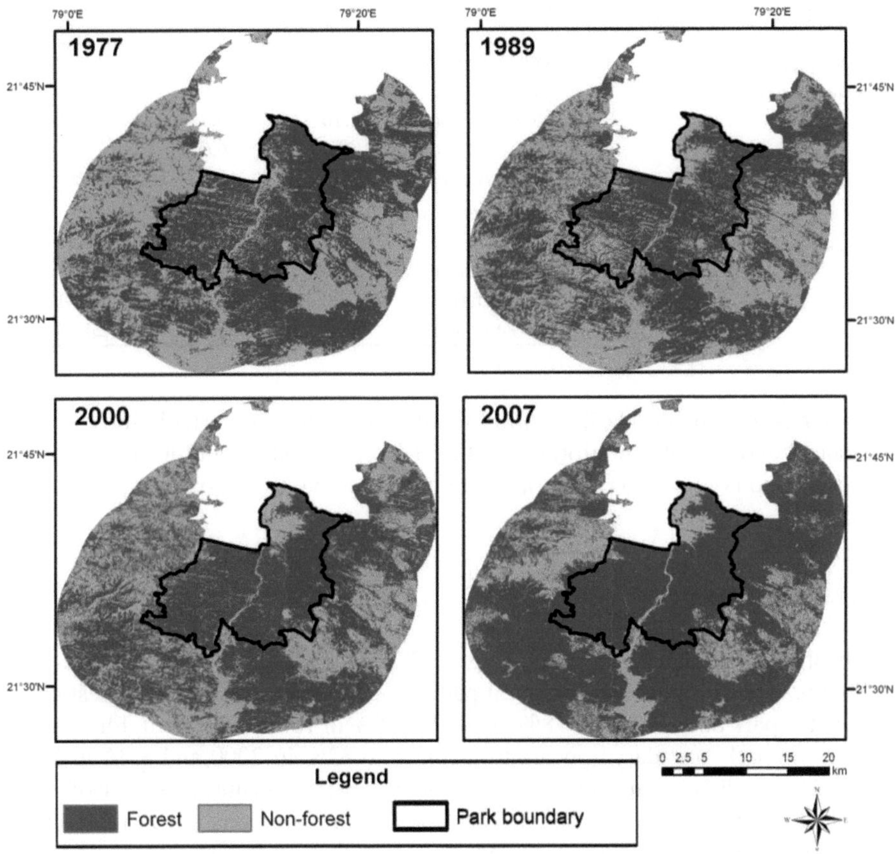

Fig. 6.2 Forest-cover distribution in Pench Tiger Reserve in 1977, 1989, 2000, and 2007

mean square (RMS) error of less than 0.5 pixels (<15 m). Radiometric calibration and atmospheric corrections were also applied to the images (Green et al. 2005). All of the preprocessing steps were carried out in ERDAS Imagine 9.3 and ENVI 4.3.

Following the corrections, the images were classified using a hybrid approach. In the first step, an ISODATA clustering method was used to generate preliminary classes. Then classes were merged based on similarity in spectral signatures and training samples collected during May–June 2008. The final classes are forest and nonforest, where the forest class includes both natural forests and plantations and the nonforest class includes bare agricultural land, cleared forest, and water bodies (Fig. 6.2). The overall classification accuracy for the most recent image is 94% with a kappa value of 0.89. Post-classification change detection was performed to compare two-date change trajectories (1977–1989, 1989–2000, and 2000–2007) for the park and the buffer (Table 6.2). The classified Landsat MSS image was resampled to 30 m to maintain spatial compatibility with the other images prior to the change-detection analysis. For each of the three time spans, four change

Table 6.2 Trajectories of land-cover change (%) in Pench Tiger Reserve and the surrounding 10-km area

	1977–1989	1989–2000	2000–2007
Pench Tiger Reserve			
Stable forest	59.99	66.70	81.37
Deforestation	18.70	3.72	1.36
Reforestation	10.54	16.02	6.20
Stable nonforest	10.77	13.56	11.07
Surrounding area			
Stable forest	36.83	42.37	50.54
Deforestation	11.01	7.53	5.26
Reforestation	13.05	13.44	22.99
Stable nonforest	39.12	36.67	21.21

trajectories were reported (Table 6.2) for areas that remained in stable forest (forest on dates 1 and 2), stable nonforest (nonforest on dates 1 and 2), deforestation (forest on date 1 and nonforest on date 2), and reforestation (nonforest on date 1 and forest on date 2).

6.3.1.3 Results

Both the single-date snapshots (Fig. 6.2) and two-date change trajectories (Table 6.2) reveal a distinct temporal pattern of forest-cover change. Forest cover within the park declined from 78% in 1977 to 70% in 1989. Afterward, forest cover within the park boundary appears to increase to 82% in 2000 and 87% in 2007. Likewise, the change trajectories show an increasing trend in the proportion of stable forest (Table 6.2). Interestingly, deforestation in the park was about 18% during 1977–1989, even after Pench was declared a national park with strictly protected status. This could be attributed to two main reasons. First, a considerable amount of forested area (~20 km^2) was submerged due to construction of a hydroelectric dam, contributing to the forest loss in the northern part of the park. Second, forested tracts were cleared in the southwestern part of the park under regular forest management activities.[1] However, Pench witnessed minimal deforestation afterward: only about 3% and 1% during 1989–2000 and 2000–2007, respectively. The pre-1989 deforestation and post-1989 recovery can be linked to the national ban on tree felling within national parks implemented through the 1988 Revised Forest Policy (Mondal and Southworth 2010b). Reforestation within the park was highest during 1989–2000 at 16%, increasing from 10% during 1977–1989 and declining to 6% during 2000–2007. Stable nonforest in the park was between 10% and 13% throughout the three decades.

The nonprotected area surrounding the park has less forest cover compared to the park. Nevertheless, this region also shows a consistent increase in forest cover

[1]Pench Tiger Reserve management plan and personal communication with the Office of the Field Director, Pench Tiger Reserve.

from 47% in 1977 to 73% in 2007. Forested tracts in the buffer region are primarily managed for commercial extraction and are huge sources of revenue for both government and private sectors. These forests are also subject to human disturbances such as resource extraction and cattle grazing. All of these factors collectively contribute to the considerably lower extent of stable forest compared to the park. Stable forest in the buffer increased from 36% during 1977–1989 to 50% during 2000–2007 (Table 6.2). Deforestation and reforestation show opposing trends with deforestation decreasing from 11% to 5% and reforestation increasing from 13% to 22% during the study period. Stable nonforest denotes either agricultural land or water bodies in this landscape and covers a significant proportion of area in the buffer (Table 6.2).

These findings indicate that Pench Tiger Reserve has been reasonably effective in maintaining and regenerating forest cover. This landscape shows increasing reforestation in the nonprotected surrounding area, which can be traced to the implementation of nationally and internationally funded plantation projects in the buffer.

6.3.2 Landscape 2: Tadoba Andhari Tiger Reserve

6.3.2.1 Study Area

The Tadoba Andhari Tiger Reserve is also located in Maharashtra state (Fig. 6.1), and covers an area of 625 km^2 in a landscape largely dominated by dry tropical forests interspersed with grasslands and water bodies (Nagendra et al. 2006). This central Indian forest landscape is rich in biodiversity, with over 41 mammal species and 195 bird species (Khawarey and Karnat 1997). The park is situated in a landscape where people and forests have coexisted for centuries. Thus, after creation of the park, six villages found themselves located within the protected area and the inhabitants lost many of their traditional rights over forest products after creation of the park (Khawarey and Karnat 1997; Nagendra et al. 2006). Resettlement of these villages has long been on the agenda of the Maharashtra Forest Department, and in 2007, after much dispute, two of the villages were relocated outside the park.

The park is relatively well insulated from human disturbance on three sides to the north, south, and east by forest patches categorized as Reserve Forest and Protected Forest, lower categories of protection that act as a buffer from human disturbance. To the northeast, the road network is quite well developed and provides greater access to the forest. Fifty-three villages located around the park are dependent on the protected area for a large part of their fuel, fodder, timber, and nontimber forest requirements. The park also experiences substantial seasonal use from migrant herders, and is frequented by timber, bamboo, and wildlife poachers (Ostrom and Nagendra 2006). In addition, the park experiences substantial pressure from tourism, with over 60,000 tourists visiting the park each year (Mawdsley et al. 2009). Despite the diverse sources of pressure on the park, park authorities have

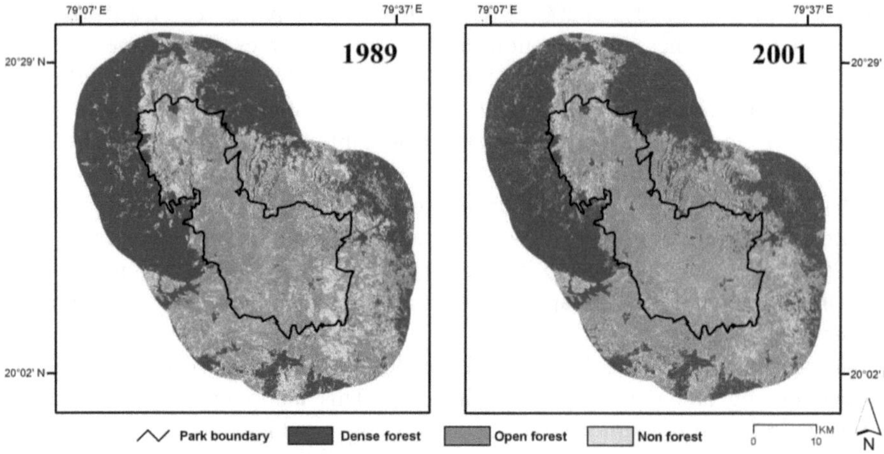

Fig. 6.3 Forest-cover distribution in Tadoba Andhari Tiger Reserve in 1989 and 2001

been focused for a remarkably long time on resettlement of the villages, considering this to provide a major solution to problems of park degradation (Ghate 2007; Ghate and Beazley 2007). Little attention has been paid to identifying different groups of users, quantifying their relative impacts on the park, or to developing approaches to differentially manage these impacts (Nagendra et al. 2010).

6.3.2.2 Methods

Two satellite images—a Landsat TM image of November 1989 and a Landsat ETM+ image of October 2001—were used to map forest-cover change within and outside the park (Fig. 6.3). The 1989 image was georeferenced to Survey of India topographic sheets, and the 2000 image was registered to the 1989 image, with RMS error of image-to-image registration retained below 0.5 pixels (Jensen 2000). Images were classified using supervised classification, based on ground training data collected in December 2004. The study area was mapped into three classes: dense forest, open forest, and nonforest (predominantly agriculture, grassland, water, or village settlements). Classification accuracy was verified using an independent dataset, with a producer's accuracy of 89.5% for the 2001 image and 86.7% for the 1989 image. Classified images were overlaid on each other to delineate land-cover change trajectories. The nine change categories produced as a result of the overlay were collapsed into a set of four classes. These represent areas of stable forest (forest in 1989 as well as 2001), stable nonforest (devoid of forest cover in 1989 and 2001), deforestation (forest in 1989 but nonforest in 2001), and reforestation (nonforest in 1989 but forest in 2001). A GIS buffer of 2 km outside the park was used to study changes in land cover in the surrounding area and compare them to changes within the park.

Table 6.3 Trajectories of land-cover change (%) in Tadoba Andhari Tiger Reserve and the surrounding 10-km area

	1989–2001
Tadoba Andhari Tiger Reserve	
Stable forest	62.64
Deforestation	15.04
Reforestation	17.22
Stable nonforest	5.10
Surrounding area	
Stable forest	27.87
Deforestation	12.21
Reforestation	10.97
Stable nonforest	48.95

During 2000–2003, we conducted surveys to assess the socioeconomic and institutional conditions of the six villages within the park, as well as household interviews with selected individuals in each village. Interviewees were selected in a stratified manner so as to sample across ethnic groups, income categories, and genders. During these interviews, we assessed the relationships of the respondents to the park by asking about the products they regularly harvest from the forest and about cattle and livestock depredations due to wildlife. We also assessed attitudes toward the park by asking a series of questions. Further details about the surveys can be found in Nagendra et al. (2010).

6.3.2.3 Results

The forest landscape within the park is quite dynamic, with as much as one-third of the landscape experiencing a mix of deforestation and reforestation (Table 6.3). The peripheral areas of the park appear to be the most depleted of forest cover (Fig. 6.3), while the park core has a greater proportion of area in stable forest. Overall, there appears to be lesser forest area where the density of village settlements and road networks is greatest, in the northeast and northwest, while forest cover is relatively well preserved toward the south, where surrounding state-owned, protected forests buffer the park from human influence (Fig. 6.3). Forest cover in the park interior also appears to be comparatively protected. There is substantial increase in density, or regrowth, within formerly open, degraded forest. This higher density can be observed both within the park toward the northern boundary and the southern end and in the adjacent protected and reserve forest areas, indicating that efforts toward park protection seem to be paying off in some areas.

Our interviews and surveys indicate that this protection has come at a high social cost for the villages located within the park. These communities face substantial difficulties due to the loss of their traditional rights over forest harvests, with many continuing to collect forest produce and to graze cattle within the park, despite efforts from forest guards to protect the park (Ostrom and Nagendra 2006). There has been an increase in park-people conflict, with increasing incidences of armed insurgency within and outside the park. Thus, in the long term, strict protection by policing from forest guards may not be sustainable unless local communities are involved. Previous experience in this region has indicated that the park can be much better protected against wildlife poachers and timber loggers when local communities are involved (Nagendra et al. 2006). Thus, ensuring a decrease in park-people conflicts in this densely inhabited protected area appears to be an essential prerequisite for its effective protection over the long term.

6.3.3 Landscape 3: Mahananda Wildlife Sanctuary

6.3.3.1 Study Area

The Mahananda Wildlife Sanctuary is located at the foothills of the eastern Himalayas with the Baikunthapur Reserve Forest to the south (Fig. 6.1). The park was declared a wildlife sanctuary in 1976 and enlarged in 1988 (Wildlife Circle 1997). The elevation in the park ranges between 350 and 1,500 m above sea level, with a wide range in forest types. The steep northern slopes contain a mix of evergreen and deciduous forest cover, while the southern region of the park, which is much flatter, harbors deciduous forests dominated by *Shorea robusta* (sal) interspersed with grassy alluvial plains along the riverbank (Shankar 2001; Wildlife Circle 1997). The park is home to a diversity of flora and fauna and forms a critical part of the elephant migration route in this region. Until the early 1990s, substantial portions of the park, especially in the riverine plains and foothills, were maintained for commercial timber extraction. The park core and hilly slopes were, however, protected from timber extraction. Subsequently, changes were made to national policies such that tree felling and fuel wood extraction was banned within all Indian parks including Mahananda (IIFM 2001).

A number of tea garden settlements and villages surround the park on all sides except the northern hilly boundary and the southern boundary, which is bordered by the Baikunthapur Reserve Forest—another government-protected area with a lower category of protection (Das and Guha 2003). The tea-estate settlement communities extract firewood and timber from the park and graze their cattle in these forests (NESPON 2000; Wildlife Circle 1997). A number of agricultural settlements in the southwestern and western parts of the sanctuary also depend on the forest for fuel wood, timber, and cattle grazing. Major vehicle access routes—National Highway 31 and the Darjeeling-Siliguri road—and a meter-gauge railway line pass through the center of the sanctuary. These routes connect to an extensive network of smaller roads outside the park, providing easy access for the transport of illegally extracted timber outside the sanctuary to the furniture shops, sawmills, and markets located in the nearby urban center of Siliguri (NESPON 2000). Siliguri and its surrounding area constitute a major regional trade center fueling the demand for timber from the forest. Despite regular patrolling by forest department guards, large amounts of timber continue to be harvested from the park (Ostrom and Nagendra 2006).

6.3.3.2 Methods

Changes in forest cover were assessed using a November 1990 Landsat TM satellite image and a December 2000 Landsat ETM+ image. The 1990 image was georeferenced to 1:50,000-scale Survey of India topographic maps, and the 2000 image was georeferenced to the registered 1990 image, keeping RMS errors below 0.5 pixel (15 m). Training information collected during May–June 2004, April–May 2005, and July 2006 was used in a supervised classification of images into dense forest, open forest, and nonforest (constituting agriculture, grassland, tea plantations, water bodies, and village settlements). One image was overlaid on the other to create a change map, classifying pixels that underwent change from nonforest to open forest, nonforest to closed forest, and open forest to closed forest as "reforestation"; pixels that changed from closed forest to open forest, closed forest to nonforest, and open forest to nonforest as "deforestation"; pixels that remained open forest or closed forest in both years as "stable forest"; and pixels that remained nonforest in both 1990 and 2000 as "stable nonforest." An accuracy assessment of the 1990–2000 change image, using field data, found a producer's accuracy of 86% and a kappa of 0.81.

A GIS buffer of 5 km beyond the sanctuary and the reserve forest was used to compare land-cover change in the surrounding landscape with the change within the park. These divisions were used to subset the raster classified images (1990, 2000) and the change trajectory image (1990–2000), allowing a comparison of land-cover change within the two protected areas to change in the surrounding landscape. This follows approaches frequently adopted for research in other protected areas (Nagendra 2008).

6.3.4 Results

Within the Mahananda Wildlife Sanctuary, natural protection in the steep northern section has led to the maintenance of large patches of stable forest here, and substantial regrowth has taken place in the southern sections where commercial timber extraction in the park was discontinued in the 1990s (Table 6.4, Fig. 6.4). Yet there is still some clearing toward the periphery, particularly in areas near the road and railway networks. Regular extraction of fuel wood and timber as well as grazing from the tea garden settlements and villages has led to considerable pressure on the park and consequent degradation (Nagendra et al. 2009; Ostrom and Nagendra 2006). Our interviews indicate the extensive transport of illegally extracted timber outside the sanctuary, primarily to fill the furniture markets in the nearby Siliguri urban center. Although the underfunded and ill-staffed forest department conducts frequent monitoring of the sanctuary, and has seized significant volumes of illegal timber from individual and large-scale poachers, there is so much pressure on the park due to subsistence and commercial extraction that they are unable to completely protect the forest.

Table 6.4 Trajectories of land-cover change (%) in Mahananda Wildlife Sanctuary, Baikunthapur Reserve Forest, and the surrounding 5-km area

	1990–2000
Mahananda Wildlife Sanctuary	
Stable forest	50.29
Deforestation	10.91
Reforestation	29.75
Stable nonforest	9.05
Baikunthapur Reserve Forest	
Stable forest	43.11
Deforestation	17.28
Reforestation	21.18
Stable nonforest	18.43
Surrounding area	
Stable forest	3.12
Deforestation	3.79
Reforestation	13.10
Stable nonforest	79.99

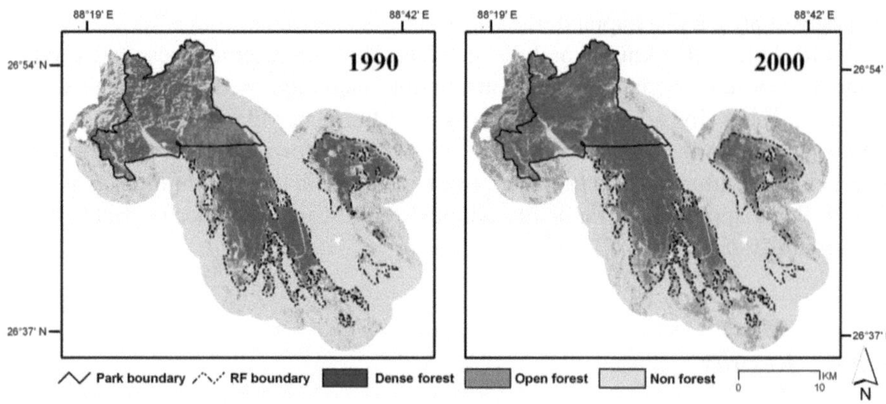

Fig. 6.4 Forest-cover distribution in Mahananda Wildlife Sanctuary and the adjacent Baikunthapur Reserve Forest in 1990 and 2000

The surrounding landscape appears to be experiencing a trend toward reforestation and regrowth, largely due to the partial or total abandonment of many of the tea gardens surrounding the park. These tea gardens are shade grown and interspersed with trees, and have witnessed substantial increases in tree canopy cover now that they are not as extensively maintained. Yet, despite these positive trends, urbanization from the Siliguri urban center and development initiatives in the region pose significant threats to the future of forest cover in this landscape. As Siliguri expands and connects to other parts of the northeast through roads and the Indian Railway's plans to convert the existing meter-gauge railway line in this area into broad-gauge, timber smuggling is likely to significantly increase. The demand from Siliguri for fuel wood for domestic use and construction is also increasing, and will impact the park significantly in decades to come.

6.3.5 Landscape 4: Bannerghatta National Park

6.3.5.1 Study Area

Bannerghatta National Park lies in the southern region of the Indian peninsula (Fig. 6.1) and is located 22 km south of the city of Bangalore. The park covers an extent of 109 km^2 in Bangalore's urban and rural districts. It is surrounded by several village communities on the east and west and is contiguous with Talli Reserve Forest to the southeast and Bilikal Reserve Forest to the south. The forest cover inside the park is mostly southern tropical dry deciduous, interspersed with southern tropical moist deciduous and dry deciduous scrub forest. The dry deciduous forests are mostly open forests with a canopy cover of 10–40% and the scrub forest has a canopy cover of 10% or less. The mixed deciduous forests are limited to a few patches in the northern and southern ends of the park with canopy covers of 35% and above (Radha Devi 2003).

The park was formed in 1971 and declared as a national park in 1974. It constitutes a major habitat for the Asian elephant and harbors a number of species of other wildlife and birds (Radha Devi 2003). The park was primarily started with the objective of bioconservation, biorecreation for tourists, and bioeducation for students and researchers. There are six large enclosures located within the park. Three enclosures contain a total of six villages dependent on agriculture and cattle grazing. Another enclosure forms a large farming estate, and the final two enclosures contain the tourism zone with a zoo, safaris, and an animal rescue and rehabilitation center. Since the enactment of the Wildlife (Protection) Act, 1972, hunting has been officially prohibited within the park (Anonymous 1994), but in practice, hunting of rodents, wild boars, and porcupines continues to take place.

The eastern and western parts of the area surrounding the park consist mostly of farming land where commercial crops like banana, coconut, vegetables, mulberry, and flowers are grown. However, in recent years, plantations of coconut palm (*Cocos nucifera*), mango (*Mangifera indica*), teak (*Tectona grandis*), and eucalyptus (*Eucalyptus cinerea*) have increased in these areas. Some of the other important land uses around the park are pastoral grassland, urban-built areas, and stone quarries. The northern area is a densely built, mostly suburban residential and commercial development connected to the city of Bangalore.

Thus, the park is well connected to Bangalore city through a network of cart and paved roads, many of which run through the national park and village enclosures, making the park accessible to village communities in and around the park. Although extraction of timber and nontimber forest products is prohibited inside all Indian national parks (Guha 1983; Saxena 1999), there is regular extraction of fuel wood for domestic cooking and to facilitate sand mining, timber for building agricultural equipment, furniture, and houses and selective poaching of high-value sandalwood trees. Frequent visits to the many religious shrines located inside the park also impact forest cover. A range of minor forest products are also extracted from the park by the indigenous and other village communities who are economically dependent on the sale of these products. The park faces constant pressure and encroachment for conversion to agriculture on its edges and near village enclosures. A highly contested grant of the park's forest land in the north-central area to village communities by revenue authorities prior to 1972 has also engendered substantial people-park conflict, contributing to deforestation (Radha Devi 2003). Other strong drivers of land-cover change are stone quarrying and sand mining for construction in Bangalore—prohibited activities that continue nevertheless due to the strong commercial demand for construction material from the city—and cattle grazing in and around the park by local communities.

6.3.5.2 Methods

Satellite images for three dates—a Landsat MSS image from February 1973, a Landsat TM image from January 1992, and an IRS LISS III image from May 2007—were used to map trajectories of forest change. All images were cloud-free and acquired in the dry season, when agriculture is easy to distinguish from forest. The park boundary was delineated using 1:50,000 topographic maps of the Survey of India and the Bannerghatta National Park management plan from the Karnataka Forest Department, Bangalore. A GIS buffer extending 5 km from the park boundary was created to compare forest change between the park and its immediate surroundings. The distance of 5 km was selected as most of the forest cover outside BNP lies within this boundary.

All satellite images were corrected for radiometric and atmospheric errors (Green et al. 2005), georectified, and resampled to a ground resolution of 30×30 m and projected to WGS 84 UTM with an RMS error of less than 0.5 pixels using the nearest-neighbor algorithm. The satellite image of 1992 was rectified using the topographic maps. The other two images were geometrically corrected to the 1992 image.

The images were classified using a hybrid approach. The ISODATA clustering method was used to generate 250 clusters, which were subsequently merged into 10 classes using visual interpretation and spectral signature analysis. Using training sites collected during two separate summer season field visits (2007 and 2008), spectral signatures were added to the 10 clusters from unsupervised classification. Supervised classification was run using Gaussian maximum likelihood method. Those 10 classes were combined to form three final forest classes of native forest, tree plantation, and nonforest (Fig. 6.5). The native forest category was formed by combining dense, open, and scrub forest as this landscape belongs to the forest category of Terminalia-Anogeissus latifolia-Tectona series (Champion and Seth 1968). The tree-plantation class included plantations of coconut trees, eucalyptus, teak, and mango. The nonforest area included agriculture, bare soil, stone, water, built area, and grassland. The overall classification accuracy for the 2007 image was 82.14% with an overall kappa of 0.70. Two-date change trajectories of 1973–1992 and 1992–2007 were created using the post-classification change detection method (Table 6.5) to create a series of possible classes of change for the two time spans.

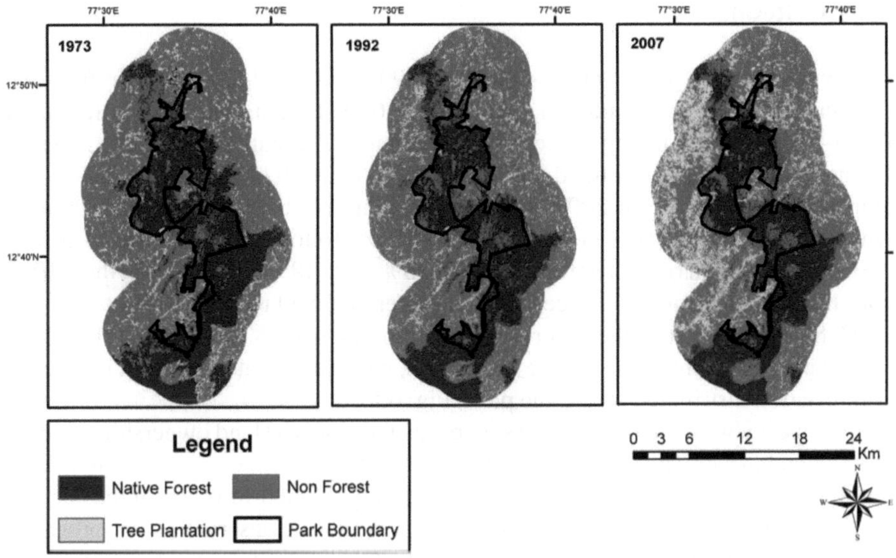

Fig. 6.5 Forest-cover distribution in Bannerghatta National Park in 1973, 1992, and 2007

Table 6.5 Trajectories of land-cover change (%) in Bannerghatta National Park and the surrounding 5-km area

	1973–1992	1992–2007
Bannerghatta National Park		
Stable forest	64.43	65.70
Deforestation	15.18	6.31
Tree-plantation harvest	0.15	0.06
Reforestation	8.67	10.97
Tree-plantation reforestation	0.09	0.87
Stable nonforest	9.86	13.33
Surrounding area		
Stable forest	16.31	15.45
Deforestation	7.40	3.08
Tree-plantation harvest	6.35	4.64
Reforestation	2.56	2.84
Tree-plantation reforestation	7.02	16.21
Stable nonforest	56.03	50.73

These classes were stable forest (forest both dates), stable nonforest (nonforest both dates), deforestation (forest on date 1 and nonforest by date 2), reforestation (nonforest on date 1 to forest on date 2), tree-plantation harvest (tree plantation on date 1 to nonforest on date 2) and tree-plantation reforestation (nonforest on date 1 to tree plantation on date 2). In-depth, open-ended interviews were also conducted with the local people living in and around the park to understand the past land-use history.

6.3.5.3 Results

The two-date change trajectories of 1973–1992 and 1992–2007 (Table 6.5) and the land-cover maps of individual dates (Fig. 6.5) show highly distinct trends of forest-cover change taking place within the park. The native forest cover inside Bannerghatta initially declined from around 80% in 1973 to 74% in 1992, and then increased to 77% in 2007. An opposite trend is observed in nonforest areas which increased from 18% in 1973 to 25% in 1992, followed by a decline to 20% in 2007. There was an overall deforestation of 15% from 1973 to 1992 even though the park had formal protection since 1974; this later declined to 6% between 1992 and 2007. Spatially, most of the deforestation within the park is located near the north-central part, adjacent to the village enclosures, with other scattered small patches of deforestation distributed across the park (Fig. 6.5).

Our interviews with local people suggest that unresolved land ownership issues in this region have created a high degree of people-park conflict and led to significant deforestation. Park-people conflicts have been highlighted in other studies as one of the biggest threats to long-term sustainability of conservation in protected areas (McNeely 2005; Nagendra et al. 2009; Ostrom and Nagendra 2006). The deforestation near the village enclosures appears to be caused by land encroachment for cultivation, timber logging, cattle grazing, and illegal sand mining, as confirmed by our field interviews and direct field observations of these activities.

It is noteworthy, however, to mention that along with deforestation, the park simultaneously experienced spatially distributed and patchy reforestation. This trajectory of change was higher in the second time period compared to the first time period (Table 6.5). Similar trends toward reforestation have been observed in many tropical and subtropical forests.

In the surrounding area outside the park, the dominant trajectories of change are deforestation of native forest cover and increase or planting of tree plantations. Most of the deforestation of native forests took place between 1973 and 1992 (Fig. 6.5, Table 6.5). Deforestation between 1992 and 2007 was quite low as very few patches of native forest cover remained outside the park and the adjacent reserve forests at this time. Deforestation was mostly located in the north-central part (Fig. 6.5) and on the edges of the national park and the reserve forests. A dense network of roads coupled with suburban residential and commercial development appears to be the major driving force behind the deforestation. The stable forest cover remaining outside the park is mostly located in the two adjacent reserve forests, which also show a steady decline from 1973–1992 to 1992–2007 (Table 6.5). Small patches of reforestation were also observed in these reserve forests.

The most dominant trajectory of land-cover change outside the park is the increase in tree-plantation area, from 8% in 1973 to 9% in 1992 and 19% in 2007. Most of this increase is due to agricultural crops like coconut and mango, and woody trees like eucalyptus and *Acacia*. The increase in tree plantation can be traced to the large-scale promotion of tree plantations around protected areas encouraged in India during the 1980s and 1990s by national and international organizations (Martin and Lemon 2001). In recent decades, this has been further encouraged by the growing demand for these agricultural products in Bangalore. Critically though, our field interviews suggest that a large part of this trend can be traced to a move away from agricultural crops like rice and sugarcane, which are severely impacted by frequent wild animal raids (especially by elephants) from the park to tree plantations like coconut and eucalyptus, which are less likely to suffer damage from wild animals. The stable nonforest area in the park surroundings declined from 56% in 1973–1992 to 51% in 1992–2007 due to the conversion to tree plantations, which is seen within the agricultural class.

Overall, this landscape has undergone two major changes: deforestation of native forest cover and its recovery due to recovery in native forests as well as an increase in tree plantations. Our findings suggest that despite significant park-people conflicts, the park has been largely able to maintain its forest cover because of its status as a protected area. The landscape surrounding the park has witnessed large-scale clearing of forest cover in privately owned lands, but is now witnessing an increase in tree cover. This is not due to reforestation, however, but driven by changes in agricultural preferences toward the planting of tree crops and tree plantations. This is in keeping with overall trends of reforestation in India, which appear to be driven mostly by increases in tree plantations rather than in native forest regrowth (Grainger 2010).

Yet, the horizontal expansion of Bangalore indicates that there will be significant pressure on the park in decades to come. Tourist visits to the park have increased tremendously in recent years, and there is a demand for housing sites located near the park, with many urban residents expressing a preference to live near the largest expanse of green space close to the city. Thus, land prices close to the park have risen sharply in recent years, and many of the formerly agricultural settlements located close to the park are being rapidly converted into high-end apartment complexes and gated communities. The impact of this land-use change can have severe repercussions on forest conservation in Bannerghatta, which represents one of the few Indian-protected areas to be located this close to a major urban settlement.

6.4 Conclusions

Globally, there has been significant interest in understanding whether protected areas have been successful in limiting habitat change and whether this appears to be an effective strategy. A number of recent studies appear to indicate that, despite incidences of "paper parks" that exist only in name, many parks have been largely effective in limiting forest clearing and habitat degradation within their boundaries (Bruner et al. 2001; Nagendra 2008; Naughton-Treves et al. 2005). Yet an equally substantial body of literature also indicates that there is increasing human pressure in the landscapes surrounding parks (DeFries et al. 2010; Nepstad et al. 2005), leading to isolation of parks, fragmentation of ecological networks, and reduction in the capacity of these parks to provide effective habitats for species or to effectively maintain the range of ecosystem services they once provided. Our analyses of 15 parks from published literature and four parks based on our own in-depth studies largely confirm this picture for protected area–embedded landscapes in India.

Most parks in India are located within a region dominated by dense human settlements and are subject to human disturbance due to grazing, lopping for fodder, and collection of leaf litter, fuel wood, timber, bamboo, thatch grass, and a range of minor nonwood forest produce. These extractions occur for both subsistence and commercial sale. The extraction of produce for commercial sale appears to be the more worrisome, especially as urbanization and concentration of settlements continue to increase, and many of the formerly small towns and cities near some protected areas expand into larger towns and metropolitan cities or become connected to other nearby large cities and towns by road and rail networks. As India continues to advance toward the twenty-second century, such change is inevitable. Industrial development fueled by the construction of ports, dams, hydroelectric projects, railways, and highways has impacted many of the parks referred to here in our study. Urbanization is impacting many others. While protected areas appear to have been largely successful at maintaining forest cover and connectivity of habitat within their core, many of them report forest clearing due to land conversion for agriculture, pastures, and grazing and habitat degradation due to extraction of forest products, fire, and grazing. This change is giving rise to an increasingly common scenario of a forested protected area that is becoming isolated and embedded in a fragmented landscape of plantations, agriculture, and expanding human settlements networked by roads and railways.

Along with a steep loss in native forest cover around most protected areas, we also find that many of those landscapes appear to have tree plantations, from commercial wood crops like teak and eucalyptus to fruit crops like mango and coconut. In Bannerghatta National Park, our interviews with landowners indicate a specific shift away from agriculture toward the plantation of tree crops was a response to crop raids by wild animals. Further, in protected areas adjoining urban centers such as Bangalore, there is an increased demand for houses located near them, due to the scarcity of green spaces within the cities. This results in an accelerated conversion of native forest land and agriculture in the surrounding landscape to apartments and residential homes, further decreasing the capacity of these larger landscapes to host native biodiversity, or provide a full suite of ecosystem services. As more areas like the Siliguri urban center near Mahananda Wildlife Sanctuary, or Chandrapur town near Tadoba Andhari Wildlife Sanctuary expand and grow from towns to cities, similar patterns of change are likely to take place, leading to increased land conversion, more tourism, and an insatiable demand for wood products and forest products that can only be met by further encroachments into the parks and the remaining native habitat outside them. Urbanization, which is a dynamic trend in India, is poised to make a powerful impact on parks, and urban areas, unlike small villages, cannot be resettled to reduce the impact on parks. This study points to the need to examine the impacts of urban growth on the protected area paradigm in India in much greater depth than is currently being investigated.

This research also indicates that we need to follow the long-term impacts that urbanization may have on forest change. Studies of forest transitions in many developing and developed economies indicate that urbanization leads to long-term forest recovery in agricultural rural locations, due to the large-scale movement of people to urban areas. This pattern may be less clear-cut in India, where many migrants continue to retain close ties with their agricultural lands and villages, returning there to farm during specific times of the year. Still, more research clearly needs to be conducted to understand the critical issue of what long-term impacts urbanization will have on forest change within and outside protected areas in the Indian context.

This study draws extensively on satellite remote sensing and GIS to address questions of whether protected areas in India have experienced distinct patterns of land-cover change—in particular, greater stability and lesser forest conversion—as compared to unprotected, privately owned lands in the surrounding landscape. Such studies are integral to providing a broader, more integrated view of the effectiveness of protected areas as an approach to conservation in peopled landscapes, by placing land-use/land-cover change in parks within the context of larger ecosystem and landscape changes within which protected areas are nested (DeFries et al. 2010; Ostrom and Nagendra 2006).

References

Anonymous. (1994). *The Wildlife (Protection) Act, 1972*. Dehradun: Natraj Publishers.

Bawa, K. S., Raven, P. H., & Seidler, R. (2004). Reconciling conservation paradigms. *Conservation Biology, 18*, 859–860.

Bruner, A. G., Gullison, R. E., Rice, R. E., & da Fonseca, G. A. B. (2001). Effectiveness of parks in protecting tropical biodiversity. *Science, 291*, 125–128.

Champion, H. G., & Seth, S. K. (1968). *A revised survey of forest types of India*. New Delhi: Government of India.

Chapin, M. (2004). A challenge to conservationists. *World Watch, Nov/Dec*, 17–31.

Cincotta, R. P., & Engelman, R. (2000). *Nature's place: Human population and the future of biological diversity*. Washington, DC: Population Action International.

Das, T. K., & Guha, I. (2003). *Economics of man-animal conflict: A lesson for wildlife conservation*. Kolkata: ACB Publications and DSA Centre for Regional Economic Studies, Jadavpur University.

DeFries, R., Karanth, K., & Pareeth, S. (2010). Interactions between protected areas and their surroundings in human-dominated tropical landscapes. *Biological Conservation, 143*, 2870–2880.

Ghate, R. (2007). Bureaucratic strategies in relocation. *Economic and Political Weekly, 42*, 21–23.

Ghate, R., & Beazley, K. (2007). Aversion to relocation: A myth? *Conservation and Society, 5*, 331–334.

Giriraj, A., Murthy, M. S. R., & Beierkuhnlein, C. (2010). Evaluating forest fragmentation and its tree community composition in the tropical rain forest of Southern Western Ghats (India) from 1973 to 2004. *Environmental Monitoring and Assessment, 161*, 29–44.

Grainger, A. (2010). The bigger picture: Tropical forest change in context, concept and practice. In H. Nagendra & J. Southworth (Eds.), *Reforesting landscapes: Linking pattern and process* (Landscape Series, No. 10, pp. 15–43). Dordrecht: Springer.

Green, G. M., Schweik, C. M., & Randolph, J. C. (2005). Retrieving land-cover change information from Landsat satellite images by minimizing other sources of reflectance variability. In E. F. Moran & E. Ostrom (Eds.), *Seeing the forest and the trees: Human-environment interactions in forest ecosystems* (pp. 131–160). Cambridge, MA: The MIT Press.

Guha, R. (1983). Forestry in British and post-British India: A historical analysis. *Economic and Political Weekly, 18*, 1882–1896.

IIFM (Indian Institute of Forest Management, Bhopal). (2001). *Public forests, fuelwood collection and migration: A case study in North Bengal*. Bangkok: Food and Agriculture Organization of the United Nations Regional Wood Energy Development Programme in Asia.

Imam, E., Kushwaha, S. P. S., & Singh, A. (2009). Evaluation of suitable tiger habitat in Chandoli National Park, India, using multiple logistic regression. *Ecological Modelling, 220*, 3621–3629.

Jenkins, C. N., & Joppa, L. (2009). Expansion of the global terrestrial protected area system. *Biological Conservation, 142*, 2166–2174.

Jensen, J. R. (2000). *Remote sensing of the environment: An earth resource perspective*. Englewood Cliffs: Prentice-Hall.

Joseph, S., Blackburn, G. A., Gharai, B., Sudhakar, S., Thomas, A. P., & Murthy, M. S. R. (2009). Monitoring conservation effectiveness in a global biodiversity hotspot: The contribution of land cover change assessment. *Environmental Monitoring and Assessment, 158*, 169–179.

Khawarey, K. N., & Karnat, M. (1997). *Management plan for Tadoba-Andhari Tiger Reserve (Tadoba National Park and Andhari Wildlife Sanctuary), plan period 1997–98 to 2006–07*. Chandrapur: Government of Maharashtra.

Martin, A., & Lemon, M. (2001). Challenges for participatory institutions: The case of village forest committees in Karnataka, South India. *Society and Natural Resources, 14*, 585–597.

Mawdsley, E., Mehra, D., & Beazley, K. (2009). Nature lovers, picnickers, and bourgeois environmentalism. *Economic and Political Weekly, 19*, 49–59.

McNeely, J. (Coord.). (2005). *Workshop stream II: Building broader support for protected areas*. IUCN, benefits beyond boundaries: Report of the Vth IUCN World Parks Congress, Durban, South Africa, 8–17 September 2003 (pp. 56–63). Gland: International Union for Conservation of Nature and Natural Resources (IUCN).

Mondal, P., & Southworth, J. (2010a). Protection vs. commercial management: Spatial and temporal analyses of land cover changes in the tropical forests of central India. *Forest Ecology and Management, 259*, 1009–1017.

Mondal, P., & Southworth, J. (2010b). Evaluation of conservation interventions using a cellular automata-Markov model. *Forest Ecology and Management, 260*, 1716–1725.

Nagendra, H. (2008). Do parks work? Impact of protected areas on land cover clearing. *Ambio, 37*, 330–337.

Nagendra, H., Pareeth, S., & Ghate, R. (2006). People within parks: Forest villages, land-cover change and landscape fragmentation in the Tadoba-Andhari Tiger Reserve, India. *Applied Geography, 26*, 96–112.

Nagendra, H., Pareeth, S., Paul, S., & Dutt, S. (2009). Landscapes of protection: Forest change and fragmentation in northern West Bengal, India. *Environmental Management, 44*, 853–864.

Nagendra, H., Rocchini, D., & Ghate, R. (2010). Beyond parks as monoliths: Spatially differentiating park-people relationships in the Tadoba Andhari Tiger Reserve in India. *Biological Conservation, 143*, 2900–2908.

Nandy, S., Kushwaha, S. P. S., & Mukhopadhyay, S. (2007). Monitoring the Chilla-Motichur wildlife corridor using geospatial tools. *Journal for Nature Conservation, 15*, 237–244.

Naughton-Treves, L., Holland, M. B., & Brandon, K. (2005). The role of protected areas in conserving biodiversity and sustaining local livelihoods. *Annual Review of Environment and Resources, 30*, 219–252.

Nautiyal, S., & Kaechele, H. (2007). Adverse impacts of pasture abandonment in Himalayan protected areas: Testing the efficiency of a natural resource management plan (NRMP). *Environmental Impact Assessment Review, 27*, 109–125.

Nayak, S., & Bahuguna, A. (2001). Application of remote sensing data to monitor mangroves and other coastal vegetation of India. *Indian Journal of Marine Sciences, 30*, 195–213.

Nepstad, D., Schwartzman, S., Bamberger, B., Santilli, M., Ray, D., Schlesinger, P., et al. (2005). Inhibition of Amazon deforestation and fire by parks and indigeneous lands. *Conservation Biology, 20*, 65–73.

NESPON (North Eastern Society for Preservation of Nature and Wildlife). (2000). State of north Bengal's forests: Background paper. Discussion on combating deforestation: The question of rural development in forest areas of North Bengal, March 17, Siliguri Tathyakendra, West Bengal, India.

Ostrom, E., & Nagendra, H. (2006). Insights on linking forests, trees, and people from the air, on the ground, and in the lab. *Proceedings of the National Academy of Sciences of the United States of America, 103*, 19224–19331.

Palria, S., Singh, S., Sharma, J. R., & Pathak, S. (2005). Habitat evaluation for sarus crane in the Keoladeo National Park using IRS LISS III and PAN merged data and GIS. *Photonirvachak: Journal of the Indian Society for Remote Sensing, 33*(2), 259–266.

Pandey, D. (1992). *Assessment of tropical forest plantation resources*. Unpublished report to Forestry Department, Food and Agriculture Organization of the United Nations, Rome. Department of Forest Survey, Swedish University of Agricultural Sciences.

Prasad, R. (2000). Sustainable forest management for dry forests of South Asia. In T. L. Cheng & P. B. Durst (Eds.), *Development of national-level criteria and indicators for the sustainable management of dry forests of Asia: Background papers* (pp. 27–74). Bangkok: RAP Publishers/Food and Agriculture Organization of the United Nations.

Radha Devi, A. (2003). *Karnataka Forest Department Master Plan for consolidation of Bannerghatta National Park boundaries and elephant corridors*. Bangalore: Kalkere.

Robbins, P. F., Chhangani, A. K., Rice, J., Trigosa, E., & Mohnot, S. M. (2007). Enforcement authority and vegetation change at Kumbhalgarh Wildlife Sanctuary, Rajasthan, India. *Environmental Management, 40*, 365–378.

Sarma, P. K., Lahkar, B. P., Ghosh, S., Rabha, A., Das, J. P., Nath, N. K., et al. (2008). Land-use and land-cover change and future implication analysis in Manas National Park, India using multi-temporal satellite data. *Current Science, 95*, 223–227.

Saxena, N. C. (1999). *Forest policy in India* (WWF India Policy and Joint Forest Management Series, No. 1. IIED Publication 7554). New Delhi: WWF India and International Institute for Environment and Development (IIED).

Schwartzman, S., Moreira, A., & Nepstad, D. (2000). Rethinking tropical forest conservation: Perils in parks. *Conservation Biology, 14*, 1351–1357.

Shahabuddin, G. (2010). *Conservation at the crossroads: Science, society, and the future of India's wildlife*. Delhi: Permanent Black.

Shahabuddin, G., & Rangarajan, M. (Eds.). (2007). *Making conservation work: Securing biodiversity in this new century*. Ranikhet: Permanent Black.

Shankar, U. (2001). A case of high tree diversity in a sal (*Shorea robusta*) -dominated lowland forest of Eastern Himalaya: Floristic composition, regeneration and conservation. *Current Science, 81*, 776–786.

Singh, G., Velmurugan, A., & Dakhate, M. P. (2009). Geospatial approach for tiger habitat evaluation and distribution in Corbett Tiger Reserve, India. *Photonirvachak: Journal of the Indian Society for Remote Sensing, 37*, 573–585.

Somashekar, R. K., Nagaraja, B. C., & Urs, K. (2008). Monitoring of forest fires in Bhadra Wildlife Sanctuary. *Photonirvachak: Journal of the Indian Society for Remote Sensing, 36*, 99–104.

Southworth, J., Nagendra, H., & Munroe, D. K. (2006). Introduction to the special issue: Are parks working? Exploring human-environment tradeoffs in protected area conservation. *Applied Geography, 26*, 87–95.

Sudhakar, S., Prasad, P. R. C., Awwaratia, M. L., Sudha, K., Babar, S., & Rao, S. K. S. V. R. (2008). Landscape analysis in Fambong Lho Wildlife Sanctuary, east district, Sikkim, India using remote sensing and GIS techniques. *Photonirvachak: Journal of the Indian Society for Remote Sensing, 36*, 203–216.

Wildlife Circle. (1997). *Management plan of Mahananda Wildlife Sanctuary, West Bengal*. Kolkata: Government of West Bengal.

World Bank. (1991). *Maharashtra forestry project*. http://web.worldbank.org/external/projects/main?pagePK=64283627&piPK=73230&theSitePK=40941&menuPK=228424&Projectid=P010390. Accessed 13 Sept 2010.

Chapter 7
Public-Private Interactions in the Conservation of Private Forests in the United States

Tatyana B. Ruseva and Burnell C. Fischer

Abstract This chapter discusses the investment decisions of private and public actors in the USA regarding the management and conservation of private forests. Managed by nearly 11 million private owners, these forestlands provide valuable ecological, economic, and social benefits to society. What happens on these lands as a result of public and private investment decisions has implications for society, both in the present and in the future. First, this chapter reviews the interactions between private and public actors in private forestry. A typology is proposed that characterizes public and private investments in forestlands based on their target audiences (e.g., a private investment decision for largely public benefit). This chapter next explores the different ownership features and uses of forests over time and space. The discussion centers on the challenges and opportunities of governing private forests as a bundle of property rights within the proposed typology of public-private interactions in private forestry.

7.1 Introduction

Much attention over the past few decades has been devoted to understanding the sources and consequences of forest change dynamics (Agrawal et al. 2008; Nagendra 2007; Persha et al. 2011; Rudel et al. 2005; Sugden et al. 2008). Human activities directly or indirectly alter the extent, health, and productivity of

---

T.B. Ruseva (✉)
Department of Government & Justice Studies, Appalachian State University, Boone, NC, USA
e-mail: rusevatb@appstate.edu

B.C. Fischer
School of Public and Environmental Affairs, Indiana University–Bloomington, Bloomington, IN, USA
e-mail: bufische@indiana.edu

E.S. Brondízio and E.F. Moran (eds.), *Human-Environment Interactions: Current and Future Directions*, Human-Environment Interactions 1,
DOI 10.1007/978-94-007-4780-7_7,

forestlands. Forested landscapes are progressively being converted to other land uses, such as agriculture, biofuel production, and urban development (Stein et al. 2009). At the same time, local communities with support from state and national programs are restoring forests in countries around the world (Chazdon 2008: 1458). This chapter reviews and synthesizes the interactions of public and private actors in the USA and their decisions to invest in the sustainable management of private forestlands.

Interest in the public value of private forests and their contributions to local communities has grown alongside improved awareness of environmental issues (Best and Wayburn 2001). Society enjoys an array of ecological, social, and economic benefits from private forests, which presently account for over half of all forestland in the USA (Butler 2008). Most of these lands are in the eastern part of the country. What happens on private forestlands as a result of the decisions and actions of their owners has implications for society, both in the present and in the future. Privately owned forests provide important goods and services, such as timber and wood products, fish and wildlife habitats, carbon storage, soil and water regulation, and scenic beauty (NRC 1998). Under direct and indirect human pressures of conversion, fragmentation, urban development, and others, and in the absence of incentives, these public goods and services are likely to be lost or undersupplied (Ruhl et al. 2008; Stein et al. 2009).

Additionally, the growing diversity of social and ecological interactions, the complex and nonlinear nature of biophysical processes, changing generational and urban profiles of America's forest owners, and increasing interest in the amenity value of private forests are shaping the changes occurring on private forestlands (Bascompte 2009; Bruyere et al. 2009; Fischer and Ruseva 2010; Ostrom 2009). To better understand the promise and pitfalls of these processes, this chapter explores the interactions between private and public actors and the manner in which they shape private forest management and conservation. As Agrawal and colleagues (2008: 1462) point out, "The challenge of understanding the coupled social and ecological systems that all forest governance represents urgently needs more emphasis and attention than it has received until now."

In this chapter, we review existing institutional arrangements for investment in private forestlands. Institutions affect the incentives and patterns of interactions among different participants, as well as the outcomes of these interactions (Ostrom 2005). First, a brief overview of private forests in the USA is provided. Second, the interconnections between private and public investment decisions are discussed. A typology is proposed that characterizes public and private investment decisions based on their target audiences or primary addressees. Next, this chapter explores different ownership characteristics and uses of forests over time and space. Different features of forests (timber, wildlife habitats, water regulation, etc.) can be linked to specific property-rights regimes and thus subject to different rights of access, use, and appropriation (Ostrom 2009; Ruhl et al. 2007). The discussion focuses on the challenges to and prospects for governing private forests as a bundle of property rights within the existing public-private interactions in private forestry. Insights from this chapter can be useful in light of the trends and evolving arrangements for natural resource management under private ownership, specifically in a developed country context.

7.2 Sustaining Private Forests in the USA: Threats and Opportunities

Privately owned forests represent a substantial share (56%) of forest resources in the USA (Butler 2008). An estimated 423 million acres of forestlands are owned by 11 million private landowners with diverse values and motivations for land ownership (Bengston et al. 2011; Butler 2008). Among them, individuals, families, estates, trusts, and unincorporated groups—known as family forest owners—are the majority. Family forest owners account for 92% of all private forest owners and over one-third of forestlands in the USA (Butler 2008). These forest owners are integral to the idea and practice of sustainable private forestry—"a model of forest management that attempts to ensure that the flow of goods and services from a land area can be sustained into the future (i.e. current management is not depleting future productivity)" (Daniels et al. 2010: 51). As a policy issue, forest sustainability becomes compounded when studied in the context of private land ownership, where the decisions of autonomous individuals overlap with dynamic social and biophysical processes.

Degradation, conversion, and fragmentation of forested landscapes present major threats to the sustainability of private forests (Schaaf and Broussard 2006; Smith et al. 2009; Stein et al. 2009). About 23 million acres of forests are at risk of being lost to developed uses by 2050 (Alig et al. 2010; Stein et al. 2005); another 57 million acres of rural private forestlands will likely have been affected by increased housing density from 2000 to 2030 (Stein et al. 2009). Urban growth, consumption pressures, and amenity-driven relocations put ever greater demands on forestlands and ecosystems (Best and Wayburn 2001; Schaaf and Broussard 2006; Stein et al. 2009). Nowak and Walton (2005) project that forestland area roughly the size of Pennsylvania (118,300 km^2) will be lost as a result of urbanization between 2000 and 2050. Eighty percent of Americans already live in cities and towns where forestland is of growing importance for the many services provided (US Census Bureau 2011). The predictions that more than 100 million acres of US forestland will be in urban forests, some of which will be managed by family forest owners, are especially noteworthy (Nowak and Walton 2005; Nowak et al. 2005). The implications for forest management in and near these urban areas are complex and ever-changing. For example, many pests from abroad are introduced first to urban areas and then expand into rural areas. The emerald ash borer (*Agrilus planipennis*) is one of the latest examples (Nowak et al. 2005; USFS 2011).

Besides urban and exurban expansion, the changing ownership, generational, and demographic profiles of America's forest owners pose additional challenges to forest health and sustainability. Results from the 2006 National Woodland Owner Survey conducted by the US Forest Service (USFS) show that 1 out of 5 acres of forestland will likely be sold or transferred in the next 5 years and that one-fifth of all family forestlands are owned by someone 75 years of age or older (Butler 2008; Smith et al. 2009). Parcelization of private landholdings into smaller forest tracts facilitates conversion to nonforest use, fragmentation of habitats, and disruption of landscape- or watershed-wide ecological processes (Rickenbach et al. 2011; Sampson and DeCoster 2000). Last but not least, threats to privately owned forest resources arise from wildfires, insect and disease outbreaks, ozone pollution, and other natural disturbances (Smith et al. 2009).

Opportunities for sustaining and conserving private forestlands are not in shortage. Studies show an overall commitment and a desire toward forest stewardship and conservation among family forest owners (Arano et al. 2004; Kendra and Hull 2005; Kilgore et al. 2007). Landowners are investing in forest plantations and restoration in pursuit of aesthetic, recreational, and other amenity values (Ross-Davis et al. 2005). Individuals and families see forests as a source of privacy, self-identity, and a legacy for future generations (Ross-Davis et al. 2005). Kendra and Hull (2005) argue that new patterns of forest ownership amid urbanizing and fragmented forest landscapes offer opportunities rather than challenges for sustaining America's private forests. Similarly, Butler and Leatherberry (2004: 9) posit that with more individuals owning forested land, "there will be more people in intimate contact with the land ... [which] affords an opportunity to educate more people about the benefits and responsibilities associated with forest stewardship." The relationship between forest owners and their land is central to sustainable forest outcomes. Equally important are their connections to other people, as evidence from peer-to-peer learning, forest cooperatives, and cross-boundary cooperation suggests (Allred and Goff 2009; Barten et al. 2001; Gass et al. 2009; Kittredge 2005; Knoot and Rickenbach 2011; Kueper and Sagor 2011; Rickenbach and Jahnke 2006; Rickenbach et al. 2011). Such findings relate to the broader literature on the role of cooperation, collaborative environmental management, and social networks in natural resource governance (Bodin and Crona 2009; Bodin et al. 2006; Crona and Bodin 2006; Crossley et al. 2009; Ernstson et al. 2008; Janssen et al. 2006).

Other barriers to, but also opportunities for, forest sustainability include differences in the life spans of people and forests, a disconnect between the ecological and market values of forestlands, and the effects of global trade and nascent ecosystem services markets (Beach et al. 2005; Best and Wayburn 2001; Dauvergne and Lister 2011). For instance, agricultural subsidies, tax incentives, and market dynamics have a major bearing on the conversion of agricultural lands to forests (Evans et al. 2010; Rudel 2010). Recent transformations and restructuring within the industrial forest sector offer another line of possibilities and opportunities (Bliss et al. 2010). Last, the complex web of federal, state, and local institutions amidst changing agency mandates and dwindling budgets proffers further questions and outlooks toward sustainable private forestry (Best and Wayburn 2001; Schroeder et al. 2011).

7.3 Public-Private Investment Decisions in the Conservation of Private Forestlands

Private and public actors interact in multiple, dynamic ways to conserve and keep private forests healthy and in doing so influence the provision of forest ecosystem goods and services. Forest stewardship and conservation practices undertaken by private landowners constitute private investment in private forestlands. Forest stewardship is defined here as the application of site level, ecologically sound practices that can positively influence the capacity of forestlands to sustain a variety of ecological, economic, and social benefits (Kilgore et al. 2008: 358).[1] Such practices, promoted through technical and financial assistance from state forestry agencies and the USFS State and Private Forestry organization, constitute social investment in the production of goods and services from private forests. Professor of natural resource economics and policy Michael Kilgore (2004: 11) explains:

> The public's interest in private forestry is grounded in two fundamental concerns. One is the need to encourage continued investment in land management for a wide range of benefits provided by private forests. The long-term nature of forestry and lack of markets for many of its products tend to discourage landowners from investing in their forests. Public policies are also used to minimize negative impacts that can be associated with forest management and timber-harvesting activities, such as the loss of wildlife habitat, diminished water quality, soil erosion, and reduced visual quality.

The conservation of private forests in the USA is promoted through private and public investments (NRC 1998). The decision of individual landowners to protect forest resources or to increase forestland by planting trees represents a *private investment decision*. Such decisions affect the level of private goods (e.g., timber) and public benefits (e.g., carbon storage, groundwater recharge) produced from private forestlands. On the other side, collective-choice decisions formulated at federal, state, and local levels determine the type and amount of *public investment* in private forest conservation. Public policies about the management, use, and protection of private forests seek to prevent the loss of and/or to perpetuate the benefits society derives from forestlands, such as the sale or donation of development rights, or the enhancement of wildlife habitats. Both private and public investment decisions have relevance for the ecological and social benefits derived from forests, specifically what quantities of land to designate to forest use, what quality standards and forest practices to apply, or how to plan and manage forestlands with an eye to the future (e.g., via a written landowner management plan).

With an inadequate level of private investment in forestlands, and declining resources for public investment, concerns about the future sustainability of private forests are growing (Rickenbach et al. 2011; Schroeder et al. 2011). The USFS

---

[1]The characteristics and definitions of "forest stewardship" vary. Drawing from others, forest stewardship is understood, here broadly, as a type of sustainable land use and management (Kilgore et al. 2008, 358).

Table 7.1 Four spheres of public-private interactions in private forestry

Primary Addressee	Investment Decision: PUBLIC	Investment Decision: PRIVATE
PRIVATE	**Public investment for (largely) private benefit:** • Educational programs • Technical assistance • Financial assistance (cost-share) • Economic incentives (tax breaks, direct payments, cost-share) • Other (free or low-cost seedlings) -box 1-	**Private investment for private benefit:** • Timber production • Food and other forest products • Cultural practices (thinning, pruning) • Certification of private forests • Conservation easements (upfront private gain) -box 3-
PUBLIC	**Public investment for public benefit:** • Regulatory programs (timber harvesting, road and trail practices) • Zoning regulations • Environmental laws (e.g. Endangered Species Act) • Purchase of rights in real property; purchase of forestland -box 2-	**Private investment for (largely) public benefit:** • Reforestation practices • Forest protection practices • Harboring and enhancing wildlife • Conservation easements (long-term, perpetual social gains)[a] -box 4-

Sources: Best and Wayburn (2001), Ellefson et al. (2007), Kilgore et al. (2008), Tompkins and Eakin (2012)

[a]Conservation easements have both private and public benefits spread across time, including (1) reduced tax burden for property owners as an upfront private benefit and (2) protection of forest from future development as a long-term social benefit

estimates that only 4% of family forest owners have written management plans, which represents 17% of family forestland (Butler 2008: 25). State-level resources and staff are also limited, and funds for landowner assistance and outreach have and likely will continue to wane. Combined with the larger socioeconomic and environmental challenges outlined above, these trends suggest a need for assessing the potential of coupling public with private investments in private forests.

The interconnections between private and public investment decisions in the context of private forests can be conceptualized as activities in four spheres of public-private interactions (Table 7.1). Table 7.1 displays, in a fairly simplistic way, four areas of private and public action. Each entails specific investment tools and has direct implications for private landowners or the public interest. In basic terms, differences exist with regard to the target audiences or addressees of investment decisions. A public policy, such as the legal requirement to obtain a permit before timber harvesting on private lands, has a direct effect on a segment of the population (i.e., private landowners and the timber buyers/producers who harvest and process the timber) and only indirect implications for society as a whole. When individual actions generate positive externalities, policies will target those individuals and their behavior in order to encourage the provision of public goods or services (Weimer and Vining 2010). Vaccination is one case illustrating how private actions can generate benefits for society in terms of improved public health (Tompkins and Eakin 2012). Private landowners who conserve or restore forests similarly create social and environmental benefits. Alternatively, public policies may provide benefits to all of us and have direct implications for society. Creating a state forest or a national park generates socially and ecologically valuable assets for present and future generations. This brief discussion allows us to characterize public and private investment decisions in relation to their primary addressees or direct benefit receivers (Table 7.1).

An important caveat is that the typology presents categories of public/private investment that vary in degree, type of forest features, and the scale and timing of effects. Certain forest investments can be directly addressed to private landowners in the short run (financial assistance) but provide medium- to long-run secondary benefits to neighbors (e.g., water regulation, recreation), local communities (e.g., scenic beauty, clean air), and society as a whole (e.g., carbon storage). This consideration is undertaken in the fourth part of this chapter. The purpose here is to distinguish between the provider and the primary benefit receiver of investment decisions, which offers practical and policy insights. In particular, the spheres of public-private interactions displayed in the upper left and bottom right of Table 7.1 are of growing policy relevance: one, because of the wealth of experience with some of these arrangements (box 1) and, two, due to the increased popularity of some partnerships to conserve forestland (box 4). Each box in Table 7.1 is described in detail below with illustrations from national- and state-level[2] contexts.

7.3.1 *Public Investment for Largely Private Benefit*

A number of government programs for technical, educational, and financial assistance are addressed directly at nonindustrial private forest (NIPF) owners.[3] The main purpose is to encourage and support the sustainable management of

[2]We use the state of Indiana, in the Midwest USA, because of its significant share of privately owned forests and similarities to other Midwest states, which display a mix of small-scale forest and agricultural land use, residential expansion, and susceptibility to invasive plants and natural disturbances (see Kauneckis and York 2009; Koontz 2001; Smith et al. 2009: 48).

[3]Nonindustrial private forest is a subset of private forestlands, where the owner does not operate wood-using plants. A difference exists between families and individuals who own forestland but do not operate a wood-processing facility (nonindustrial private forest owners) and private groups and corporations who own and operate a primary wood-processing facility (industrial private forest owners) (Butler 2008: 3; Smith et al. 2009: 144).

private forestlands. While a stream of forest goods and services is expected, the direct benefits accrue principally to individuals or families who own forest and participate in these programs (Tompkins and Eakin 2012). The public does not and cannot manage private forestlands. The government, on behalf of the public, invests financial and human resources in private forestlands because they are identified as important to society (NRC 1998). Public investment in private forests takes the form of federal and state assistance programs that promote forest stewardship through information and educational programs (e.g., publications, workshops, training courses), professional advice and technical assistance from extension foresters to forest owners ("walking the forest" or writing a forest management plan), financial assistance through grants and cost sharing for forest practices, property tax incentives, and provision of low-cost or subsidized tree seedlings.

Most states in the USA rely on a mix of education and incentive programs tailored to local needs and landscapes (Ellefson et al. 2007; Kilgore et al. 2008). In Indiana, where 86% of forests are privately owned, most government programs address private forestlands. A tax incentive program with a long and successful history in Indiana is the Classified Forest and Wildlands program, which places an assessed tax value of one dollar per acre on enrolled properties. In return, landowners must meet specific forest conservation standards, such as protection from fire and grazing, property development (e.g., construction of a building), and following a written management plan (IDNR n.d.). Similar voluntary programs exist in other states, for instance, Vermont's Agricultural and Managed Forest Use Value Program, Iowa's Forest Reserve Laws, and Wisconsin's Forest Crop Law and Managed Forest Law (Greene et al. 2010). Practitioners argue that with sound tax policies reasonable returns from forestland investments are possible (Kimbell et al. 2009: 93). Examples of incentive programs at the federal level include the Wildlife Habitat Incentive Program, Environmental Quality Incentives Program, and the Conservation Reserve Program, among others. A grant under the Wildlife Habitat Incentive Program primarily and directly benefits the landowner but also encourages activities on private lands, for instance, the enhancement of habitats for migratory birds, the enjoyment of which is joint and nonexcludable. The expectation is that such public investment (e.g., in wildlife habitat) on private forestlands will yield public benefits in the future.

The activities described in box 1 in Table 7.1 are where a significant share of public-private interactions related to private forest conservation currently takes place. These arrangements suggest several important trends and empirical lessons. Federal funding for NIPF assistance programs has fluctuated over time and declined considerably in recent years (Best and Wayburn 2001; NASF 2010; Schroeder et al. 2011). Shrinking investment and agency capacities, paralleled by a growing number of NIPF owners and threats to forests (wild fires, declining wetlands), has important distributional implications for forest stewardship. Other considerations include changing landowner demographics and ownership structure, parcelization, rising land values, and development, which collectively challenge the effectiveness and acceptance of NIPF assistance programs (Davis et al. 2010; Knoot et al. 2010). Over the years, researchers have sought to examine the extent, to which public programs lead to additional private investment in forest management and their impact on forest land use and management (Beach et al. 2005; Kauneckis and York 2009; Kilgore et al. 2007; Munroe and York 2003; Nagubadi et al. 1996; Schaaf and Broussard 2006; York et al. 2005). Studies show that the involvement of NIPF owners is often conditioned on ownership characteristics (preferences, motivations, ownership size, and duration), land conditions, market drivers, and public policies (Beach et al. 2005; Kauneckis and York 2009). In all, the engagement of private landowners is paramount because they are the primary addressees of the public investment decisions in box 1 in Table 7.1.

7.3.2 Public Investment for Public Benefit

Box 2 in Table 7.1 captures public investment decisions for primarily public benefit. Decisions by state and federal governments to conserve forest resources take several forms, regulation being the prime example. Other approaches include the purchase of land for conservation or public use and purchase of development rights.

Regulation is "generally used to set minimum standards for forest practices to ensure protection of public trust resources on private lands" (Best and Wayburn 2001: 130). A variety of state-level regulatory programs seek to control activities on private land, albeit in a limited fashion. A comprehensive study by Ellefson et al. (2007) indicates that only 15 states have prominent regulatory programs.[4] These regulatory programs target different forestry practices on private lands, namely, timber harvesting, road and trail practices, forest protection and reforestation, cultural practices, chemical substance application, and administrative practices. Regulatory programs seem to be most effective for controlling the application of chemicals and least effective with regard to cultural practices like pruning, thinning, and timber stand improvement (Ellefson et al. 2007). Generally, state governments have jurisdiction over the regulation and zoning of different types of activities on private lands. On average, five state-level agencies are involved in the design, implementation, and enforcement of regulatory programs (Ellefson et al. 2007). Research, however, indicates that the effectiveness of these programs is "modest when compared to non-regulatory programs" while the investment cost remains considerable (Ellefson et al. 2007: 629).[5]

In addition, governments can directly purchase forestland and create state and national parks, forests, wildlife areas, and wilderness lands, in the interest of present

[4]Most states rely on education and incentive programs to encourage sustainable forestry practices. Ellefson et al. (2007) examined statewide programs for all private lands. Certain targeted programs, such as the Indiana Classified Forest and Wildlands program, do set standards and regulate forest use on lands enrolled *voluntarily* in the program. This is an important differentiation since many states have such voluntary programs.

[5]The estimated annual state government investment in regulatory programs targeted at private forestlands was $57.6 million dollars in 2003 (Ellefson et al. 2007: 629).

and future generations. Where private forests harbor and serve as habitats for endangered or threatened species, government agencies administer rules regarding the taking of such species from private lands (ESA 2005: Sec. 9). A hybrid partnership between government actors, namely, the US Fish and Wildlife Service, and private landowners arises in the context of developing and adopting habitat conservation plans. The primary objective of the plans is to protect endangered or threatened plants and wildlife on private lands, which "are of aesthetic, ecological, educational, historical, recreational, and scientific value to the Nation and its people" (ESA 2005: Sec. 2(a)(3)).[6] A statutory requirement, thus, initiates public-private collaboration to preserve biological diversity as a benefit to current and future generations.

Last, the government can acquire development rights through a conservation easement as a way to protect private forestland from future development. Conservation easements are "legally binding agreements transferring a negotiated set of property rights from one party to another, without removing the property from private ownership" (USFS 2008). While directly benefiting and serving public purposes, easements under the federal Forest Legacy Program depend on private landowners' choices to donate or sell development rights.[7] Conservation easements produce another hybrid public-private partnership for largely public benefit but with an added short-term gain for property owners from reduced taxes. A recent study of private rural lands in Massachusetts suggests that easements (whether through a public or private conservation organization) may prove to be a critical mechanism for maintaining forest benefits from private lands, especially where timber revenues are inadequate to meet escalating property taxes (D'Amato et al. 2010). Acknowledging both the financial and conservation benefits from easements, D'Amato et al. (2010: 36) note that conservation easements can protect forestland from conversion, preserve forest management, and "ensure continued flow of ecosystem services and public benefit."

7.3.3 Private Investment for Private Benefit

Both industrial and nonindustrial forest owners can make investment decisions about the production of timber and other forest products. As Best and Wayburn (2001) argue, private forestlands are some of the most productive lands in the country. About 58% of the nation's timber supply comes from private lands (Smith et al. 2009). Traditionally, industrial forest owners have possessed the requisite

[6]Habitat conservation plans in the Endangered Species Act fall under the responsibility of the US Fish and Wildlife Service (http://www.fws.gov/endangered/what-we-do/hcp-overview.html. Accessed 18 Sept 2011).

[7]The Forest Legacy Program (FLP) "complements private, Federal and State programs focusing on conservation in two ways. First, FLP directly supports property acquisition. Additionally, FLP supports efforts to acquire donated conservation easements. FLP funded acquisitions serve public purposes identified by participating states and agreed to by the landowner" (USFS 2008).

resource stocks and sawmills to process forest products and by and large have engaged in the activities described in box 3 of Table 7.1. Large-scale transformations since the 1990s, however, have led to the disintegration of the forest industry in the USA (Bliss and Kelly 2008; Smith et al. 2009). Significant shares of timberland are now in the hands of timber investment management organizations, real estate investment trusts, family forest owners, land trusts, and other institutional investors (Bliss et al. 2010).

At the state and local levels, key players are sawmill companies, such as Pike Lumber Company in Indiana which owns and operates sawmill facilities, manufactures and distributes lumber, and engages in market transactions with other private landowners. Private-private partnerships, such as the Family Forest Legacy Program offered by Pike Lumber Company,[8] currently manage over a 100 tracts of private forestland in Indiana, Michigan, and Ohio in order to ensure a "healthy vibrant forest free from development pressure." A study examining the effects of the forest industry divestiture suggests three potential paths of development: intensive timber production, continued parcelization and conversion, and conservation (Bliss et al. 2010: 53). Researchers further note that new forms of ownership, rising land values, smaller ownership parcels, and the costs of harvesting timber are likely to change the cost structure of private forest production in the future (Bliss and Kelly 2008; Bliss et al. 2010; Sampson 2004: 12; Sampson and DeCoster 2000).

Presently, 62% of family forest owners manage forestland parcels less than 10 acres (Butler 2008). Their decisions about how to use forestlands are based on the costs and benefits of the goods supplied by forests. Existing markets for private goods, such as timber, food, and medicinal products, offer incentives to landowners that inform their investment decisions. The benefits of such decisions accrue largely to private interests—corporate or noncorporate private forest owners. Timber harvests on smaller NIPFs are episodic and driven by market prices for timber, financial, or family need, as well as the health of the forest (Davis et al. 2010; Sampson 2004). Research indicates that harvesting decisions are often part of the multiple objectives landowners have for their land. Their management goals often blend financial gain with an interest in the noncommodity and amenity values of forests (Best and Wayburn 2001; Egan 1999; Knoot et al. 2010; Koontz 2001).

In this context, forest certification programs on small forest landholdings have gained popularity as a form of private-private partnership, for instance, between family forest owners and the American Tree Farm System.[9] Third-party certification of family forests and audits for sustainable forest standards seek to provide landowners with access to unique markets and a better return on their investments.

---

[8]http://www.pikelumber.com/Partner.php. Accessed 23 Oct 2010.

[9]American Tree Farm System Certification Program (http://www.treefarmsystem.org/atfscertification. Accessed 23 Oct 2010).

Developing markets for certified forest products, ecosystem services, and carbon may offer potential income gains, thus stimulating innovative private investments in private forestry.

7.3.4 Private Investment for Largely Public Benefit

Private landowners, nongovernment conservation organizations, and land trusts help preserve private forestlands for the benefit of everyone. These represent private investment decisions for largely public benefits (Table 7.1, box 4). Notwithstanding the upfront or immediate private gain embedded in some of these investment decisions, the ultimate purpose is social, for the addressees—present and future generations.

The role of land trusts for the conservation of forestlands is growing in importance. The efforts and investments of land trusts, such as the Nature Conservancy, the Trust for Public Land, and many state and local land trusts, such as Sycamore Land Trust in south-central Indiana, account for 37 million acres of family forestland currently under conservation easement (Aldrich and Wyerman 2005). The Nature Conservancy completed the largest conservation purchase of private land in 2006—700,000 acres across 10 southern states, Maine, and Wisconsin (Woodard 2006 cited in Bliss et al. 2010: 61). A landowner's decision to sell or donate development rights to a private conservation organization, such as a land trust, is an important investment decision likely to support a stream of ecosystem goods and services from private lands. While also providing a financial incentive and a short-term profit to landowners, a conservation easement has a broad public intent—to perpetuate benefits from forests—which outweighs the short-term tax alleviation.

Direct social and ecological benefits result from privately undertaken activities to maintain and conserve forest resources. Research suggests that private individuals increasingly invest in their land and trees (Davis et al. 2010). About seven billion dollars were spent by private households in 2006 on forest care services, such as professional assistance, pruning, fertilizing, pest management, and removal (Butterfield 2007 cited in Davis et al. 2010: 322). According to a national survey,[10] this represents 16% of the total investment of households in the care of their trees and gardens. These data, while incomplete and context specific, provide information about the types of activities landowners engage in (e.g., enhancing beauty, building roads, planting trees) and the kind of positive externalities expected from private spending. Such practices are reflective of the motivations and values of most family forest owners today, which focus on aesthetic enjoyment, recreation, privacy, and investment (Butler 2008; Davis et al. 2010; Koontz 2001; Kroeger and Casey 2007).

[10]Results from a 2006 survey conducted in conjunction with the National Gardening Association of America estimated the total investment of American households on landscaping and tree care services (Butterfield 2007 cited in Davis et al. 2010: 322).

Planting trees is associated with the real cost of seedlings, landowner's time, and effort. Keeping forests as forests represents a cost to private landowners, who may consider a different land use. A number of other social and institutional factors may undermine private decisions to invest in forests as a source of largely social and environmental benefits. Where such efforts exist, scholars note, coordination and cooperation among landowners may be needed to optimize landscape- and watershed-scale benefits from forestlands (Best 2004; Butler and Leatherberry 2004; Vokoun et al. 2010). Researchers and practitioners observe that important landscape-wide benefits from forests will be lost in the face of increasing parcelization, fragmentation, and lack of coordination among family forest owners (Best 2004; Bliss 2001; Butler and Leatherberry 2004; Vokoun et al. 2010). From a policy standpoint, the private provisioning of forest benefits might be encouraged through improved communication, cooperation, and coordination among forest owners and other private and public actors.

To summarize, this section reviewed four spheres of private-public interactions as a way to better understand the sufficiency and utilization of investment in private forests. It proposed a typology of decisions by private and public actors to invest in the provision of forest benefits. These investment decisions were organized based on the direct recipient or addressee of the investment arrangement. Such a categorization, however, fails to escape the fact that all forest ecosystems, as well as human systems, are multidimensional and dynamic. We can start to incorporate the multiple dimensions of forests as a socioecological system by looking at its individual components, namely, the nature of the resource system (forests) and its outputs (forest benefits), and the decisions and interactions of resource owners at multiple scales.

7.4 Forests as a Bundle of Property Rights: Implications for Forest Ownership and Governance

Forests are complex socioecological systems that provide a range of ecosystem goods and services and are subject to different property-rights regimes (Ostrom 2009). Some features of forests are common-pool resources (CPRs), such as hunting and habitats for wild plants and animals; some have public good characteristics, for instance, water purification, pollination, carbon sequestration, and aesthetics, and other features are private goods, such as timber, mushrooms, and land for homesites (de Groot et al. 2002; Fischer and Ruseva 2010; Fisher et al. 2009) (Table 7.2). The effectiveness of many of the private-public interactions described above is challenged by a lack of existing mechanisms to separate these features over time and space. In particular, the public goods and CPRs provided by private forests are hard to quantify, price, and exchange; they are consumed jointly and simultaneously, and their meaning is dependent on the decision context and social actors involved (Costanza and Farber 2002; Lubchenco 1998). Other important characteristics of

Table 7.2 Possible governance arrangements for forest goods and services

Forest ecosystem functions	Forest features (goods and services)	Type of good	Public-private spheres of interaction	Possible governance arrangements
Regulation	Clean air, water, and soil regulation, carbon storage	Public goods and services	State forestry assistance, regulation (*Table* 7.1, *boxes 1, 2*)	Government programs Polycentric and multiscale governance
Habitat	Hunting	Common-pool resource	Habitat enhancement (*Table* 7.1, *box 4*)	Community organizations Private-public partnerships at parcel or multiple scale
Production	Food, timber, raw materials, energy resources	Private good	Forest certification, working forests (*Table* 7.1, *boxes 1, 3*)	Market-based arrangements
Information	Spiritual enrichment, recreation and aesthetic enjoyment	Public good Toll/club good	Conservation easements (*Table* 7.1, *boxes 2, 4*)	Public-private partnerships Polycentric and multiscale governance

Sources: de Groot et al. (2002), Fisher et al. (2009): 647, Gatzweiler (2006): 299

forests include the frequency of use or consumption of forest products and services; varying spatial and temporal scales; joint investment decisions (e.g., conservation easements offered by both public and private conservation organizations); complexity due to feedbacks, time lags, and nested processes; and dependence as well as interactions among all these characteristics (Fisher et al. 2009). All in all, the multiple features of forests presuppose the existence of a bundle of property rights.

Property-rights systems are key components in the use and misuse of forest resources (Costanza and Farber 2002; Ostrom 2005). Political scientists Edella Schlager and Elinor Ostrom (1992: 256) write, “Different bundles of property rights, whether they are de facto or de jure, affect the incentives individuals face, the types of actions they take, and the outcomes they achieve.” Various aspects of forests, such as timber, wildlife habitat, and carbon sequestration, can be associated with different property-rights regimes and users and subject to different rights of access, management, allocation, and appropriation (Fischer and Ruseva 2010; Ruhl et al. 2007). Additionally, the difference between private landowners, who typically hold a complete set of rights, and all others, who do not have complete rights, and their respective discount rates shape public and private investment decisions and their interactions (Schlager and Ostrom 1992).

A public-private interaction aimed at a specific forest good, such as timber, can be fairly uncomplicated due to the existence of private markets for timber products.

An interaction between private actors focused on a set of forest features spread across space and time, for instance, carbon credits, soil formation, and clean water credits, may entail a detailed designation of different users/consumers. While soil formation is produced locally and used directly for agricultural cultivation, water regulation services produced by forests at the top of a hill or mountain are consumed as benefits downstream (Fisher et al. 2009: 648). Similar disconnects between the scale of production and scale of consumption of forest features (e.g., carbon sequestration) underscore the spatial and temporal heterogeneity of forest features. A designation of the type of forest feature, its providers, and users/consumers may be essential for enhancing private, as well as public, investment in private forests.

As a bundle of property rights, private forests may become subject to multiple regimes of ownership and governance. A typology of forest ecosystem functions includes regulation, habitat, production, and information functions (de Groot et al. 2002; Gatzweiler 2006). Each function is associated with a particular type of forest good or service. In turn, these features can be characterized according to their joint use/rival consumption and exclusion, i.e., as a public, private, or toll good or CPR (Fisher et al. 2009: 647; Ostrom and Ostrom 1999: 78). This private-public good aspect of ecosystem services guides potential public-private investment decisions and governance arrangements for their delivery (Table 7.2).

Forest governance provides a foundation for understanding the implications of managing forests as a bundle of property rights within existing public-private interactions in private forestry. Institutions, understood as rules, guide human behavior and influence the use and misuse of forest resources (Costanza and Farber 2002; Ostrom 2005). Markets, government programs, and property rights are key institutional mechanisms that enable or hinder outcomes on private forestlands. By providing information to resource owners and users, for instance, through price signals, educational programs, title, or property boundaries, these institutions shape the nature and scope of private-public interactions. Different institutional regimes are associated with different forest features (e.g., wildlife enhancement, timber production), policy levels (federal, state, local), actors (private, public), and scales of interactions (ownership parcel, neighborhood, watershed, ecoregion, or other) (Koontz et al. 2004; Rickenbach et al. 2011; Wondolleck and Yaffee 2000). In brief, these institutional arrangements affect the incentives and patterns of interactions among different actors, as well as the outcomes of their interactions (Ostrom 1998).

Governance is a useful concept when we seek to understand the institutions that support and constrain the conservation of private forests. Governance is characterized by the decisions, actions, and interactions of actors in the private forest sector (Tucker 2010: 690). These actors have authority to make decisions that shape activities and social and ecological outcomes on forestlands. Governance also entails "continuing interactions between [actors], caused by the need to exchange resources and negotiate shared purposes" (Rhodes 2007: 1246). In the context of private forestry, landowners, land trusts, professional foresters, timber buyers, developers, and state and federal agencies engage in conversations and consultations with each other. They share and process information and form beliefs and preferences, which affect their behavior and decisions about forest use and management (Fishbein and Ajzen 2010). Thus, *private forest governance* can be conceptualized as the decisions and interactions of private landowners and other public and private actors, caused by the need to exchange resources, and bound toward different activities and outcomes on private forestlands (Oakerson 1999: 18; Rhodes 2007: 1246; Tucker 2010). Such interactions, across levels and sectors of governance, are reliant on information, knowledge, and other resource exchanges.

Information and knowledge have direct and indirect effects on natural resource management decisions (Dietz et al. 2003; Ostrom 1998). As an investment, the decisions of public and private actors (described in Table 7.1) are contingent upon available and relevant information. Information and knowledge are key components of decision-making processes and are central to the choices private and public actors make with regard to private forests. Information can directly influence individual choices (e.g., what and how much land to keep as forest) and indirectly affect the sustainability of private forestlands. Learning occurs in the process of information and resource sharing among actors. Learning also affects the attitudes, preferences, and choices of individuals with regard to forest practices and land use (Ostrom 1998; Smith et al. 1995 cited in Tompkins and Eakin 2012).

Private forest governance may also vary based on the communication patterns and social ties of private forest owners. Interactions with others, as observed by sociologist Bernice Pescosolido (1992: 1101), form "an essential element in the dynamics of decision-making processes." Landowners make decisions about the care and management of their forests within particular social environments. An environment encompasses the connections a landowner has, both with the land and with others, such as family members, neighbors, and forestry professionals. Informal and subjectively meaningful social interactions form a network of social relationships (Crossley et al. 2009; Pescosolido 2007). Social networks, such as this, provide the "context for embedded, rather than individualized, decision-making" (Heath et al. 2009: 657). Conceptual and empirical work indicates that social contacts are positively associated with information and knowledge sharing, the adoption of best management practices, and willingness to coordinate forest practices toward ecological and economic benefits (Gass et al. 2009; Olsson et al. 2007; Prokopy et al. 2008; Warriner and Moul 1992). Information channeled through informal conversations and interpersonal contacts may facilitate learning and diffusion of knowledge and enhance the governance of private forests (Knoot and Rickenbach 2011; Nybakk et al. 2009; Rickenbach 2009).

In brief, social interactions form a basic mechanism through which landowner attitudes, behaviors, and practices are shaped (Borgatti et al. 2009; Freeman 2004; Pescosolido 2007). This mechanism has direct implications for private forest governance, where parcel-scale decisions are linked to social interactions at multiple scales. Figure 7.1 summarizes the social and ecological challenges and opportunities that cumulatively shape the governance of private forests. Items in each rectangle have the capacity to operate as risks or windows of opportunity for sustainable forest governance. Arrows represent feedback mechanisms among human-environment interactions at different scales. The importance of private ownership is recognized

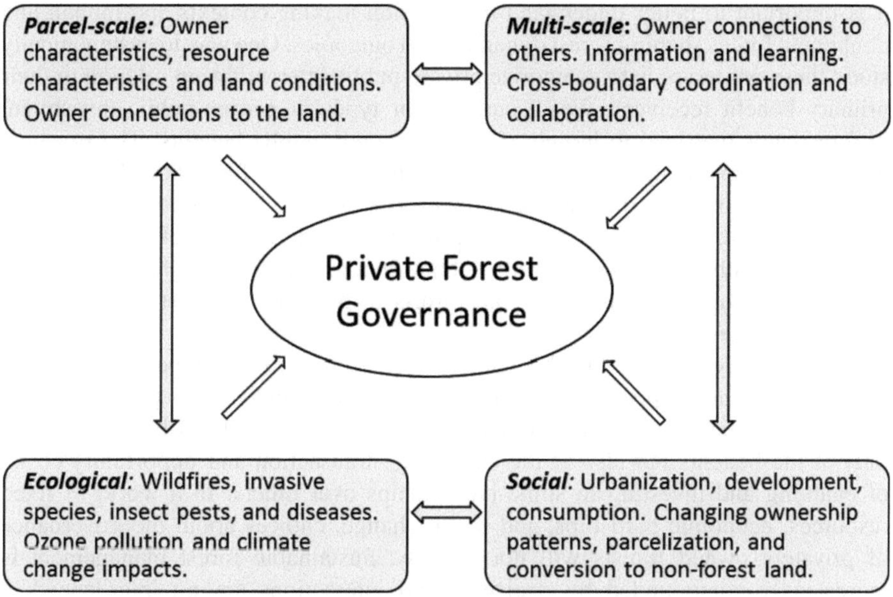

Fig. 7.1 Challenges and opportunities in the governance of private forests

by the separation of parcel scale as opposed to other multiple-scale processes within definable socioecological units (e.g., ownership parcels, watersheds, ecoregions) (Rickenbach et al. 2011). The diversity of actors involved in forest management decisions and practices presupposes multiple scales. Coordination and collaboration among landowners, neighbors, professional foresters, loggers, land trusts, and others are important dimensions of forest governance with implications for forest sustainability. Such partnerships that cross boundaries are able to address issues that cannot be solved or easily solved by a single actor or entity (Agranoff and McGuire 2003). Thus, effective interactions among landowners and other private and public actors have the potential to lead to multisector and multiscale partnerships that facilitate the long-term management of private forestlands.

7.5 Conclusion

This chapter reviewed existing scholarship on private forest management with an eye to identifying and synthesizing the processes, risks, and challenges to forest governance. It discussed the interactions among private and public actors involved in the management and conservation of privately owned forests in the USA. With more than half of all forestlands in the country under private ownership, it is important to better understand the decision-making contexts and human and nonhuman forces shaping forest dynamics and outcomes. One way to systematically study these processes is to categorize private-public interactions in relation to their primary benefit receivers. We discussed four types of private-public investment in forestlands that lead to largely private or largely public benefits. The practical difficulty of separating the private from the public value of forests can be aided by a categorization of the different features (goods and services) produced by forest ecosystems. We believe that integrating the proposed typology of private-public interactions with a refined and legally defined bundle of property rights over forests, as a socioecological system, may aid efforts to preserve and sustain the wealth produced by these valuable ecosystems.

The typology presented in this chapter may need to be developed further, both conceptually and empirically, in order to respond to and inform policy-relevant decisions at local, state, and national levels. Attention needs to be directed not only at the benefits but also at the costs (e.g., transaction and opportunity costs) of enabling and investing in some partnerships over others. In a world of fixed resources, economic hardships, and social change, choices about the governance of privately owned forests will not be easy. Sustainable forest management is increasingly compounded by decisions and interactions among interdependent individuals within diverse and dynamic socioeconomic settings. From such a broad socioecological perspective, this chapter offers some ideas for the sustainable management of natural resources under private ownership.

References

Agranoff, R., & McGuire, M. (2003). Inside the matrix: Integrating the paradigms of intergovernmental and network management. *International Journal of Public Administration, 26*(12), 1401–1422.

Agrawal, A., Chhatre, A., & Hardin, R. (2008). Changing governance of the world's forests. *Science, 320*(5882), 1460–1462.

Aldrich, R., & Wyerman, J. (2005). *2005 National land trust census report*. Washington, DC: Land Trust Alliance.

Alig, R., Latta, G., Adams, D., & McCarl, B. (2010). Mitigating greenhouse gases: The importance of land base interactions between forests, agriculture, and residential development in the face of changes in bioenergy and carbon prices. *Forest Policy and Economics, 12*, 67–75. doi:10.1016/j.forpol.2009.09.012.

Allred, S., & Goff, G. (2009). The power of peer learning programs in natural resources. *Rural New York Minute, 32*(August 2009) (Published by Human Dimensions Research Unit and Community & Rural Development Institute, Cornell University, Ithaca, NY).

Arano, K. G., Munn, I. A., Gunter, J. E., Bullard, S. H., & Doolittle, M. L. (2004). Comparison between regenerators and non-regenerators in Mississippi: A discriminant analysis. *Southern Journal of Applied Forestry, 28*(4), 189–195.

Barten, P. K., Damery, D., Catanzaro, P., Fish, J., Campbell, S., Fabos, A., et al. (2001). Massachusetts family forests: Birth of a landowner cooperative. *Journal of Forestry, 99*(3), 23–30.

Bascompte, J. (2009). Disentangling the web of life. *Science, 325*(5939), 416–419.

Beach, R., Pattanayak, S., Yang, J., Murray, B., & Abt, R. (2005). Econometric studies of non-industrial private forest management: A review and synthesis. *Forest Policy and Economics, 7*(3), 261–281.

Bengston, D. N., Asah, S. T., & Butler, B. J. (2011). The diverse values and motivations of family forest owners in the United States: An analysis of an open-ended question in the National Woodland Owner Survey. *Small-Scale Forestry, 10*, 339–355.

Best, C. (2004). Non-governmental organizations: More owners and smaller parcels pose major stewardship challenges. *Journal of Forestry, 102*(7), 10–11.

Best, C., & Wayburn, L. (2001). *America's private forests: Status and stewardship*. Washington, DC: Island Press.

Bliss, J. C. (2001). *Extension and the future of family forests: Making connections*. Published by International Union of Forestry Research Organizations, provided by The Regional Institute Ltd. www.regional.org.au/au/iufro/2001/bliss.htm. Accessed 15 Nov 2010.

Bliss, J. C., & Kelly, E. C. (2008). Comparative advantages of small-scale forestry among emerging forest tenures. *Small-Scale Forestry, 7*, 95–104. doi:10.1007/s11842-008-9043-5.

Bliss, J. C., Kelly, E. C., Abrams, J., Bailey, C., & Dyer, J. (2010). Disintegration of the U.S. industrial forest estate: Dynamics, trajectories, and questions. *Small-Scale Forestry, 9*, 53–66. doi:10.1007/s11842-009-9101-7.

Bodin, Ö., & Crona, B. (2009). The role of social networks in natural resource governance: What relational patterns make a difference? *Global Environmental Change, 19*(3), 366–374.

Bodin, Ö., Crona, B., & Ernstson, H. (2006). Social networks in natural resource management: What is there to learn from a structural perspective? *Ecology and Society, 11*(2), resp. 2 [online]. http://www.ecologyandsociety.org/vol11/iss2/resp2/

Borgatti, S. P., Mehra, A., Brass, D. J., & Labianca, G. (2009). Network analysis in the social sciences. *Science, 323*(5916), 892–895.

Bruyere, B., Teel, T., & Newman, P. (2009). Response to more kids in the woods: Reconnecting Americans with nature. *Journal of Forestry, 107*, 378–379.

Butler, B. J. (2008). *Family forest owners of the United States, 2006* (General Technical Report NRS-27). Newtown Square: U.S. Department of Agriculture, Forest Service, Northern Research Station.

Butler, B. J., & Leatherberry, E. C. (2004). America's family forest owners. *Journal of Forestry, 102*, 4–14.

Chazdon, R. L. (2008). Beyond deforestation: Restoring forests and ecosystem services on degraded lands. *Science, 320*(5882), 1458–1460.

Costanza, R., & Farber, S. (2002). Introduction to the special issue on the dynamics and value of ecosystem services: Integrating economic and ecological perspectives. *Ecological Economics, 41*(3), 367–373.

Crona, B., & Bodin, Ö. (2006). What you know is who you know? Communication patterns among resource users as a prerequisite for co-management. *Ecology and Society, 11*(2), art. 7 [online]. http://www.ecologyandsociety.org/vol11/iss2/art7/

Crossley, N., Prell, C., & Scott, J. (2009). Social network analysis: introduction to special edition. *Methodological Innovations Online, 4*, 1–7, http://www.pbs.plym.ac.uk/mi/pdf/17-04-09/1.%20Crossley%20et%20al%20intro%201-7.pdf

D'Amato, A. W., Catanzaro, P. F., Damery, D. T., Kittredge, D. B., & Ferrare, K. A. (2010). Are family forest owners facing a future in which forest management is not enough? *Journal of Forestry, 108*, 32–38.

Daniels, S. E., Kilgore, M. A., Jacobson, M. G., Greene, J. L., & Straka, T. J. (2010). Examining the compatibility between forestry incentive programs in the US and the practice of sustainable forest management. *Forests, 1*, 49–64.

Dauvergne, P., & Lister, J. (2011). *Timber*. Hoboken: Wiley.

Davis, M., Steiner, L. E., & Fly, J. M. (2010). Do you hear what i hear: Better understanding how forest management is conceptualized and practiced by private forest landowners. *Journal of Forestry, 108*(6), 321–328.

de Groot, R. S., Wilson, M. A., & Boumans, R. M. J. (2002). A typology for the classification, description and valuation of ecosystem functions, goods and services. *Ecological Economics, 41*(3), 393–408.

Dietz, T., Ostrom, E., & Stern, P. C. (2003). The struggle to govern the commons. *Science, 302*, 1907–1912.

Egan, A. F. (1999). Reducing forest road erosion: Do foresters and logging contracts matter? *Journal of Forestry, 97*(8), 36–39.

Ellefson, P. V., Kilgore, M. A., & Granskog, J. E. (2007). Government regulation of forestry practices on private forest land in the United States: An assessment of state government responsibilities and program performance. *Forest Policy and Economics, 9*(6), 620–632.

ESA (Endangered Species Act of 1973 As Amended through the 108th Congress). (2005). http://www.nmfs.noaa.gov/pr/pdfs/laws/esa.pdf

Ernstson, H., Sörlin, S., & Elmqvist, T. (2008). Social movements and ecosystem services—The role of social network structure in protecting and managing urban green areas in Stockholm. *Ecology and Society, 13*(2), art. 39 [online]. http://www.ecologyandsociety.org/vol13/iss2/art39/

Evans, T. P., Donnelly, S., & Sweeney, S. (2010). Threats to the forest transition in the Midwest United States. In H. Nagendra & J. Southworth (Eds.), *Reforesting landscapes: Linking pattern and process* (pp. 175–203). Dordrecht: Springer.

Fischer, B. C., & Ruseva, T. B. (2010). What is happening in and outside America's private woodlands? *Journal of Forestry, 108*(6), 304–306.

Fishbein, M., & Ajzen, I. (2010). *Predicting and changing human behavior: The reasoned action approach*. East Sussex: Psychology Press.

Fisher, B., Turner, R. K., & Morling, P. (2009). Defining and classifying ecosystem services for decision making. *Ecological Economics, 68*(3), 643–653.

Freeman, L. C. (2004). *The development of social network analysis: A study in the sociology of science*. Vancouver: Empirical Press.

Gass, R., Rickenbach, M., Schulte, L., & Zeuli, K. (2009). Cross-boundary coordination on forested landscapes: Investigating alternatives for implementation. *Environmental Management, 43*(1), 107–117.

Gatzweiler, F. W. (2006). Organizing a public ecosystem service economy for sustaining biodiversity. *Ecological Economics, 59*(3), 296–304.

Greene, J., Daniels, S., Jacobson, M., Kilgore, M., & Straka, T. (2010). *Financial incentive programs for non-industrial private forest owners*. USDA Forest Service, Southern Research Station. http://www.srs.fs.usda.gov/econ/data/forestincentives/. Accessed 26 Sept 2011.

Heath, S., Fuller, A., & Johnston, B. (2009). Chasing shadows: Defining network boundaries in qualitative social network analysis. *Qualitative Research, 9*(5), 645–661.

IDNR (Indiana Department of Natural Resources). (n.d.). IDNR Division of Forestry strategic plan 2008–2013. http://www.in.gov/dnr/forestry/files/fo-Forestry-Strategic-Plan-2008-2013.Final.pdf. Accessed 11 Oct 2010.

Janssen, M. A., Bodin, Ö., Anderies, J. M., Elmqvist, T., Ernstson, H., McAllister, R. R. J., et al. (2006). Toward a network perspective of the study of resilience in social-ecological systems. *Ecology and Society, 11*(1), art. 15 [online]. http://www.ecologyandsociety.org/vol11/iss1/art15/

Kauneckis, D., & York, A. (2009). An empirical evaluation of private landowner participation in voluntary forest conservation programs. *Environmental Management, 44*(3), 468–484.

Kendra, A., & Hull, R. B. (2005). Motivations and behaviors of new forest owners in Virginia. *Forest Science, 51*(2), 142–154.

Kilgore, M. A. (2004). Public forest policies and the family forest. *Journal of Forestry, 102*(7), 11–12.

Kilgore, M. A., Greene, J. L., Jacobson, M. G., Straka, T. J., & Daniels, S. E. (2007). The influence of financial incentive programs in promoting sustainable forestry on the nation's family forests. *Journal of Forestry, 105*(4), 184–191.

Kilgore, M. A., Snyder, S., Taff, S., & Schertz, J. (2008). Family forest stewardship: Do owners need a financial incentive? *Journal of Forestry, 106*(7), 357–362.

Kimbell, A. R., Schuhmann, A., & Brown, H. (2009). More kids in the woods: Reconnecting Americans with nature. *Journal of Forestry, 107*, 373–377.

Kittredge, D. B. (2005). The cooperation of private forest owners on scales larger than one individual property: International examples and potential application in the United States. *Forest Policy and Economics, 7*(4), 671–688.

Knoot, T., & Rickenbach, M. (2011). Best management practices and timber harvesting: The role of social networks in shaping landowner decisions. *Scandinavian Journal of Forest Research, 26*(2), 171–182.

Knoot, T., Schulte, L., & Rickenbach, M. (2010). Oak conservation and restoration on private forestlands: Negotiating a social-ecological landscape. *Environmental Management, 45*(1), 155–164.

Koontz, T. M. (2001). Money talks – But to whom? Financial versus nonmonetary motivations in land use decisions. *Society and Natural Resources, 14*, 51–65.

Koontz, T. M., Steelman, T. A., Carmin, J. A., Smith Korfmacher, K., Moseley, C., & Thomas, C. W. (2004). *Collaborative environmental management: What roles for government?* Washington, DC: RFF Press.

Kroeger, T., & Casey, F. (2007). An assessment of market-based approaches to providing ecosystem services on agricultural lands. *Ecological Economics, 64*(2), 321–332.

Kueper, A. M., & Sagor, E. S. (2011). *Peer learning and landowner networks: Lessons from five case studies*. St. Paul: University of Minnesota Extension. http://api.ning.com/files/iFd3RfosUw6JpOgZbJ3*BZPf7KAIPoJRKZLv5j*QvfNBZSXQPXrmS2acIWfSkkJMnGFyLInG6hqbSnuFJWtvhgX9DL0*Y7B3/AMKresearchoverviewfinalPDF.pdf. Accessed 15 Sept 2011.

Lubchenco, J. (1998). Earth's unruly tenant: Credible science and human impacts on global ecology. *Open Spaces, 2*(1) [online]. http://www.open-spaces.com/article-v2n1-lubchenco.php

Munroe, D. K., & York, A. M. (2003). Jobs, houses, and trees: Changing regional structure, local land-use patterns, and forest cover in southern Indiana. *Growth and Change, 34*(3), 299–320.

Nagendra, H. (2007). Drivers of reforestation in human-dominated forests. *Proceedings of the National Academy of Sciences of the United States of America, 104*(39), 15218–15223.

Nagubadi, V., McNamara, K. T., Hoover, W. L., & Mills, W. L., Jr. (1996). Program participation behavior of nonindustrial forest landowners: A probit analysis. *Journal of Agricultural and Applied Economics, 28*(2), 323–336.

NASF (National Association of State Foresters). (2010). *State foresters concerned that president's budget will fall short in meeting conservation goals on all lands*. Press release. Washington, DC: NASF. http://www.stateforesters.org/files/02112010-NASF-prez-budget.pdf. Accessed 17 Sept 2011.

Nowak, D. J., & Walton, J. T. (2005). Projected urban growth (2000–2050) and its estimated impact on the US forest resource. *Journal of Forestry, 103*(8), 383–389.

Nowak, D. J., Walton, J. T., Dwyer, J. F., Kaya, L. G., & Myeonng, S. (2005). The increasing influence of urban environments on US forest management. *Journal of Forestry, 103*(8), 377–382.

NRC (National Research Council), (1998). *Forested landscapes in perspective: Prospects and opportunities for sustainable management of America's nonfederal forests*. Washington, DC: National Academies Press.

Nybakk, E., Crespell, P., Hansen, E., & Lunnan, A. (2009). Antecedents to forest owner innovativeness: An investigation of the non-timber forest products and services sector. *Forest Ecology and Management, 257*(2), 608–618.

Oakerson, R. J. (1999). *Governing local public economies: Creating the civic metropolis*. Richmond: ICS Press.

Olsson, P., Folke, C., Galaz, V., Hahn, T., & Schultz, L. (2007). Enhancing the fit through adaptive co-management: Creating and maintaining bridging functions for matching scales in the Kristianstads Vattenrike Biosphere Reserve Sweden. *Ecology and Society, 12*(1), art. 28 [online]. http://www.ecologyandsociety.org/vol12/iss1/art28/

Ostrom, E. (1998). The comparative study of public economies. *The American Economist, 42*(1), 3–17.
Ostrom, E. (2005). *Understanding institutional diversity*. Princeton: Princeton University Press.
Ostrom, E. (2009). A general framework for analyzing sustainability of social-ecological systems. *Science, 325*(5939), 419–422.
Ostrom, V., & Ostrom, E. (1999). Public goods and public choices. In M. D. McGinnis (Ed.), *Polycentricity and local public economies: Readings from the workshop in political theory and policy analysis* (pp. 75–106). Ann Arbor: University of Michigan Press.
Persha, L., Agrawal, A., & Chhatre, A. (2011). Social and ecological synergy: Local rulemaking, forest livelihoods, and biodiversity conservation. *Science, 331*(6024), 1606–1608.
Pescosolido, B. A. (1992). Beyond rational choice: The social dynamics of how people seek help. *The American Journal of Sociology, 97*(4), 1096–1138.
Pescosolido, B. A. (2007). Sociology of social networks. In C. D. Bryant & D. L. Peck (Eds.), *21st century sociology: A reference handbook* (Vol. 1, pp. 208–217). Thousand Oaks: Sage.
Prokopy, L. S., Floress, K., Klotthor-Weinkauf, D., & Baumgart-Getz, A. (2008). Determinants of agricultural best management practice adoption: Evidence from the literature. *Journal of Soil and Water Conservation, 63*(5), 300–311.
Rhodes, R. A. W. (2007). Understanding governance: Ten years on. *Organization Studies, 28*(8), 1243–1264.
Rickenbach, M. (2009). Serving members and reaching others: The performance and social networks of a landowner cooperative. *Forest Policy and Economics, 11*, 593–599.
Rickenbach, M., & Jahnke, A. (2006). Wisconsin private sector foresters' involvement in nonindustrial private forestland cross-boundary forestry practices. *Northern Journal of Applied Forestry, 23*(2), 100–105.
Rickenbach, M., Schulte, L. A., Kittredge, D. B., Labich, W. G., & Shinneman, D. J. (2011). Cross-boundary cooperation: A mechanism for sustaining ecosystem services from private lands. *Journal of Soil and Water Conservation, 66*(4), 91A–96A.
Ross-Davis, A. L., Broussard, S. R., Jacobs, D. F., & Davis, A. S. (2005). Afforestation motivations of private landowners: An examination of hardwood tree plantings in Indiana. *Northern Journal of Applied Forestry, 22*(3), 149–153.
Rudel, T. (2010). Three paths to forest expansion: A comparative historical analysis. In H. Nagendra & J. Southworth (Eds.), *Reforesting landscapes: Linking pattern and process* (pp. 45–57). Dordrecht: Springer.
Rudel, T. K., Coomes, O. T., Moran, E., Achard, F., Angelsen, A., Xu, J., et al. (2005). Forest transitions: Towards a global understanding of land use change. *Global Environmental Change Part A, 15*(1), 23–31.
Ruhl, J. B., Kraft, S. E., & Lant, C. L. (2007). *The law and policy of ecosystem services*. Washington, DC: Island Press.
Ruhl, J. B., Kraft, S. E., & Lant, C. L. (2008). The tragedy of ecosystem services. *Bioscience, 58*(10), 969–975.
Sampson, N. (2004). Responses to "America's family forest owners": Implications for forest production. *Journal of Forestry, 102*(7), 12–13.
Sampson, N., & DeCoster, L. (2000). Forest fragmentation: Implications for sustainable private forests. *Journal of Forestry, 98*, 4–8.
Schaaf, K. A., & Broussard, S. R. (2006). Private forest policy tools: A national survey exploring the American public's perceptions and support. *Forest Policy and Economics, 9*, 316–334.
Schlager, E., & Ostrom, E. (1992). Property rights regimes and natural resources: A conceptual analysis. *Land Economics, 68*(3), 249–262.
Schroeder, J., Becker, D. R., & Kilgore, M. A. (2011). *State forestry programs, budget strategies, and the recession* (Staff Paper Series, No. 213). St. Paul: Department of Forest Resources, University of Minnesota.

Smith, W. B., Miles, P. D., Perry, C. H., & Pugh, S. A. (Coords.). (2009). *Forest resources of the United States, 2007* (General Technical Report WO-78). Washington, DC: U.S. Department of Agriculture, Forest Service, Washington Office.

Stein, S. M., McRoberts, R. E., Alig, R. J., Nelson, M. D., Theobald, D. M., Eley, M., et al. (2005). *Forests on the edge: Housing development on America's private forests* (General Technical Report PNW-GTR-636). Portland: U.S. Department of Agriculture, Forest Service, Pacific Northwest Research Station.

Stein, S. M., McRoberts, R. E., Mahal, L. G., Carr, M. A., Alig, R. J., Comas, S. J., et al. (2009). *Private forests, public benefits: Increased housing density and other pressures on private forest contributions* (General Technical Report PNW-GTR-795). Portland: U.S. Department of Agriculture, Forest Service, Pacific Northwest Research Station.

Sugden, A., Smith, J., & Pennisi, E. (2008). The future of forests. *Science, 320*(5882), 1435–1460.

Tompkins, E., & Eakin, H. (2012). Managing private and public adaptation to climate change. *Global Environmental Change Part A: Human & Policy Dimensions, 22*(1), 3–11.

Tucker, C. M. (2010). Learning on governance in forest ecosystems: Lessons from recent research. *International Journal of the Commons, 4*(2), 687–706.

U.S. Census Bureau. (2011). *GCT-P1. Urban/rural and metropolitan/nonmetropolitan population: 2000. Data set: Census 2000 summary file 1 (SF 1).* http://factfinder.census.gov/servlet/GCTTable?_bm=y&-geo_id=01000US&-_box_head_nbr=GCT-P1&-ds_name=DEC_2000_SF1_U&-format=US-1. Accessed 3 Oct 2011.

USFS (USDA Forest Service). (2008). *Forest legacy program.* http://www.fs.fed.us/spf/coop/programs/loa/aboutflp.shtml. Accessed 19 Sept 2011.

USFS (USDA Forest Service). (2011). *Northeastern area, forest health protection—Emerald ash borer*. http://na.fs.fed.us/fhp/eab/. Accessed 2 Oct 2011.

Vokoun, M., Amacher, G. S., Sullivan, J., & Wear, D. (2010). Examining incentives for adjacent non-industrial private forest landowners to cooperate. *Forest Policy and Economics, 12*(2), 104–110.

Warriner, G. K., & Moul, T. M. (1992). Kinship and personal communication network influences on the adoption of agriculture conservation technology. *Journal of Rural Studies, 8*(3), 279–291.

Weimer, D., & Vining, A. (2010). *Policy analysis: Concepts and practice* (5th ed.). New York: Pearson Longman.

Wondolleck, J. M., & Yaffee, S. L. (2000). *Making collaboration work: Lessons from innovation in natural resource management*. Washington, DC: Island Press.

York, A. M., Janssen, M. A., & Ostrom, E. (2005). Incentives affecting land use decisions of nonindustrial private forest landowners. In P. Dauvergne (Ed.), *Handbook of global environmental politics* (pp. 233–248). Cheltenham: Edward Elgar.

[illegible]

Chapter 8
The Monitoring of Land-Cover Change and Management across Gradient Landscapes in Africa

Cerian Gibbes, Lin Cassidy, Joel Hartter, and Jane Southworth

Abstract Understanding the interactive effects of land management decisions and socioecological functioning is central to the study of human-environment interactions. Strategies such as designating or physically bounding parks are commonly used to conserve biodiversity and mitigate direct human impact on the environment. Remote sensing is an attractive source of data for monitoring such parks, as it provides a continuous source of consistent data across broad spatial extents. The current challenge to the field is its application in gradient landscapes where shifts from one land-cover class to another are subtle, as is the case in many savanna regions across Africa. This chapter explores implications of landscape monitoring and management strategies employed in eastern and southern Africa. We examine the suitability of various remote sensing approaches for quantifying land-use and land-cover change and how such studies can be used to monitor and inform the management of conservation areas in the broader African landscape.

C. Gibbes (✉)
Department of Geography and Environmental Studies, University of Colorado, Colorado Springs, Colorado Springs, CO 80918, USA
e-mail: cgibbes@uccs.edu

L. Cassidy
Okavango Research Institute, University of Botswana, Maun, Botswana
e-mail: lcassidy@orc.ub.bw

J. Hartter
Department of Geography, University of New Hampshire, Durham, NH 03824, USA
e-mail: joel.hartter@unh.edu

J. Southworth
Department of Geography, University of Florida, Gainesville, FL 32611, USA
e-mail: jsouthwo@geog.ufl.edu

E.S. Brondízio and E.F. Moran (eds.), *Human-Environment Interactions: Current and Future Directions*, Human-Environment Interactions 1,
DOI 10.1007/978-94-007-4780-7_8, © Springer Science+Business Media Dordrecht 2013

8.1 Introduction

8.1.1 Human-Environment and Land-Use/Land-Cover Changes

Human-environment interactions and their study, especially under changing conditions, pose challenges to research that require a *multidisciplinary approach*, with contributions from several science disciplines. Documents such as *Grand Challenges in the Environmental Sciences* (NRC 2001) and the *Global Land Project Science Plan* (GLP 2005) emphasize the development of land change science (LCS), that is, strategies that incorporate the range of the sciences from the biophysical to the social. Social, environmental, and geographical information–remote sensing sciences are combined in an interdisciplinary effort to examine and model human-environment interactions and their implications for global environmental change and sustainability (Turner et al. 2007).

The need to understand the spatial and temporal dynamism inherent in human-environment systems drives the analyses of land-use and land-cover change (LULCC) (Foley et al. 2005; McMichael et al. 1999; Turner et al. 1995). Much of human-environment research focuses on the ways in which human uses of landscapes, that is, land use, interact with the earth's biophysical conditions, as reflected by land cover. Changes in land use and land cover have important implications for multiple issues from the function and state of ecosystems to socioeconomic processes and to socioecological systems (DeFries et al. 2004; Geist and Lambin 2001; Steffen et al. 2004). LULCC is often examined to gain a baseline understanding of how human decisions and actions are affecting the environment, which in turn can influence ecosystem functioning, biodiversity, and climate (Southworth 2004). The monitoring of LULCC is concerned with addressing the following central questions: What kinds of changes are taking place? Where are such changes occurring? What are the rates of these changes? What are the factors influencing each of the above? The assumption underlying the use of remote sensing for LULCC analyses is that key environmental variables that relate to human-environment interactions can be remotely detected (Southworth and Gibbes 2010). Satellite imagery offers repeat data at varying spatial and temporal scales, thereby enabling multiscalar assessments of change in the quantity and distribution of land-use and land-cover patterns. These LULCC patterns can then be linked to human-environment interactions and processes.

The empirical nature of many LULCC studies has been one of the underlying strengths of human-environment research. Case studies tend to consist of quantitative assessments of change. Many also tend to be spatially explicit, allowing for the location of the system in space and time to be taken into consideration and for the patterns and processes of that system to be linked to conditions related to a given geographic position. This means that change over time can also be linked to spatial variability in processes across multiple scales (Lausch and Herzog 2002;

Turner et al. 2001). Digital remote sensing offers an attractive source of land-use and land-cover data as it provides a representation of the earth's surface, in a consistent, spatially continuous, and repeatable manner. As such, remote sensing is heavily relied upon in LULCC analyses, and the current availability of data, which spans a wide range of spatial and temporal scales, makes this data source even more attractive for multi-scale analyses.

8.1.2 Dynamic Data: Moving Beyond Reductionist Classifications

Current attention to the multi-scale processes interacting with LULCC, for example, NASA's Land-Cover and Land-Use Change (LCLUC) Program, the Global Land Project, and the International Geosphere-Biosphere Programme (IGBP) (GLP 2005; Gutman et al. 2004; Lambin and Geist 2006; Steffen et al. 2004), has emphasized that understanding and modeling dynamic systems demands information about these systems that is itself dynamic. However, variability in human-environment interactions is not adequately captured by traditional remote sensing approaches to LULCC. In particular, the overreliance on traditional land-cover classifications limits remote sensing-based LULCC research.

The incorporation of land-cover classifications in LULCC analyses has enabled a basic understanding of discrete, or categorical, land-cover changes. However, this approach is limited by the subjectivity associated with classifications, the pure pixel assumption, insufficient inclusion of spatial data, and an inability to evaluate within-class variability (Southworth et al. 2004). As simplified representations of landscapes, classifications are subjective and reduce the inherent variability within landscapes (Di Gregorio and Jansen 1998). The subjectivity and local/regional context of classifications has challenged the development of a global land-cover taxonomy and resulted in meta-analyses of land-cover change that rely on multiple, independently defined land-cover classifications (Geist and Lambin 2001; Rudel 2008). Furthermore, as is demonstrated in the case studies to follow, the use of traditional classifications is not necessarily appropriate for all human-environment analyses.

The limitations of classifications are frequently addressed through the incorporation of vegetation indices and enhanced methodologies. The use of vegetation or spectral indices (e.g., the Normalized Difference Vegetation Index or NDVI) emerged in response to the inability of land-cover classification analyses to explain within-class changes and as a means to objectively link the data regarding emitted and reflective energy to the use of energy by vegetation. Based on an established understanding of the relationship between spectral reflectance and biomass, vegetation indices are commonly used to examine the interannual and intra-annual changes in quantities and distributions of green biomass (Carlson and Ripley 1997; Jiang et al. 2006; Wang et al. 2009). Although NDVI is one of the most popular vegetation indices, there is concern regarding its suitability for all landscapes, particularly those with high biomass where this index is shown to saturate (Huete et al. 2002; Sellers 1985), or very low biomass such as savannas, where mineral surfaces mask out the vegetation reflectance (Ringrose et al. 1989), as well as those where the land-cover changes of interest have similar spectral reflectance patterns.

In addition to the use of indices, the incorporation of more advanced remote sensing methodologies is used to address the limitations of traditional classifications for studies of human-environment interactions (Blaschke and Hay 2001; Castilla 2003; Hay et al. 2001, 2003, 2005), a shift that has resulted from technological developments within remote sensing. These technological developments include increases in commercially available high-resolution image products, for example, Quickbird and IKONOS (Hay et al. 2005; Wulder 1998, 2004), and hyperspectral imagery like Hyperion and AVIRIS (Filippi and Jensen 2006). Coupled with these developments is the push to move away from pixel-only-based quantifications of landscapes and toward quantification of actual geographical objects (Hay et al. 2005). In other words, in order to capture the complexity of a given system, we need to consider the true entities or units we wish to quantify and analyze (e.g., tree cover, landscape patch) (Wang et al. 2004).

8.1.3 Gradient Landscapes in Africa

Gradient landscapes are characteristic of much of Africa, in part due to extensive areas of natural vegetation. Though gradient landscapes are found in developing regions such as South America, Africa, unlike any of the other continents, is distributed across approximately 70° of latitude. As a result, the continent exhibits substantial variation in climate and, consequently, in land cover, with gradients of cover density decreasing from the closed-canopy tropical forests at the equator, through woodland savannas, to grassland savannas, and deserts. While such gradients can be assessed at the continental scale, other gradients at finer scales, such as grazing effects and desiccation, are overlooked (see, e.g., Lambin and Ehrlich 1997). At more regional levels, other factors, such as topography, fire, and land use, interact with climate to create dynamic and complex landscapes from which it is difficult to tease short-term variability from long-term trends or predict social and ecological consequences of those interactions (Walther et al. 2002; Zeng and Neelin 2000).

Some of the main land-cover changes underway globally include tropical deforestation, cropland expansion and contraction, dryland degradation, and urbanization (Geist and Lambin 2001: 3). The negative socioecological impacts of deforestation have likely prejudiced LULCC research to focus on deforestation. However, more recent awareness of the complexity of LULCC within forested landscapes has shifted research foci to include analyses of patterns and processes associated with reforestation. The regeneration and conservation of forests is an increasingly important component of global LULCC patterns (Nagendra and Southworth 2009).

In this chapter, we focus on reforestation, forest conservation, and land degradation in forested landscapes, and semiarid savanna or drylands, with particular emphasis placed on these processes as they occur within African landscapes.

The regeneration and conservation of forests is an increasingly important component of global LULCC patterns. Forest restoration and expansion have significantly reduced the net loss of the forest area in sub-Saharan Africa, though only 16.4% (74,585 ha) of forest area on the continent is designated primarily for conservation. Generally, forested areas in the tropics are converted to other land uses, particularly agricultural land uses (Brown and Lugo 1994) and in some African contexts to urban growth. Protection of forests (or other landscapes) within this context is often undertaken in the form of protected areas. In some regions, these protected area landscapes are more successful than others; across our three case studies, the role of protection of landscapes and their associated biodiversity can be evaluated using remotely sensed analyses. As the number of protected areas continues to increase worldwide, evaluation of their effectiveness and monitoring becomes increasingly important. In this chapter, we evaluate different types of landscapes and mosaics of protected-unprotected lands and land-cover types.

LULCC in drylands is estimated to affect 250 million people, an estimate that is likely to increase considerably with climate change (Reynolds et al. 2007). With the adoption of programs such as the Convention to Combat Desertification (CCD) and frameworks like the Drylands Development Paradigm (DDP) to tackle economic, social, and environmental issues associated with arid and semiarid areas, studies of human-environment interactions in dry and semiarid regions of the world are increasingly relevant.

Widespread LULCC is expected in the semiarid savannas of Africa as a result of changes in climate patterns and increasing human populations (Archer et al. 2001). The discussion of LULCC in African savannas centers on shifts in vegetation composition, specifically a decline in tree cover. Changes in tree cover within savanna systems impacts the productivity of the system, modifies the availability of resources for both wildlife and humans, and holds the potential to greatly impact earth-atmosphere interactions (Beerling and Osborne 2006; Ludwig et al. 2008; Ringrose et al. 2002). Unlike forests, savannas have a discontinuous tree canopy that is fundamental to savanna functioning as it increases structural complexity and influences biomass allocation and microhabitats necessary for the success of other species (Belsky et al. 1989; Scholes and Archer 1997). Declines in tree cover, therefore, reduce availability of herbaceous resources, which in turn can result in degradation of habitat for wildlife and loss of direct (trees themselves) and indirect (presence of wildlife used to attract tourism to the region) resources for human populations.

As already mentioned, throughout Africa, parks have been the principal strategy used to manage and conserve landscapes, both forested (see Sect. 8.2) and savanna (see Sects. 8.3 and 8.4). More recently, considerable attention has been paid to intertwining "win-win" conservation and development schemes into the management portfolios of African countries. One such scheme widely used in African savanna landscapes is community-based natural resource management (CBNRM).

The incorporation of CBNRM in landscape management within Africa reflects the shift in management approaches to strategies that combine natural resource management with development by incorporating local communities in the resource management process (Child 2009; Nelson and Agrawal 2008). CBNRM is currently being combined with park establishments to manage the savannas and address shifts from tree-dominated savanna habitat to a largely shrub-dominated landscape (Ringrose et al. 2002).

8.1.4 What This Chapter Addresses

The remainder of this chapter will address three case studies. The first one examines the monitoring of land-cover change and management strategies used in a forested landscape in Uganda. The second case study describes the impacts of a fence designed to separate wildlife from livestock, which cuts across a moisture and vegetation gradient in the southern Okavango Delta region in Botswana. The third one explores the utility of advanced remote sensing methodologies to quantify vegetation components and monitor the effect of management strategies in Caprivi, Namibia. Together, these three studies represent a variety of land cover and management types across this region and are useful examples of the types of human-environment questions in which we can use remote sensing in Africa. The landscape studies range from forested parks to savanna regions in complex human-environmental situations and management schemes. Correspondingly, the remote sensing techniques used to address these issues also vary from traditional classification techniques and vegetation indices commonly found in the literature to more advanced techniques of thermal analyses and object-oriented classifications. Hence, as we progress through the case studies, a suite of situations, methods, and approaches is presented within the African context.

8.2 Case Study: Uganda

8.2.1 Land-Use and Land-Cover Change in and Around Parks in Uganda

Uganda's forests are under continuous threat of conversion due to population growth, in-migration, and intensive agriculture. Over 80% of the land in Uganda is used for small-scale farming (Mukiibi 2001), and continued population growth leads to added pressure on the land. Forests in Uganda are widespread and complex, representing one of the most dominant forms of land cover in Uganda, covering approximately 4.9 million ha of the country's total land surface (24.1 million ha, NEMA 2001). Nearly 3 million ha of forests and woodlands remain unprotected and are used by nearby communities for various purposes (NEMA 2001). They are important to the livelihoods of neighboring communities in several ways. They provide thatch, handcraft materials, medicines, food, fuel, building poles, timber, and other resources (Hartter 2010). Nationwide, unsustainable domestic tree harvesting for firewood and non-timber forest products continues (Kayanja and Byarugaba 2001). Forest conversion continues to transform landscapes into other land uses such as farming, woodlots, and pasture. Although estimates vary, Uganda continues to lose between 0.8 (NEMA 2001) and 3% of closed-canopy forest annually. In particular, the Albertine Rift region, one of the world's hotspots for biodiversity (Cordeiro et al. 2007; Plumptre 2002; Plumptre et al. 2003, 2007) is also one of the most threatened in the world due to dense, intensive smallholder agriculture, high levels of land and resource pressures, and high rates of habitat loss (Brooks et al. 2001; Plumptre 2002).

Patterns of land-cover change in most tropical developing countries relate significantly to anthropogenic impacts occurring across multiple spatial and temporal scales. Therefore, landscapes around parks, particularly in tropical forested areas, are important because they represent reservoirs of land, resources, and economic opportunities for people and simultaneously are often viewed as buffers for parks by managers. Whereas human populations around savanna parks usually are limited by low and sporadic rainfall, which strongly constrains agriculture, forest parks in the tropics often occupy and/or are surrounded by land that is highly suitable for agriculture (Goldman et al. 2008). As such, landscapes around parks in eastern and southern Africa have become mosaics of interacting natural and human-influenced patches.

In forest park landscapes, forest is an important land cover since the local human population has been restricted from park resources (timber, building poles, and firewood). Without access to the park, we would expect that population growth and resource and land needs would lead to pressure on land and resources outside the park. In particular, unprotected forests are vulnerable to exploitation and agricultural encroachment. This increased pressure leads to local action where continuous landscapes are disaggregated: conversion of land to agriculture, pasture, and woodlots. On their own, land-use decisions and management manifested on 10-, 20-, or 40-ha parcels result in a relatively small footprint that often is invisible at broader scales. However, together the choices and behaviors of many households break up continuous landscapes into smaller parcels. Gradient landscapes that once transitioned from forest to grassland to savanna have been converted to small farm plots. The accumulation of many small land-use decisions can have dramatic and long-lasting ecological impacts, such as fragmented wildlife habitat, decreased seed source, and diminished ability to provide resources.

While vegetation in parks often remains intact, population growth and resource pressures lead to further conversion. Continued use and conversion of these forests leads to further landscape fragmentation. In turn, this puts additional pressures onto the park as its rich resources become an increasing temptation in a resource-poor environment. The impacts that fragmentation has on both wildlife and vegetation within a fragment and, perhaps more importantly, the impact of loss of intact habitat and wildlife on the people relying on the remaining fragments are important to understanding and protecting against future decline. As fragments decline and become more degraded, encroachment into the park may increase to collect resources, while at the same time the number and severity of human-wildlife incidences may also increase as well.

In this section of this chapter, we examine land-cover change and forest-productivity change in a forest park landscape. The landscape in and around Kibale National Park (Kibale) in the Albertine Rift of Uganda is emblematic not only of population and resource pressures—issues threatening other forest parks—but is a particularly good example of a natural gradient landscape that has been interrupted and replaced with discrete transitions and clear boundaries. Due to anthropogenic pressures, the natural gradient has been replaced by a landscape of discrete land-cover classes and abrupt edges between land-cover types. The transition from moist mid-altitude forest to savanna has been halted due to large-scale land conversion and intensive agriculture, thus transforming the configuration and extent of land-cover classes. As a result, natural variation and the recognizable gradient of natural vegetation cover types over space have been removed.

8.2.2 Kibale National Park

Kibale National Park is a mid-altitude tropical moist forest covering approximately 795 km^2 in western Uganda. This transitional forest (between lowland rainforest and montane forest) has an average elevation of 1,110–1,590 m and is a remnant of a previously larger mid-altitude forest region (Struhsaker 1997). The climate is warm throughout the year, with an average range of 15–23°C and a mean annual rainfall of over 1,600 mm (1970–2009, T. Struhsaker and C. Chapman, unpublished data). The bimodal rainfall pattern produces two major rainy seasons (long: late February to early May; short: late August to late November) (see Fig. 8.1).

Prior to the elevation to a national park in 1993, Kibale was designated a forest reserve since 1932, with the main objective to provide a sustained production of hardwood timber (Struhsaker 1997). Logging was conducted at regular intervals between 1950 and 1975 at variable intensities (low to high), but mostly between 1967 and 1969, resulting in large forest gaps and variable forest disturbance in parts of Kibale (Hartter et al. 2009). Harvested areas were allowed to regenerate naturally. During Uganda's political upheaval in the 1970s and 1980s, the plantations were not managed (Chapman et al. 2002), and as early as 1971, illegal destruction and encroachment occurred in the corridor, which led to forest conversion to farms. Management plans changed when Kibale became a national park in 1993. In 1964, the Kibale Forest Corridor Game Reserve, along the western flank of the reserve that connects to the northern edge of Queen Elizabeth National Park, was gazetted (formally published as government land) and administered as a separate entity until it was merged with the reserve to create Kibale National Park

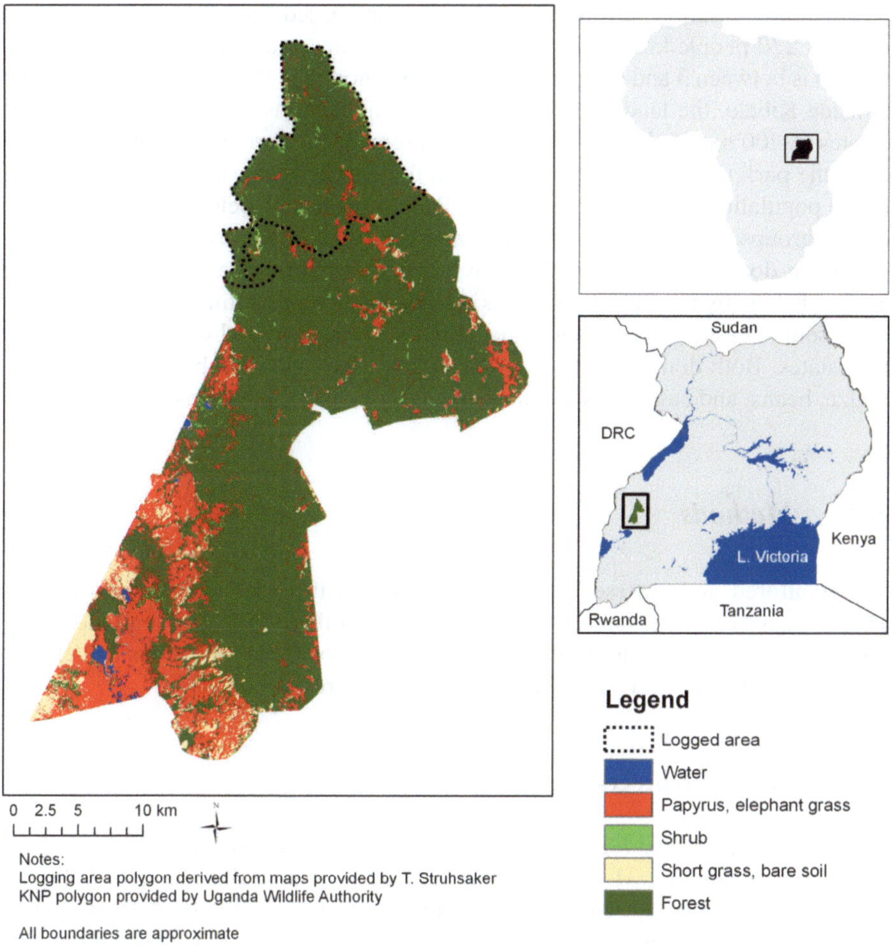

Fig. 8.1 Study area map of Kibale National Park and Corridor region and its location in Uganda

(Struhsaker 1997). The estimates of people who had settled in the corridor during Uganda's 20 years of political unrest vary considerably (8,800–170,000), but all were evicted by 1993 (Chapman et al. 2010). This corridor is a mixture of forest and grassland, as it transitions from mid-altitude forest to savanna woodland and then to savanna of Queen Elizabeth National Park. As a result of this settlement and abandonment, some of the forest was converted to farmland, which was in turn abandoned following eviction. Much of the abandoned farmland reverted to shrub and bush (dominated by pioneer species such as *Acanthus pubescens*) but also grassland. Since 1995, Uganda Wildlife Authority and the Forests Absorbing Carbon Emissions (FACE) Foundation have established a program to reforest some of the former game corridor (Omeja et al. 2011).

Human population surrounding Kibale has increased sevenfold since 1920 and exceeds 270 people/km^2 at the western edge (Hartter and Ryan 2010). Population growth is between 3 and 4% per year, which is among the highest rates in the world. Outside Kibale, the landscape is a mosaic of small farms (most <5 ha), large tea estates (>200 ha), and interspersed forest fragments and wetlands that are isolated from the park (Hartter and Ryan 2010). In the Kibale landscape, the vast majority of the population is subsistence farmers. People in the area belong primarily to two ethnic groups: the Batoro (west side) and the Bakiga (southwest and east side). The once dominant Batoro population has been surpassed in many communities around Kibale by the Bakiga. The Bakiga have been immigrating to the Kibale area from southwestern Uganda since the 1950s seeking land and employment on the tea estates. Both dominant ethnic groups plant a mixture of subsistence (bananas, maize, beans, and cassava as the main staple foods) and cash crops.

8.2.3 Methods

We monitored and assessed forest change within the park and the surrounding landscape over the previous three decades, using both discrete and continuous data analyses of satellite imagery. We augmented land-cover classifications and change trajectories of land use with NDVI as a proxy for forest productivity.

8.2.3.1 Image Preprocessing

Three Landsat TM and ETM+ scenes were obtained for this study: 26 May 1984, 17 January 1995, and 31 January 2003. All were acquired at the end of the dry season when fallow agricultural lands can be easily distinguished from forests, except for 1984, which was acquired at the end of the rainy season since this was the only available haze- and cloud-free image. Images were geometrically registered to 1:50,000 scale survey topographic maps of the region, followed by radiometric calibration and atmospheric correction. All images were georegistered to within a root mean square error of <0.5 (below 15 m accuracy).

8.2.3.2 Land-Cover Classification

Land-cover maps were derived for each date by independent supervised classification for three Landsat TM and ETM+ scenes: 1984, 1995, and 2003 (Hartter et al. 2009; Hartter and Southworth 2009; Southworth et al. 2010). The 1984 image provides baseline data prior to official park establishment, the 1995 image captures conditions close to park establishment, and the 2003 image corresponded to training data. The most useful classification for this study identified five land-cover types: (1) forest, (2) crops/bare land (including pastures, cultivated crops, and kitchen gardens), (3) papyrus (*Cyperus papyrus L.*) and elephant grass (*Pennisetum purpureum*), (4) tea, and (5) water. The accuracy assessment for the 2003 image, using field samples collected in June and July of 2005, indicated an overall accuracy of 89.1%, with a Kappa of 0.867 (Hartter and Southworth 2009). The classifications were done independently for each image and were not based on the same spectral signatures due to a lack of exact anniversary dates, although the pattern of land-cover signatures was consistent across all image dates.

The 1984, 1995, and 2003 land-cover composites were used to create change trajectories, that is, sequences of successive changes in land-cover types (Petit et al. 2001). This technique is used to determine the change between two or more time periods of a particular region or for a particular land cover and provides quantitative information on spatial and temporal distribution of categories of land-cover change and landscape fragmentation (Brockington 2006; DeFries et al. 2005; Ferraro and Kiss 2002; Parks and Harcourt 2002; Schelhas and Greenberg 1996; Walther et al. 2002). We wanted to understand the dominant land-cover trajectories, so we limited the number of possible trajectories to only those main landscape conversions. Land-cover classes examined here are areas forested on all three image dates (stable forest), areas of wetland and grasses on all three image dates (wetland or grassland), and areas of agriculture on all three image dates (agriculture), which represent the three stable land-cover classes in the region. Next, we have two conversion classes of interest: reforestation (when agriculture, wetland, or grassland on date one or date two is forested by date three) and deforestation (any area of forest on date one or date two that is non-forest—agriculture, wetland, or grassland—on date three). Water was excluded from this analysis as the crater lakes predominant in this region do not vary in area.

8.2.3.3 NDVI Change

Given the absence, presence, and extent of forest, we can now use the NDVI analysis to examine productivity. To understand within-class variability and change in forest productivity, additional cloud-free, dry season Landsat TM and ETM+ scenes supplemented those used for the land-cover classification: 4 August 1986, 20 August 1989, 9 January 2001, and 11 September 2008. The same preprocessing protocol was followed. We calculated the NDVI value for each image and adjusted NDVI values for the 2001 and 2003 ETM+ images to TM values (Stevens et al. 2003).

8.2.3.4 Results

Land-cover classifications from the three time periods show extreme contrast between the park and the surrounding area. Forest remains the dominant land-cover type within Kibale (Fig. 8.2, Table 8.1). Over time, park boundaries have been maintained, with no evidence of large-scale encroachment. Most of the park is stable forest (79%), some has been regenerating since 1984 (11%), and only

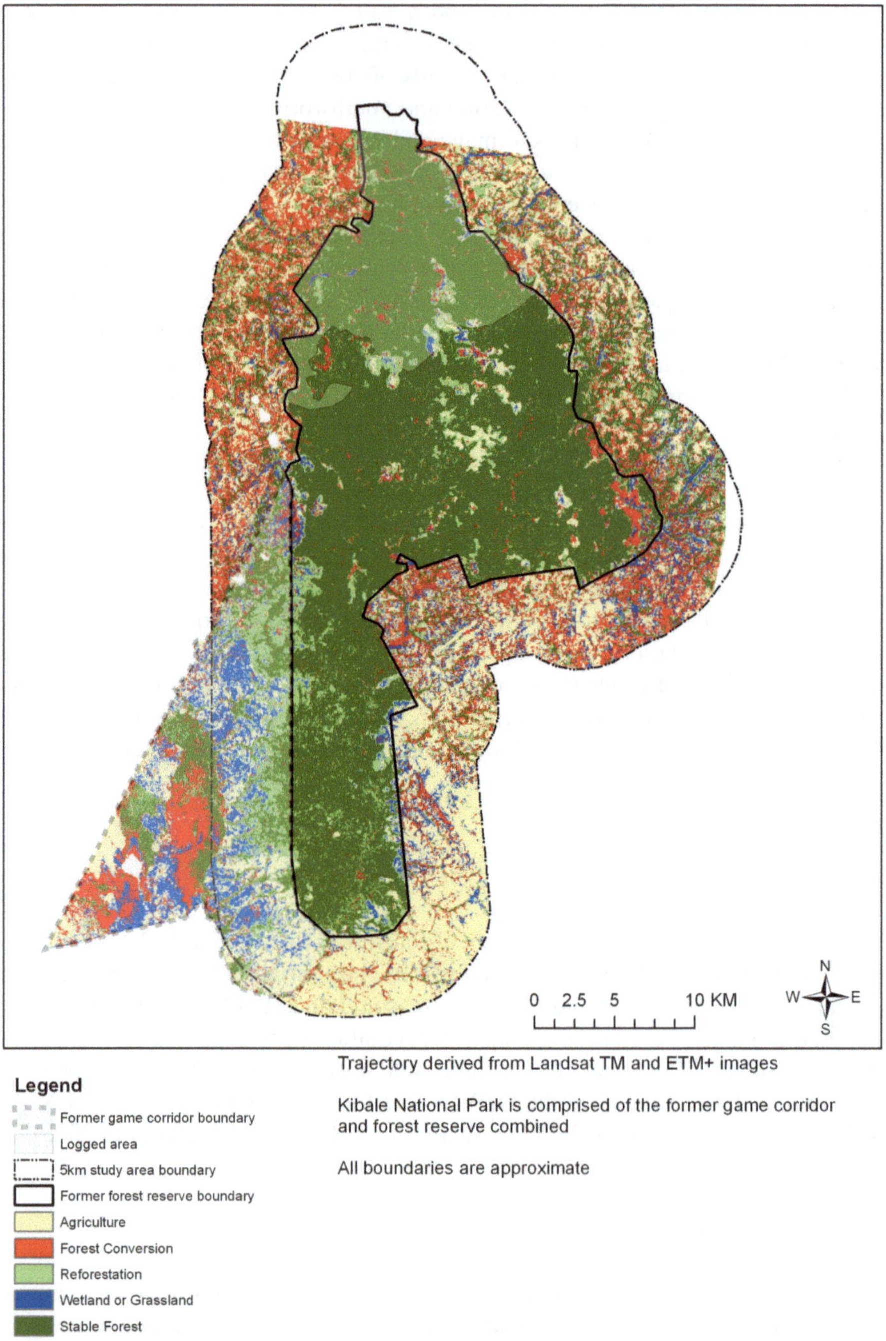

Fig. 8.2 Land-cover classifications resultant change trajectory

Table 8.1 Forest-cover change over time across the different landscape components

		Former forest reserve		Former game corridor		Surrounding landscape	
Land cover	Year	%	ha	%	ha	%	ha
Forest	1984	86.6	48,825	36.0	8,525	31.9	23,563
	1995	86.9	48,528	39.2	9,293	34.3	23,696
	2003	90.6	51,096	37.2	8,819	29.2	21,324

Table 8.2 NDVI values and proportion of forest for each image date

Unlogged area in former forest reserve				Logged area in former forest reserve			
	Forest	NDVI			Forest	NDVI	
Year	(% area)	Mean	Std Dev	Year	(% area)	Mean	Std Dev
1984	86.1	0.570	0.071	1984	88.5	0.669	0.050
1986	85.8	0.581	0.053	1986	89.6	0.638	0.050
1989	85.8	0.604	0.051	1989	89.6	0.650	0.046
1995	85.8	0.533	0.046	1995	89.6	0.600	0.049
2001*	90.3	0.336	0.067	2001*	91.0	0.386	0.066
2003*	90.3	0.448	0.049	2003*	91.0	0.488	0.050
2008	90.3	0.609	0.049	2008	91.0	0.644	0.046
Surrounding landscape				Former game corridor			
	Forest	NDVI			Forest	NDVI	
Year	(% area)	Mean	Std Dev	Year	(% area)	Mean	Std Dev
1984	31.9	0.645	0.063	1984	36.0	0.630	0.068
1986	34.3	0.603	0.071	1986	39.2	0.584	0.070
1989	34.3	0.600	0.085	1989	39.2	0.575	0.091
1995	34.3	0.575	0.045	1995	39.2	0.548	0.045
2001*	29.2	0.413	0.069	2001*	37.2	0.407	0.078
2003*	29.2	0.503	0.051	2003*	37.2	0.376	0.050
2008	29.2	0.629	0.078	2008	37.2	0.637	0.068

Proportions for 1985 and1989 were estimated using the 1995 land-cover classification; proportions for 2001 and 2008 were estimated using the 2003 land-cover classification.
*Adjusted NDVI values (Stevens et al. 2003)

about 4% of park area is deforested (Fig. 8.2, Table 8.2). Unlike in the park, the corridor maintains much more of this transitional landscape to savanna. In the corridor, only 20% of the forest has remained since 1984, while the corridor also sustained 16.4% deforestation, with almost 11% of this being recent (since a year before being gazetted). Reforestation has increased over the period of record, with overall reforestation in the corridor at 16.6%. Grassland is the dominant land cover. While the land-cover classification classified some areas in the park and corridor as agriculture, there is no agriculture within the park, and these misclassified areas actually represent areas of short grasses, shrubs, and regenerating stands that remain from previously harvested pine plantations.

Outside the park, however, the domesticated landscape is in a state of flux in order to support local livelihoods. The surrounding landscape is a fine-grained mosaic of all the land-cover classes, with crops/bare and forested land covering nearly equal areas, papyrus/elephant grass and tea plantations around the northern boundaries of the park, and a network of riparian or bottomland forests and papyrus and other wetlands interspersed throughout. Conversion of forest continues at a much higher rate, while there is some effort toward reforestation. Although there is an overall loss of 2% total forest, only about 16% of the forest outside Kibale has remained stable since 1984. This area has experienced higher reforestation rates as the park but has also had nearly 25% deforestation.

In examining the natural forest landscape without the anthropogenic signal, we have removed the agriculture land-use classes (tea and crops). In doing so, NDVI composites were created for forest within the park (including the former game corridor) and the surrounding landscape (Fig. 8.3). Within the park, the highest NDVI values are found in the northern section of the park, which were logged between 1950 and 1975 and are now regenerating forest. The rest of the park, which has not been commercially logged, shows NDVI values below the mean. The logged area within the park has higher mean NDVI values compared to the entire park and the surrounding landscape for all image years (Fig. 8.3). There is also an increasing gradient of NDVI in the park from the lower elevations in the east and south to the higher northern forests, which may also be related to the logged-unlogged areas. Over time, there is a general trend of decreasing NDVI (Table 8.2, Fig. 8.3). So while the park has maintained and increased forest cover, amount or area, quality or productivity of that forest cover has actually declined over time.

Within the corridor, NDVI values varied about the landscape average within a similar range to the fragments and logged area, with the striking exception of 2003, when the area logged (Fig. 8.3) had a lower NDVI value than any other forest category in that year and the largest difference from the intact forest of any category in any year. Outside the park, forest fragments have a higher-than-average NDVI value in all years except 1989, and in 2001 and 2003, the highest mean NDVI values overall. In addition, most individual fragments had above-mean NDVI values for each image year.

Establishing parks is the primary mechanism used to protect tropical forest biodiversity (Oates 1999; Terborgh 1999; Terborgh et al. 2002), particularly in regions with high human densities. Park landscapes protect and maintain threatened or endangered, often endemic, flora and fauna (Hartter et al. 2009). But if they are going to do so to the detriment of the surrounding landscape, perhaps a different approach to conservation is necessary. Since most parks are ecosystem remnants of a limited size, it is important to consider each park as a component of a larger landscape (DeFries et al. 2005; Parks and Harcourt 2002), and the fates of biodiversity in parks and the surrounding landscapes are inextricably linked (Chazdon 2008; Chazdon et al. 2009; Schelhas and Greenberg 1996). By considering whole landscapes, we can more broadly assess and monitor the long-term and multiscalar changes in and around parks. Whether or not the park has caused the drastic landscape transformation is certainly debatable, but it may have played a role.

The presence and use of forest fragments outside parks illustrate the interconnected nature between a park and the surrounding domesticated landscape. The remaining fragments outside the park may serve as a buffer to the park,

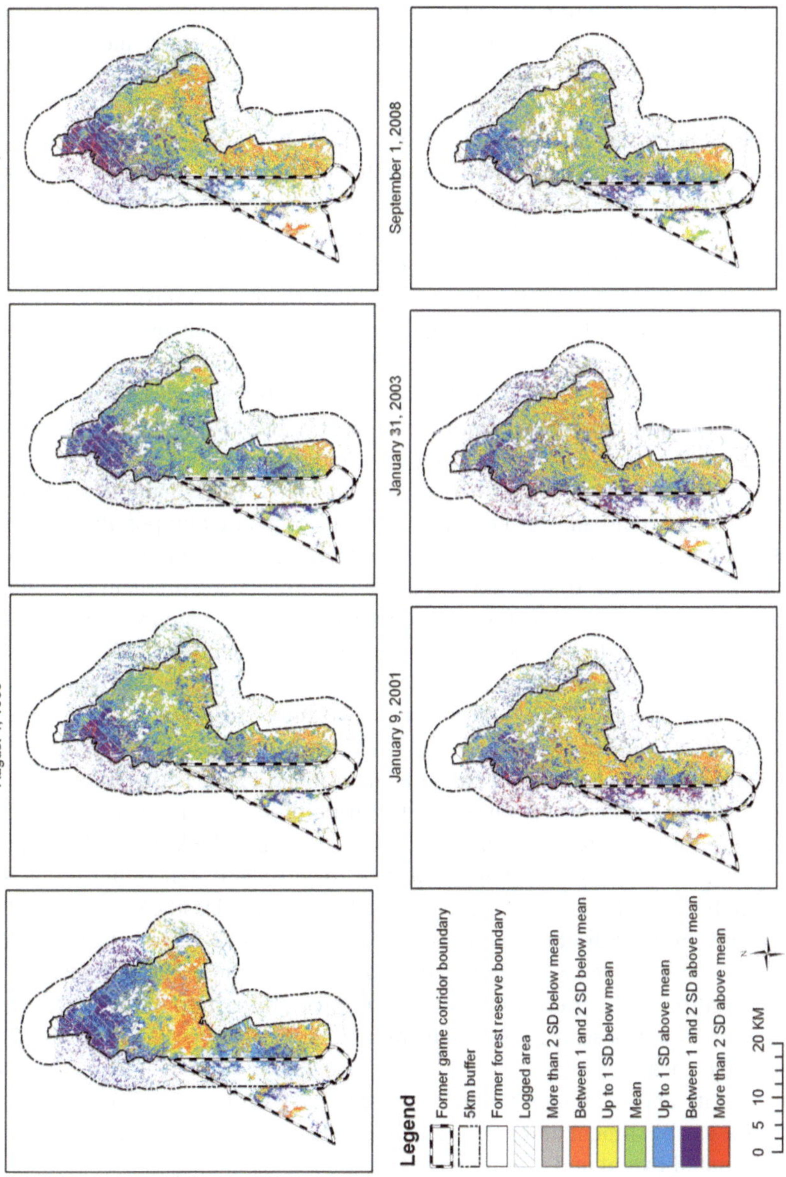

Fig. 8.3 Standardized NDVI images showing variation over space and time for these vegetated forest regions

by providing resources to people that would otherwise be extracted from the park. Knowledge of the impacts that fragmentation has on both the wildlife and vegetation within a fragment and, perhaps more important, the impact of loss of intact habitat and wildlife on the people relying on the remaining fragments is essential to understanding and protecting against future decline. As fragments decline and become more degraded, encroachment into the park may increase to collect resources, while at the same time the number and severity of human-wildlife incidences may also increase as well.

The Kibale landscape is a good example of park landscapes that have been partitioned and where extensive population growth and settlement has led to large-scale land-cover transformation. In particular, forest and grassland conversion has removed much of the natural gradient from one land-cover class to another. What has been produced is a collection of parcels of discrete land-cover types. In landscapes such as this, the wildlife corridor may be an important mechanism to restore this natural gradient that has been abruptly halted. The corridor was initially gazetted to allow for the free movement of animals, particularly elephants (*Loxidonta africana*), between forest areas to the north and Queen Elizabeth National Park to the south (Baranga 1991). This corridor is important not only because it facilitates recovery and maintenance of tree and animal species of conservation concern but also because it represents a purposeful effort to maintain and restore the natural gradient. In this setting, where land and resource pressures accompanied by population growth are in fierce competition with conservation objectives, little natural gradient will exist unless it is created. In the case of Kibale, the man-made gradient (corridor) is important because it may allow for connectivity over space, which will become ever more important under future global environmental change. Restoration efforts are underway to restore degraded forests, which may further improve the corridor to act as a transition between mid-altitude forest and savanna ecosystems.

Land-cover classifications provide a particularly effective tool for an analysis of landscape dynamics in Kibale. Much of the natural variability and gradient between forest and savanna has been removed due to agricultural conversion and intensification, which in turn has created a landscape of discrete, interlocking patches. Forests do not transition to scrub woodland and then to grassland, but rather there is an abrupt edge between forest and crops. Thus, a land-cover classification can be used to assess and monitor direct loss of specific cover types (e.g., forest or wetlands) over time. As this breakup continues, these spatially explicit classifications allow further evaluation of the transition out of natural variability into these discrete units to examine land-cover arrangement and fragmentation, factors with critical effects on park landscapes.

The viability and sustainability of forested ecosystems depends not only on successful park establishment and management but also on the ability to monitor their change in spatial extent and productivity. Continuous data analysis can supplement discrete analyses to examine more subtle, within-class variability (Southworth et al. 2004). In the heterogeneous landscapes around many parks, where land parcels are relatively small and land use is highly diverse, continuous data analyses can provide more revealing spatial analyses focused on biophysical indicators such as is the case above for forest health and in the following case studies (Sects. 8.3 and 8.4) for actual land-cover analysis; the next two cases represent regions where traditional land-cover classification approaches have only limited utility. This Ugandan case study thus sets the stage as an exemplar of both continuous and discrete approaches used in unison to better represent the landscape dynamics. However, as we continue to move south across the African landscape and our focus on savanna increases, so too must our reliance on continuous remote sensing approaches and newer analysis tools within the field of remote sensing.

8.3 Case Study: Southern Okavango, Botswana

8.3.1 Introduction

The challenges to studying and managing human impacts on savannas lie partially in that savannas are gradient landscapes and the changes in the ecological functioning tend to be subtle and steady along a given geographical axis rather than abrupt. Savannas are essentially one land-cover type, but because they cover such large spatial extents, this categorization is not useful to management activities. Further, though the variations may not have distinct boundaries, they are very real gradations (Scanlon et al. 2007), such as canopy closure (closed woodland, open woodland) and plant structure (woodland, scrub, grassland), which lead to strong differences in the availability of different ecosystems services at the opposite ends of the gradient. Within the same category, for example, grazing might be a dominant land use at one end of the gradient, while timber harvesting is most important at the other. Additionally, savannas tend to be patchy, with combinations of woody vegetation and grass distributed heterogeneously across the landscape, even within a given gradation range (Scanlon et al. 2007).

In addition, it is important to be able to capture and interpret the interactions between the spatial gradients of vegetation cover and between the temporal gradients in influential factors, such as successively drier rainfall years for an extended period (Ringrose et al. 2005). Scale is a key consideration as well. While the savanna level itself is too large for monolithic management decisions, the smallness of the patch level, which in Botswana's savannas is on the order of magnitude of meters squared, is challenging to focus on over large areas using readily available, but coarse-resolution, remote sensing data (see, e.g., the methodological concerns of Saura 2002). This challenge of scale means that analytical tools themselves should focus on the savanna landscape as a dynamic and continuous surface rather than in terms of discrete classes (Southworth et al. 2004). The impact on landscape-level environmental functioning of fencing off different forms of land use has had a high profile in the ongoing conservation-development debate (Berkes 2004; Brandon et al. 1998; Brechin et al. 2002; Brockington 2002; Chapin 2004; Chomitz 1994; Redford and Sanderson 2000; Spierenburg and Wels 2006; Wilshusen et al. 2002). Fences remain a contentious issue, both socially due to loss of access to resources and ecologically because of the impact of curtailed animal movements on herbivory and soils (James 2003; Pollard et al. 2003). In Botswana, fence line studies have largely highlighted degradation due to livestock and the impact of fences on wildlife populations (Albertson 1998; Mbaiwa and Mbaiwa 2006; Perkins and Thomas 1993; Williamson and Williamson 1981), reinforcing the idea that wilderness areas need to be "protected" from agricultural expansion. In 1982–1983, the Southern Buffalo Fence was erected toward the distal end of the Okavango Delta to separate Cape buffalo (*Syncerus caffer*), known carriers of foot-and-mouth disease, from the cattle that formed an important part of Ngamiland District's economy and cultural heritage (Campbell 1976; Government of Botswana 2003; Ngamiland District Council 2002; Tlou 1972, 1976). With the implementation of the 1991 Ngamiland Land Use Plan, this fence (consisting of two parallel fences 50 m apart) has also become a boundary separating different land-use zones: wildlife management areas and communal grazing areas. The vegetation differences between the two sides of the Southern Buffalo Fence along the Matsibe sandveld tongue are so strong that they are clearly visible on vegetation maps, even at a scale encompassing the whole of Africa. There is now an abrupt boundary of land use cutting across the land-cover gradient. Understanding the nature of the change, and its likely causes and consequences, is critical if the Botswana government is to be able to meet its obligations of sound, science-based management of the Okavango Delta Ramsar site as a signatory to the convention. In this section of this chapter, we present findings related to changes in land cover over time on the two sides of the fence.

In terms of human-environment interactions, the fence itself does not impact the landscape. Instead, the effect is more indirect, through its impact on animal movements, which then affect the environment (Albertson 1998). Scale plays an important part in assessing the overall impact of the fence as well, as any changes in ecosystem functioning and services, such as biodiversity, may involve a change that brings about reduction on one side of the fence but an overall increase for the landscape as a whole (cf. Crist et al. 2003). Further, policies like the decision to fence wildlife off from livestock are themselves an indirect interaction of humans with the environment and tend to operate at broad, regional scales, while the effects on the landscape are often more localized due to environmental variation. We hope to inform policy makers by exploring the causes and consequences of the fence's presence. Here, we focus on an initial assessment and ask, What is the general nature and degree of change in vegetation associated with the Southern Buffalo Fence, after controlling for underlying factors such as soils and the history of the fence alignment?

8.3.2 Study Area

Botswana is a sparsely populated but politically stable country lying between 18° and 27° south and between 20° and 29° east. Much of its surface area comprises semiarid savanna, and as such, the wetlands of Botswana's Okavango Delta, and the wildlife they support, are among the country's most important economic and ecological features (Murray-Hudson and Crisman 2003). As a permanent source

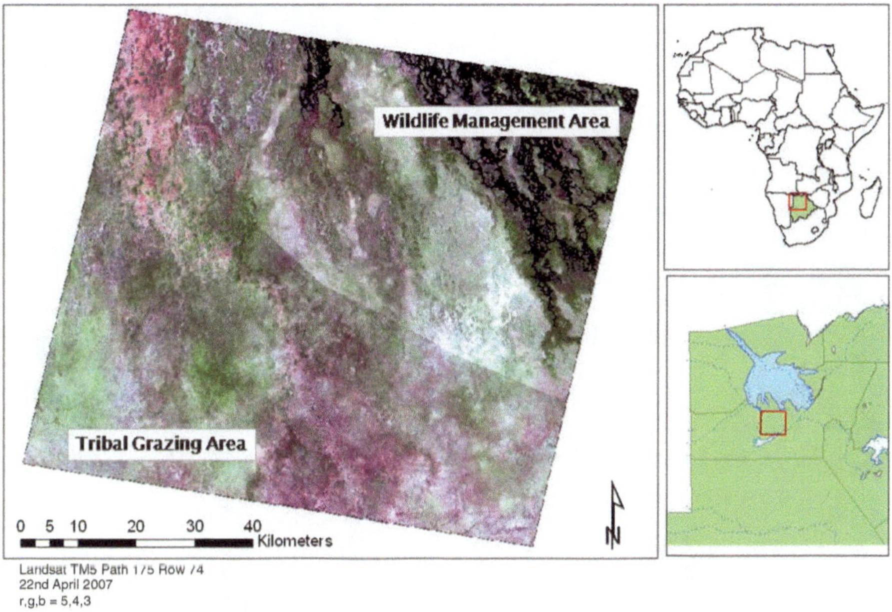

Fig. 8.4 Map showing the location of the Southern Buffalo Fence, visible as a boundary between areas of high and low reflectance in Landsat imagery from 2007. The fence separates two forms of land use—the economically important wildlife management areas where safari tourism is conducted and the socially important tribal grazing areas where rural people raise cattle, goats, and sheep

of freshwater, the system attracts and supports an abundance of plant and animal life. However, the presence of vector-borne diseases that affect both humans and livestock has until recently kept human population levels low (Tlou 1985). Economic growth associated with the burgeoning wildlife-based tourism industry conflicts with the traditional livestock economy, primarily through the need to separate wildlife, as disease vectors, from cattle (Campbell 1976; Ngamiland District Council 2002; Tlou 1976).

The Southern Buffalo Fence is located in the southwestern distal end of the Okavango Delta wetlands. It bisects a sandveld tongue that runs between the seasonal Matsibe and Kweeni river systems (Fig. 8.4). Due to its remoteness, and the historical presence of tsetse flies, the area is largely uninhabited. A time series of satellite imagery of the Okavango Delta reveals a marked reduction in the standing productivity of sandveld vegetation to the north of the Southern Buffalo Fence relative to the south side, where the fence crosses the sandveld tongue (Fig. 8.5). What is important is that the area that has exhibited most vegetation change—an apparent loss in tree cover and diversity and a reduction in standing plant biomass—is *inside* the protected wildlife management area, while the open-access tribal grazing area appears little changed.

In the tribal grazing area, two small villages sit within 30 km of the fence. However, most livestock are kept out at cattleposts, most of which are equipped

Fig. 8.5 False-color composite Landsat imagery showing pre- and post-fence vegetation conditions along the sandveld tongue. Vegetated areas appear *dark* to *bright red*, water appears *black*, and areas of senescent or no vegetation appear *green-blue* to *white*. (**a**) shows the location of the study area relative to the entire Okavango Delta. (**b**) shows the study area in 1979, prior to the construction of the Southern Buffalo Fence, while (**c**) and (**d**) respectively show 8 and 25 years post-construction

with boreholes to access groundwater. Land is communally owned, and with the centralization of government since independence, local headmen have lost much of their power, and resource use tends to be under an open-access regime (Cassidy 2000; Ostrom et al. 1999). Within the part of the wildlife management area addressed here, three luxury safari lodges have been developed. Access to the area is restricted under the conditions of a concession lease accorded to tour operators. Although concessionaires are obliged to produce management plans, this is more in terms of their infrastructure and activities, and decisions about management and off-take of wildlife tend to be dictated by the central government. Importantly, management decisions at the broad scale, such as the decision to build disease-control fences, are all made by the central government, which has implications for the management choices that those residing in the area can make.

8.3.3 Methods

Two analyses were conducted. The first was a series of paired-point land-cover assessments along the fence. The second was a time-series analysis of continuous

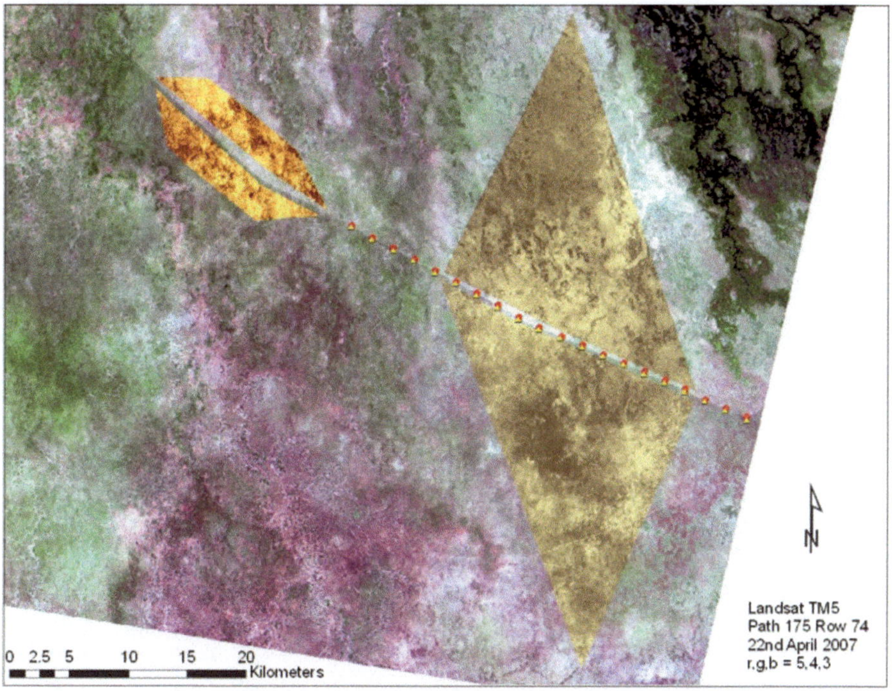

Fig. 8.6 Locations of paired points and polygons for analyses. The paired points (*red* and *yellow dots*) follow along the fence, with points 250 m apart across the fence and pairs 2 km apart along the fence. The *polygons* are shaped to avoid the influence of riparian zones and to cover comparable soil units

surface characteristics derived from satellite imagery data within paired polygons over large areas on each side of the fence (Fig. 8.6). First, however, in order to control for underlying factors, the potential roles of soils and land-use factors in influencing vegetation differences were explored. The Southern Buffalo Fence's alignment does not follow the boundary of any soil units and in fact bisects two large soil units. These are a predominantly arenic zone to the east, comprising the sandveld tongue proper, and a more calcic and sodic area to the west (Fig. 8.7). This means that any differences between the two sides of the fence are unlikely to be due to underlying soil conditions. The history of the fence alignment is interesting. The fence was placed along an old hunting concession boundary (established in 1979). However, the concession boundary itself was not arbitrary—it followed the old track between Maun, the district capital, and Habu village. The factor determining the alignment of the track lies in its main purpose—the moving of cattle from these remote bush areas to the district center for sale. The track has been in existence for over 70 years, and its alignment is probably related to the southern limit of the tsetse fly, a disease vector for livestock, as shown in Stigand's 1923 map. Stigand's map also shows the southern limit of mopane (*Colophospermum mopane*) to be further north than the alignment of the road and therefore the current alignment of the fence.

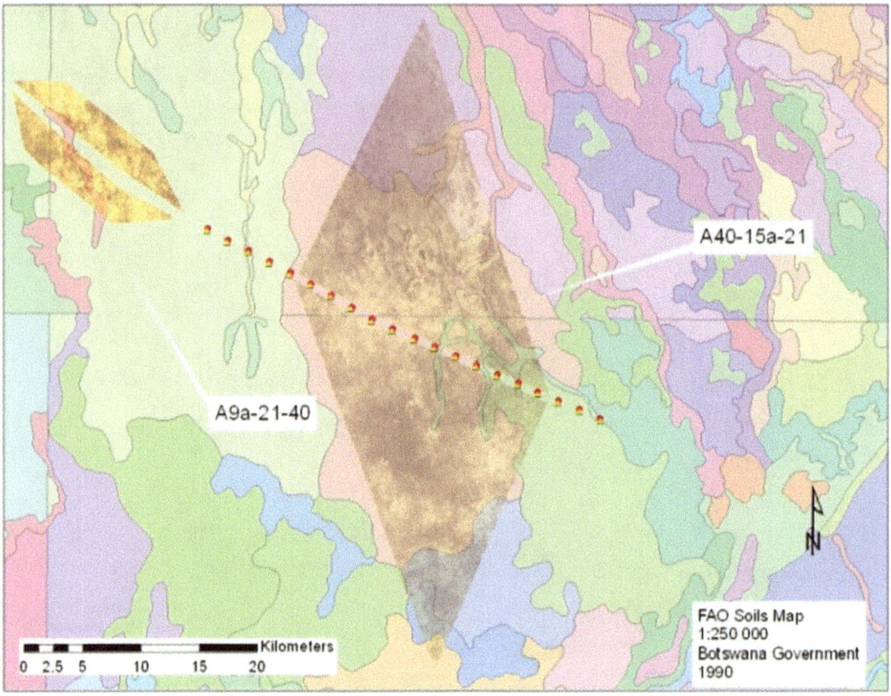

Fig. 8.7 Overlay of sampling *points* and *polygons* on soils map, showing large *brown polygons* in the Matsibe area (eastern pair) covering predominantly arenic soils on the sandveld tongue and *narrow orange polygons* in the Habu area (western area) on calcic to sodic soils

Starting at the southernmost bend in the fence, 20 paired-point assessments of land-cover types were done, each pair 2 km apart along the fence. The points were located 100 m perpendicularly away from each side of the fence, making the points of each pair approximately 250 m apart. At each point, land-cover characteristics for an area of 100-m radius surrounding the point were noted. The number of pairs obtained was limited due to challenges of accessibility and drinking water availability. Two-sided, paired sample t-tests were run for key land-cover characteristics.

Landsat imagery was obtained, georeferenced, and radiometrically and atmospherically calibrated (Green et al. 2002). The imagery covered both rainy season (April/May) and dry season (Aug/Sep) scenes for 6 years: MSS2 in 1979, prior to the fence being constructed; MSS5 in 1984, shortly after the fence was constructed; TM5 in 1990 and 1997; ETM+ in 2000; and TM5 in 2007, representing successive post-construction time steps. By conducting separate analyses on the wet and dry imagery, phenological variation due to seasonal changes could be accounted for. Tasseled cap transformations were run to extract soil brightness (and moistness from 1990 onward), and the TM and ETM+ thermal bands were calibrated to degrees Celsius (also only from 1990 onward). Although all vegetation indices, including soil-adjusted vegetation indices, are known to be problematic in semiarid savannas

Table 8.3 Differences in general land-cover characteristics at 20 paired points 250 m apart on either side of the buffalo fence

Characteristic for each 100 m × 100 m sample point area	Mean value for cattle side of fence	Mean value of wildlife side of fence	Result of paired sample t-test/X^2 test
Percentage exposed soil	42.2%	63.0%	$t = -3.908$ $p = 0.001$
Percentage covered by tree canopy	9.8%	6.2%	Not sig.
Average canopy height	4.0 m	2.8 m	$t = 2.5968$ $p = 0.018$
Average height of emergent trees	6.2 m	5.5 m	Not sig.
Average rank for canopy dbh (from 1 = 2–10 cm to 6 = 75–100 cm)	1.3	1.0	Not sig.
Average rank for emergent dbh (from 1 = 2–10 cm to 6 = 75–100 cm)	2.3	2.1	Not sig.
No. of points with thornveld as main vegetation association	9	4	Pearson = 2.849 $p = 0.091$
No. of points with woodland, not scrub, as main vegetation structure	14	7	Pearson = 4.912 $p = 0.027$

Note: *dbh* diameter at breast height

(Ringrose et al. 1989, 1998), transformed NDVIs (TNDVIs) were run for all years. The resultant surfaces were subset to pairs of polygons—named here Matsibe (to the east) and Habu (to the west)—in order to remove the influence of the riparian zones on the north side of the fence (Fig. 8.6). For each polygon pair, and for each surface of interest, the mean and standard deviation values within the livestock polygon were subtracted from those for the wildlife polygon. These values were then compared over time and for the different seasons.

8.3.4 Results

At the paired points, there were distinct differences between the two sides. For example, at 16 of the 20 pairs, the points on the wildlife side had more exposed soil than on the cattle side. Canopy height was lower at more than half of the wildlife points, with mean height significantly lower for that side (Table 8.3). Even though the paired points were only 250 m apart, there were significant differences in vegetation association and structure as well.

The Landsat thermal band, which is only present on TM and ETM+ platforms, shows that there is considerable interannual variation in mean surface temperature. However, since 1990, the surface temperatures within the wildlife area have been consistently hotter than in the livestock area by as much as 2°C during dry season 2007 in the Habu area, where the soils are darker (Fig. 8.8).

Tasseled cap band 3, which is found to represent soil moisture in TM and ETM+ imagery (Crist and Cicone 1984), does not have a comparable band in tasseled cap transformations derived from MSS imagery, so as with temperature, results are shown for 1990 onward only. The wildlife areas show slightly higher variation in moisture values, with the difference between the two sides particularly marked in the Matsibe region, probably due to the sandy conditions there, where exposed soils have low retention capabilities (Fig. 8.9). Note that the differences between the means of the two sides have increased for 2000 and 2007 in the Habu area. However, in all instances except the first year in the Matsibe area, soil moisture values are higher on the cattle side. Note too the strong inverse correlation with surface temperatures, whose plotted data in Fig. 8.8 form a mirror image of the moisture trends in Fig. 8.9.

The satellite imagery confirms the findings of the paired points, showing that the mean soil brightness has increased on the wildlife side more than on the cattle side since the construction of the fence (Fig. 8.10). This is the case on both the lighter arenic soils in the Matsibe region and in the darker sodic soils in the Habu area. The variability in wet season soil brightness has also increased inside the wildlife area since the construction of the fence. The high soil brightness values in the wet season on the wildlife side show the extent to which, even at the end of the growing season, there is a high proportion of exposed soil, suggesting little vegetation cover. This fact is confirmed by the values of the TNDVIs, which, though not shown here because there is very little variability or trend, nevertheless show that in all years for both seasons there is more vegetation present on the cattle side than on the wildlife side.

8.3.5 Discussion

The paired point analysis suggests both composition and structural changes to vegetation on the wildlife side of the fence, through a replacement of thornveld by mopane and the removal of taller trees, are turning woodland into shrubland. The lower tree heights and fewer thornveld trees on the wildlife side are likely due to the impact of the large population of elephants, as there are many ring-barked and felled acacia on the wildlife side (see reasons for this dietary preference in, e.g., Hiscocks 2008) and because mopane responds to elephant browse by hedging (Lewis 1986; Smallie and O'Connor 2000). The higher amounts of exposed soil, particularly at the end of the growing season at the points within the wildlife area, are reflected in

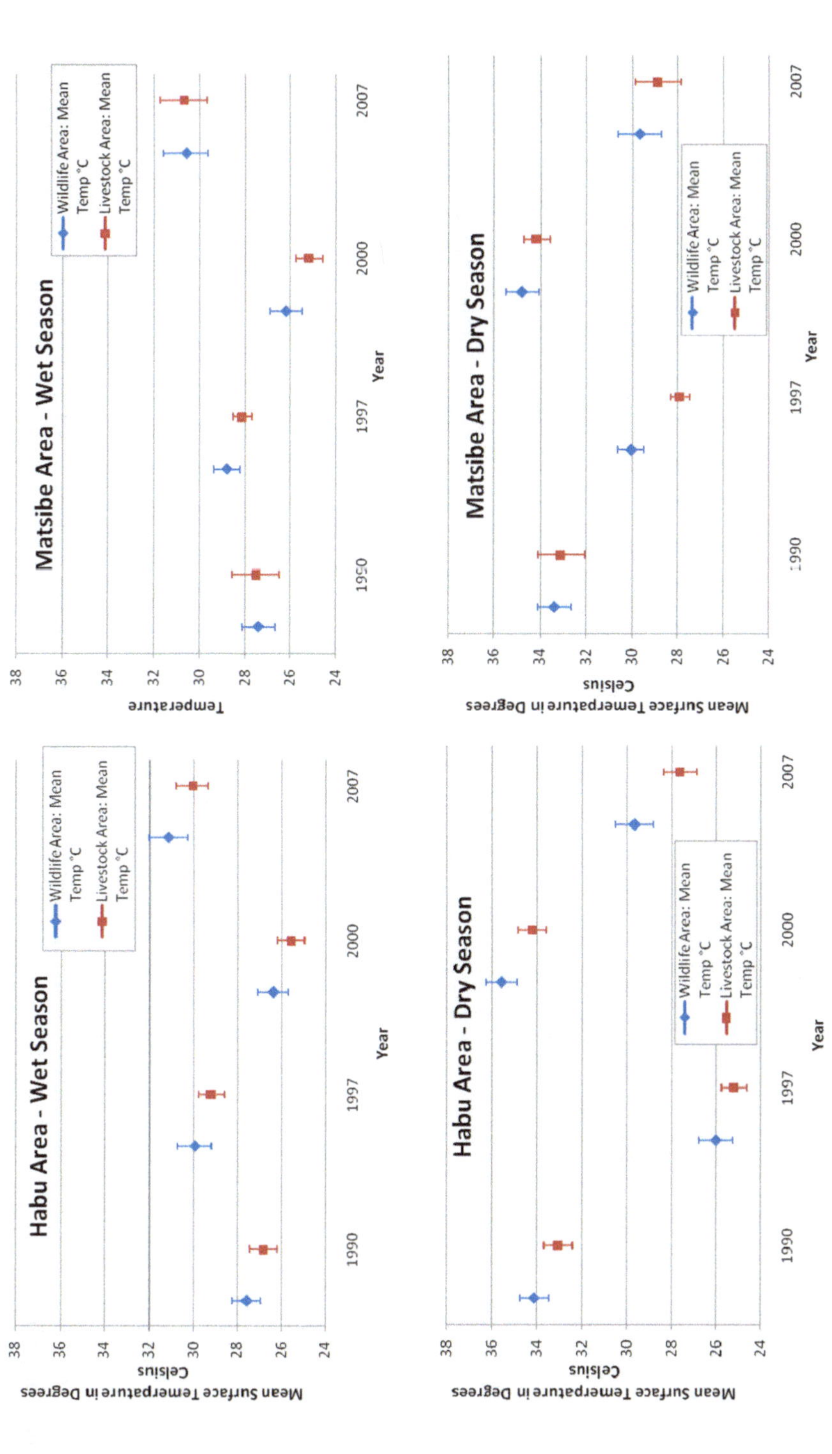

Fig. 8.8 Differences in mean Landsat-derived surface temperature values between wildlife and livestock areas for wet and dry seasons from 1990 to 2007, showing one standard deviation. The low standard deviations (compared to data in other figures below) are in part an artifact of the coarser resolution of the thermal band

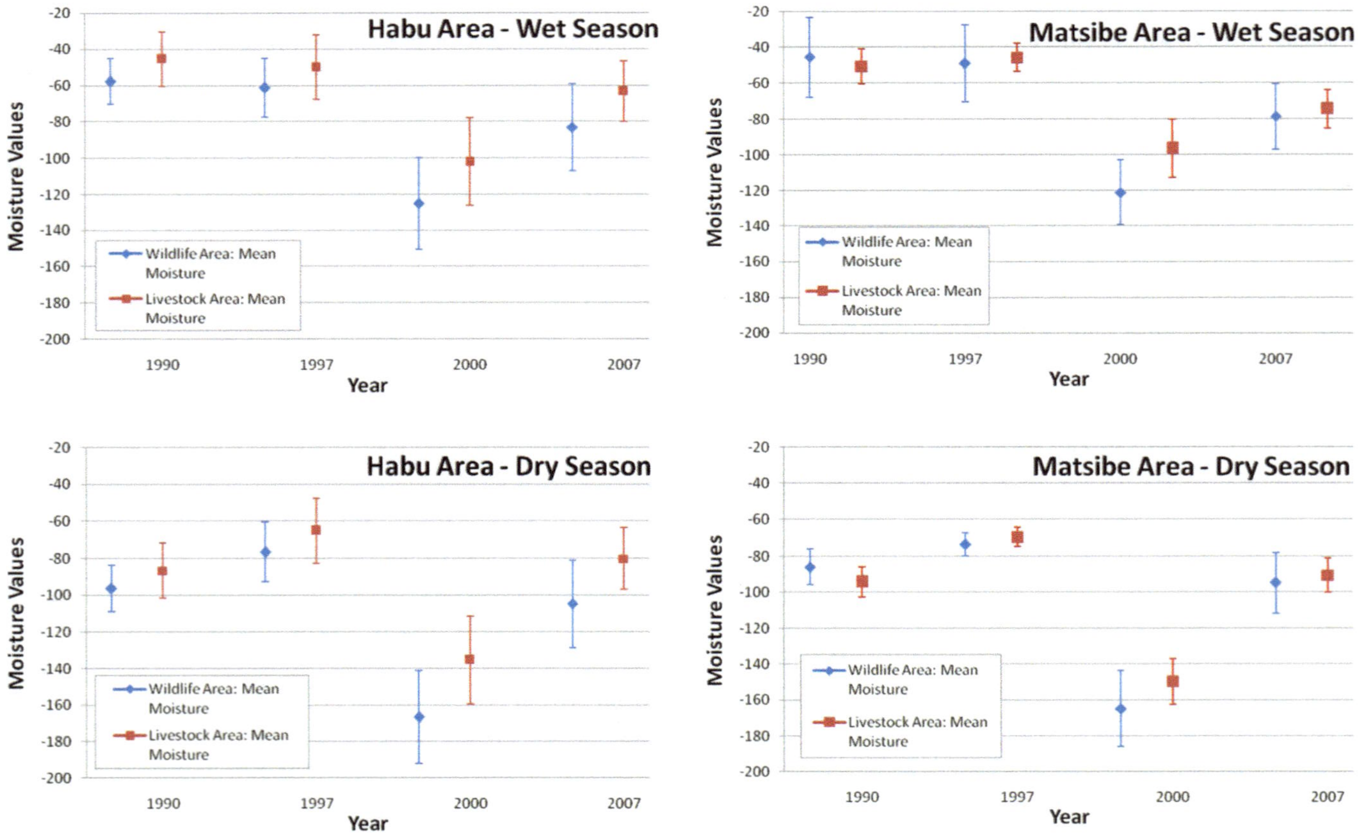

Fig. 8.9 Differences in mean Landsat-derived surface moisture values between wildlife and livestock areas for wet and dry seasons from 1990 to 2007, showing one standard deviation

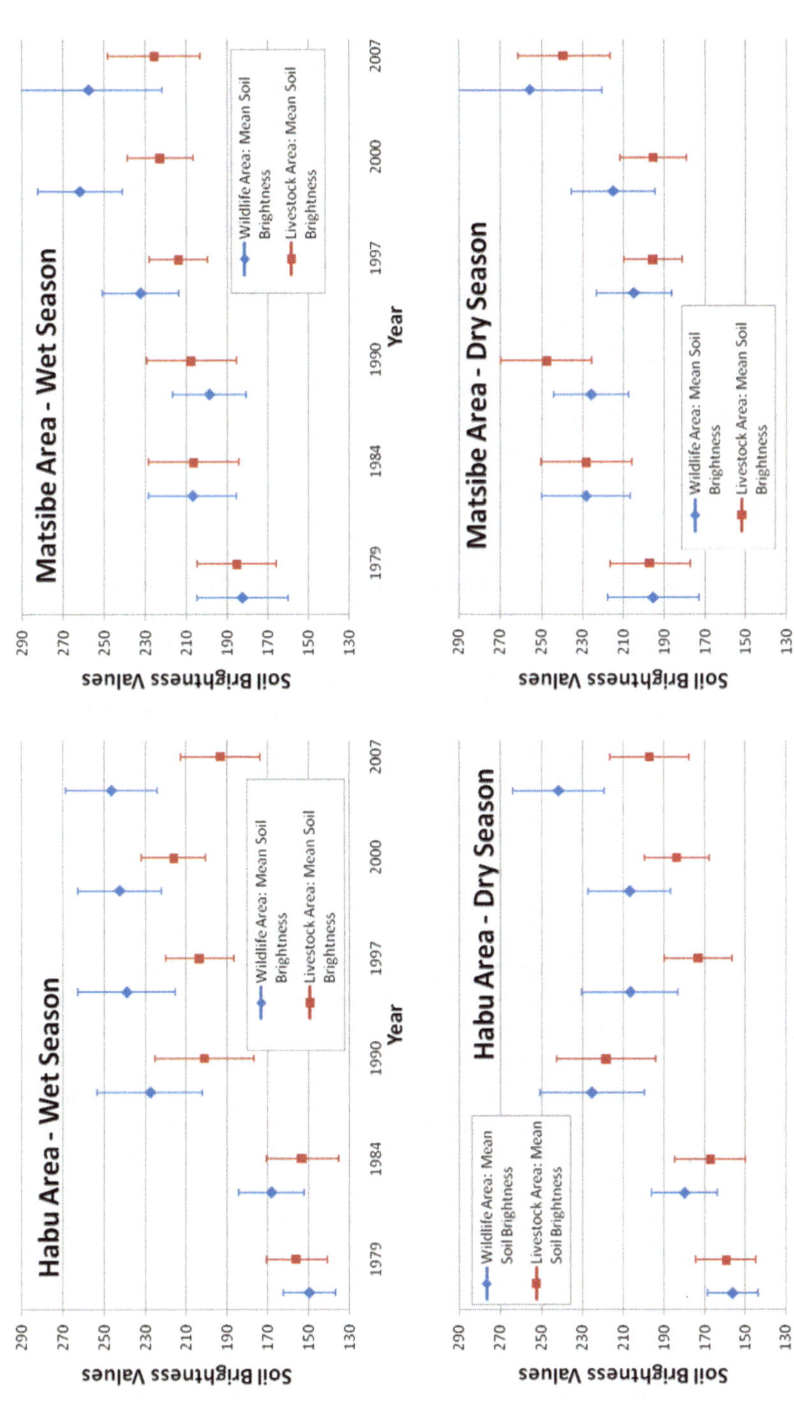

Fig. 8.10 Differences in mean Landsat-derived soil brightness values between wildlife and livestock areas for wet and dry seasons from 1990 to 2007, showing one standard deviation

the tasseled cap brightness values and give weight to the suggestion that there is a significant reduction in standing biomass due to heavier grazing rates as a result of truncated animal movements.

These landscape analyses provide corroborative evidence for the reduction in standing biomass inside the fence (i.e., within the wildlife management area). Further work is being undertaken to relate this to any changes in plant diversity and to the distribution of animal biomass (through current and past aerial survey counts). Even without these data, however, the research raises further questions: Is the reduction in standing vegetative biomass actually a problem? While some levels of grazing have been shown in other areas to enhance range quality and change diversity levels (cf. Cingolani et al. 2005), there is also the possibility that by creating two adjacent but different habitat types, the overall biodiversity of the area has been increased, for example, by creating niches for birds that prefer grasslands to woodlands.[1] Has the vegetation change in the wildlife area crossed a threshold into a different state, and if so, what are the implications for future use options? Certainly, as the landscape is now, the increased surface albedo and temperatures have environmental consequences and, in this region, contribute to creating a hotter, drier environment (Foley et al. 2005).

At the landscape level, the fence has also introduced a sharp boundary into what was previously a gradient of change. Currently, given that neither area is being actively managed in terms of stocking densities, this contrast is likely to remain intense. Regular monitoring into the future and a finer temporal resolution for satellite imagery may reveal the shape of the trend line of change, where it asymptotes, for example. The question is whether a critical threshold has been passed, such that a simpler, lower stable state has been reached (Walker et al. 1981), reducing options for future use even if good management interventions were introduced (Daily et al. 2000). It is difficult to quantify what range of the landscape's vegetative gradient has been lost, although it is clear that the fragmentation imposed by the fence's presence is reducing wildlife-carrying capacities (see also Boone and Hobbs 2004). At a more local scale, there are of course implications for management of wildlife populations. This comes at a time when Botswana's government has decided to restrict safari hunting in this area, meaning that artificial population controls, such as through culling, are unlikely to be an option. An alternative would be movement of the fence alignment. Cost implications notwithstanding, it is likely that moving the fence southward to increase the wildlife area would be culturally unacceptable, since many people see the tourism industry as being exclusionary (Mbaiwa 2005). There is also the risk of increasing wildlife populations and systematically reducing standing vegetative biomass over a greater area. If the fence were moved northward to decrease the wildlife management area, there would be intensification of vegetation loss on the wildlife side of the fence but a reduction of pressure on the existing area. However, this would have a consequent effect on

---

[1]Ongoing studies are assessing diversity of grass, woody vegetation, and birds on both sides of the fence.

the wildlife populations (cf. Wallgren et al. 2009) and decrease the lucrative returns from tourism in this area. Discussions of these options may be moot given current proposals to develop a string of fenced game ranches on the southern side of the fence to further increase the buffer between disease-carrying buffalo and cattle, causing even greater fragmentation (Western et al. 2009). The luxury of being able to project or plan far enough into the future to explore a scenario where wildlife and livestock can coexist (Western 1989) tends to evade planners and policy makers who are tied to a short-term planning horizon by election results.

In the next section, we continue to address the challenges of studying savanna environments and building from the work presented in this section to try and link to the standing vegetation type and structure, as seems necessary in this landscape. This section has addressed well the role of continuous analyses in studying landscape change. Next, we increase our spatial resolution (decrease extent) and link our landscape components to remote sensing imagery more closely, again for savanna land cover.

8.4 Case Study: Caprivi, Namibia

8.4.1 Land-Use and Land-Cover Change in Southern African Savannas

In southern African savannas, the land-cover change discourse is centered on changes in vegetation composition, specifically the distribution of trees, grasses, and shrubs across the landscape. The concern is that increased herbivore populations (specifically elephants) in combination with other local drivers of land-cover change (fire regimes, land-use patterns) are causing a detrimental loss of trees, resulting in an overall reduction in biodiversity and alterations in ecosystem functioning (Kalwij et al. 2010). Tree cover is central to savanna functioning, and one of the challenges for monitoring and managing savanna landscapes resides in understanding the spatial dynamics of trees. Thus, monitoring the spatial characterization and quantification of tree canopies is essential for landscape management and the development of policies aimed at ensuring sustainable land management practices (Holdo et al. 2009; Ludwig et al. 2001).

The gradient nature of savanna landscapes, the discontinuous tree canopy, and the subtle changes in the vegetation composition limit the use of categorical pixel-based land-cover classifications to quantify changes in vegetation composition. Yet, the spatial extent of savanna landscapes within Africa necessitates the use of remotely sensed data to characterize land-cover change and link such changes to social and ecological processes. While field-based analyses establish baseline understandings of the role of trees in savannas, they are limited in spatial and temporal scale. Remotely sensed data offer an alternative to field-based data (which are often prohibitively expensive to conduct over larger spatial extents or higher temporal frequencies) and simultaneously capture landscape-scale measures of vegetation (Scanlon et al. 2007). Therefore, it is necessary to explore the use of non-pixel-based remote sensing approaches combined with high spatial resolution imagery to characterize the distribution of savanna vegetation, specifically trees. Additionally, non-pixel-based classifications of high-resolution imagery hold the potential to provide a scaling tool, allowing a link to be created between field-based measures and traditional classifications (Thenkabail 2004).

8.4.2 Study Area

The study area is located in the Mudumu North Complex in Caprivi, Namibia (Fig. 8.11). Most of the Caprivi region is a part of the larger Kalahari woodlands landscape and as such has similar vegetation composition to lands in surrounding countries. Like much of southern Africa, the savanna landscape in Caprivi is being managed using a variety of land management strategies, including government-managed areas (parks) and community conservation areas. The study area is divided almost equally between the Bwabwata National Park Kwando Core Area (KCA) to the west of the Kwando River and the communal conservation areas to the east. People were removed from KCA beginning in the 1940s due largely to the decimation of cattle by sleeping sickness. During this period, the human population in the study area was low, and people supplemented cattle with harvested resources such as veld fruits. In 1968, the Caprivi Game Park was declared, and people were restricted from returning to what is now KCA (Mayes 2008). Struggles for national independence in Caprivi led to the South African Defence Force occupying KCA, during which time land and resource use within KCA was limited to conservation officials. After Namibian independence, Bwabwata National Park was declared replacing Caprivi Game Park. Land and resource use within KCA has since been limited to photographic tourism.

In the community conservation areas, improvements in the treatment of cattle diseases enabled human populations to rebound beginning in the 1980s (Mendelsohn and Roberts 1997). Human population has continued to increase as a result of immigration from surrounding countries, in particular Angola. This has resulted in increased grazing pressure, as well as an increase in clearing of land for agriculture aided by cattle draught power. The community conservation areas are multipurpose zones in which certain parts of the landscape are designated wildlife conservation corridors and other areas are used for resource collection and agriculture. Land uses in KCA and community conservation areas therefore contrast each other, with KCA having (both historically and currently) less anthropogenic conversion, while much of the communal conservation lands are directly altered for settlement, agriculture, and grazing. In KCA, management practices are determined by local and regional government officials, while savanna management practices in the community conservation areas are decided by committees consisting of local community members.

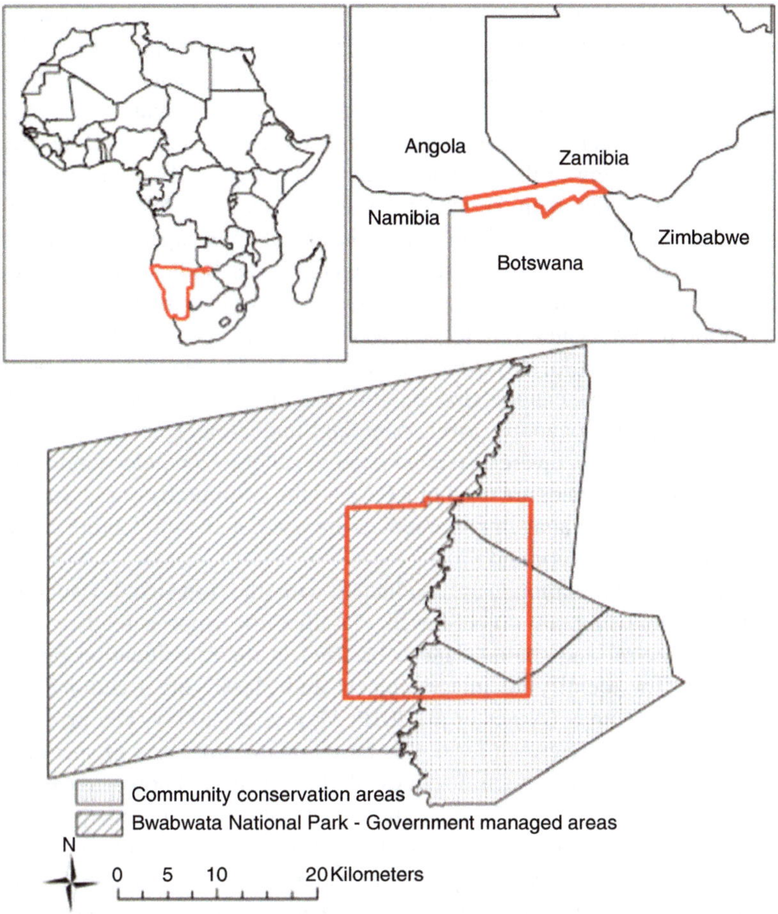

Fig. 8.11 Study region showing the larger regional context of the study site, the study area as determined by the extent of the IKONOS imagery (outlined in *red*) and the two dominant land management types used in the region

However, recent implementation of a collaborative management scheme for the study area has generated the need for a standardized approach to monitoring land cover across the larger landscape.

8.4.3 Methods

High-resolution IKONOS imagery (4 m × 4 m visual infrared bands) and medium-resolution Landsat TM imagery (30 m × 30 m) were acquired for the study area. Dry season imagery (IKONOS on May 21, 2006; Landsat on May 1, 2007) was selected for both image sources in order to minimize cloud cover. As with many studies that incorporate remotely sensed data, image selection is partly based on data availability; as such, the 1-year lag between the image sources was unavoidable. The extent of the IKONOS imagery was used to determine the extent of the study area; the Landsat TM imagery was then subset to match the spatial extent of the IKONOS imagery. Preprocessing of the imagery, including image calibration and geometric rectification, was undertaken. Water bodies and clouds were removed from images using a binary mask generated from a combination of an unsupervised classification and ancillary data sources (shapefiles identifying water bodies). This was done to reduce the likelihood of misclassifications during the object-oriented classification.

Vegetation transects and training samples were collected to enable in-field identification of the spatial locations of individual trees, demographic data associated with each tree (species, canopy size, height, diameter at breast height), land-cover measurements (percent canopy cover, dominant understory), and land-use history. Key informant interviews were also conducted to generate an understanding of what land management practices are used in KCA and the community conservation areas, who determines what management practices are used, and what factors contributed to tree growth and development.

Tree crown identification was conducted with the four-band visual infrared IKONOS image using an object-based classification approach. This approach allows both spectral and spatial data to be used in the classification process and is appropriate for identifying image objects and classifying fractional cover in heterogeneous savannas (Asner et al. 2010; Hay et al. 2003). The probability of being identified as a tree crown (or not) is based on the similarity of spectral values of the given pixel to the spectral values of the pixels of known trees identified during the training process. Segmentation was then used to incorporate spatial characteristics (e.g., shape, connectivity) and further partition of the image into objects with similar spectral and spatial characteristics. The segmented image was then converted to a vector file, and geometric characteristics (area, a perimeter-area ratio, and shadow) were used to refine the identification of trees. The location of tree polygons as identified in the final vector layer from the object-based classification was compared to the actual spatial location of individual trees as determined in the field.

The object-based classification was used to assess the current differences in tree cover across the landscape. Spatial clustering of trees was assessed using the Getis-Ord Gi* statistic which measures local association among features (Ord and Getis 1995). Bonferroni's adjustment ($\beta = \alpha/n$) was applied to identify the critical values for determining significant clustering of trees. To explore the utility of using an object-oriented classification of high-resolution imagery as a scaling tool for linking field observations to the commonly used NDVI measures determined from Landsat imagery, the proportion of each Landsat pixel covered by tree polygons was calculated and compared to the corresponding NDVI value of the Landsat image (Fig. 8.12). Classes of tree coverage were then determined by creating bins based on amount of tree crown coverage. For example, tree class 1 consists of all pixels with 0–20% tree coverage, and class 2 consists of all pixels with 21–40% coverage,

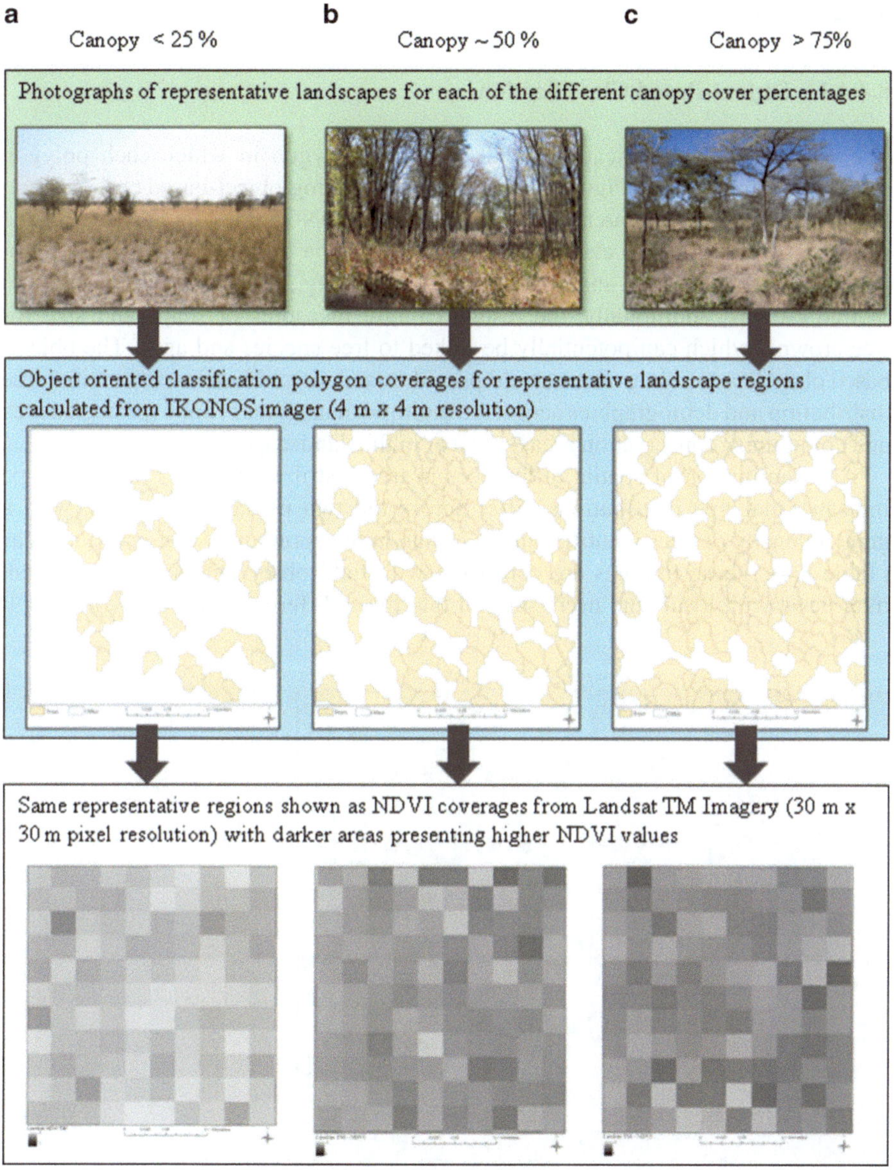

Fig. 8.12 Schematic showing the relationships between the actual tree coverage, the idealized object-oriented classification results in *polygon* form, and NDVI for the same area at the Landsat TM scale (Figure modified from Gibbes et al. 2010)

ultimately resulting in five tree-cover classes. The variation in NDVI as a function of proportion of trees is quantified by examining the difference among NDVI values for each tree-cover class and using graphical outputs and the Kruskal-Wallis test.

8.4.4 Results

8.4.4.1 Object-Based Classification

The object-based classification yields a vector polygon in which each polygon represents a tree crown. Figure 8.13 shows the resulting object-based classification for a subset of the study area as well as both the IKONOS and Landsat data for the subset. The object-based classification improves upon a pixel-based classification by incorporating shapes and spatial associations (e.g., association with shadow) of tree crowns; additionally, the results capture measures of shape and size for tree crowns, which can potentially be linked to tree species and ages. The object-based classification shown here could provide standardized baselines for initial tree distribution and demographics across the multiple management units (protected area and community conservation areas) present in this landscape.

An examination of spatial differences in tree distribution across the landscape indicates that tree distributions differ across the two management areas, with a greater density of trees identified in KCA than in the community conservation areas (Table 8.4). These findings are consistent with field observations. Key informant interviews with local land users suggest that these differences are likely the result

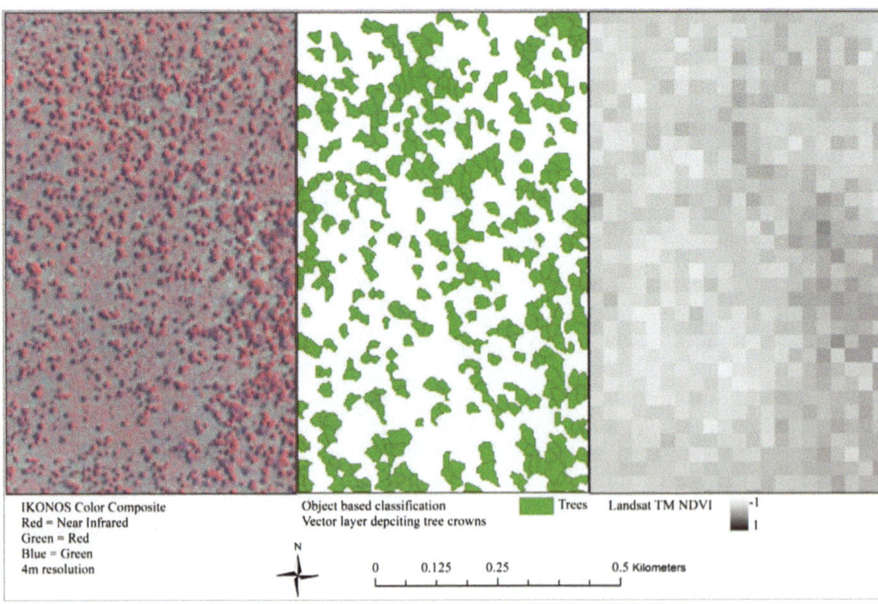

Fig. 8.13 Results of the object-based classification. Figures shown are (*left*) the IKONOS imagery for a focus area (color composite RGB = near infrared, red, green where *red* represents vegetation), (*middle*) the corresponding object-based classification for the focus area and (*right*) Landsat NDVI also for the focus area, in which the difference in scale is apparent (modified from Gibbes et al. 2010)

Table 8.4 Density of large trees within each of the two land management types—KCA, government-managed (~132.1 km^2) versus community-managed lands (~146.48 km^2)

	KCA (government-managed, protected area)	Community conservancy areas (community managed)
Total no. of trees/km^2	580.2	352.7
No. of large trees*/km^2	11.5	3.4

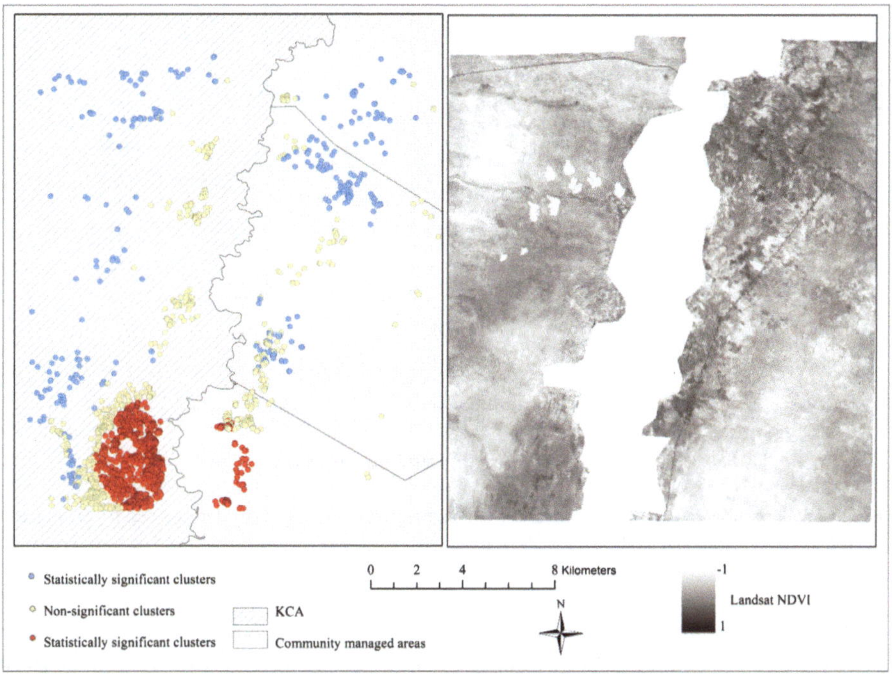

Fig. 8.14 Spatial clustering of trees (*left*) results of the Getis-Ord Gi* statistic with Bonferroni correction applied to determine statistically significant tree clusters, where *red dots* indicate clusters of trees with large crown size, while *blue dots* indicate the statistically significant clusters of trees with relatively smaller crown size (*right*) Landsat TM NDVI shown for the same study region

of variations in management practices used in the protected area and community conservation areas. Clearing land for agriculture and/or homesteads and collecting timber resources reduce the presence of trees in the community conservation areas.

The cluster analysis of trees shows significant clustering of trees with large crown sizes in the southern portions of the study area, while trees with smaller crowns have clusters that are more evenly distributed across the landscape (Fig. 8.14). The spatial locations of tree clusters are likely the result of both historical stand development processes and the influence of variation in other biotic parameters, such as soil moisture and herbivory, which influence tree growth. The Bonferroni adjusted critical value of 1.68 was used to determine significant clustering. As spatial

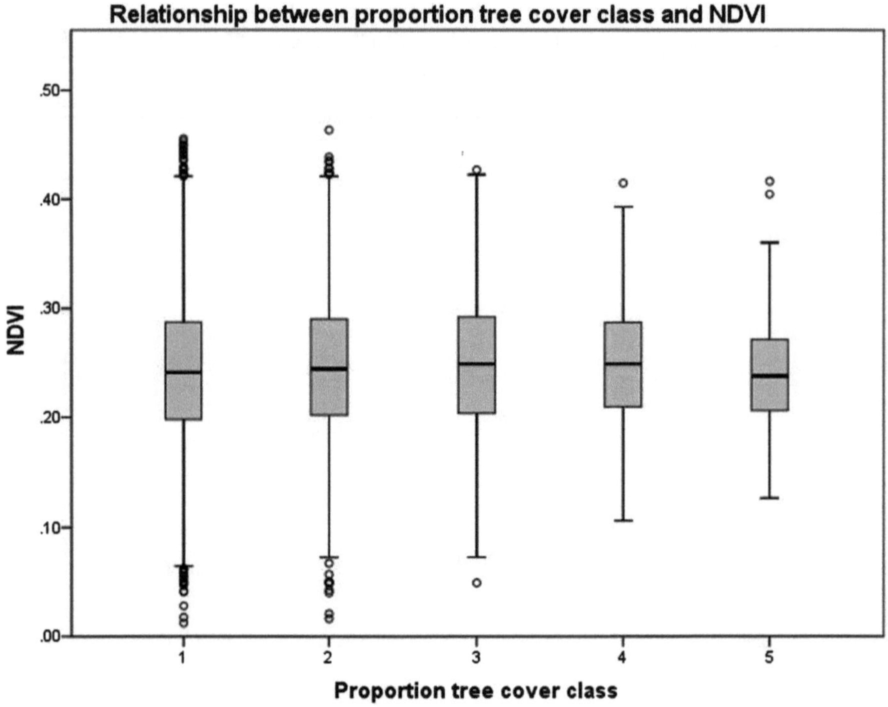

Fig. 8.15 Boxplots of the NDVI values for each tree-cover class (1–5) as determined by proportion of tree cover per Landsat TM 30 m × 30 m pixel

processes gain increasing attention in ecology and ecosystem management (Levin 1992; Turner 1989; Weisberg et al. 2007), the detection of tree clusters within the landscape contributes to developing an understanding of spatial processes and heterogeneity in savanna vegetation (Wiens 1989). Standardized characterizations and quantifications of tree clustering offer useful ecological information for monitoring and managing the spatially heterogeneous processes (e.g., herbivory, fire, anthropogenic land use) that influence the maintenance of savanna vegetation.

8.4.4.2 Scaling from Field to Landsat TM

The Kruskal-Wallis test to evaluate the relationship between the proportion of Landsat pixel covered by tree crowns and that Landsat pixel's NDVI values. The results ($H = 4.39$, significance level 0.05) indicate no significant difference in the distribution of NDVI values across tree-cover classes (Fig. 8.15). The lack of significant difference in NDVI values for pixels with high tree coverage and those with low tree coverage suggests that the land covers in the background matrix contribute greatly to NDVI values observed for the study region. This would support

Moleele et al. (2001) and Ringrose et al. (1989) who find that NDVI offers a good measure for overall amounts of vegetation but is not useful for differentiating vegetation structure.

8.4.5 Case Summary

An object-based classification is used to address the limitations of characterizing savanna vegetation structure. This methodology, when applied to high-resolution satellite imagery, is useful for identifying tree crowns within the gradient savanna landscape. Tree distribution and clustering correspond to field data collected for this area, and the spatial patterning across the land management units is as observed and expected, with larger quantities of trees and larger-sized trees found within the park. In addition to characterizing tree location and demographics, the results from the object-based classification prove useful for point-pattern analyses, offering an assessment of the spatial location and characteristics of the trees relative to surrounding vegetation. This analysis reinforces the need to incorporate both spectral and spatial data in the classification of savanna vegetation.

The results of the Kruskal-Wallis test suggest that the utility of the object-based classification as a scaling tool for linking field observations to Landsat NDVI measurements is limited. Similarity in spectral reflectance of shrubs and trees in this region makes discriminating structure solely from spectral information challenging, especially at the coarser Landsat scale. The scale at which environmental change is observed and measured in savanna systems is critical. While quantification of overall changes in biomass is possible with coarse-resolution NDVI, partitioning biomass into functional types requires finer-scale analysis, such as the object-based classifications presented here.

Current limitations of using remote sensing for ecological studies include the overreliance on pixel-based maximum likelihood classifications (Weisberg et al. 2007). Although pixel-based approaches are appropriate when landscape components of interest have very different spectral signatures, this case study demonstrates the use of an alternative approach that focuses on the identification of individual vegetation components within the landscape as opposed to a broad measure of biomass within individual pixels. One challenge of the object-based approach is the ability to discriminate between polygons of individual trees as opposed to those representing patches of trees. We attempted to address this challenge by calibrating the area within the vector object processing using individual tree crown locations as determined in the field. This ensured that rather than generating large polygons of similar spectral values, the output consisted of multiple smaller polygons. Although this does not completely remove the presence of polygons that represent multiple trees as opposed to individual trees, it certainly reduces the occurrence.

The object-based classification offers a potentially useful way for standardizing the monitoring of vegetation across large spatial extents and across multiple land management units than could be captured with field measurements. It integrates well a multiscalar approach often needed in ecological remote sensing data, and while initial linkages to Landsat NDVI data were not very successful, linking to more ecologically relevant remote sensing data, such as subpixel classifications, holds promise and is currently being validated. This research highlights the need for linkages to field data, a multiscalar-based approach and need for meaningful landscape objects for future study. With its finer spatial scale and association with individual objects—trees in this case—it also provides a clear example of how explicit remote sensing data can be and direct linkages to field data. As such, the progression from Uganda (Sect. 8.2), with traditional classification and an NDVI approach, to Botswana (Sect. 8.3), which adds more explicit variable extraction, and finally to Namibia in this (Sect. 8.4), where we can assess individual trees and their patterns across space, provides a clear trajectory of complexity, landscape mosaics, and remote sensing data. And we can see how different uses—associated with different research questions—can be extracted and utilized across this African landscape.

8.5 Conclusions

While all three studies include a clear park/protected focus, all three addressed the research in different ways in order to best evaluate and represent the type of landscape under study. Locating a park or management zone within the larger landscape, in order to better evaluate the impacts of these human decisions on the landscape over time, is essential. For African landscapes in particular, the role of gradients—and in all three case studies presented here the disruption of natural gradients by human management decisions—is a key area for human-environment research. Decisions can often have unintended consequences, such as in the Botswana example (Sect. 8.3) where the wildlife management area is much more degraded than the cattle region. All three studies clearly highlight the importance of good management decisions and monitoring of impacts both for the zone of management or impact but also the larger regional context over time. For many remote areas, such as in these African case studies, remote sensing is clearly a valuable tool. However, these tools can also be quite limited, if restricted to more traditional remote sensing techniques, for example, pixel-based land-cover change, especially across these gradient regions. As we develop a better understanding of these human-environment arenas, so too must we continue to develop appropriate measurement, monitoring, and analyses tools. The techniques presented in this chapter—from land-cover change and NDVI-based analyses in Uganda (Sect. 8.2) to vegetation indices linked to surface temperatures derived from satellites along transects and linked to vegetation plots in Botswana (Sect. 8.3) and then finally downscaled to individual tree analyses in Namibia (Sect. 8.4)—present a pathway of increasing complexity and show a path of increased linkages and connectivity to field-based research, ecological components, and real information representing the landscapes in question. The tighter linkage from plot to pixel significantly strengthens our environmental understanding and, as more studies address similar concepts, will build a better understanding of these systems. Overall, the gradient landscapes still in evidence across much of this African landscape warrant gradient or continuous methods of study. This chapter hopefully serves to introduce some of these concepts, their necessity, and their usefulness in addressing real-world human-environment-based research questions.

References

Albertson, A. (1998). *Northern Botswana veterinary fences: critical ecological impacts*. Maun: Okavango People's Wildlife Trust.

Archer, S., Boutton, T., & Hibbard, K. A. (2001). Trees in grasslands: Biogeochemical consequences of woody plant expansion. In D. Schulze, M. Heimann, S. Harrison, E. Holland, J. Lloyd, I. Prentice, & D. Schimel (Eds.), *Global biogeochemical cycles in the climate system*. San Diego: Academic.

Asner, G., Levik, S., & Smit, I. (2010). Remote sensing of fractional cover and biochemistry in savannas. In M. Hill & N. Hanan (Eds.), *Ecosystem function in savannas*. Boca Raton: CRC Press.

Baranga, J. (1991). Kibale forest game corridor: Man or wildlife? In D. A. Saunders & R. J. Hobbs (Eds.), *Nature conservation: The role of corridors* (pp. 371–376). London: Surrey Beatty and Sons.

Beerling, D., & Osborne, C. (2006). The origin of the savanna biome. *Global Change Biology, 12*, 2023–2031.

Belsky, A., Amundson, R., Duxbury, J., Riha, S., Ali, A., & Mwonga, S. (1989). The effects of trees on their physical, chemical, and biological environment in a semi-arid savanna in Kenya. *Journal of Applied Ecology, 26*, 1005–1024.

Berkes, F. (2004). Rethinking community-based conservation. *Conservation Biology, 18*, 621–630.

Blaschke, T., & Hay, G. J. (2001). Object-oriented image analysis and scale-space: Theory and methods for modeling and evaluating multi-scale landscape structure. *International Archives of Photogrammetry and Remote Sensing, 34*, 22–29.

Boone, R., & Hobbs, N. (2004). Lines around fragments: Effects of fencing on large herbivores. *African Journal of Range & Forage Science, 21*, 147–158.

Brandon, K., Redford, K. H., & Sanderson, S. E. (Eds.). (1998). *Parks in peril: People, politics, and protected areas*. Washington, DC: Island Press for The Nature Conservancy.

Brechin, S. R., Wilshusen, P. R., Fortwangler, C. L., & West, P. C. (2002). Beyond the square wheel: Toward a more comprehensive understanding of biodiversity conservation as social and political process. *Society and Natural Resources, 15*, 41–64.

Brockington, D. (2002). *Fortress conservation: The preservation of the Mkomazi Game Reserve, Tanzania*. Oxford: James Currey.

Brockington, D. (2006). An anthropological perspective on some unexpected consequences of parks. *Conservation Biology, 20*, 609–616.

Brooks, T., Balmford, A., Burgess, N., Fjeldsa, J., Hansen, L. A., Moore, J., et al. (2001). Toward a blueprint for conservation in Africa. *BioScience, 51*, 613–624.

Brown, S., & Lugo, A. (1994). Rehabilitation of tropical lands: A key to sustaining development. *Restoration Ecology, 2*, 97–111.

Campbell, A. C. (1976). Traditional utilisation of the Okavango Delta. In The Botswana Society (Ed.), *Proceedings of the symposium on the Okavango Delta and its future utilization* (pp. 163–173). Gaborone: The Botswana Society.

Carlson, T. N., & Ripley, D. A. (1997). On the relation between NDVI, fractional vegetation cover, and leaf area index. *Remote Sensing of Environment, 62*, 241–252.

Cassidy, L. (2000). *CBNRM and legal rights to resources in Botswana*. Gaborone: SNV/IUCN CBNRM Support Programme.

Castilla, G. (2003). *Object-oriented analysis of remote sensing images for land cover mapping: Conceptual foundations and a segmentation method to derive a baseline partition for classification*. Ph.D. thesis, Polytechnic University of Madrid, Madrid, Spain. http://oa.upm.es/133/01/07200302_castilla_castellano.pdf. Last date accessed 15 Jan 2008.

Chapin, M. (2004). A challenge to conservationists. *World Watch Magazine, 17*, 17–31.

Chapman, C. A., Chapman, L. J., Zanne, A., & Burgess, A. (2002). Does weeding promote regeneration of an indigenous tree community in felled pine plantations in Uganda? *Restoration Ecology, 10*, 408–415.

Chapman, C. A., Chapman, L. J., Jacob, A. L., Rothman, J. M., Omeja, P., Reyna-Hurtado, R., et al. (2010). Tropical tree community shifts: Implications for wildlife conservation. *Biological Conservation, 143*, 366–374.

Chazdon, R. L. (2008). Beyond deforestation: Restoring forests and ecosystem services on degraded lands. *Science, 320*, 1458–1460.

Chazdon, R. L., Harvey, C. A., Komar, O., Griffith, D. M., Ferguson, B. G., Martínez-Ramos, M., et al. (2009). Beyond reserves: A research agenda for conserving biodiversity in human-modified tropical landscapes. *Biotropica, 41*, 142–153.

Child, B. (2009). Rights based natural resource management. In H. Suich & B. Child (Eds.), *Evolution and innovation in wildlife conservation*. London: Earthscan.

Chomitz, K. M. (1994). Transferable development rights and forest protection: An exploratory analysis. *International Regional Science Review, 27*, 348–373.

Cingolani, A., Noy-Meir, I., & Díaz, S. (2005). Grazing effects on rangeland diversity: A synthesis of contemporary models. *Ecological Applications, 15*, 757–773.

Cordeiro, N. J., Burgess, N. D., Dovie, D. B. K., Kaplin, B. A., Plumptre, A. J., & Marrs, R. (2007). Conservation in areas of high population density in sub-Saharan Africa. *Biological Conservation, 134*, 155–163.

Crist, E., & Cicone, R. (1984). Application of the tasseled cap concept to simulated Thematic Mapper data (transformation for MSS crop and soil imagery). *Photogrammetric Engineering and Remote Sensing, 50*, 343–352.

Crist, T., Veech, J., Gering, J., & Summerville, K. (2003). Partitioning species diversity across landscapes and regions: A hierarchical analysis of alpha, beta, and gamma diversity. *American Naturalist, 162*, 734–743.

Daily, G. C., Söderqvist, T., Aniyar, S., Arrow, K., Dasgupta, P., Ehrlich, P. R., et al. (2000). The value of nature and the nature of value. *Science, 289*, 395–396.

DeFries, R., Foley, J., & Asner, G.P. (2004). Land Use Choices: Balancing Human Needs and Ecosystem Function. *Frontiers in Ecology and the Environment, 2*(5), 249–257.

DeFries, R., Hansen, A., Newton, A. C., & Hansen, M. C. (2005). Increasing isolation of protected areas in tropical forests over the past twenty years. *Ecological Applications, 15*, 19–26.

Di Gregorio, A., & Jansen, L. J. M. (1998). *Environment and natural resources service*. GCP/RAF/287/ITA Africover – East Africa Project and Soil Resources, Management and Conservation Service. Rome: Food and Agriculture Organization of the United Nations.

Ferraro, P. J., & Kiss, A. (2002). Direct payments to conserve biodiversity. *Science, 298*, 1718–1719.

Filippi, A. M., & Jensen, J. R. (2006). Fuzzy learning vector quantization for hyperspectral coastal vegetation classification. *Remote Sensing of Environment, 100*, 512–530.

Foley, J. A., DeFries, R., Asner, G. P., Barford, C., Bonan, G., Carpenter, S. R., et al. (2005). Global consequences of land use. *Science, 309*, 570–574.

Geist, H. J., & Lambin, E. F. (2001). *What drives tropical deforestation? A meta-analysis of proximate and underlying causes of deforestation based on subnational case study evidence* (LUCC Report Series, No. 4). Louvain-la-Neuve: Land Use and Cover Change International Project Office.

Gibbes, C., Adhikari, S., Rostant, L., Southworth, J., & Qui, Y. (2010). Application of object based classification and high resolution satellite imagery for savanna ecosystem analysis. *Remote Sensing, 2*, 2748–2772.

GLP (Global Land Project). (2005). *Science plan and implementation Strategy* (IGBP Report No 53/IHDP Report No 1. 19). Stockholm: International Geosphere-Biosphere Programme.

Goldman, A., Hartter, J., Southworth, J., & Binford, M. (2008). The human landscape around the island park: Impacts and responses to Kibale National Park. In R. Wrangham & E. Ross (Eds.), *Science and conservation in African forests: The benefits of longterm research* (pp. 129–144). Cambridge: Cambridge University Press.

Government of Botswana. (2003). *National development plan 9, 2003/04–2008/09*. Gaborone: Ministry of Finance and Development Planning.

Green, G., Schweik, C. M., & Hanson, M. (2002). *Radiometric calibration of LANDSAT multispectral scanner and thematic mapper images: Guidelines for the global change community* (CIPEC Working Paper CWP-02-03). Bloomington: Center for the Study of Institutions, Population, and Environmental Change (CIPEC), Indiana University.

Gutman, G., Janetos, A., Justice, C., Moran, E., Mustard, J., Rindfuss, R., et al. (2004). *Land change science: Observing, monitoring, and understanding trajectories of change on the earth's surface*. Dordrecht: Kluwer Academic Publishers.

Hartter, J. (2010). Resource use and ecosystem services in a forest park landscape. *Society and Natural Resources, 23*, 207–233.

Hartter, J., & Ryan, S. J. (2010). Top-down or bottom-up? Decentralization, natural resource management, and usufruct rights in the forests and wetlands of western Uganda. *Land Use Policy, 27*(3), 815–826.

Hartter, J., & Southworth, J. (2009). Dwindling resources and fragmentation of landscapes around parks: Wetlands and forest fragments around Kibale National Park, Uganda. *Landscape Ecology, 24*, 643–656.

Hartter, J., Southworth, J., & Binford, M. (2009). Parks as a mechanism to maintain and facilitate recovery of forest cover: Examining reforestation, forest maintenance and productivity in Uganda. In H. Nagendra & J. Southworth (Eds.), *Reforesting landscapes: Linking pattern and process* (Landscape Series, Vol. 10, pp. 275–296). New York: Springer.

Hay, G. J., Marceau, D. J., Dube, P., & Bouchard, A. (2001). A multiscale framework for landscape analysis: Object-specific analysis and upscaling. *Landscape Ecology, 16*, 471–490.

Hay, G., Blaschke, T., Marceau, D., & Bouchard, A. (2003). A comparison of three object oriented methods for the multiscale analysis of landscape structure. *Journal of Photogrammetry and Remote Sensing, 57*, 327–345.

Hay, G. J., Castilla, G., Wulder, M. A., & Ruiz, J. R. (2005). An automated object-based approach for the multiscale image segmentation of forest scenes. *International Journal of Applied Earth Observation and Geoinformation, 7*, 339–359.

Hiscocks, K. (2008). The impact of an increasing elephant population on the woody vegetation in southern Sabi Sand Wildtuin, South Africa. *Koedoe-African Protected Area Conservation and Science, 42*, 47–55.

Holdo, R. M., Holt, R. D., & Fryxell, J. M. (2009). Opposing rainfall and nutrient gradients best explain the wildebeest migration in the Serengeti. *American Naturalist, 173*, 431–445.

Huete, A., Didan, K., Miura, T., Rodriguez, E. P., Gao, X., & Ferreira, L. G. (2002). Overview of the radiometric and biophysical performance of the MODIS vegetation indices. *Remote Sensing of Environment, 83*, 195–213.

James, C. (2003). Response of vertebrates to fenceline contrasts in grazing intensity in semi-arid woodlands of eastern Australia. *Austral Ecology, 28*, 137–151.

Jiang, Z., Huete, A., Chen, J., Chen, Y., Li, J., Yan, G., & Zhang, X. (2006). Analysis of NDVI and scaled difference vegetation index retrievals of vegetation fraction. *Remote Sensing of Environment, 101*, 366–378.

Kalwij, J. M., De Boer, W. F., Mucina, L., Prins, H., Skarpe, C., & Winterbach, C. (2010). Tree cover and biomass increase in a southern African savanna despite growing elephant population. *Ecological Applications, 20*, 222–233.

Kayanja, F. I. B., & Byarugaba, D. (2001). Disappearing forests of Uganda: The way forward. *Current Science, 81*, 936–947.

Lambin, E., & Ehrlich, D. (1997). Land-cover changes in sub-Saharan Africa (1982–1991): Application of a change index based on remotely sensed surface temperature and vegetation indices at a continental scale. *Remote Sensing of Environment, 61*, 181–200.

Lambin, E. F., & Geist, H. (Eds.). (2006). *Land use and land cover change: Local processes, global impacts*. New York: Springer.

Lausch, D. P., & Herzog, F. (2002). Applicability of landscape metrics for the monitoring of landscape change: Issues of scale, resolution and interpretability. *Ecological Indices, 2*, 3–15.

Levin, S. (1992). The problem of pattern and scale in ecology. *Ecology, 63*, 1943–1967.

Lewis, D. (1986). Disturbance effects on elephant feeding: Evidence for compression in Luangwa Valley, Zambia. *African Journal of Ecology, 24*, 227–241.

Ludwig, F., de Kroon, H., Prins, H. H. T., & Berendse, F. (2001). Effects of nutrients and shade on tree-grass interactions in an East African savanna. *Journal of Vegetation Science, 12*, 579–588.

Ludwig, F., De Kroon, H., & Prins, H. (2008). Impacts of savanna trees on forage quality for a large African herbivore. *Oecologia, 155*, 487–496.

Mayes, S. (2008). Special edition: Bwabwata National Park. Newsletter for the Strengthening the Protected Area Network (SPAN) project. *Sandpaper, 10*, 6–9.

Mbaiwa, J. (2005). Enclave tourism and its socio-economic impacts in the Okavango Delta, Botswana. *Tourism Management, 26*, 157–172.

Mbaiwa, J., & Mbaiwa, O. (2006). The effects of veterinary fences on wildlife populations in Okavango Delta, Botswana. *International Journal of Wilderness, 12*, 17–41.

McMichael, A. J., Bolin, B., Costanza, R., Daily, G. C., Folke, C., Lindahl-Kiessling, K., et al. (1999). Globalization and sustainability of human health: An ecological perspective. *BioScience, 49*(3), 205–210.

Mendelsohn, J., & Roberts, C. (1997). *An environmental profile and atlas of Caprivi*. Windhoek: Directorate of Environmental Affairs/Ministry of Environment and Tourism.

Moleele, N., Rinrose, S., Arnberg, W., Lunden, B., & Vanderpost, C. (2001). Assessment of vegetation indexes useful for browse (forage) prediction in semi-arid rangelands. *International Journal of Remote Sensing, 22*, 741–756.

Mukiibi, J. K. (Ed.). (2001). *Agriculture in Uganda: General information* (Vol. 1). Kampala: Fountain Publishers.

Murray-Hudson, M., & Crisman, T. (2003). Ecotourism as a sustainable land-use in African wetlands: The Okavango Delta and Kwando Wildlife management areas of Botswana. In T. Crisman, L. Chapman, C. Chapman, & L. Kaufman (Eds.), *Conservation, ecology and management of African freshwaters* (pp. 403–429). Gainesville: University Press of Florida.

Nagendra, H., & Southworth, J. (Eds.). (2009). *Reforesting landscapes: Linking pattern and process* (Landscape Series, Vol. 10). New York: Springer.

Nelson, F., & Agrawal, A. (2008). Patronage or participation? Community-based natural resource management reform in sub-Saharan Africa. *Development and Change, 39*, 557–585.

NEMA (National Environment Management Authority). (2001). *State of environment report for Uganda 2000/2001*. Kampala: National Environment Management Authority.

Ngamiland District Council. (2002). *Ngamiland district development plan 6: 2003–2009*. Gaborone: Ministry of Local Government, Lands and Housing.

NRC (National Research Council). (2001). *Grand challenges in environmental sciences*. Washington, DC: National Academies Press. http://www.nap.edu/books/0309072549/html/

Oates, J. F. (1999). *Myth and reality in the rain forest*. Berkeley: University of California Press.

Omeja, P. A., Chapman, C. A., Obua, J., Lwanga, J. S., Jacob, A. L., Wanyama, F., et al. (2011). Intensive tree planting facilitates biomass and biodiversity accumulation. *Forest Ecology and Management, 261*, 703–709.

Ord, J. K., & Getis, A. (1995). Local spatial autocorrelation statistics: Distributional issues and an application. *Geographical Analysis, 27*, 287–305.

Ostrom, E., Burger, J., Field, C. B., Norgaard, R. B., & Policansky, D. (1999). Revisiting the commons: Local lessons, global challenges. *Science, 284*, 278–282.

Parks, S. A., & Harcourt, A. H. (2002). Reserve size, local human density and mammalian extinctions in U.S. parks. *Conservation Biology, 16*, 800–808.

Perkins, J., & Thomas, D. (1993). Spreading deserts or spatially confined environmental impacts? Land degradation and cattle ranching in the Kalahari Desert of Botswana. *Land Degradation and Development, 4*, 179–194.

Petit, C., Scudder, T., & Lambin, E. (2001). Quantifying processes of land-cover change by remote sensing: Resettlement and rapid land-cover changes in south-eastern Zambia. *International Journal of Remote Sensing, 22*, 3435–3456.

Plumptre, A. J. (2002). *Extent and status of the forests in the Ugandan Albertine Rift*. New York: Wildlife Conservation Society for WWF under contract to Republic of Uganda Ministry of Water, Lands and Environment.

Plumptre, A. J., Behangana, M., Ndomba, E., Davenport, T. R. B., Kahindo, C., Kityo, R., et al. (2003). *The biodiversity of the Albertine Rift* (Albertine Rift Technical Report No. 3). Kampala: Wildlife Conservation Society.

Plumptre, A. J., Davenport, T. R. B., Behangana, M., Kityo, R., Eilu, G., Ssegawa, P., et al. (2007). The biodiversity of the Albertine Rift. *Biological Conservation, 134*, 178–194.

Pollard, S., Shackleton, C., & Carruthers, J. (2003). Beyond the fence: People and the lowveld landscape. In J. T. du Toit, K. H. Rogers, & H. C. Biggs (Eds.), *The Kruger experience: Ecology and management of savanna heterogeneity* (pp. 422–446). Washington, DC: Island Press.

Redford, K., & Sanderson, S. (2000). Extracting humans from nature. *Conservation Biology, 14*, 1362–1364.

Reynolds, J. F., Stafford Smith, D. M., Lambin, E. F., Turner, B. L., II, Mortimore, M., Batterbury, S. P., et al. (2007). Global desertification: Building a science for dryland development. *Science, 316*, 847–851.

Ringrose, S., Matheson, W., Mogotsi, B., & Tempest, F. (1989). The darkening effect in drought affected savanna woodland environments relative to soil reflectance in Landsat and SPOT wavebands. *Remote Sensing of Environment, 30*, 1–19.

Ringrose, S., Matheson, W., & Vanderpost, C. (1998). Analysis of soil organic carbon and vegetation cover trends along the Botswana Kalahari Transect. *Journal of Arid Environments, 38*, 379–396.

Ringrose, S., Kampunzu, A., Vink, B., Matheson, W., & Downey, W. (2002). Origin and palaeo-environments of calcareous sediments in the Moshaweng dry valley, southeast Botswana. *Earth Surface Processes and Landforms, 27*, 591–611.

Ringrose, S., Jellema, A., Huntsman-Mapila, P., Baker, L., & Brubaker, K. (2005). Use of remotely sensed data in the analysis of soil-vegetation changes along a drying gradient peripheral to the Okavango Delta, Botswana. *International Journal of Remote Sensing, 26*, 4293–4319.

Rudel, T. (2008). Meta-analyses of case studies: A method for studying regional and global environmental change. *Global Environmental Change, 18*, 18–25.

Saura, S. (2002). Effects of minimum mapping unit on land cover data spatial configuration and composition. *International Journal of Remote Sensing, 23*, 4853–4880.

Scanlon, T., Caylor, K., Levin, S., & Rodriguez-Iturbe, I. (2007). Positive feedbacks promote power-law clustering of Kalahari vegetation. *Nature, 449*, 209–212.

Schelhas, J., & Greenberg, R. S. (Eds.). (1996). *Forest patches in tropical landscapes*. Washington, DC: Island Press.

Scholes, R. J., & Archer, S. R. (1997). Tree-grass interactions in savanna. *Annual Review of Ecology and Systematics, 28*, 517–544.

Sellers, P. J. (1985). Canopy reflectance, photosynthesis, and transpiration. *International Journal of Remote Sensing, 6*, 1335–1372.

Smallie, J., & O'Connor, T. (2000). Elephant utilization of *Colophospermum mopane*: Possible benefits of hedging. *African Journal of Ecology, 38*, 352–359.

Southworth, J. (2004). An assessment of Landsat TM Band 6 thermal data for analyzing land cover in tropical dry forests. *International Journal of Remote Sensing, 25*, 689–706.

Southworth, J., Munroe, D. K., & Nagendra, H. (2004). Land cover change and landscape fragmentation: Comparing the utility of continuous and discrete analyses for a western Honduras region. *Agriculture, Ecosystems and Environment, 101*, 185–205.
Southworth, J., & Gibbes, C. (2010). Digital remote sensing within the field of land change science: Past, present and future directions. *Geography Compass*, *4*(12), 1695–1712.
Southworth, J., Hartter, J., Binford, M. W., Goldman, A. C., Chapman, C. A., Chapman, L. J., et al. (2010). Landscape effects of an East African national park on the protected tropical forest and its surrounding area. *Tropical Conservation Science, 3*, 122–142.
Spierenburg, M., & Wels, H. (2006). "Securing space": Mapping and fencing in transfrontier conservation in southern Africa. *Space and Culture, 9*, 294.
Steffen, W., Sanderson, A., Tyson, P. D., Jager, J., Matson, P. M., Moore, B., et al. (2004). *Global change and the earth system: A planet under pressure*. New York: Springer.
Stevens, M. D., Malthus, T. J., Baret, F., Xu, H., & Chopping, J. (2003). Intercalibration of vegetation indices from different sensor systems. *Remote Sensing of Environment, 88*, 412–422.
Stigand, A. (1923). Ngamiland. *The Geographical Journal, 62*, 401–419.
Struhsaker, T. T. (1997). *Ecology of an African rain forest: Logging in Kibale and the conflict between conservation and exploitation*. Gainesville: University Press of Florida.
Terborgh, J. (1999). *Requiem for nature*. Washington, DC: Island Press.
Terborgh, J., van Schaik, C., Davenport, L., & Rao, M. (2002). *Making parks work: Strategies for preserving tropical nature*. Washington, DC: Island Press.
Thenkabail, P. S. (2004). Inter-sensor relationships between IKONOS and Landsat-7 ETM+ NDVI data in three ecoregions of Africa. *International Journal of Remote Sensing, 25*, 389–408.
Tlou, T. (1972). *A political history of northwestern Botswana to 1906*. Dissertation, University of Wisconsin.
Tlou, T. (1976). The peopling of the Okavango Delta. In The Botswana Society (Ed.), *Proceedings of the symposium on the Okavango Delta and its future utilisation* (pp. 49–53). Gaborone: The Botswana Society.
Tlou, T. (1985). *A history of Ngamiland 1790 to 1906: The formation of an African State*. Gaborone: Macmillan Botswana.
Turner, M. G. (1989). Landscape ecology: The effect of pattern on process. *Annual Review of Ecology and Systematics, 20*, 171–197.
Turner, B. L., II, Skole, D., Sanderson, S., Fischer, G., Fresco, L., & Leemans, R. (1995). *Land-use and land-cover change science/research plan*. Louvain-la-Neuve: LUCC International Project Office/University of Louvain.
Turner, M. G., Gardner, R. H., & O'Neill, R. V. (2001). *Landscape ecology: In theory and practice*. New York: Springer.
Turner, B. L., II, Lambin, E. F., & Reenberg, A. (2007). The emergence of land change science for global environmental change and sustainability. *Proceedings of the National Academy of Sciences of the United States of America, 104*, 20666–20671.
Walker, B. H., Ludwig, D., Holling, C. S., & Peterman, R. M. (1981). Stability of semi-arid savanna grazing systems. *Journal of Ecology, 69*, 473–498.
Wallgren, M., Skarpe, C., Bergström, R., Danell, K., Bergström, A., Jakobsson, T., et al. (2009). Influence of land use on the abundance of wildlife and livestock in the Kalahari, Botswana. *Journal of Arid Environments, 73*, 314–321.
Walther, G., Post, E., Convey, P., Menzel, A., Parmesan, C., Beebee, T., et al. (2002). Ecological responses to recent climate change. *Nature, 416*, 389–395.
Wang, L., Sousa, W., & Gong, P. (2004). Integration of object based and pixel based classification for mapping mangroves with IKONOS imagery. *International Journal of Remote Sensing, 25*, 5655–5668.
Wang, H., Li, X., Long, H., Gai, Y., & Wei, D. (2009). Monitoring the effects of land use and cover changes on net primary production: A case study in China's Yongding River basin. *Forest Ecology and Management, 258*, 2654–2665.
Weisberg, P., Lingua, E., & Pillai, R. (2007). Spatial patterns of pinyon-juniper expansion in central Nevada. *Rangeland Ecology & Management, 60*, 115–124.

Western, D. (1989). Conservation without parks: Wildlife in the rural landscape. In D. Western & M. Pearl (Eds.), *Conservation for the twenty-first century* (pp. 158–165). New York: Wildlife Conservation International.

Western, D., Groom, R., & Worden, J. (2009). The impact of subdivision and sedentarization of pastoral lands on wildlife in an African savanna ecosystem. *Biological Conservation, 142*, 2538–2546.

Wiens, J. A. (1989). Spatial scaling in ecology. *Functional Ecology, 3*, 385–397.

Williamson, D., & Williamson, J. (1981). An assessment of the impact of fences on large herbivore biomass in the Kalahari. *Botswana Notes and Records, 13*, 107–110.

Wilshusen, P. R., Brechin, S. R., Fortwangler, C. L., & West, P. C. (2002). Reinventing a square wheel: Critique of a resurgent "protection paradigm" in international biodiversity conservation. *Society and Natural Resources, 15*, 17–40.

Wulder, M. A. (1998). Optical remote sensing techniques for the assessment of forest inventory and biophysical parameters. *Progress in Physical Geography, 22*, 449–476.

Wulder, M. A., Hall, R. J., Coops, N. C., & Franklin, S. E. (2004). High spatial resolution remotely sensed data for ecosystem characterization. *BioScience, 54*, 511–521.

Zeng, N., & Neelin, J. (2000). The role of vegetation-climate interaction and interannual variability in shaping the African savanna. *Journal of Climate, 13*, 2665–2670.

Part III
Institutions and Political Ecology Approaches

Part III brings together five cases discussing the evolution of institutional arrangements affecting natural resource management and urbanization in different parts of the Americas. They discuss trajectories of change in the context of interactions between local social-historical conditions, national policies, development projects, infrastructure change, and economic pressures and opportunities experienced by different social groups, from farmers and fishery communities to peri-urban and urban residents. They illustrate the sophistication of institutional approaches in incorporating multiple methodologies and research tools, including longitudinal and cross-sectional assessments, experimental economics, remote sensing and GIS, surveys, and institutional analysis frameworks (e.g., the Institutional Analysis and Development framework or IAD).

In Chap. 9, Castro analyzes the political and institutional history leading to recent and ongoing implementation of agro-extractive settlement projects (PAEs) in the lower Amazon floodplain. PAEs represent, in principle, a form of legal recognition of local resource management systems in the floodplain. This study analyzes the implementation process by focusing on the roles of four main stakeholders involved in 15 projects: federal agencies, NGOs, regional fisheries unions, and local residents. Castro sets this analysis by providing an overview of different lines of research on common-pool resource (CPR) management. He brings a unique perspective that benefits from long-term, longitudinal, and historical research dating back to the initial stages of comanagement systems in the floodplain. He analyzes how policies aiming at including local communities indirectly exclude other user groups, spurring tensions previously incipient in the area. He calls attention to the way conflicts develop as related to temporal and spatial dynamics of the resource system, i.e., marked by uncertainties regarding seasonal variations, land-water boundaries, and physical stability of the settlement and the opportunities and limitations created by cooperation among stakeholders to the performance of the comanagement systems in the region.

In Chap. 10, Lopez brings an experimental economics perspective to examine the role of incentives and sanctions on rural populations' behavior toward natural resources in Colombia. Her experiments on public goods and CPRs reveal the limitations and pitfalls of top-down sanctions on the management of natural resources In particular, Lopez calls attention to the limits and potentially counter-productive results of monetary penalties on small-scale farmers' land-use decisions.

In Chap. 11, Futemma takes a comparative perspective to examine small-scale farmers living within and around national forests in Brazil. She analyzes conflicts faced by the farmers to meet their economic goals and expectations to conserve forest resources. She also benefits from longitudinal research examining institutional and land-use change at household and community levels in the lower Amazon. The two distinct Brazilian regional realities, i.e., national forests in the states of São Paulo and Pará, offer the opportunity to contrast trajectories of agriculture and cattle raising, forest management and non-timber forest products, rural tourism, and the differential ways small-scale farmers search for alternative solutions to improve their livelihoods.

Another excellent example of the value of longitudinal research, in Tucker's Chap. 12, provides a careful review of phases of institutional evolution in rural Honduras and its implications for local well-being and forest environment. Tucker examines the evolution of local resource management institutions amid changes prompted by market expansion, infrastructure, and local social and cultural changes. The chapter offers richly detailed analysis of local changes. Longitudinal and systematic ethnographic and institutional analysis approaches are integrated to remote sensing analysis and socioeconomic indicators to examine the (mixed) outcomes of economic and institutional changes to the local population. In general, as in other cases discussed in the book, Tucker shows how local efforts to improve economic conditions and to develop institutions for forest management are hampered by lack of access to basic services, such as health and education.

Closing Part III, in Chap. 13, York and Munroe examine the process of urbanization and exurbanization in the United States, in particular the role of federal, state, and local land-use policies. The chapter considers the expansion of urban areas into agricultural and forest lands and discusses the implications of fragmenting urban and rural spaces to rural communities. Contextualized within a broader literature review and empirical research, it integrates policy analysis within a spatial framework to examine urban settlement cases in the states of Ohio, Indiana, and Arizona. It considers zoning, development impact fees, and public land designations. It calls particular attention to the impact of these policies on local land markets and, in turn, their role on the rate and intensity of expansion of urban settlements. York and Munroe highlight the differential role of federal and state institutions on urban land use in the American West and the importance of local zoning in the Midwest.

Chapter 9
Between Cooperation and Conflict: The Implementation of Agro-Extractive Settlements in the Lower Amazon Floodplain

Fabio de Castro

Abstract In recent years, numerous territorial categories have been created in order to integrate local management systems into a broader legal framework. Despite the political advance of this institutional innovation to include marginalized groups, different perceptions and motivations among key stakeholders in the implementation process may challenge the performance of such initiatives. This chapter is focused on the implementation of a new territorial model for the Amazonian floodplain aimed at combining conservation, social inclusion, and local development. A diachronic analysis of the engagement of four main stakeholders reveals a tension between cooperation and competition among them in which power relations, leadership, and economic incentives play central roles. This study illustrates the importance of coupling cooperation and conflict analysis in co-management studies.

9.1 Introduction

In recent decades, natural resource management has gone through major transformations in many developing countries. From centralized, top-down, blueprint strategies, a large range of collaborative initiatives including state and non-state actors have begun to be developed (Armitage et al. 2008; Borrini-Feyerabend et al. 2007). This shift in the natural resource management paradigm is a result of changes in perception and motivations of researchers, practitioners, and policy makers.

This research has been financed by WWF-Brazil, CEDLA, and the European Commission (FP7-SSH-CT-2010-266710). I thank the staff at the Institute for Environmental Research in Santarém for their logistic support during my visits to the field.

F. de Castro (✉)
Centre for Latin American Research and Documentation (CEDLA), Keizersgracht 395-397, 1016 EK Amsterdam, The Netherlands
e-mail: f.decastro@cedla.nl

E.S. Brondízio and E.F. Moran (eds.), *Human-Environment Interactions: Current and Future Directions*, Human-Environment Interactions 1,
DOI 10.1007/978-94-007-4780-7_9,

Since the seminal conference on the "commons" (NRC 1986), a large number of disciplines have been drawn to a joint research agenda regarding the relationship between property rights and patterns of natural resource use. The collection of numerous case studies of local management systems covering several resources and regions worldwide supported the development of a robust "theory of the commons." The ultimate recognition of this effort came with the Nobel Prize in economic sciences being awarded to its leading theorist, Elinor Ostrom, in 2009. A large body of theoretical, empirical, and methodological research supports the claim for the potential of long-standing informal institutions to promote sustainable production and improved local livelihoods.[1]

In addition to the academic community, both state and non-state actors underwent major transformations in policy-making and advocacy work. The democratization process in many developing countries favored the revitalization of social movements and their active participation in the decision-making process through decentralization measures (Ribot 2004). At the same time, alliances between local and international organizations strengthened the political and economic power of local actors, creating new paths for advocacy work (Keck and Sikkink 1998). As a result, from being virtually invisible in conservation policies, local management practices have gradually been incorporated into broader co-management systems worldwide (Armitage et al. 2008).

Co-management systems usually combine social and ecological goals, which can be conflictive. Moreover, it often assumes homogeneous communities (Gibson and Agrawal 2001) with conservation goals (Castro et al. 2006). However, co-management systems bring together a large number of state and non-state actors (Berkes and Pomeroy 1997). The complexity of social relations among different actors leads to cross-scale connections where power plays a major role in the configuration of social networks (Cash et al. 2006).

The configuration of social networks in this process is a key contextual component that reveals how interactions at different scales can promote or prevent cooperative behavior (Dietz and Henry 2008) and how winners and losers are defined in this process (Adger et al. 2005). While most co-management studies focus on factors influencing cooperative behavior (Berkes and Folke 1998; Jentoft 2005; Jentoft et al. 1998; Ostrom 1990), I argue in this chapter that a coupled analysis of conflict-cooperation is needed in order to better understand the social dynamics among actors in shaping sustainable co-management systems and evaluate their potential and shortcomings for sustainable use of natural resources.

The performance of social networks in a co-management system depends on the features of the stakeholders involved in the process and structural factors influencing their behavior (Adger et al. 2005). Although each stakeholder holds common history and position in the social network, internal differences in motivations and perceptions among individuals strongly influence their level of cohesion. Therefore,

[1]A database of research on the commons is freely available at http://dlc.dlib.indiana.edu/cpr/index.php

a diachronic and disaggregated analysis of stakeholders helps to reveal the motivations behind their conflictive and cooperative behaviors in co-management systems.

The aim of this chapter is to understand the interplay between internal and external conflict and cooperation among stakeholders and how it may influence the performance of co-management systems. This analysis will be based on the establishment of agro-extractive settlement projects (PAEs[2]) in the Lower Amazon floodplain in Brazil. PAE is a settlement category created by the Brazilian Institute for Agrarian Reform (INCRA) for traditional communities to fulfill the demand for exclusive-use rights for local residents. Based on concession of collective rights, the residents are entitled to a small private lot for residence and farming activities and a large area where natural resources (e.g., fish, grazing area, and flooded forest) are collectively managed.

This chapter focuses on the role of four main actors involved in the implementation of 15 settlements: the government agency INCRA, the nongovernment organization (NGO) Institute for Environmental Research (IPAM), the grassroots organization Fishers' Union, and the floodplain dwellers. Despite the involvement of numerous other organizations in different stages of this process, the scope of this chapter is limited to the social interactions of these four stakeholders. First, I briefly describe the methodological strategy of my research and my special position in this long-term project. The second section presents the historical context of the creation of PAEs and how each actor has been engaged in the process. In the third section, I present the historical background, current structure, role, and motivations of each actor and analyze how internal social dynamics and individual leadership have influenced their conflictive and cooperative behaviors. Two main aspects raised in this analysis are discussed in the fourth section: the role of social heterogeneity driving conflictive behavior and the role of individual leaders' promoting collaborative behavior. In the last section, I present some concluding remarks.

9.2 Methodological Strategy

Although PAE is a recent initiative in the Lower Amazon, it represents a final stage of a process of local and regional organization for sustainable production initiated in the region during the early 1990s. I have been able to follow this process from its early stages. Between 1990 and 1994, I was part of the NGO-led *Várzea* Project and initiated my research in the region. Between 1994 and 1999, during my doctoral program, I became collaborating researcher to the *Várzea* Project, which allowed me to maintain contact with the activities in the region. From 2000 to 2002, I was consultant for the bilateral cooperation research program of PPG7 *Provarzea* and was directly involved in discussions of the institutional advances

[2]From the Portuguese Projeto de Assentamento Agroextrativista.

of local management projects. Finally, since 2008, I have been consultant for the World Wildlife Fund (WWF) to support the Monitoring and Evaluation program of the *Várzea* Project.

My continuous presence in the area in almost all phases of this 20-year process, access to a rich source of data, and a view "from within" of the NGO activities have given me the unique opportunity to assess the everyday life aspects of design and implementation process of PAEs. The analysis is based on data collected throughout the whole process. It combines an extensive quantitative and qualitative analysis of the communities in the region as part of my dissertation in the mid-1990s with more recent data from interviews carried out with representatives of all four stakeholders. Archival data, observations at several meetings, and informal conversations with several individuals complement the contextual analysis.

9.3 From Local Management to Co-Management

Local management systems in the Amazonian floodplain date back to pre-Columbian times (Castro 2002). However, a more conscious management system based on proactive collective action and clear rules of resource use emerged in the 1980s in response to the fishing intensification in the region (McGrath et al. 1993). The "fishing accord," developed with support from the Catholic organizations in the region, is a management concept based on a lake zoning system (e.g., no take, managed, open access) and a set of rules for natural resource use and monitoring systems (Castro 1999). Until the mid-1990s, the fishing accords were mostly fragmented, each created by a single community, and focused primarily on fishing resources (Castro and McGrath 2003). More recently, the floodplain has experienced several important changes. Politically, communities became more organized, and "glocal" alliances between local and international organizations for sustainable production were consolidated. Economically, cattle ranching intensified in the floodplain (Merry et al. 2004), commercial fishing activities increased from both community and urban fishers, and the flow of money into the community was improved via several governmental social benefits (e.g., pension, unemployment salary, conditional cash transfer program, or *Bolsa Familia*). As a result, new socioenvironmental demands arose in the region. The conflict between fishing management and other land-use activities became more evident, and the tension between community residents and cattle ranchers amplified (McGrath et al. 2007).

In the mid-1990s, a gradual process of integrated co-management began to take shape in Santarém through the collaboration of the Brazilian Institute of Environment and Natural and Renewable Resources (IBAMA) and the Public Ministry. The change was based on three pillars: (1) formalization of the fishing accords and elaboration of a local monitoring system involving voluntary environmental agents of IBAMA, (2) formulation of agreements between local residents and cattle ranchers regarding land-use activities, and (3) the creation of regional fisheries councils represented by different stakeholders to discuss and define solutions for sustainable production in the floodplain (McGrath et al. 2008).

Despite the advances of this co-management system, the lack of exclusive property rights for community residents, limited economic opportunities, and organizational constraints remained major bottlenecks for the sustainability of the co-management system. As a result, only a few communities that combined favorable socioenvironmental conditions to support the fishing accords were able to maintain their local management systems (Almeida 2004; Castro 1999). Aquatic systems being state property in Brazil limited the ability of local residents to monitor use and sanction violators and to benefit from credit lines (Benatti 2005). In addition, the co-management system was limited to conservation goals and did not address economic demands necessary to support the management system (McGrath et al. 2008). It became clear that fishing accords were not only about fishing but also property rights and food security (Castro 1999). In sum, the lack of an integrated policy system for the floodplain led to punctuated, disconnected initiatives, some more successful than others due to local contextual factors (Castro et al. 2002).

A first attempt of the government to address those issues as a pilot project took place in four communities in the early 2000s. In this initiative, the informal land tenure system was to be regularized as exclusive-use rights and integrated with a collective management system of the floodplain resources (e.g., fish, grassland, flooded forest). This project was supported by both community residents and cattle ranchers as exclusive rights to their informal properties were expected. However, the complex bureaucratic process, limited human and financial resources, and the short-term contract for concession of exclusive use limited the success of this initiative. In addition, most of the informal properties fell under (for community residents) or above (for cattle ranchers) the rural module of 80–100 ha established by INCRA (Benatti 2005). As the first experience of INCRA's involvement in the floodplain, it became evident that the socioenvironmental reality did not fit well into the predefined standards for the upland, and the project was discontinued.

In 2005, government organizations, NGOs, grassroots organizations, and floodplain dwellers resumed the discussion in search of a co-management system for the region, and PAE was chosen as the most appropriate land tenure model for the floodplain. In 2006, a technical cooperation was established among three government agencies—the agency in charge of floodplain tenure governance (GRPU), the agency in charge of agrarian reform (INCRA), and the agency in charge of environmental protection (IBAMA). In this partnership, INCRA was designated as the lead organization in charge of the creation and implementation of approximately 50 PAEs across the region.

Despite the promising prospects of the PAE, the steps taken in the first stage of this initiative were far from ideal. What was supposed to be a process to fulfill local demands became a top-down, blueprint, technocratic process. Although the creation of a PAE is triggered by a formal request from the local residents, the Fishers' Union, as the representative organization of the floodplain communities, issued the request to INCRA for the whole region without previous consultation with local residents. Between July and December 2006, approximately 50 PAEs were demarcated in eight municipalities, encompassing an area of approximately 750,000 ha. Ambiguous information, lack of transparency, and distrust in government agencies by floodplain dwellers triggered opposition from both community members and cattle ranchers who felt threatened by this top-down process. They questioned the difference between the individual and collective land-use concessions, eligibility criteria for the program, economic benefits, accountability, and the position of cattle ranchers.

INCRA's strategy to convince community residents was twofold. First, they stressed the economic advantages of improved infrastructure, small grants, and access to credit lines that would follow. Second, the participatory management plan would empower community residents since grazing areas traditionally held by cattle ranchers would turn into collective areas. As expected, the support gained from most of the communities was confronted by major opposition from cattle ranchers, who tried to obstruct the process several times with no success. In 2007, INCRA underwent an internal restructuring process, and the implementation of PAEs resumed in 2008 with a new team of recently hired staff. This phase was initiated in collaboration with IPAM, which had been working closely with communities in the region since the early 1990s. INCRA and IPAM signed a cooperation agreement for the elaboration of the management plan for 15 PAEs in the Lower Amazon. IPAM proposed to bring its knowledge of the region, experience with community work, and mobilization capacity to develop a participatory management plan in all 15 PAEs.

In sum, the creation of a PAE is an outcome of a gradual process of incorporation of informal local practices into an integrated floodplain tenure framework. The community-based, fishing-oriented management systems of the 1980s are now part of a regional-based, floodplain-oriented conservation and development program including land tenure, sustainable production, and rural development. Community residents have been included as key players in the management plan by integrating their local knowledge, ruling system, and social practices. Cattle ranchers have been excluded from the process and are expected to be dispossessed of their traditional private ownership. In this process, where numerous actors are involved with more or less relevance in different stages and procedures, the social interactions are far from smooth. Actors differ in their motivations, perceptions, and expectations in this endeavor. Moreover, each actor faces internal social dynamics which adds to this complex network.

9.4 The Four Key Stakeholders

The implementation of a PAE brings together four heterogeneous actors in terms of type and level of organization, role in the process, and level of interaction with one another. This section highlights the internal and external social interactions of each actor, namely, INCRA, the *Várzea* Project of IPAM, the Santarém Fishers' Union, and the floodplain communities.

9.4.1 INCRA

INCRA is the federal agency in charge of the creation and implementation of the PAEs in the region. It was created in 1970 during the military dictatorship to carry out the agrarian reform, manage the national cadastral record of rural properties, and administrate the national land property. INCRA has traditionally carried out top-down, blueprint, upland settlement based on individual land property for migrant peasants and provision of infrastructure, credit lines, and technical assistance (Alston et al. 1999). Only recently has INCRA incorporated conservation and social inclusion of traditional populations in its policy agenda, with creation of three special settlement categories—PAE, Sustainable Development Project, and Forestry Settlement Project.

In general, INCRA settlement projects are motivated by political pressure such as land conflicts, demands from social movements, and election campaigns. Pará is the only state with three regional offices due to its size (second-largest state with 1,247,689 km^2) and being the stage of major land conflicts (Simmons 2004). The Santarém regional office was created in 2005 in response to the aggravated land conflict in Pará that had led to the murder of Catholic nun Dorothy Stang. In order to fulfill the political demand of settlement creation in the region, the new regional office set high targets. In the first 2 years (2005–2006), more than 51,000 families were settled in the region, which corresponded to 20% of the total beneficiaries in the whole country and 50% in the state of Pará at that time. The fast pace led to an intervention of the Public Ministry due to irregularities in the process, causing the resignation of the head of the INCRA Regional Office.

While the main political focus of INCRA is on the upland, the regional office covers many floodplain communities where traditional populations have lived for generations but hold no formal title to their land. This area represents a potential focus for establishment of new settlements to be included in the office's target. Creation and implementation of settlements in the floodplain has many technical advantages over upland settlements. First, the floodplain represents fewer political challenges than the upland, violent events are less frequent, and conflicts between social movements and rural elite are less polarized. Second, settlements in the upland are implemented from scratch, whereas people have been living in the floodplain for a long time, and social organization and basic infrastructure are already in place. Therefore, transaction and implementation costs are relatively low. Third, the main implementation cost of upland settlements (e.g., roads) is not applicable to the floodplain, where the transportation system is mainly aquatic. Therefore, from INCRA's perspective, creation and implementation of floodplain settlements is a relatively easy task when compared to the upland reality.

What seemed to be easy turned into a major fiasco. INCRA staff's lack of previous experience in the floodplain led to a direct transfer of its upland settlement model. However, specific socioenvironmental features of the floodplain demand context-specific solutions for infrastructure, economic alternatives, and technical assistance. In addition, the human occupation pattern, diversified socioeconomy, and political organization require special attention on the participatory mechanisms and design of integrated management (Futemma et al. 2002; McGrath et al. 2007; WinklerPrins 2002). When the implementation process started in 2008, the picture had changed remarkably. More than 150 new staff members had joined INCRA as part of a decision to expand regional offices nationwide. The search for and selection of new employees at the national level had two major impacts in the organization. First, it constrained the influence of local elite in the selection process. Second, it attracted a number of young professionals committed to social justice.

Despite opposition from the older generation, who had been in office for more than 20 years, the "new blood" quickly turned the regional office into a vibrant group committed to develop new alternatives to the top-down, blueprint procedures formerly adopted in the region. Interesting enough, this new generation was opposed to partnerships with NGOs. For them, the neoliberal policies and "glocal" alliances of the 1990s had set the state aside, and the current repositioning of INCRA depended on state control over the process. For this reason, the proposal of INCRA's PAE coordinator to team up with the *Várzea* Project was not well received by his colleagues. Only after a long process of discussion and persuasion was the coordinator able to gain support from the National Coordination in Brasilia and some colleagues in Santarém. The resistance to engage in partnerships with the NGO remains among many INCRA staff. However, the collaboration between the two organizations has been successful due to the active role of the local coordinators at INCRA and IPAM.

In sum, INCRA's motivation to create the PAE in the floodplain was mainly driven by the pressure to boost the creation of settlements in the region. The team was initially composed of staff trained in technocratic, blueprint-based, upland-oriented procedures designed during the military government. Their perception of a top-down approach as the appropriate way to create and implement settlement projects and their convincing methods based on financial incentives (e.g., infrastructure, grants, and access to credit) accentuated the clientelist approach and underestimated the social capital built from years of local and regional organization in the region. Likewise, the argument toward exclusion of cattle ranchers in the process ignored the social connection and potential conflict between the two local actors. Only upon the arrival of new staff was an alternative approach developed. Conflicts at INCRA between older and newer staff groups over top-down procedures, among new staff over partnerships with NGOs, and between upland-oriented and floodplain-oriented staff were some of the challenges for the development of an alternative settlement model for the floodplain. A unique aspect of this process was the action of the PAE coordinator to seek support from a local NGO to improve the institutional learning process, communication among stakeholders, and participation of the community residents in the process (Table 9.1).

Table 9.1 Features of the four main actors involved in the implementation of the PAE in the Lower Amazon

	Level	Role	Motivation	Internal conflicts	External conflicts
INCRA	National	Legal framework Economic support Infrastructure Technical assistance	Political pressure National target Institutional innovation Inclusive policy	Old/New staff Upland/Floodplain perspective Pro/Against NGO partnership	Local rural elite
Várzea *Project*	Regional	Participatory management plan Legal advice	Financial input Local commitment Political influence Professional recognition	Research/ Advocacy work International/ Government donors	Local rural elite
Santarém Fishers' Union	Municipal	Formal request to create PAE Unemployment salary	Political prestige Increased representativeness	Urban/Rural fishers	Large-scale fishers
Community Residents	Local	Local knowledge input Management activities (council)	Exclusive use rights Financial input Empowerment Sustainable production	Religion Access to resource Violation of local rules	Urban fishers Ranchers

9.4.2 *Várzea* Project

The *Várzea* Project is in charge of the management plan of 15 PAEs in the region. The presence of the *Várzea* Project in the region dates back to early 1990s and can be divided into three main phases. The first phase (1991–1994) of the *Várzea* Project was a small project of a nonlocal NGO focused on three floodplain communities. This phase was characterized by diagnostic-oriented research to identify the key aspects of the floodplain management system in the region. Limited funding, a small research team, and the semiprofessionalized character of the NGO limited the research to a few sites and reliance on close contact with community members for data collection. The small-scale approach of the research and frequent contact with several community members helped to build trust relationships between the studied communities and the project team and supported the move to the second phase (1994–2007). With a large grant from WWF to focus on one site (Ituqui Island), the initial stage of this phase (1994–1997) was marked by a sudden change in the project structure and organization. The project staff mushroomed to a couple dozen members divided into five subprojects on production, management, education, health, and capacity building. The *Várzea* Project became part of IPAM in 1995 upon the foundation of the NGO, with an office in Santarém. The wide range of professionals, sizeable funding, and multiple goals (e.g., research, capacity building, advocacy) led the project to internal conflicts. Different components of the project ran parallel, with little cross-fertilization, and staff members often engaged in conflicts stemming from different conceptual perspectives and academic statuses.

The research component had limited direct return to the community. Apart from the numerous academic publications, community residents had little feedback on data collected from numerous interviews, questionnaires, and measurements of production systems. On the other hand, the capacity building and advocacy components developed more smoothly in response to local demands such as local training programs, political organization, and legal orientation. This period (late 1990s) coincided with a vibrant debate among stakeholders on alternatives for co-management and sustainable production in the floodplain (McGrath et al. 2008). Driven by this new demand, the *Várzea* Project gradually moved to a strong focus on environmental education, capacity building, and policy-making, and expanded its activities to other communities.

One important figure in this process was the lawyer in charge of the policy-making component of the project. Based on a legalist approach, she worked closely with community leaders, the Fishers' Union, and government agencies. She played a key role in the design and creation of community (e.g., community associations) and regional (e.g., regional fisheries councils) organizations, reformulation of the structure of the Fishers' Union, and formalization of fishing agreements in the region. Born in the region, and with strong communication skills, her ability to interact with different stakeholders (e.g., community residents, grassroots organizations, researchers, policy makers) helped to facilitate the connection among them, leading to a broad social network in the region.

By 2007, the *Várzea* Project enjoyed a reputation as a key organization in supporting local democracy and increased participation in sustainable production of floodplain resources. However, as a professionalized NGO, its institutional stability was vulnerable to lack of continuity in funding. The third phase (2007–present) is marked by the phasing out of WWF funding after 15 years of partnership with the project and the beginning of partnership with INCRA in the implementation process of the PAEs. The reduced funding led to a major drop in their staff. Unable to continue with the large volume of activities, many initiatives were discontinued. With no other funding alternative at hand, and a continued interest in playing an active role in this process, the *Várzea* Project took on the challenge to team up with INCRA in the implementation process of the 15 PAEs.

The partnership with INCRA had major impacts in the *Várzea* Project work. Understaffed and overcommitted to the targets set by INCRA, the former structure of specialized teams working independently in subprojects shifted to a single multidisciplinary team working together in multiple tasks. For the first time, the *Várzea* Project enjoyed cross-fertilization of different perspectives, well-coordinated work with the communities, and more integrated research/advocacy work. The process went relatively smoothly in the four PAEs of Santarém, where the project has had a long-term presence. However, the limited organizational capacity and experience in working with external agencies by local communities created major barriers for the *Várzea* project to replicate their participatory approach as efficiently in other municipalities.

In addition, shifting from an independent organization funded by international agencies such as WWF and the United Kingdom's Department for International Development (DFID), the *Várzea* Project had to cope with the rigid, bureaucratic institutional framework of INCRA in this new partnership. The conflict between WWF-DFID and INCRA goals directly reflects on their activities. Like INCRA, WWF-DFID functions under political pressure and targets and depends on expected outcomes from the *Várzea* Project to report back to their own sponsors. While WWF seeks to partner with the project to contribute with improved livelihood, sustainable production, and institutional organization in the region, INCRA seeks partnership to finalize the implementation of 15 PAEs. Squeezed between both funders, the *Várzea* Project juggles its limited time, budget, and personnel to combine both (sometimes conflicting) goals.

In short, the *Várzea* Project has gone a long way to build trust and recognition in the region as a key partner of the local communities. However, the project has undergone major internal conflicts due to different motivations for and perceptions of the project's role in the region. More recently, as the team became more integrated, it has faced conflict between the goals of its two main funding agencies. The leading role of the local coordinator in focusing the project toward capacity building, policy-making, and regional network building opened new channels of active participation in the co-management system in the region. The partnership with INCRA provides the financial stability and an active role in the implementation of the PAEs but under major constraints and risks. Its partnership with WWF gave it recognition as an independent organization; however, it opened it up to risks of funding instability.

9.4.3 Fishers' Union

The Fishers' Union had a key role in the creation of the PAEs as the representative organization that issued the formal request to INCRA on behalf of the community residents (see Table 9.1). The Fishers' Union represents, therefore, the legitimization of the PAE as a bottom-up process. However, as discussed earlier, the Fishers' Union has assumed the creation of the PAE as the realization of an "old dream" of all communities and, therefore, that it had no need for local consultation. In order to understand the role of the Fisher's Union in Santarém in this process, it is important to look into the historical trajectory.

The Fishers' Union is a nationwide organization created in 1912 by the federal government with the goal to protect the national territory, reduction of class antagonism, and consolidation of Brazilian nationalism. Through a top-down process and compulsory registration, the Fishers' Union served as a depoliticizing machine, controlled by local political and economic elites (Breton et al. 1996). From the 1940s, when compulsory registration was abolished, the Fishers' Union was consolidated as an organization of large-scale fishers. After the democratization process in the 1980s, some local units of the Fishers' Union were gradually taken over by small-scale fishers. In Santarém, where the Catholic Church has had a strong political involvement in favor of minorities, the Fishers' Union was one of the first in this process. However, the Fishers' Union did not become a well-represented organization until 1994, when there was an economic incentive to affiliate to the Union. As of 1994, professional fishers affiliated with the Fishers' Union became eligible for unemployment benefits during the annual closing of the fishing season (3 months). In 1 year, registrations with the union grew manifold. As part of an agreement between the Fishers' Union and the Rural Workers' Union, it was established that the former would represent floodplain communities and the latter would represent upland communities. This period coincided with the institutional development of regional management of the fisheries and the creation of Regional Fisheries Councils, in which the Fishers' Union played a major role.

Although the Santarém Fishers' Union has become a political representative of small-scale fishers in the municipality, a strong subclass division characterizes this organization—community and urban fishers. The former prioritized their local management systems due to conflicts with outsider fishers. According to them, urban fishers violate community rules and, therefore, should not be allowed to fish in the community lakes. They claim that urban fishers have free access to the river where they can fish. The latter condemned the local management systems and argued that they lack their own lakes to manage and means to farm in the city. Therefore, they claim that community lakes should be accessible to all fishers. The conflict between these two groups regarding the local management system is reflected in the Fishers' Union organization. Until the mid-1990s, urban fishers controlled the Fishers' Union direction, and local management systems received little attention from the union. The increased political and economic visibility of the Fishers' Union created incentives for floodplain community leaders to organize themselves and compete in the union elections.

Today, community fishers are politically stronger in the organization, and the division between urban and community fishers is evident. The Fishers' Union has approximately 8,000 registered members, from which 20% are urban fishers. While seven Regional Fisheries Councils were created in the floodplain in the 1990s, including 180 communities, the Urban Fisheries Council was not created until 2005. According to its coordinator, urban fishers are underrepresented because they lack incentive to join the Fishers' Union. In general, they are not eligible for unemployment salary because they need to prove major income from fishing to be considered a "professional fisher." However, considering the limited access to fishing grounds, lack of access to farming systems, and high living costs in the city, urban fishers are forced to combine other activities in their economic portfolio (e.g., construction work, informal business).

In sum, the Santarém Fishers' Union has grown to become an important political organization for the small-scale fishers. The main motivation of the Fishers' Union in the implementation of the PAEs was to increase its political influence as representative of community residents. However, the exclusion of other constituencies (e.g., urban fishers) in the design and implementation of the PAEs only increased the internal conflict in the organization.

9.4.4 Floodplain Dwellers

The floodplain dwellers are the main beneficiaries of the PAEs. The 15 PAEs implemented in the Lower Amazon correspond to an area of approximately 230,000 ha covering five municipalities, including around 100 communities and more than 4,000 dwellers (Fig. 9.1, Table 9.2). Each PAE is composed of riparian communities (between 1 and 14 communities) and cattle ranches (from one to several large properties). A typical riparian community is composed of 30–60 families, living on high levees along side channels. A few collective buildings can be found in the community core, such as churches, a school, and health and community centers. The communities are organized in associations or leadership groups, and social activities include soccer teams, mothers' clubs, and catechist groups, giving a sense of local identity to their residents (Castro 1999).

Despite the relatively standard social structure and economy, variability across communities (Castro 2009) and across households within communities (Futemma et al. 2002) is rather common. Internal differences among community residents can lead to conflict regarding access, use, and management of the floodplain resource. One main source of social division in communities is religion. From a traditionally Catholic orientation, most of the communities have experienced a rapid growth of Pentecostal churches in the region. The emergence of new religious segments has created another identity layer based on "brotherhood in faith" among Pentecostals. In some cases, it has affected the community identity by triggering internal conflicts between followers of different religions regarding their perceptions toward natural resource use. On one hand, the leadership work of the Catholic Church has strongly influenced the posture of Catholics toward conservation issues, defending their rights to access and control their resources, while Pentecostals tend to believe that the resource supply is protected by divine forces. On the other hand, Catholics are more flexible on their duties, including their commitment to collective action, which can threaten the covenant, while Pentecostals tend to adhere more closely to their personal commitment to their "brothers and sisters" (Castro 1999). Their conflictive behavior and different perceptions toward natural resources can have major impacts on how these two groups can collaborate. In extreme cases, the split of the community due to religion-related conflict has happened in a few communities in the region.

Differences among communities sharing a PAE are also common. Communities enjoy different accesses to natural resources according to their physical locations.

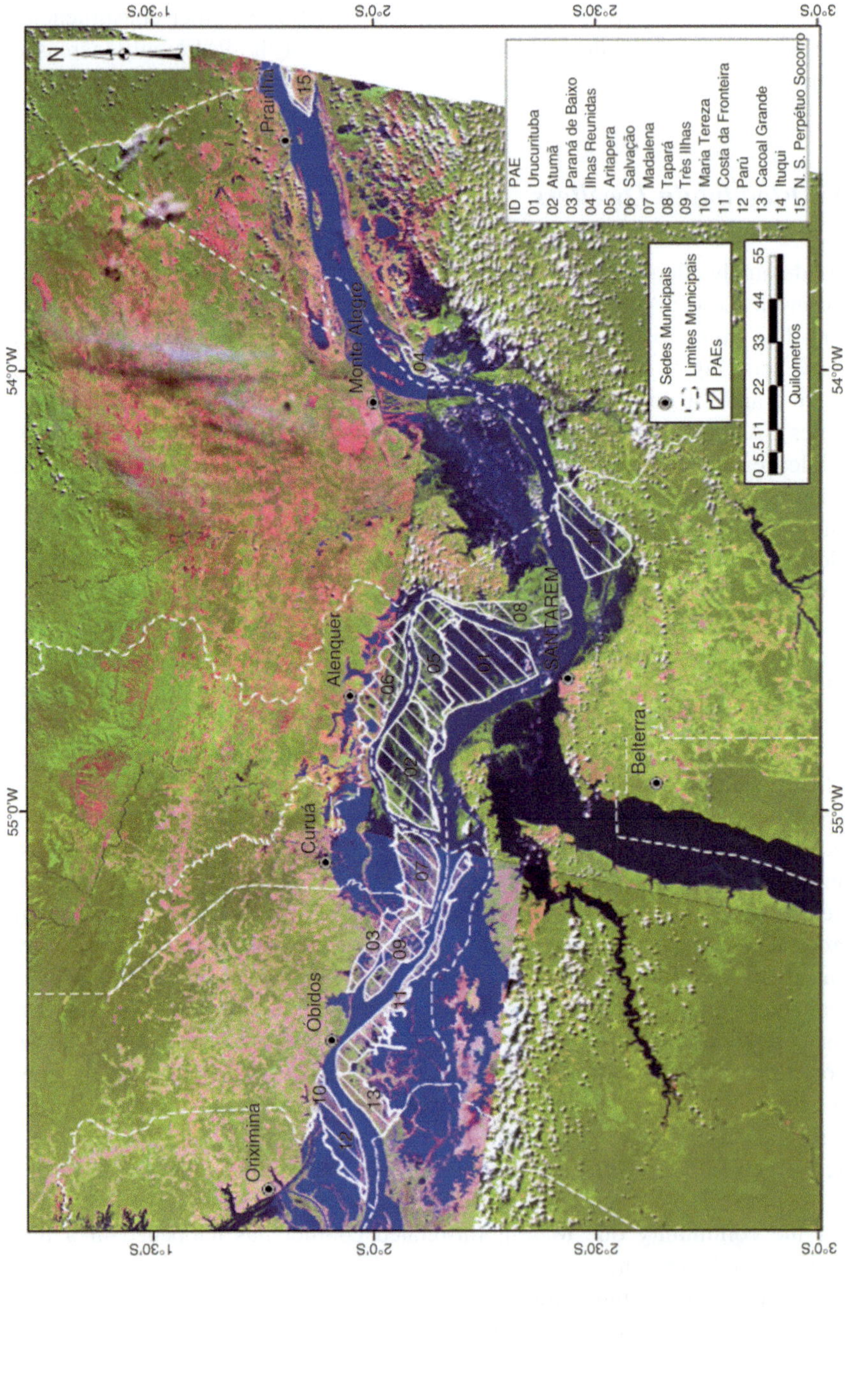

Fig. 9.1 Spatial distribution of 15 PAEs in the Lower Amazon

Table 9.2 List of the PAEs in implementation phase in the Lower Amazon

Municipalilty	PAE	Area (ha)	Communities (#)
Obidos	Paraná de Baixo	9, 999	7
Obidos	Três Ilhas	8, 960	8
Obidos	Maria Tereza	13, 400	1
Obidos	Costa Fronteira	15, 900	9
Obidos	Paru	15, 920	4
Obidos	Cacoal Grande	10, 480	4
Curua	Madalena	15, 850	11
Alenquer	Salvação	19, 000	3
Alenquer	Atumã	33, 500	12
Santarem	Aritapera	25, 000	14
Santarem	Urucurituba	19, 500	6
Santarem	Tapará	11, 700	9
Santarem	Ituqui	21, 920	5
Prainha	Ilhas Reunidas	6, 000	3
Prainha	N.S. Perpétuo Socorro	6, 527	2
		227, 557	100

Source: IPAM

Island communities are usually limited by the flooding season whereas riverbank communities enjoy access to uplands year round (Castro 2009; Futemma et al. 2002). Economic alternatives are more available in riverbank communities, including cattle ranching due to minimized transportation costs between floodplain and upland (Futemma et al. 2002). The island communities, on the other hand, are more vulnerable to environmental risks.

Finally, political organization of communities is also diverse. Due to a long-term presence of NGOs and government-led projects in Santarém, communities in this municipality are usually organized into associations, have representation in the regional councils, and have experience with NGO work. Community leadership has developed particularly well in Santarém as a result of the activities of different projects. Some of those leaders left the community to occupy key political positions such as president of the Fishers' Union, city representative, and regional secretary of the state fisheries office. Their influence in bringing the floodplain residents' demands to the agenda of those organizations was fundamental to this process.

Interesting enough, communities that were more resistant to joining a PAE were the most organized ones. In two cases of strong resistance to the PAE, community leaders were well prepared in the meetings with representatives of INCRA and the *Várzea* Project and questioned the reasons for the project. They stressed their lack of trust in the government organizations and their pride on their community work, of which they did not want to lose control. The financial promises from INCRA worked more easily on less organized communities that had less ability to promote discussion at the community level. In sum, political, economic, cultural, and environmental differences within and among communities can lead to conflictive behavior and influence the willingness of individuals to collaborate.

An important level of interaction in the floodplain is the social relation between community residents and cattle ranchers. Cattle ranchers are local elite who (informally) own large tracks of grassland to raise cattle and water buffaloes. Often, large-scale owners are politicians, own commercial establishments in the city, or hold relevant social position in the region (e.g., doctors, lawyers). The division between those two local actors is not completely clear. Community residents have gradually become more involved in cattle ranching since 1980s (Castro 2002). The insertion of the local residents in this activity has been facilitated by a partnership system developed in the region, based on informal agreements with outside cattle owners who send their herd to the floodplain for local residents to take care of in exchange for some of the calves (Merry et al. 2004). This informal agreement creates a dependency of some local residents on large cattle ranchers to start their herds. In addition, those residents often feel personally committed to large cattle ranchers who also provide personal help in odd situation (e.g., small loans, health, contacts in the city).

Cattle play a key economic and cultural role in the household economy and could become the major activity of some local residents (Castro 2009; Merry et al. 2004). It works as a way to save money for eventual use in case of emergency (Merry et al. 2004). In addition, the economic prosperity of raising cattle drives residents to regard cattle ranchers as having a higher economic status when compared to fishers.

The close relationship between cattle ranchers and some community residents contrasts with the growing conflict between these two groups in the last years. In addition to the impact on farming and fishing systems, the large ranchers represent a social class that limits local residents' access to fishing grounds. Despite numerous conflicts, residents tend to favor management rules over banning the care of outsiders' livestock (except for water buffaloes). In the last 10 years, more than 50 cattle management agreements have been created in the region (McGrath et al. 2008). Therefore, more than controlling the cattle ranching activities, the residents are more prone to increase their power over the cattle ranchers in the region.

A preliminary analysis of the rules proposed by community residents in the four PAEs in Santarém reveals an emphasis of the management plan proposed by community residents on activities carried out by large cattle ranchers. Many community residents stressed in their interviews their expectation to "finally stop the expansion of large ranchers in their lands." The threat felt by cattle ranchers in the region led them to react through their own union (SIRSAM) to lobby with local and state politicians. Until 2007, they promoted several attempts to stop or redesign the implementation of the PAEs in the region, with no success. Since then, cattle ranchers have stopped attending the participatory meetings. Interviewees from other groups are unanimous in the opinion that the silence of cattle ranchers will be broken as soon as the management plans are approved and major conflicts will emerge. In sum, from an empowered position in the floodplain, community residents expect cattle ranchers to become a mere occupant with limited access to the local resources dictated by PAE rules. However, the long and complex social interaction between large cattle ranchers and community residents may limit the power of the formal rules to prevail. How much the management plans will play a role in this new picture will depend on the historical and current social contexts of each PAE.

9.5 Discussion

In the last decade, international donors, the national government, and grassroots organizations embraced the concept of co-management motivated by different goals and incentives. For international agencies, co-management represented a way around the national state by providing funds directly to NGOs; for the national government, decentralization of management tasks represented saving costs from their limited budget and staff; for the grassroots organizations, active participation in management decisions represented a step forward in social inclusion and local empowerment. The mixed goals and motivations lead to complex social relations and diverse outcomes.

The assessment of a large number of cases has pointed out three main sides of this process. An institutional-oriented approach highlights the role of institutional design in creating opportunities to control perverse incentives, tackle new challenges on time, and connect with multiscale arrangements (Agrawal and Ostrom 2001; Gibson et al. 2000). A political-oriented approach highlights the role of power structure and everyday life politics turning promising initiatives into a political game with little change on the ground (Larson and Soto 2008; Ribot 2004). A leadership-oriented approach highlights the importance of individuals in facilitating or preventing cooperation (Andersson et al. 2006; Hochstetler and Keck 2007). All three approaches reveal the importance of contextual factors in different phases of the process of the institutionalization of local practices. The required decentralization to include local users, for example, can lead to perverse outcomes such as elite capture (Larson and Soto 2008) and incomplete implementation due to limited institutional capacity from both state and non-state actors (Brown and Rosendo 2000; Chuenpagdee and Jentoft 2007; Graser and Oliveira 2004).

Stakeholder analysis is a useful method to identify gaps between theory and practice. However, focusing on current processes and treating stakeholders as homogeneous groups limit the understanding of internal dynamics of each group and the role of individuals in promoting or limiting collaborative behavior. The analysis of the implementation process of co-management systems with a historical perspective of social interactions among heterogeneous groups has major advantages to reveal nuances in the potential and shortcomings of this negotiation process.

In Brazil, numerous co-management system initiatives have been developed since the 1990s, as part of a strong environmental movement and institutional change (Hochstetler and Keck 2007). In the Amazon, this process started with designing extractive reserves as part of alliances between international organizations and local NGOs (Brown and Purcell 2004; Brown and Rosendo 2000), followed by a more active role of the state in providing new legal mechanisms to promote

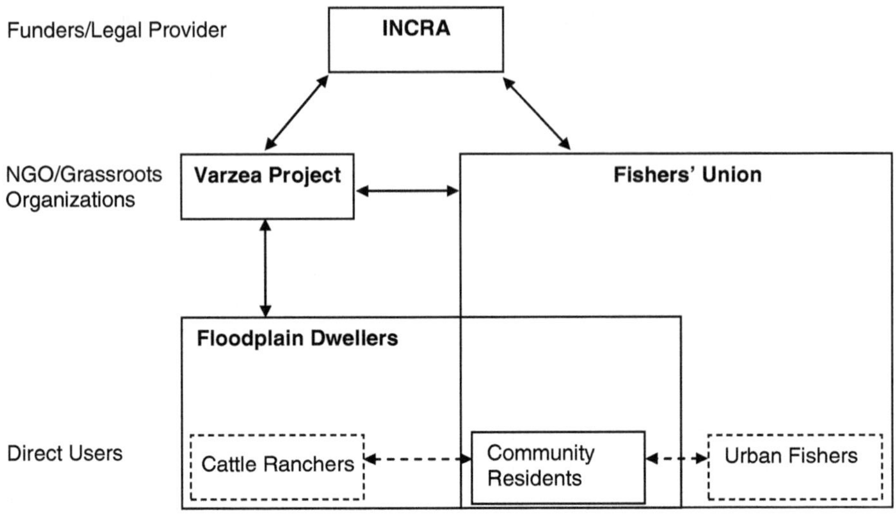

Fig. 9.2 Social interactions between the four main actors in the implementation of the PAEs in the Lower Amazon (*solid line box*: included actor; *solid line arrow*: collaborative interaction; *dashed line box*: excluded actor; *dashed line arrow*: conflictive interaction)

state and non-state partnerships (Hochstetler and Keck 2007). Numerous settlement categories combining conservation, social inclusion, and rural development goals were created in Brazil (Paz et al. 2006). This new model promotes collaborative interactions among state and non-state actors through participatory and inclusive design of institutional arrangements for sustainable production and monitoring systems. Based on concession agreements, the national state grants local residents with temporary exclusive use rights under conditions of following the management plan developed through a participatory process. However, the level of participation from local communities, the institutional arrangement, and procedures vary remarkably from case to case. More than one event, those initiatives must be analyzed as processes of continuous tension between collaboration and conflictive behavior among stakeholders.

Implementation of the PAEs shed some lights on the potential and limitation of co-management initiatives. The analysis of the historical involvement of the relevant stakeholders and their social interaction before and during implementation of the PAEs in the Lower Amazon reveals two important sides of the same processes: (1) different motivations and perceptions driving conflictive behavior among individuals of the same group and (2) similar motivations and perceptions of individual leaders in each group bringing their organizations into a social network. The analysis of the tension between these two processes is fundamental to understand both the institutionalization dimension and the everyday life politics playing out in the shaping of a co-management system. In this process, major changes have occurred in the positions of relevant stakeholders and in their interaction in the process, defining new winners and losers (Fig. 9.2).

In this new social network, the participatory process has been biased toward the community residents as part of a strategy supported by the other three actors—INCRA, the *Várzea* Project, and the Fishers' Union—at the expense of excluding two important local actors—cattle ranchers and urban fishers. A second feature of the social network is the central position of the *Várzea* Project, which is connected to all stakeholders except the two excluded groups. Through the *Várzea* Project, connections between direct users, grassroots organizations, and the government agency have been created and strengthened. The role of individuals in each of those organizations has been instrumental to overcoming bureaucratic constraints and institutional weaknesses of INCRA. In this regard, the *Várzea* Project has not only helped to build the network but also to prepare community and grassroots organization leaders to fulfill this role. Finally, a third feature of this social network is the key position of INCRA, the federal agency, in the implementation of the co-management system. While decentralization strategies and "glocal" configurations to bypass the weak state have been suggested as ways to promote social inclusion, in this case, the federal agency has been fundamental to providing the legal framework and financial support needed and avoiding elite capture. After 20 years of alliance with international donors, the *Várzea* Project was able to move to a full institutionalization of the co-management system in the region.

This new social configuration, however, has many implications. In the complex politics among the stakeholders, the sustainability of natural resources is perhaps the last item on their wish list. For INCRA, the *Várzea* Project represents a key partner to facilitate its entrance in the communities due to its long-term presence in and deep knowledge of the region, reputation, and close links with several communities and grassroots organizations. For the *Várzea* Project, the partnership represented a unique opportunity to participate actively in this implementation of an initiative that originated from its own work of nearly two decades. Furthermore, the financial support from INCRA gave the so needed organizational stability to the project. On the other hand, as an INCRA partner, the *Várzea* Project has become a service provider, confusing its image toward local residents to appear like a state organization.

By the same token, the empowerment of local communities through their legal exclusive use rights has triggered different interests among them regarding their aspirations for their new political position. Both excluded actors dominated the decision-making process at the local (ranchers in the floodplain) and regional (urban fishers in the Fishers' Union) levels until recently. Community residents became more empowered with the Fishers' Union in the late 1990s, and more recently, the implementation of the PAE opened a new window for empowerment in the floodplain region. Therefore, their motivation to support the implementation of the PAE is not only from economic incentives (e.g., grants, access to credit, improved infrastructure) but also political to increase their power over ranchers and outside fishers.

Therefore, the implementation of PAEs has become an arena for power contestation and a source for funding. Whether those goals will be translated into the claimed multiple goals of social inclusion, sustainable production, and rural development will depend on how open individual leaders and their respective organizations will be to renegotiate some of their claims.

9.6 Conclusions

After a few years of experience in implementation of co-management systems, we are now in the position to assess some of these processes in more detail. The recent implementation of the agro-extractive settlement projects in the floodplain is only the final stage of a process that evolved over 20 years. This process reveals the challenges to institutionalize local informal practices into a national legal framework. In this process, new social configuration is shaped with actors being included and excluded in the process. A complex tension between cooperation and conflict within and between stakeholders' groups plays a major role in the way the social network is shaped. The contradictions between multiple goals of social inclusion, conservation, and rural development of co-management systems and the goals of each stakeholder are in the center of this dilemma. In the effort to include formerly excluded actors, there is a danger of excluding relevant actors in the process, creating room for conflicts in the future and promoting a socially biased process. In the effort to use the momentum for the institutionalization of informal ruling systems, implementing agencies may rush into a process of top-down participation with little accountability by local users. As the process is underlined by nuanced, everyday life politics, only through a diachronic and disaggregated analysis of each stakeholder can the motivations and perceptions of each stakeholder be revealed in detail.

References

Adger, W. N., Brown, K., & Tompkins, E. L. (2005). The political economy of cross-scale networks in resource co-management. *Ecology and Society, 10*(2), 9 [online].

Agrawal, A., & Ostrom, E. (2001). Collective action, property rights, and decentralization in resource use in India and Nepal. *Politics and Society, 29*(4), 485–514.

Almeida, O. (2004). *Fisheries management in the Brazilian Amazon*. Ph.D. thesis. Imperial College, London.

Alston, L. J., Libecap, G. D., & Mueller, B. (1999). *Titles, conflict, and land use: The development of property rights and land reform on the Brazilian Amazon frontier*. Ann Arbor: University of Michigan Press.

Andersson, K. P., Gibson, C. C., & Lehoucq, F. (2006). Municipal politics and forest governance: Comparative analysis of decentralization in Bolivia and Guatemala. *World Development, 34*(3), 576–595.

Armitage, D., Berkes, F., & Doubleday, N. (Eds.). (2008). *Adaptive co-management: Collaboration, learning, and multi-level governance*. Vancouver: University of British Columbia Press.

Benatti, J. H. (Ed.). (2005). *A questão fundiária e o manejo dos recursos naturais da Várzea: Análise para elaboração de novos modelos jurídicos*. Manaus: Provárzea-Ibama.

Berkes, F., & Folke, C. (Eds.). (1998). *Linking social and ecological systems: Management practices and social mechanisms for building resilience*. Cambridge: Cambridge University Press.

Berkes, F., & Pomeroy, R. S. (1997). Two to tango: The role of government in fisheries co-management. *Marine Policy, 21*(5), 465–480.

Borrini-Feyerabend, G., Pimbert, M., Farvar, M. T., Kothari, A., & Renard, Y. (2007). *Sharing power: A global guide to collaborative management of natural resources*. London: Earthscan Publications.

Breton, Y., Benazera, C., Plante, S., & Cavanagh, J. (1996). Fisheries management and the *Colonias* in Brazil: A case study of a top-down producers' organization. *Society and Natural Resources, 9*(3), 307–315.

Brown, J. C., & Purcell, M. (2004). There's nothing inherent about scale: Political ecology, the local trap, and the politics of development in the Brazilian Amazon. *Geoforum, 36*(5), 607–624.

Brown, K., & Rosendo, S. (2000). The institutional architecture of extractive reserves in Rondonia, Brazil. *The Geographical Journal, 166*(1), 35–48.

Cash, D. W., Adger, W. N., Berkes, F., Garden, P., Label, L., Olsson, P., Pritchard, L., & Young, O. (2006). Scale and cross-scale dynamics: Governance and information in a multilevel world. *Ecology and Society, 11*(2), 8 [online].

Castro, F. (1999). *Fishing accords: The political ecology of fishing intensification in the Amazon* (CIPEC Dissertation Series, No. 4). Bloomington: Center for the Study of Institutions, Population, and Environmental Change (CIPEC), Indiana University.

Castro, F. (2002). From myths to rules: The evolution of the local management in the Lower Amazonian floodplain. *Environment and History, 8*(2), 197–216.

Castro, F. (2009). Patterns of resource use by *caboclo* communities in the Middle-Lower Amazon. In C. Adams, W. Neves, R. Murrieta, & M. Harris (Eds.), *Amazonian peasant societies: Modernity and invisibility*. Dordrecht: Springer.

Castro, F., & McGrath, D. (2003). Moving towards sustainability in the local management of floodplain lake fisheries in the Brazilian Amazon. *Human Organization, 62*(2), 123–133.

Castro, F., McGrath, D., & Crossa, M. (2002). Adaptándose a los cambios: la habilidad de las comunidades ribereñas en el manejo de sistemas de lagos de la Amazonia Brasileña. In R. C. Smith & D. Pinedo (Eds.), *Nuestros bosques, nuestros lagos: la gestión comunitaria de los bienes comunes en la Amazonía* (pp. 272–302). Lima: Instituto del Bien Común and Instituto de Estudios Peruanos.

Castro, F., Siqueira, A., Brondizio, E., & Ferreira, L. (2006). Use and misuse of the concepts of tradition and property rights in the conservation of natural resources in the Atlantic Forest (Brazil). *Ambiente e Sociedade, 9*(1), 23–39.

Chuenpagdee, R., & Jentoft, S. (2007). Step zero for fisheries co-management: What precedes the implementation. *Marine Policy, 31*(6), 657–668.

Dietz, T., & Henry, A. D. (2008). Context and the commons. *Proceedings of the National Academy of Sciences of the United States of America, 105*(36), 13189–13190.

Futemma, C., Castro, F., Silva-Forsberg, M. C., & Ostrom, E. (2002). The emergence and outcomes of collective actions: An institutional and ecosystem approach. *Society and Natural Resource, 15*(6), 503–522.

Gibson, C. C., & Agrawal, A. (Eds.). (2001). *Communities and the environment: Ethnicity, gender, and the state in community-based conservation*. New Brunswick: Rutgers University Press.

Gibson, C. C., McKean, M. A., & Ostrom, E. (2000). *People and forests: Communities, institutions, and governance*. Cambridge, MA: MIT Press.

Graser, M., & Oliveira, R. S. (2004). Prospects for the co-management of mangroves on the North Brazilian coast: Whose rights, whose duties and whose priorities? *Natural Resources Forum, 28*, 224–233.

Hochstetler, K., & Keck, M. E. (2007). *Greening Brazil: Environmental activism in state and society*. Durham: Duke University Press.

Jentoft, S. (2005). Fisheries co-management as empowerment. *Marine Policy, 29*(1), 1–7.

Jentoft, S., McCay, B. J., & Wilson, D. C. (1998). Social theory and fisheries co-management. *Marine Policy, 22*(4–5), 423–436.

Keck, M. E., & Sikkink, K. (1998). *Activists beyond borders: Advocacy networks in international politics*. Ithaca: Cornell University Press.

Larson, A. M., & Soto, F. (2008). Decentralization of natural resource governance regimes. *Annual Review of Environment and Resources, 33*, 213–239.

McGrath, D., Castro, F., Futemma, C., Amaral, B. D., & Calabria, J. (1993). Fisheries and the evolution of resource management on the Lower Amazon Basin. *Human Ecology, 21*(2), 167–195.

McGrath, D., Almeida, O. T., & Merry, F. D. (2007). The influence of community management agreements on household economic strategies: Cattle grazing and fishing agreements on the Lower Amazon floodplain. *International Journal of the Commons, 1*(1), 67–87.

McGrath, D., Cardoso, A., Almeida, O. T., & Pezzuti, J. (2008). Constructing a policy and institutional framework for an ecosystem-based approach to managing the Lower Amazon floodplain. *Environment, Development, and Sustainability, 10*, 677–695.

Merry, F. D., Sheikh, P. A., & McGrath, D. (2004). The role of informal contracts in the growth of small cattle herds on the floodplain of the Lower Amazon. *Agriculture and Human Values, 21*, 377–386.

NRC (National Research Council). (1986). *Proceedings of the conference on common property resource management*. Washington, DC: National Academies Press.

Ostrom, E. (1990). *Governing the commons: The evolution of institutions for collective action*. Cambridge: Cambridge University Press.

Paz, R. J., Freitas, G. L., & Souza, E. A. (2006). *Unidades de conservação no Brasil: história e legislação*. João Pessoa: Ed. Universitária/UFPB.

Ribot, J. C. (2004). *Waiting for democracy: The politics of choice in natural resource decentralization*. Washington, DC: World Resources Institute.

Simmons, C. S. (2004). The political economy of land conflict in the Eastern Brazilian Amazon. *Annals of the Association of American Geographers, 94*(1), 183–206.

WinklerPrins, A. (2002). Seasonal floodplain-upland migration along the Lower Amazon River. *Geographical Review, 92*, 415–431.

Chapter 10
Conservation of Natural Resources: Which Matters – Having a Regulation or the Size of the Penalty Imposed?

Maria Claudia Lopez

Abstract In this chapter, I investigate through a public good in rural Colombia how different institutional arrangements, particularly different external regulations, with a common characteristic—an external regulator with a weak enforcement capacity—derive into different contributions and compliances. I tested two monetary regulations and found that high penalties very often do more harm than good. I also found that low penalties and public and private reminders were very effective at increasing cooperation and rule compliance.

10.1 Introduction

Since the seminal paper by Becker (1968) several authors have studied the different factors that determine rule compliance. According to Becker, in his model of crime, an individual will "commit an offense if the expected utility to him exceeds the utility he could get by using his time and other resources at other activities" (ibid.: 176). Thus, an individual will take into account among other factors the probability of conviction, the punishment, and the income he could make to decide whether or not to commit a crime. In other words, an individual will commit a crime if the

---

I am particularly grateful to Juana Murillo from Instituto Colombiano de Desarrollo Rural, (INCODER) who made these experiments possible. The funding to do these experiments came from INCODER and WWF-Colombia. Carmen Candelo, Julian Caicedo, and Ana Maria Roldan from WWF-Colombia gave important insights about the fishery in Buenaventura and helped design the survey. The fieldwork was done with the support of Ana Maria Roldan, Juan Carlos Rocha, and Pablo Ramos. The experiments would not have been possible without the help of the community councils of Buenaventura.

M.C. Lopez (✉)
Facultad de Estudios Ambientales y Rurales, Pontificia Universidad Javeriana, Bogotá, Colombia
e-mail: mclopezperez@gmail.com

E.S. Brondízio and E.F. Moran (eds.), *Human-Environment Interactions: Current and Future Directions*, Human-Environment Interactions 1,
DOI 10.1007/978-94-007-4780-7_10, © Springer Science+Business Media Dordrecht 2013

utility to commit the crime is higher than the utility to comply with the law and therefore to follow a legal activity. According to this argument, implementing high penalties and/or high probabilities of conviction should be enough to guarantee rule compliance. Additionally, Becker also suggests that the optimal form of deterrence is to have high penalties since it is very costly to have high probabilities of conviction.

Other literature on rule compliance has shown that individuals follow rule prescription not only when a monetary fine is imposed but also when other intrinsic motivations come into account (Sutinen and Kuperan 1999). Sutinen and Kuperan recreate a fishery model where agents can spend time fishing legally and fishing illegally. The authors find that "if the typical illegal gains are greater than the gains from legal fishing, the expected penalty should be large enough to offset the difference between legal and illegal gains" (ibid.: 177). The authors argue that in general, in the case of fisheries, neither the size of the penalties nor the enforcement are high enough to explain fishers' compliance with rules. Therefore, they may have other motivations such as moral obligations to do the right thing; thus, these intrinsic motivations influence the economic outcomes.

Likewise, the literature in sociology (e.g., Tyler 1990) suggests that compliance can be due to two different causes: either an instrumental or a normative perspective. The instrumental view follows Becker's (1968) argument about individuals maximizing their utility; the normative argument, on the other hand, states that individuals comply with rules when they perceive them as fair and consistent with their norms (Nielsen and Mathiesen 2003). In that sense, individuals' personal values, perceptions of norms, and social influence may determine their compliance behavior (Sutinen and Kuperan 1999).

The subject of rule compliance becomes particularly important when talking about natural resource management in developing countries: first, because these countries have most of the natural resources of the world and second, because they do not have the economic and technological means to implement perfect enforcement (Sutinen and Kuperan 1999). According to Sutinen and Kuperan (1999), enforcement in natural resource settings is so expensive that it could account for half of all public expenditures. Following Becker's (1968) argument, the environmental authority should then impose high penalties. However, authors have shown that the penalties need to be realistic, first being set according to the wealth of the population facing the penalties and second, by being politically viable (Robinson et al. 2010; Rodriguez-Ibeas 2002).

The subject of rule compliance gains more importance in the context of collective territories or in communities where dwellers are supposed to craft institutional arrangements to manage their natural resources. This is the case of the Pacific coast of Colombia, where the experiments presented in this chapter were conducted. The communities living in these territories are Afro-Colombians who obtained after 1991 property rights to the land they had occupied for centuries. In general, it is difficult to measure the effects of an institutional arrangement on people's behavior in advance. These are the cases when economic experiments become a handy way to determine how an institution may affect people's behavior once that institution is in place. In this chapter, I investigate the effects of different institutional arrangements, particularly different external regulations with a common characteristic—a weak enforcement capacity. In particular, I investigate four types of regulations, two pecuniary and two nonpecuniary. Results suggest that a monetary regulation with a high penalty may do more harm than good. I also found that low penalties and public and private reminders were very effective at increasing cooperation and rule compliance.

10.2 Economic Experiments

Economic experiments create a controlled environment in which researchers can explore how agents make decisions and how different institutions may influence these decisions (Friedman and Sunder 1994; Smith 1994, 2010). The experimenter uses monetary payments or other important rewards to elucidate people's preferences. Experiments have been conducted exclusively with undergraduate students[1] in laboratory settings mainly in the United States and Europe until recently (Vollan and Ostrom 2010). According to List (2006: 3), "in an ideal laboratory experiment, this very sterility allows an uncompromising glimpse at the effect of exogenous treatments on behavior in the lab. Of course, making generalizations outside of the lab domain might prove difficult in some cases."

Different researchers from diverse disciplines decided to bring these experiments from the lab to the field (Cardenas et al. 2000; Henrich 2000; Henrich et al. 2004, 2005). The reason to go out of the lab was to run the experiments in a natural environment (Carpenter et al. 2005; Harrison and List 2004). According to a taxonomy done by Harrison and List (2004), the experiments used for this chapter can be classified as "framed field experiments." This classification implies that the experiment is conducted with "nonstandard" participants, also that they faced in their daily lives the situation represented by the experiment. In this case, the experiments were conducted with users of natural resources, more specifically with fishers, and they were framed in such a way that the participants could bring their daily experience into the experiment.

In past decades, the experimental economics literature on environmental issues has generated very valuable and interesting results that cover issues from economic valuation to the creation of economic incentives (Cardenas 2011; Cherry et al. 2007; Sturm and Weiman 2006). In the studies related to resource management, the focal point of attention has been how to solve the social dilemmas (either provision or appropriation) incurred by the use of these resources.

[1]The first reported experiments in rural settings were conducted by Hans Binswanger in the late 1970s in six villages in India. The next field experiments reported in the literature were from Henrich (2000) in the Peruvian Amazon and Cardenas et al. (2000) in Colombia.

Economic experiments with external monetary regulations are not very common. Among these studies in the lab, Beckenkamp and Ostmann (1999) and Ostmann (1998) found that subjects react more to the severity of sanctions than they do to increases in the probability of sanction. In the field, Cardenas et al. (2000: 1719) found that regulation with imperfect monitoring does not lead to a social optimum and may even crowd out cooperation: "individuals confronted with the regulation began to exhibit less other-regarding behavior and made choices that were more self-interested; that is, the regulation appeared to crowd out other-regarding behavior." Also using field experiments, Cardenas (2004) found that low penalties generate too much cooperation, while high penalties create too much free riding. Maria Alejandra Velez et al. (2010) conducted an experiment in three regions of rural Colombia, and their findings suggest that the effect of monetary regulations must be studied on a community-by-community basis because the results varied from one community to the next.

Exploring other kinds of regulations in the field, Velez and Lopez have been studying the effect of a nonmonetary external regulation that resembles a sanction prohibiting fishing for commercial purposes if someone is caught overharvesting. Their results show that older participants behave more cooperatively than younger ones because older stakeholders comply more with rules to sustain the system. In a different setting, Lopez et al. (2010) examine the effectiveness of exogenous regulatory pressure and pro-social emotions in promoting cooperative behavior in a public goods context. The public revelation of an individual's contribution and its consequences for the rest of the group leads to significantly higher public good contributions and social welfare than the regulatory pressure. This is the case even under regulations that are designed to motivate fully efficient contributions.

In this chapter, I replicate the public good game by Lopez et al. (2010) in order to investigate the effects of having four types of regulations, two pecuniary and two nonpecuniary. By using these institutions, I am testing one of the biggest social dilemmas a rural community must face when using natural resources: the provision of the resource.

10.3 Context

In 1991, Article 55 of the new Colombian Constitution[2] created a new collective-property-rights regime for the territories in the Pacific region. This region, peopled mostly by Afro-Colombians, (i.e., 90% of the inhabitants), was for centuries seen as an unoccupied territory by the Colombia government (Plant and Hvaalkof 2001). This new collective-property-rights regime requires the communities to organize into community councils in order to get the land titles. Once the title is obtained, the community councils have the right to manage their territories and to exclude outsiders from extracting their resources.

[2]This article was operationalized in 1993 with the implementation of Law 70.

Studies have shown that these Afro-Colombian communities now have the sense that the territory belongs to them, and that the new property-rights regime has empowered them (Velez 2011). Velez (2011) points out that because the main livelihood of these communities is related to the extraction of natural resources such as fishing, hunting, mining, and timber exploitation (Grueso et al. 1998), it is a priority for these dwellers to craft institutional arrangements to manage their resources; however, they do not have the economic and technological means to do so effectively.

In order to cope with this situation, the Instituto Colombiano de Desarrollo Rural (INCODER) has been advising communities on how to create new institutional arrangements to manage their common-pool resources, with the idea that the management of the common-pool resources generates two dilemmas—one related to the appropriation of the resources and the second related to the provision of the resources and the rules to manage those resources. Thus, I was invited in 2007 by INCODER to conduct a public good experiment to explore the provision dilemma, with fishers from different community councils from the municipality of Buenaventura. The experiments aimed to investigate the effects of different institutional arrangements with a common characteristic: an external regulator with a weak enforcement capacity. The idea was to conduct an experiment recreating one situation of primordial importance for the community councils.

This study was done with fishers from the coastal communities of the municipality of Buenaventura. Fishing is a very important economic activity in the municipality and is done by at least one member of each family. Due to the importance of fishing, finding the correct institutions to manage the fishery is relevant since studies (Jenny et al. 2007; Nielsen and Mathiesen 2003) have revealed that compliance with rules will improve if fishers find the rules meaningful.

The recruitment for these experiments focuses on fishers from different communities within the municipality of Buenaventura; thus, a total of 130 fishers participated in these experiments. The invitations to participate in the experiments were handled by officials of INCODER and WWF Colombia, well known among the different communities. The invitation was only for one family member whose most important economic activity was fishing. Thus, 96% of the participants in this sample were fishers (main economic activity) and 93% were male. A striking statistic showing the poverty of this region is the fact that on average the participants only had 3.7 years of education. The average age among the participants was 40.5 years old (ages ranging from 17 to 77). Finally, 74.62% of the sample have been living in the area for more than 10 years.

10.4 Experimental Design

For this chapter, I followed the experimental design reported by Lopez et al. (2010). A total of 26 groups participated in these experiments, each group having five participants. All groups participated in a 15-round frame field experiment with one of five possible treatments. The experiment is based on a standard linear public good game with five homogenous players being members of a group with identical monetary payoffs. Each player received an initial endowment of 25 tokens, then he/she had to decide how many to contribute to a group project and how many to keep for him/herself. The sum of contributions to the group account is multiplied by 2 and then distributed equally among the group members. The experiment was set in such a way that the dominant Nash strategy[3] for each individual was to contribute zero tokens to the group account, but the social optimum was reached when each person contributed all of his/her 25 tokens to the group project. The participants were told that the group project was similar to a situation where fishers are asked to contribute to a group project such as cleaning the mangrove or the beaches. Additionally, before starting to play, each group received the following information:

> Before we begin playing for real money, I would like to point something out: As you may have noticed, the earnings for the group are the highest when everybody contributes 25 tokens to the group project. If you decide to keep tokens for yourself, you can increase your individual earnings but you are reducing the earnings of the group.

This treatment is called the *Baseline*. The reasoning behind this treatment follows Sutinen and Kuperan's (1999) argument that some rules may elicit some moral obligations and therefore drive people to do the right thing. With this *Baseline*, I wanted to test whether only stating how the group will increase its earnings, i.e., explaining how to do the "right thing," would derive higher contribution levels than what is found in the public goods literature.

In order to investigate the effects of different regulations on people's behavior, I conducted the same public good game experiment with four variances, two nonpecuniary sanctions and two pecuniary sanctions. In all of these treatments at the end of each round, one participant was randomly inspected. The inspection was done by choosing a ballot from a bag containing five ballots; each ballot had a participant's number.[4]

In the case of the nonpecuniary sanctions, the participant inspected got either a private reminder (*External Private Reminder* treatment) or a public reminder (*External Public Reminder* treatment) of how his/her decisions were affecting the group. This reminder is shown in Table 10.1.

If the inspected participant was fully contributing to the group project, he/she received the following message: "You contributed all of your 25 tokens to the project, which means you did all you could to make the earnings for the group the highest."

The two nonpecuniary treatments varied in the sense that in the *External Private Reminder* treatment, the inspected participant received a piece of paper that he/she

---

[3]*Dominant Nash strategy* is a term used in game theory to describe a situation in which each player in a game who only considers his or her self-interest is making the best decision that he or she can, while considering the decisions of others.

[4]Each participant in the group was identified for the whole section with a participant number of 1–5.

Table 10.1 Reminder received if player was inspected at the end of a round

The earnings of the group are the highest when everybody contributes all of his or her 25 tokens to the group
Tokens you contributed to the group project
Total tokens contributed to the project
Total tokens contributed to the project if you had contributed your 25 tokens
Losses for the group because you did not contribute all of your 25 tokens

could read privately. So everybody else in the group knew who was inspected, but none of them knew how many tokens that person had contributed to the group project. In the *External Public Reminder* treatment, the inspected participant received the information publicly when the monitor read the information aloud in front on the whole group.

For the pecuniary treatments, two different kinds of monetary external regulations were tested, a low and a high external regulation. Thus, through both external regulation treatments, I kept constant the level of enforcement and changed the size of the penalty. These two treatments were built on the *External Private Reminder* treatment; thus, the inspected participant received all the information in private. Additionally, at the beginning of the experiment, the participants were told that a monetary regulation was in place; if the inspected person was not fully contributing his/her 25 tokens to the group account, he/she had to face a penalty. In the *Low Regulation* treatment, the penalty was one token for each token the participant did not contribute to the group project. Whereas in the *High Regulation* treatment, the penalty was four tokens for each token not contributed to the group project. In both treatments, the result from the inspection was kept private. For these two treatments, I wanted to test Becker's (1968) argument that people will follow rules depending on the size of the penalty and the level of enforcement.

For the *External Private Reminder*, *External Public Reminder*, and *Low Regulation* treatments, the dominant Nash strategy and the social optimum were the same as for the *Baseline* treatment, thus, a zero-token contribution for the group project for the former and a contribution of 25 tokens for the latter. However, the experiment was set up in such a way that for the *High Regulation* treatment, the dominant Nash strategy and the social optimum were the same, i.e., a full contribution of 25 tokens to the group project.

This design was supposed to have the same number of groups for each treatment; however, once in the field, it was impossible to run all the sections planned by the research team and therefore the different groups we were able to run are not evenly divided among treatments. Table 10.2 presents the number of groups we had for each treatment.

The experiments were conducted in school classrooms, or any other space that was available for us. The experiment started by having the entire group (up to four groups participated at the same time) together facing a blackboard. The monitor first explained that each participant was going to be asked to make economic decisions.

Table 10.2 Number of groups in each treatment

Treatment	Number of groups
Baseline	4
External Private Reminder	6
External Public Reminder	4
Low Regulation	6
High Regulation	6

She mentioned that during the experiment, tokens were going to be used instead of money, but she also informed them that at the end of the experiment, the research team would add all the tokens each participant had earned during the 15 rounds and pay 25 Colombian pesos[5] for each token. Then we read the instructions[6] aloud and showed them posters with all the forms they would have to fill individually during the experiment. Participants were asked to remain silent and to raise their hands in case they had questions. Because the level of education was so low, we went over the instructions more than once and did multiple practice rounds to assure comprehension of the game. Participants with difficulties filling the forms were helped by assistants from our team. After the instructions were finished, the participants within each group were asked to sit facing each other's backs to prevent them from seeing other participants' forms.

In the instructions, I stated that the experiment was going to last 15 rounds. I also specified that the decision task was the same for all rounds. Thus, at each round, each participant had to decide how many tokens he/she would contribute to the group project and how many to keep for him/herself. These decisions were written in two forms, a decision card to be handled to the monitor and a decision sheet kept by the participant with the record of all his/her decisions throughout the experiment. Once the monitor collected all the decision cards, she announced in public the group total contributions to the project and the amount of tokens each participant got (the total contributions multiplied by 2 and then divided by 5). Then each participant had to calculate individually his/ her own earnings and wait to see the result from the inspection at that round. At the end of the 15 rounds, the research team calculated all the participants' earnings and paid each one in cash. Individuals' earnings ranged between 11.350 and 20.450 pesos with an average of 16.599 pesos (about 6.63 US dollars). A show-up payment was not provided, but transportation was guaranteed for all participants; we also provided lunch since the experiments lasted, on average, 3 h.

[5]In July 2006, when the experiments were conducted, one US dollar was equivalent to 2,500 Colombian pesos. A day's wage in the fishery industry in Buenaventura was about 12,000 pesos.

[6]Instructions and forms are available upon request.

10.5 Results

The first graphical approximations to the data are presented in Figs. 10.1 and 10.2. Figure 10.1 shows the round average for individual contributions to the group project by treatment, and Fig. 10.2 shows round average for individual earnings by treatment. The figures show that the results both in terms of contributions and earnings are different across treatments. However, the results obtained from these graphs deserve a more careful analysis.

In Table 10.3, I present the average group contributions and earnings for different rounds of the experiment. A first result consistent with Lopez et al. (2010) is

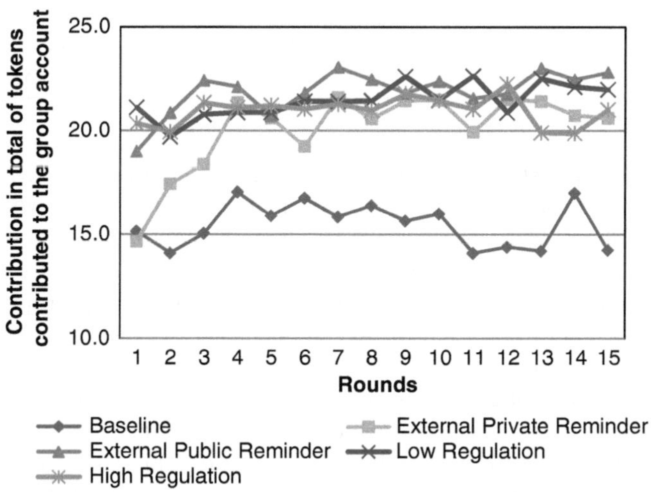

Fig. 10.1 Round average individual contribution levels by treatment

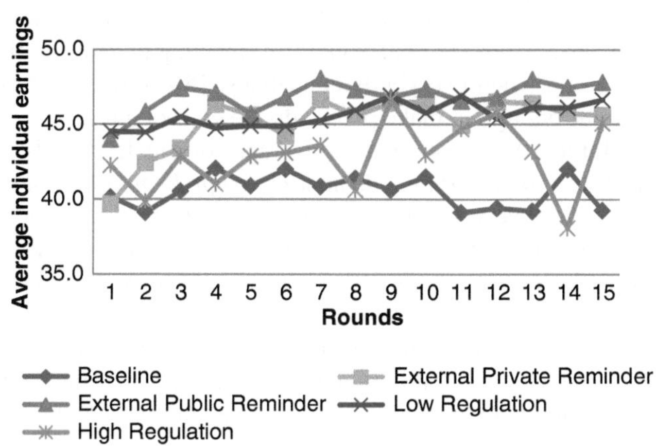

Fig. 10.2 Average round individual earnings by treatment

Table 10.3 Average group contributions and average group earnings for the 15 rounds, the first round, and the last round (standard deviations in parentheses)

	Average contributions 15 rounds	Average contributions round 1	Average contributions round 15	Average earnings 15 rounds	Average earnings round 1	Average earnings round 15
Baseline	15.46	15.15	14.25	40.54	40.15	39.25
	(6.2)	(6.12)	(6.82)	(6.2)	(6.12)	(6.82)
External Private Reminder	20.06	14.67	20.6	45.06	39.67	45.6
	(6.74)	(7.53)	(6.93)	(6.74)	(7.53)	(6.93)
External Public Reminder	21.87	19	22.8	46.86	44	47.8
	(4.88)	(4.83)	(3.8)	(4.87)	(4.83)	(3.79)
Low Regulation	21.44	21.1	21.97	45.6	44.53	46.63
	(6.45)	(6.02)	(5.96)	(6.42)	(5.78)	(5.94)
High Regulation	20.96	20.33	21	42.84	42. 26	45.07
	(5.82)	(5.81)	(8.1)	(11.69)	(9.62)	(8.81)

the fact that the *Baseline* treatment (in their case, called Frame) generates lower contributions and earnings than the rest of the treatments. The *Baseline* treatment is slightly different from the standard baseline of a linear public good game due to the fact that I informed participants in the instructions that the earnings were the highest when everybody contributes 25 tokens to the group project. However, this information does not seem to vary dramatically the result from the standard baseline of a linear public good game reported in the literature (Ledyard 1995) since the contributions in that case go from 40% to 60% of the endowment. In this case, the contributions for the 15 rounds were around 62% of the endowment, a little bit above the interval found in the public good literature. This result is also supported by Lopez et al. (2010: 8), since they find that "informing subjects that contributing all tokens to the group project maximizes aggregate earnings did not affect average contributions or earnings." This first result is interesting because it eliminates the hypothesis by Sutinen and Kupernan (1999) that people will follow the rules just because it is the correct thing to do. Explicitly explaining that the earnings were the highest for the group when everybody contributes 25 tokens to the group project does not seem to be enough to make people do the correct thing.

The rest of Table 10.3 shows that on average for the 15 rounds, the *External Private Reminder*, *External Public Reminder*, *High Regulation*, and *Low Regulation* treatments were very similar. However, it is interesting to note that at the beginning of the experiment (round 1), these treatments had very different results. The contributions were on average 19 tokens for the *External Public Reminder* treatment, 14.67 for the *External Private Reminder* treatment, and 21.1 and 20.33, respectively, for the *Low Regulation* and *High Regulation* treatments. Another result from Table 10.3 that deserves further analysis is the fact that the *Low Regulation* treatment had in general not only better earnings than the *High Regulation* treatment but also higher contributions.

Table 10.4 Random effects Tobit estimation of individual extraction (standard error in parentheses)

Variable	Model 1 (contributions)	Model 2 (earnings)
Constant (default is Baseline treatment)	12.15***	38.44***
	(2.85)	(1.45)
External Private Reminder	13.44***	7.8***
	(4.09)	(2.01)
External Public Reminder	13.20***	8.37***
	(4.04)	(2.01)
Low Regulation	10.12**	6.57***
	(4.02)	(2.01)
High Regulation	6.41	−0.52
	(4.01)	(2.01)
Round	0.28***	0.144***
	(0.04)	(0.03)
Fixed effects by groups (not reported)	26 groups	26 groups
Rho	0.44	0.13
Prob > χ^2	0.000	0.000
N	1,950	1,950

Note: The dependent variable in model 1 is the contribution to the group account; thus, the dependent variable is truncated between 0 and 25. The dependent variable in model 2 is the individual earnings; thus, the dependent variable is truncated between −26 and 65
***Denotes $p < 0.01$; **denotes $p < 0.05$; *denotes $p < 0.10$

To further explore these results, I conducted an econometric analysis. Since every player made decisions for 15 rounds and because contribution decisions were truncated between 0 and 25 tokens, I did a random Tobit effect model. In model 1 presented in Table 10.4, the constant is the average contribution for the *Baseline* treatment; additionally, I included fixed-effect variables for each treatment and for each group, and I included the variable round in the regression. Model 2 in Table 10.4 is the same model, but for the earnings, this variable is truncated from −26 to 65.[7] All the variables in Table 10.4 should be interpreted as deviations from the constant, the *Baseline* treatment. Two additional models, not reported in this chapter, were conducted with the same variables but adding socioeconomic variables such as age, education, time living in the area, main activity of the fisher, and gender, but these variables were not significant.

From the models in Table 10.4, we can see that all the treatments except *High Regulation* performed better in terms of contributions and earnings than the *Baseline* treatment. This second result is interesting since this was the only treatment in which theoretical economic prediction would call for full contributions to the public good. This also contradicts Becker's argument that a larger penalty will yield more compliance.

[7] In the regulations treatments, it was possible to get negative earnings. However, since the probability of inspection was low, we did not encounter any case of a participant having negative earnings at the end of the experiment.

I did a Wald test to investigate the statistical differences among the other treatments. I found that *External Private Reminder* and *High Regulation* as well as *External Public Reminder* and *High Regulation* were statistically different at the 5% level of significance. Thus, we can say that in terms of amount of contributions to the group account, the *External Private Reminder*, *External Public Reminder*, and *Low Regulation* treatments have very similar effects.

For both model 1 and model 2, I found that the round variable is positive and significant at the 1% level of significance, meaning that contributions to the group project increased over time. This result can be understood as a learning effect to cooperate.

In the case of earnings, model 2 shows that the *High Regulation* treatment generates the same earnings as the *Baseline* treatment, therefore supporting the result that a high regulation is not the best option to manage this fishery. Again, as for the contributions, the differences between the *External Private Reminder*, *External Public Reminder*, and *Low Regulation* treatments are not statistically significant. Additionally, all earnings from these treatments were different from the *High Regulation* treatment at 1% level of significance.[8]

This last result is different from what Lopez et al. (2010) found in their study, where the *External Public Reminder* treatment led to the highest level of both contributions and earnings, followed by the *Low Regulation* and *High Regulation* treatments. In their case, the *External Private Reminder* treatment did not have any differences with respect to the *Baseline* treatment reported in this chapter. These different results across sites verify what others have studied empirically (e.g., Acheson 2006) and lie in the fact that different people will behave in different ways under different institutions. In some sense, it is an example that the same institutional arrangement will not fit all cases and therefore that it is necessary to investigate the effects of different institutions in different contexts. In the Buenaventura fishery, actors react in the same way to nonpecuniary regulations as to regulations with a very low penalty. Additionally, it is clear that this result does not support Becker's (1968) model of crime; instead, it seems that other kinds of motivations are behind participants' following the rules.

In order to understand better the compliance behavior for all these treatments, I start by presenting in Fig. 10.3 the different contribution distributions for the different treatments. From the figure, it is clear that the distribution of the *Baseline* treatment is different from the other distributions. Kolmogorov-Smirnov tests were conducted to verify that all distributions of the treatments were different from the distribution of the *Baseline* treatment, and in all cases they were statistically different ($p = 0.00$).

An additional model presented in Table 10.5 investigates a model for compliance. A random logit model was created (the variable is 1 when 25 tokens were contributed to the group account and 0 otherwise). As the previous models, this model has fixed effects for the different treatments and for the different groups.

[8] Performing a Wald test.

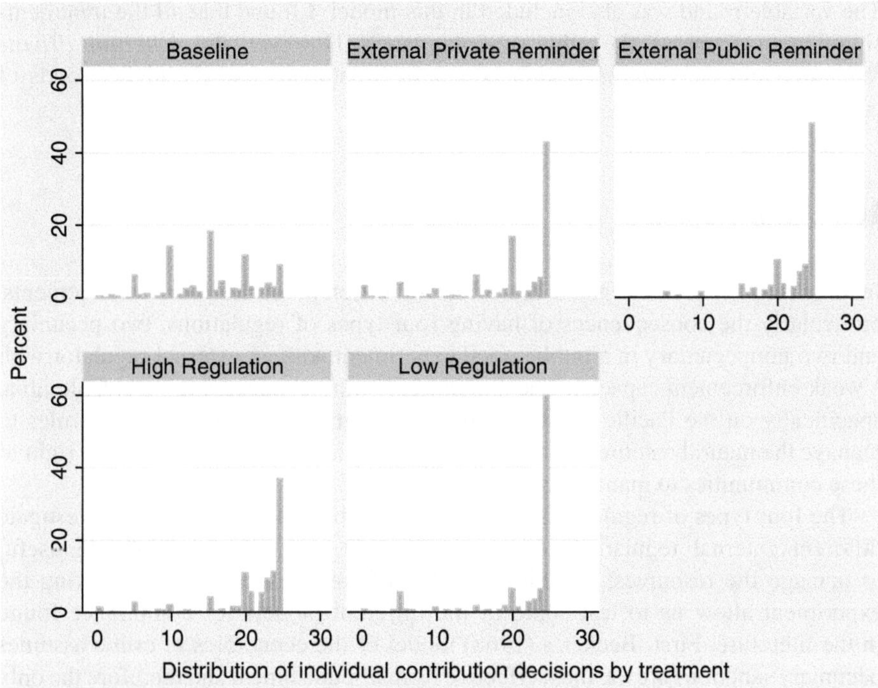

Fig. 10.3 Distribution of individual contribution decisions by treatment

Table 10.5 Random effects logit estimation of individual compliance (standard error in parentheses)

Variable	Model 3
Constant (default is the Baseline treatment)	
External Private Reminder	6.29***
	(1.74)
External Public Reminder	6.27***
	(1.72)
Low Regulation	5.14***
	(1.72)
High Regulation	4.26**
	(1.73)
Round	0.14***
	(0.016)
Fixed effects on groups (not shown)	26 groups
$Prob > \chi^2$	0.000
N	1,950

Note: The dummy dependent variable equals 1 if 25 tokens were contributed to the group account; otherwise, it is 0

***Denotes $p < 0.01$; **denotes $p < 0.05$; *denotes $p < 0.10$

The variable round was also included in this model. I found that all the treatments significantly increase the odds of compliance. However, the *External Private Reminder* and the *External Public Reminder* treatments had the highest odds of compliance.

10.6 Conclusion

In this chapter, I investigated the effect of different institutional arrangements, particularly the consequences of having four types of regulations, two pecuniary and two nonpecuniary in a public good experiment with an external regulator with a weak enforcement capacity. The experiments were conducted in rural Colombia, specifically on the Pacific coast, a region where it is fundamental to find rules to manage the natural resources, because the Colombian Constitution gave the right to these communities to manage their own resources.

The four types of regulations I tested through the experiment aim to investigate different external regulations that a community council may think are useful to manage the resources. Additionally, the different institutions used during the experiment allow us to test some of the different models for compliance found in the literature. First, Becker's (1968) model of the economics of crime assumes pecuniary sanctions are the most effective form of punishment and therefore the only policy mechanism to avoid rule-breaking behavior. Second, Sutinen and Kuperan's (1999) model states that individuals may have other motivations to follow rules, such as moral obligations to do the right thing, and these intrinsic motivations influence rule compliance. Finally, Tyler (1990) suggests that individuals do not follow only extrinsic motivations when deciding to comply with a rule. He identifies intrinsic motivations such as legitimacy and fairness of rules to explain rule compliance.

In this chapter, I found in the *Baseline* treatment that just by telling people what was the correct thing to do (thus testing the hypothesis by Sutinen and Kupernan (1999)) does not seem to be enough to make people generate more contributions than in other public good settings reported in the literature. I also found that people who live on the Pacific coast of Colombia did not seem to follow external regulations with high penalties (with a theoretical prediction of full cooperation). This result is consistent with Cardenas and colleagues' (2000) findings suggesting that high monetary regulations do not lead to a social optimum and may even crowd out cooperation. Further, I reject Becker's (1968) hypothesis that this treatment will be the most efficient in terms of contributions to the public good. Paradoxically, the participants in these experiments reacted positively and similarly in the *External Private Reminder*, *External Public Reminder*, and *Low Regulation* treatments. These three treatments generated an increase in cooperation and rule compliance with respect to the *Baseline* treatment and to the *High Regulation* treatment. First, consistent with Cardenas (2004), the *Low Regulation* treatment with a low penalty (with a theoretical prediction of noncooperation) generates a lot of cooperation. Second, consistent with Velez et al. (2010), I found that the same institutions may have different results in different communities. This result supports the findings by

Ostrom (2007), Poteete et al. (2010), and Vollan and Ostrom (2010) that individual and group decisions are influenced by context and "micrositational" variables. It may be the case that these three treatments make explicit some extrinsic motivations and therefore enhance rule compliance, as Tyler (1990) suggests. This result is important for the members of the community councils in charge of the provision of rules to manage their natural resources. But it also evident that it deserves further research.

The results from these experiments show the importance of finding institutional arrangements where communities are involved in rule crafting. It seems that for these communities of the Pacific coast, the institutions that are meaningful and therefore increase compliance should include moral obligations.

References

Acheson, J. M. (2006). Institutional failure in resource management. *Annual Review of Anthropology, 35*, 117–134.

Beckenkamp, M., & Ostmann, A. (1999). Missing the target? Sanctioning as an ambiguous structural solution. In M. Foddy, M. Smithson, S. Scheneider, & M. Hogg (Eds.), *Resolving social dilemmas: Dynamic, structural, and intergroup aspects* (pp. 165–180). Philadelphia: Psychology Press.

Becker, G. S. (1968). Crime and punishment: An economic approach. *Journal of Political Economy, 76*, 169–217.

Cardenas, J. C. (2004). Norms from outside and from inside: An experimental analysis on the governance of local ecosystems. *Forest Policy and Economics, 6*, 229–241.

Cardenas, J. C. (2011). Social norms and behavior in the local commons as seen through the lens of field experiments. *Environmental and Resource Economics, 48*, 451–485.

Cardenas, J. C., Stranlund, J. K., & Willis, C. E. (2000). Local environmental control and institutional crowding-out. *World Development, 28*(10), 1719–1733.

Carpenter, J., Harrison, G., & List, J. (2005). Field experiments in economics: An introduction. In J. Carpenter, G. Harrison, & J. List (Eds.), *Field experiments in economics* (pp. 1–16). Greenwich/London: JAI/Elsevier.

Cherry, T. L., Kroll, S., & Shogren, J. F. (2007). *Environmental economics, experimental methods*. New York: Routledge.

Friedman, D., & Sunder, S. (1994). *Experimental methods: A primer for economists*. Cambridge: Cambridge University Press.

Grueso, L., Rosero, C., & Escobar, A. (1998). The process of Black community organizing in the southern Pacific coast of Colombia. In S. Alvarez, E. Dagnino, & A. Escobar (Eds.), *Cultures of politics/politics of cultures: Re-visioning Latin American social movements* (pp. 430–447). Boulder: Westview Press.

Harrison, G. W., & List, J. (2004). Field experiments. *Journal of Economic Literature, 42*(4), 1009–1055.

Henrich, J. (2000). Does culture matter in economic behavior? Ultimatum game bargaining among the Machiguenga of the Peruvian Amazon. *The American Economic Review, 90*, 973–979.

Henrich, J., Boyd, R., Bowles, S., Camerer, C., Fehr, E., & Gintis, H. (2004). *Foundations of human sociality: Economic experiments and ethnographic evidence from fifteen small-scale societies*. Oxford: Oxford University Press.

Henrich, J., Boyd, R., Bowles, S., Camerer, C., Fehr, E., Gintis, H., McElreath, R., Alvard, M., Barr, A., Ensminger, J., Henrich, N. S., Hill, K., Gil-White, F., Gurven, M., Marlowe, F. W., Patton, J. Q., & Tracer, D. (2005). "Economic man" in cross-cultural perspective: Behavioral experiments in 15 small-scale societies. *The Behavioral and Brain Sciences, 28*, 795–855.

Jenny, A., Hechavarria Fuentes, F., & Mosler, H. J. (2007). Psychological factors determining individual compliance with rules for common pool resource management: The case of a Cuban community sharing a solar energy system. *Human Ecology, 35*, 239–250.

Ledyard, J. (1995). Public goods: A survey of experimental research. In J. Kagel & A. Roth (Eds.), *The handbook of experimental economics* (pp. 111–194). Princeton: Princeton University Press.

List, J. A. (2006). Field experiments: a bridge between lab and naturally occurring data. *Advances in Economic Analysis and Policy, 6*(2), 8 [online]. URL: http://www.bepress.com/cgi/viewcontent.cgi?article=1747&context=bejeap

Lopez, M. C., Murphy, J. J., Spraggon, J. M., & Stranlund, J. K. (2010). Comparing the effectiveness of regulation and individual emotions to enhance cooperation: Experimental evidence from fishing communities in Colombia. *Economic Inquiry* Online First. URL: http://onlinelibrary.wiley.com/doi/10.1111/j.1465-7295.2010.00344.x/pdf

Nielsen, J. R., & Mathiesen, C. (2003). Important factors influencing rule compliance in fisheries lessons from Denmark. *Marine Policy, 27*(5), 409–416.

Ostmann, A. (1998). External control may destroy the commons. *Rationality and Society, 10*(1), 103–122.

Ostrom, E. (2007). A diagnostic approach for going beyond panaceas. *Proceedings of the National Academy of Sciences of the United States of America, 104*, 15181–15187.

Plant, R., & Hvaalkof, S. (2001). *Land titling and indigenous peoples*. Washington, DC: Inter-American Development Bank.

Poteete, A. R., Janssen, M., & Ostrom, E. (2010). *Working together: Collective action, the commons and multiple methods in practice*. Princeton: Princeton University Press.

Robinson, E. J. Z., Mahaputra, A. K., & Albers, H. J. (2010). Protecting developing countries' forests: Enforcement in theory and practice. *Journal of Natural Resources Policy Research, 2*(1), 25–38.

Rodriguez-Ibeas, R. (2002). Regulatory enforcement with discretionary fining and litigation. *Bulletin of Economic Research, 54*, 105–118.

Smith, V. L. (1994). Economics in the laboratory. *Journal of Economic Perspectives, 8*(1), 113–131.

Smith, V. L. (2010). Theory and experiment: What are the questions? *Journal of Economic Behavior and Organization, 73*(1), 3–15.

Sturm, B., & Weiman, J. (2006). Experiments in environmental economics and some close relatives. *Journal of Economic Surveys, 20*, 419–457.

Sutinen, J. G., & Kuperan, K. (1999). A socio-economic theory of regulatory compliance. *International Journal of Social Economics, 26*(1–3), 174–193.

Tyler, T. (1990). *Why people obey the law*. New Haven: Yale University Press.

Velez, M. A. (2011). Collective titling and the process of institution building: The new common property regime in the Colombian Pacific. *Human Ecology, 39*(2), 117–129.

Velez, M. A., Murphy, J. J., & Stranlund, J. K. (2010). Centralized and decentralized management of local common pool resources in the developing world: Experimental evidence from fishing communities in Colombia. *Economic Inquiry, 48*(2), 254–265.

Vollan, B., & Ostrom, E. (2010). Cooperation and the commons. *Science, 330*(6006), 923–924.

Chapter 11
Small-Scale Farmers and the Challenges of Environmental Conservation and Rural Development: Case Studies from the State of São Paulo and the Amazon Region

Célia Futemma

Abstract In the past 10 years, small-scale farmers have been the target of both environmental and rural development concerns at national, state, and municipal levels in Brazil. At the federal level, public policies aim at enhancing family-based farming systems (*agricultura familiar*) through increasing their participation in the market and guaranteeing their food security. Simultaneously, environmental policies restrict small-scale farmers from using 100% of their property, since they have to conserve a minimum of forest area—forest reserve—as well as a gallery forest. Some government officials and scholars argue that small-scale farmers are not able to follow such environmental rules and should be forgiven, because of the small size of their properties, which limit their choices in production. The question is, Considering government programs for rural development and environmental conservation, are small-scale farmers who live inside or surrounding protected areas

This chapter was originally presented at the Workshop "Interdisciplinary Approaches to Research on Human-Environment Interactions," Indiana University, Bloomington, February 25–27, 2010. I gratefully acknowledge support from FAPESP for research funds (No. 07/53308-1 and No. 01/11473-0) and from the Federal University of São Carlos for academic support. I owe gratitude to Indiana University on behalf of Emilio Moran and Eduardo Brondízio for providing me with support to conduct some of the data analysis presented in this chapter. I am grateful to IBAMA, on behalf of Ângelo de Lima Francisco and Fabiana Bertoncini, for giving me permission to conduct research at Tapajós Flona and Ipanema Flona, respectively. I thank the fieldwork team who helped collect data in 2008 in the two Ipanema settlements and Rosângela Calado da Costa for helping collect data in both places. Finally, I am deeply thankful to all the small-scale farmers and their families from the Tapajós communities and the Ipanema settlements and all other informants for making the collection of data possible. I assume full responsibility for the content presented in this chapter.

C. Futemma (✉)
Center for Environmental Studies and Research (NEPAM), University of Campinas (UNICAMP), Campinas, São Paulo, Brazil
e-mail: cfutemma@unicamp.br

E.S. Brondízio and E.F. Moran (eds.), *Human-Environment Interactions: Current and Future Directions*, Human-Environment Interactions 1,
DOI 10.1007/978-94-007-4780-7_11,

able to increase their participation in the market, produce for their livelihoods, and meet environmental conservation rules? To address this question, decisions and activities of small-scale farmers in the state of São Paulo and the Amazon region will be analyzed.

11.1 Introduction

Human ecologists and scholars from several related disciplines have been trying to elucidate the challenging human-environment interactions faced in the new millennium (Moran and Ostrom 2005), wherein changes in natural systems are among the biggest concerns the world faces. The interaction between local human populations and protected areas (PAs) is characterized by a conflicting situation that is hard to solve (Almeida and Cunha 2000; Benatti 2001; Terborgh and Peres 2002). The complexity of this relationship includes a number of variables—social, cultural, economic, and political. Public policies play a major role in several places around the world with regard to performance of local development and conservation. Political ecology offers some analytical tools to analyze these situations, in which multiple stakeholders, access to natural resources, poverty, and asymmetry of power are in place.

Since the early 1990s, political ecology started to appear as a potential analytical tool for studying human-environment interactions (Durham 1988; Painter 1988; Schmink 1987; Stonich 1993) and has gained the interest of several scholars (see discussion in Fox et al. 2009). Gerber et al. (2009) propose a combination of political ecology and ecological economics for studying conflicts related to tree plantations in Cameroon and Ecuador. Doolittle (2010) points out a lack of adequate methodological tools for analyzing political ecology cases and proposes a holistic research strategy based on qualitative data by integrating historical, ethnographic, and spatial information. This multimethod and multiscalar approach allowed her to analyze questions of conservation and rural livelihoods in Sabah, Malaysia.

Political ecology also has been a target of criticisms. Vayda and Walters (1999) argue that political ecologists fail by considering only one factor (i.e., political) to explain complex human-environment relations. These authors called "green romanticism" the belief among political ecologists that local communities will always manage properly their natural resources once the control of resources is passed on to them. Forsyth (2003) criticizes political ecologists for not defining properly the term "ecology" within the political ecology frame and proposes a "critical political ecology" in which environmental issues should be addressed as an environmental politics or the politics of ecology.

Despite these criticisms and their diversified approaches and methods, in the past 20 years, political ecology has contributed to uncovering some puzzles in political arenas that involve human populations and environments. Multiple scales of analysis help to address more systemic relationships from local (rural properties, households, communities) to global scales (Stonich 1993). Social arenas that involve conflicting public policies, unstable property regimes, and power disputes among different stakeholders are present in several places (Sheridan 1988), especially in developing and emergent countries. Throughout the world, conflicts involving the presence of local people and conservation policies are frequent.

In countries of Latin America, such as Honduras and Nicaragua, government programs for replacing fire (used in shifting cultivation) with new technologies and guaranteeing food security and soil conservation have failed in not considering local knowledge and local realities. For instance, ignoring the effects of the cost of labor to manage the soil and the cost of pest and weed controls, among others, ended up favoring wealthier local people rather than the poor (Shriar 2007). Costa Rica presented a history of land speculation similar to Brazil's experience (Alston et al. 1999), in which, land was cleared (deforested) to give it an economic value as a result of government incentives to occupy these lands for agricultural purposes (Schellas and Sánchez-Azofeifa 2006). Schellas and his colleagues argue that a mosaic of different forms of land use in the buffer zone of the Braulio Carrillo National Park made conservation or even recovery of the forest difficult.

Several countries from Southeast Asia suffered changes in policies regarding shifting cultivation and the rise of conservation programs (Fox et al. 2009). The introduction of institutionalized conservation funded the wealthy sectors of the society and contributed to enhance the implementation of authoritarian national policies in Malaysia, Indonesia, Thailand, the Philippines, Vietnam, China, and Lao PDR. The creation of national parks was forcibly imposed on lands that were formerly the swidden fields and fallows of local people in northern Thailand, Indonesia, and Malaysia. In China, Vietnam, and Lao PDR, the top-down nature of government authority led to conservation agenda within policy and law, where nature reserves and national parks were imposed. Tree plantations affected the capacity of swidden cultivators to grow their own food as less land was available for agriculture (Fox et al. 2009).

In Brazil, incentives boost family-based agriculture, thereby encouraging a larger participation of small-scale farmers in the market; yet, at the same time, they must comply with conservation goals. The reconciliation between market-oriented agriculture and conservation can become harsher when PAs are concerned. PAs present sets of rules that restrict farmers' options regarding types of land use inside and around them. The question is, considering government programs for rural development and environmental conservation, are small-scale farmers who live inside or surrounding a PA able to increase their participation in the market, to produce for their livelihoods, and to meet environmental conservation rules? To address this question, this chapter presents the following objectives: (1) to investigate the main public policies established by the government that affect decisions of small-scale farmers and (2) to analyze the main economic activities carried out by small-scale farmer families who live within or near a PA. Two cases will be studied: a rural settlement in the State of São Paulo and communities in the State of Pará. In both places, the local populations reside either within the PA or in the buffer zone.

11.2 Sociocultural Settings

My study focuses on one type of PA in Brazil—the national forest—located in different regions and occupied by different types of local human populations. The National Forest of Tapajós (Tapajós Flona) in the lower Amazon region, State of Pará, is inhabited by a variety of social groups: native nonindigenous people, native indigenous people, and migrants. The National Forest of Ipanema (Ipanema Flona), in the Atlantic Forest region of the State of São Paulo, is peopled by old-resident migrants and new ones (from the movement of the landless people, or MST).

11.2.1 The Tapajós National Forest and the Native Communities (in the Amazon Biome)

Tapajós Flona was created on February 19, 1974, by Decree 73.684. The federal agency Chico Mendes Institute for Biodiversity Conservation (ICMBio) has been responsible for its administration since 2007. Local residents only knew about the creation of the national forest PA around 1980, when the federal government tried to force them to leave through an expropriation process. However, residents resisted, and they are still living in the area. There were 28 communities, approximately 1,438 families and 7,181 inhabitants (FLONA/IBAMA 2006). Tapajós Flona has a total area of approximately 545,000 ha. This PA was created for one main reason: to support research on logging exploitation.

Tapajós Flona occupies territories of four municipalities: Belterra, Aveiros, Placas, and Rurópolis. This PA is located between Tapajós River to the west and Cuiabá-Santarém (BR-163) highway to the east. Along the highway, Tapajós Flona is surrounded by colonization settlements organized by the Brazilian Agency for Land and Agrarian Reform (INCRA) as well as large farms that were established in the 1970s. Inconsistent with its original goals, during the 1970s and 1980s, INCRA distributed 52 land titles to farmers inside the PA. The logging concession was given by the Brazilian Institute of Environment and Renewable Natural Resources (IBAMA) to the intergovernmental International Tropical Timber Organization in the mid-1990s (Veríssimo et al. 2002).

Most of the local communities are composed of native nonindigenous people—caboclos—who have been living in this area over 100 years. Traditionally, they grew annual crops (manioc, corn) for their livelihood and eventually sold some forest products such as vines, vegetal oil, woods, rubber, and wild fruits to obtain cash for their household economy (Faria 2004; Soares 2004). Manioc was their main staple food and source of calories, and fishing or hunting was the principal source of protein.

Tapajós Flona is characterized by a heterogeneous environment in terms of topography, vegetation, and soils. This diversified environment leads to different forms of land and forest use (Flona-Tapajós Report 1996). It is important to take into account this diversity to understand the presence of assorted decisions across communities within this PA. Some patches are characterized by sandy soils, located by the riverbank. As altitude increases, soils become more clayish and fertile. There are several spots of black soil, locally called "indigenous soil" because it is believed to be a result of ancient organic dumping in former indigenous villages, therefore more suitable for farming. Some areas are hilly, presenting a difference of 200 m between plateau and upland zones. At least six types of forests are found in this area: (1) tropical humid dense forest in the plateau, (2) tropical humid dense forest in the upland, (3) tropical flooded forest, (4) tropical open forest (palm trees and lianas) in the plateau, (5) tropical open forest in the upland, and (6) secondary forests in both altitudes (Dubois 1976). Differences in soil and altitude cause heterogeneous distribution of plant species, including useful plants for medicinal, building, and food purposes. Therefore, commercial wood and good soil for agriculture are important assets across communities and shape the forms of land and forest use.

11.2.2 Ipanema National Forest and Rural Settlements in the Atlantic Forest Biome

Ipanema Flona was created on May 20, 1992, by Decree 530 (MMA 2003). It covers an area of 5,069.73 ha and occupies territories of three municipalities: Iperó, Capela do Alto, and Araçoiaba da Serra in the State of São Paulo.

The focus of this study is on two rural settlements comprised of former landless people from the MST: Ipanema 1 and Ipanema 2. Ipanema 1 presents a larger area of 1,120 ha and is comprised of 86 individual lots. Ipanema 2 presents 647 ha with 62 individual lots. The total number of inhabitants at the time of the study was approximately 800 people based on a socioeconomic survey I and my colleagues carried out in July 2008. The two settlements and Ipanema Flona are in an area of the former, private Ipanema Ranch.

Ipanema Ranch covered a total area of 6,827 ha, and part of this territory was converted into national forest by the federal government (Machado 1998), specifically Ipanema Flona (Fávero et al. 2004). Four days before Ipanema Flona was formally established, several people from the MST were camped there, according to their own report. They also reported that they already had a plan to occupy the large patches of cleared (deforested) area after Ipanema Ranch was abandoned.

After a long process of negotiation with IBAMA (the agency responsible for PAs in Brazil until 2007, when ICMBio took over the charge), INCRA created the Ipanema Settlement through Ordinance 342 on December 12, 1995. The total area of the official settlement was 1,200 ha and was divided into small lots from 3 to 15 ha. Ipanema Ranch was divided into Ipanema Flona with 5,069.25 ha, the Marine Ministry with 858 ha, the Ministry of Agriculture with 495 ha, and miscellaneous areas with 402 ha (Machado 1998). By 1997, the settlements had been demarcated by INCRA as a result of a settlers' petition. In 2008, the total area covered by the two settlements was 1,613 ha.

The division of land is still contentious, as parts of the rural settlements overlap with areas of ICMBio, Marine Ministry, and Agriculture Ministry; thus, local settlers are constantly concerned with land security. Also, in 2008, more than 10 new families "invaded" the settlements, but they were not welcomed by the first settlers. During 2008 and 2009, INCRA tried to define the future of these new residents but, thus far, the situation remains undefined.

Since the beginning of the two settlements, the houses have been provided with electricity, allowing them to have refrigerators, televisions, and washing machines, in contrast to their Amazonian counterparts, who face serious problems of infrastructure (Brondízio et al. 2009; Futemma and Brondízio 2003). The main and secondary roads were in good condition for traffic and presented easy access to the main urban centers (e.g., Sorocaba and Iperó) and made it possible for local farmers to sell their products (e.g., crops and milk) and purchase household supplies. Children of family farmers can go to school or work off-farm jobs because of the availability of roads.

In the last 6–7 years, families had also gained access to reception for their mobile phones and Internet. Most families had more than one mobile phone per household. Few families had access to Internet because it required a computer and a fixed phone line. All these types of communication facilitated their contact with the external world, reducing their isolation and the distance between local residents and outsiders.

11.3 Methods

In 2002, for 45 days, I visited 16 riparian, so-called "traditional" communities located inside Tapajós Flona. I analyzed in depth three communities—Jamaraquá, Pedreira, and Nazaré—by interviewing heads of households and conducting group meetings. To assess spatial and temporal changes in land use and land cover within each Tapajós community, Landsat TM images for 1986 and 1999 were used; and in situ data (field reconnaissance to collect training samples used for image classification) and interviews with local, small producers were used for image classification of land cover (Futemma and Brondízio 2004).

Between 2007 and 2009, I and three other researchers made several field trips to the Ipanema settlements located inside and near Ipanema Flona. In 2007, we carried out exploratory visits to reconnoiter and obtain local permission to conduct the study. In 2008, a different team of researchers (seven people) conducted a socioeconomic survey by providing questionnaires to 96% of the households (160 households). In 2009, from July to August, semistructured interviews were conducted with 14 families who were involved in cattle ranching and cultivation (annual and perennial crops). Local leaders were also interviewed in order to collect data on social and political organizations. Another set of data that dealt with conservation of gallery forests and rural tourism was also collected in 2009.

11.4 The National Forests and the Main Economic Activities

Human populations that live within or surrounding PAs depend on multiple economic activities to fulfill their subsistence and income needs. Some of them invested more in cropping, others in livestock or even in off-farm jobs. Small-scale farmers from both Tapajós Flona and Ipanema Flona were involved in agriculture, cattle raising, tourism (rural and ecotourism), forest management, and off-farm jobs.

11.4.1 Agriculture

Conventional agriculture based on high energy inputs, chemical fertilizers, and weed killers has been considered damaging for biodiversity and for human health (Altieri 1999). Natural scientists, for at least 30 years, have brought to the attention of public officials and decision makers that we need to replace this conventional system with more ecologically oriented production (Altieri 1999; Altieri et al. 1987). Shifting or swidden cultivation (here these two terms are used as synonyms) has also been considered deleterious to the environment in that it uses fire to burn old fields and thus increases deforestation rates throughout tropical regions (Geist and Lambin 2002; Myers 1993; Ranjan and Upadhyay 1999).

Small-scale farmers from Tapajós Flona continued to grow their traditional annual crops (Table 11.1), such as manioc (20% of the families, 288 families), corn (23%, 331 families), and rice (20.5%, 295 families) through shifting cultivation. They also grew several other cultivars as shown in Table 11.1. Amazonian farmers always had practiced a type of farming similar to what we know today as agroforestry, in which they mixed annual with perennial crops. For instance, in their gardens, one could frequently find banana (*Musa sp.*), cupuaçu (*Theobroma grandiflorum*), urucum (*Bixa orellana*), acerola (*Malpighia emarginata*), and a variety of palm trees (e.g., tucumã [*Astrocaryum aculeatum*] and açaí [*Euterpe oleracea*]). A more systematic production of agroforestry was proposed by researchers and extensionists from a nongovernment organization (NGO), and some Tapajós families were involved in that project. By 2002, the agroforestry project was not very successful, as several problems had been encountered during the execution phase. The NGO was trying to overcome those constraints and failures to keep the project going.

Small-scale farmers in the Ipanema settlements grew a diversity of annual and perennial crops for both subsistence and commercial purposes. They cultivated a variety of vegetables (lettuce, tomatoes, cucumber, eggplants, among others), beans, and sweet manioc (*Manihot utilissima*), but they planted manioc much less than their Amazonian counterparts (Table 11.1). Different from Amazonian farmers, southern farmers did not make manioc flour, they did not even know or grow the bitter manioc (*Manihot esculenta*) commonly used to make flour. Their main staple foods were rice and beans. Corn was cultivated by 60% of Ipanema families, because they feed

Table 11.1 Frequency of families from the Tapajós communities ($n = 1{,}438$) and from the Ipanema settlements ($n = 167$) that cultivated annual crops

Crop	Tapajós Flona[a] (% families)	Ipanema Flona[b] (% families)
Pumpkin	0.4	23.0
Açaí	0.3	0.0
Peanut	0.3	2.4
Rice	20.5	4.8
Potatoes	0.4	4.2
Yam	0.1	0.0
Onion	0.1	0.6
Manioc flour	21.1	0.0
Beans	12.0	37.7
Manioc	20.0	48.0
Watermelon	0.2	0.0
Corn	23.5	60.0
Black pepper	0.1	0.0
Soybeans	0.1	0.0

[a]Secondary data from census 2006 (Flona/Ibama)
[b]Primary data

small animals (chickens and pigs) with corn. They cultivated many types of fruits, such as orange, tangerine, acerola, lemon, lime, guajava, avocado, and mango, to list a few. They sold their products, which provided food for social programs, such as childcare, elder care and nursing homes, and orphanages, with support from the federal government's program to help them commercialize their products (Brazil Zero Hunger Program, see Sect. 11.6.1). Farmers reported that they were proud of producing food for social purposes.

Some families were trying to change from conventional agriculture to production free of chemicals (i.e., organic) in both annual (e.g., corn, beans, vegetables) and perennial (e.g., passion fruits, guajava, acerola) crops. They were concerned about their own and consumers' health, and they reported that organic production was more compatible with environmental conservation.

11.4.2 Cattle Raising

Despite being considered one of the main villains to conservation, cattle are still attracting the interest of rural producers, including small-scale farmers throughout Brazil (Hetch 1993; Porro et al. 2004; Veiga et al. 2004; Walker et al. 2000). Cattle present several advantages to farmer's families. They require low labor inputs, present low investment risk, provide milk and several other by-products (e.g., cheese, butter, creams), function financially as a savings account for the family because of their liquidity, represent a symbol of strength and courage among young men (cowboy culture), and hold the possibility to be better off, among other benefits (Porro et al. 2004; Veiga et al. 2004). In both Tapajós Flona and Ipanema Flona, several families invested in cattle.

A census conducted by the Flona/Ibama team in 2006 indicated a total of 6,350 heads created by communities located inside Tapajós Flona (FLONA/IBAMA 2006). The three communities analyzed here presented average heads of cattle per family of 0.56 (Jamaraquá, 9 heads), 0.63 (Pedreira, 27 heads), and 1.79 (Nazaré, 70 heads). Despite the small number of cows per family, some communities held a larger herd, for instance, São Jorge (upland community) with an average of 4.26 heads per family (totaling 950 cows) and Aveiros (riparian community) with 4.42 per family (totaling 2,073 cows). São Jorge and Aveiros were atypical communities. São Jorge was composed of recent migrants (mainly from northeastern and southern regions of Brazil) and represented more a rural settlement than caboclo communities; Aveiros was an urban area but entirely within Tapajós Flona's boundaries. Only 4 out of 31 communities had no cattle at all. These numbers confirm the attraction of cattle for small-scale Amazonian farmers.

In the Ipanema settlements, a local environmental factor influenced the settlers' decisions on patterns of land use: fertility of soil. Ipanema 1 has sandy soils and is less fertile; thus, the size of lots was larger than in Ipanema 2 so that a family could invest in cattle raising to compensate for the poor quality of land. The decision regarding sizes of lots based on fertility of the soil was made in the initial phase of the settlement (1992–1994). The 2008 survey indicated that families in Ipanema 2 had smaller lots than families in Ipanema 1 and they invested mostly in annual and perennial crops; none of them had cattle as an important source of income. In contrast, 15% of families (16) in Ipanema 1 declared that from 50 to 100% of their household income was based on cattle production. Approximately 63.5% of 94 families from both settlements had cattle—68 families (71.6%) in Ipanema 1 and 26 (27.4%) families in Ipanema 2. The average was 11 heads per family, but six families held between 30 and 60 heads each.

In general, annual and perennial crops were the main source of income for 50% of the families in Ipanema 1 and for 57% of the families in Ipanema 2. Retirement income received was another important source of money for 21% of families in Ipanema 1 and 10% in Ipanema 2.

11.4.3 Rural Tourism or Ecotourism

Tourism has been playing a key role in the economy of Brazil (Lage and Milone 1991). The socioenvironmental movement brought in a more rural-oriented or natural-oriented tourism. Both rural tourism and ecotourism, generally speaking, tried to satisfy both social inclusion and environmental conservation (Luchiari 2002; Ruschmann 2000). These categories of tourism fall right into place when one considers another source of income to small-scale rural producers.

By 2002, few families in Tapajós Flona were involved in ecotourism. Informally and without structure, these families received tourists interested in visiting the Amazonian native forests and learning about traditional local cultures. Some residents were trained to work with ecotourism; however, better capacity building was needed. The Tapajós Flona office with an information desk was in the city and needed better infrastructure to receive tourists and provide them with good information about the forest and surrounding area, lodging, and transportation by boat, bus, or taxi to assure a visitor a safe and enjoyable visit.

Since 2007, some families in the Ipanema settlements have been receiving tourists from São Paulo city (approximately 150 km away) (Martins and Futemma 2009). Predominantly, tourists were teachers and students from private elementary and high schools (upper class) who wanted to know the history of the MST and the setters' lifestyle. They spent 1 or 2 days in the settlement, but they stayed in nearby hotels, not in the settlements. Families delegated different tasks to different individuals from both settlements. Some women were responsible for preparing lunches, some young people helped to guide the tourists to see some rural properties and cultivating areas. Some community leaders were in charge of telling the history of the MST and how they arrived and settled in the area. This type of tourism is considered education-oriented tourism within a rural tourism (Beni 2002). There was potential to combine visits to Ipanema Flona and its settlements, once the tourists became interested in both places. However, to make it possible, the manager of the national forest had to be willing to do so. The current director reported that they wished to reconcile tourism in both places but that the legalization of the settlements (legal rights of possession) had to come first. In other words, until legalization of the land occupied by the settlers happens, partnerships will not happen. Legalization of land is issued by the federal government and is a step toward land titling.

11.4.4 Forest Management

The Ipanema settlements had no dense or sufficient natural forest in which to exploit either timber or non-timber products. In fact, their lands were traditionally industrial and pasture areas so-called the Ipanema Rancher (see Sect. 11.2.2 for more information); therefore, basically, no natural forest remained in the area. Only a narrow strip of gallery forests existed along some parts of the rivers. Flona Ipanema's officers intended to plant eucalyptus trees in order to reforest degraded areas and extract timber to sell so they could invest in infrastructure, according to their own reports.

In contrast, the Tapajós communities were covered by a dense, native, tropical humid forest. There were a number of timber and non-timber species that could be explored. However, by federal law, to extract any resource from a forest, an individual needed to have a forest management plan. Since the mid-1990s, IBAMA (now ICMBio) and several NGOs have been trying to put into practice a community-based forest management plan. To do so, they tried to help local residents elaborate and execute a management project. Up to 2002, only a few members of four communities were involved in the project.

Any group of people interested in exploring forest resources for commercial purposes had to write and submit a project plan to IBAMA. A team of consultants hired by IBAMA evaluated projects to give final approvals. By 2002, only three projects had been approved and executed by local residents with direct support of either IBAMA or an NGO (Amazon Environmental Research Institute and/or Health and Joy Project). These three projects had the following targets: (1) extraction of vegetal oil products, (2) wood debris for small furniture, and (3) vegetal leather (ecoleather) for handbags (from planted rubber trees).

11.4.4.1 Extraction of Vegetal Oils

Vegetal oils'—andiroba (*Carapa guianensis*) and copaíba (*Copaifera* sp.)—extraction began with a group of women, who had been collecting and preparing those oils for medicinal and cosmetic purposes. Local Amazonian people used these oils on a regular basis. Extraction of vegetal oils requires local ecological knowledge and techniques to find a tree, to extract the oils, to know what season to collect them, and how to prepare and use them.

A group of women proposed a project with external help from IBAMA-ProManejo. ProManejo (Forest Resource Management Project) refers to a project that was financed by G7 (seven richest countries) within PPG7 (Pilot Program to Conserve the Brazilian Rain Forest) and was launched in 1992 and ended in 2009. The main target of ProManejo was protection of the Amazonian tropical rain forests through sustainable management. IBAMA was in charge of implementing this project in the Brazilian Amazon region. After some initial successes in selling non-timber forest products and increasing cosmetic companies' demand from southern Brazil, some men started to participate in this endeavor. By 2002, three communities were involved in vegetal oil projects with male and female participants.

11.4.4.2 Use of Wood Debris for Small Furniture

In 2002, only one community was involved in this project, which was carried out by the Amazon Environmental Research Institute. The NGO worked directly with local residents and provided them with some training in crafting wood debris for making small furniture and selling it in local markets and São Paulo.

11.4.4.3 Vegetal Leather

Similar to the wood debris project, only one community was involved in the production of leather made from rubber trees. Local art crafters were assisted and financed by national and international sponsors and agencies (e.g., IBAMA-ProManejo). These external agents helped local people in several steps from elaboration of the project to its execution and commercialization. This group was called Eco-Leather Producers Group. Men and women were involved in this group.

11.4.5 Off-Farm Jobs

The decision of one member or several members of a family to work outside the rural properties had to do with local perceptions of lack of opportunities within the rural property or that farming tasks were too hard or simply that they had no skills or willingness to be a farmer. Particularly, young individuals (farmers' children) desired to leave the properties to work off farm and be in contact with the external world, which they perceived as more attractive and modern. The 2008 survey indicated more than 50 types of jobs—carpenters, cooking assistants, plumbers, and housekeepers, among others—employed the residents of the Ipanema settlements. Most of them worked nearby, so they could go and come back every day. Some worked far away and did not stay in the settlements during weekdays. Off-farm work contributed to family income, 20% among families in Ipanema 1 and 10% in Ipanema 2.

11.5 Environmental Outcomes

11.5.1 Tapajós Flona and Its Communities

Analysis of land-cover transitions between 1986 and 1999 (Futemma and Brondízio 2004) indicates that the total Tapajós Flona area presented a large area of mature forest (93.4%) and most communities had low rates of deforestation. Between 1974, when Flona-Tapajós was created, and the 1990s, the institutional structure of communities inside the national forest was unclear regarding land tenure and rights to the forest. Although their tenure status has not been fully defined, in the last 10 years, they have officially gained the rights to use and manage the forest on a community basis.

The riparian communities presented lower rates (6.5%) of clearing than the upland communities after 1986 (Table 11.2). When considering the communities together, forest cover presented little variation, ranging from 80 to 90%, although when one includes outliers they correspond to about 68%. In general, the communities presented low levels of deforestation and high levels of forest cover (>80%) with the exception of one community accessible by road and subjected to the land market and land invasions (Fig. 11.1). Communities along the river represented areas of long-term occupation by caboclo populations. They typically grew manioc,

Table 11.2 Absolute (area) and relative (percent) distribution of land-cover changes from 1986 to 1999 according to different types of institutional arrangements: public and community

Institutional arrangement		*N*	Area (ha) Areas in use	Clearance of secondary forest	Deforestation of mature forest	Regeneration	Mature forest	Total
Public/government	FLONA-Tapajós[a] total	1	5,519.25	4,311.81	7,501.5	9,472.05	184,338.3	406,151.5
Communal[b]	FLONA-Tapajós riparian community	14	801.27	837.36	3,748.77	1,586.79	48,684.69	57,458.88
	FLONA-Tapajós upland community	1	943.38	688.95	2,177.64	668.52	9,502.65	14,040.36

Institutional arrangement		*N*	Area (%) Areas in use	Clearance of secondary forest	Deforestation of mature forest	Regeneration	Mature forest	Total
Public/government	FLONA-Tapajós total	1	1.4	1.1	1.8	2.3	93.4	100
Communal	FLONA-Tapajós riparian community	14	1.4	1.5	6.5	2.8	84.7	100
	FLONA-Tapajós upland community	1	6.7	4.9	15.5	4.8	67.7	100

[a]The total area of FLONA-Tapajós is approximately 600,000 ha, but we consider only the northern 406,000 ha in our analysis. This includes community and logging concession areas and forest reserves for direct use and indirect use (research only)

[b]Despite FLONA-Tapajós being under public governance, we treated communities as having communal institutional arrangements, because local residents have rights to use forest resources

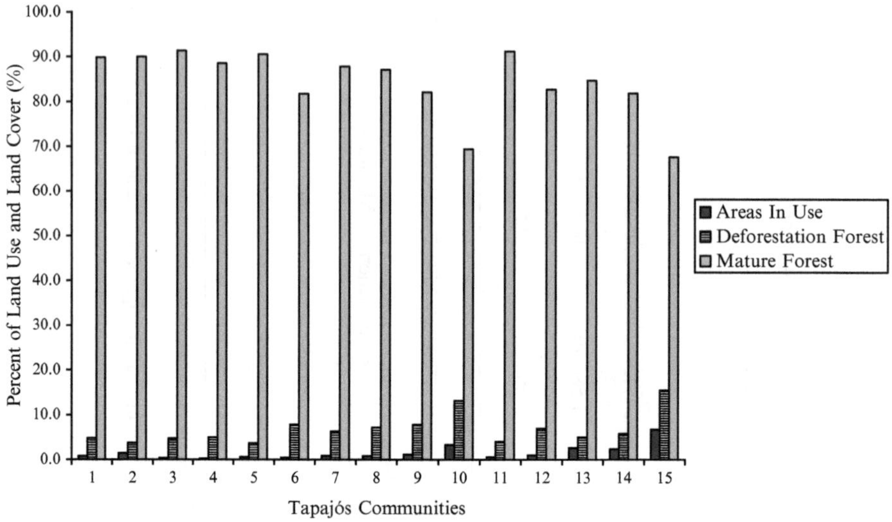

Fig. 11.1 Distribution of land use and land cover among the Tapajós communities, 1986–1999. Numbers 1–14 refer to riparian communities. Number 15 refers to the upland community

corn, and rice and collected some forest products, including roots, vegetable oils, vines, and timber, for household use and to sell for income. Thirteen communities accessible only by river had more than 80% of their land in forest cover. The upland community accessible by road had 65% in forest cover (Fig. 11.1).

11.5.2 Ipanema Flona and Its Settlements

A sample of 30 properties (15 in Ipanema 1 and 15 in Ipanema 2) were studied (Venturino and Futemma 2009) to evaluate the presence or not of a minimum area of gallery forest required by law (Forest Code 1965; see Machado 1989) as a permanent preservation area (APP) (Machado 1989; see also Sect. 11.6.2). Regarding forms of land use, 100% of the sample invested in cropping and 66.7% had cattle in their properties. Only 6.7% produced for subsistence; the majority of 93.3% produced for markets. More than half, or 63.3%, of the market-oriented families participated in the local cooperative in order to sell their output to the Zero Hunger Program, promoted by the federal and municipal governments.

Pasture areas were observed in 83.3% of the properties sampled near the river, that is, cattle activity took place within the APP boundaries due to easy access to the water. The presence of APP was counted when gallery vegetation covered a width of at least 30 m from the riverside, according to the APP rules. Vegetation with less than 30 m was not considered an APP. Of 30 properties, 12 (40%) followed the APP rule, and 18 (60%) did not comply (Table 11.3).

Table 11.3 Distribution of property lots according to size of properties, size of APP (gallery forest) per property at settlements Ipanema 1 and Ipanema 2 (2009)

Property ID	Size of existing gallery forest within a property (ha)	Size of APP according to legal rule (ha)	Compliance to the APP rule (yes/no)
Settlement Ipanema 1			
1.1	0.55	1.11	No
1.2	0.25	0.64	No
1.3	0.10	0.13	No
1.4	1.06	0.52	Yes
1.5	0.87	0.78	Yes
1.6	–	0.26	No
1.7	1.50	2.44	No
1.8	0.55	0.52	Yes
1.9	0.11	0.58	No
1.10	3.75	2.50	Yes
1.11	1.85	2.10	No
1.12	0.10	0.90	No
1.13	0.42	1.05	No
1.14	1.60	1.20	Yes
1.15	0.93	0.90	Yes
Settlement Ipanema 2			
2.1	1.25	1.57	No
2.2	1.31	1.01	Yes
2.3	0.38	1.29	No
2.4	–	0.450	No
2.5	0.43	0.39	Yes
2.6	0.32	0.45	No
2.7	0.99	2.11	No
2.8	0.63	0.75	No
2.9	1.34	1.21	Yes
2.10	0.56	0.06	Yes
2.11	1.51	1.53	Yes
2.12	1.25	1.79	No
2.13	1.12	1.79	No
2.14	0.28	0.58	No
2.15	0.42	0.42	Yes

11.6 Public Policies: Rural Development vs. Environmental Conservation

The new millennium started with strong debates on both environmental conservation and social justice (also included were food security, guaranteed source of income, social inclusion, among others). In Brazil, officials tried to include some of these new demands toward conservation and rural development into the political agenda, especially within the Ministry of Environment (MMA) and the Ministry of Agrarian

Development (MDA, policies paid particular attention to small-scale rural producers www.mda.gov.br). In addition, to encouraging families to stay in rural areas and to be better off, the federal government, in 2003, created incentives to boost family-based agriculture throughout Brazil. The Ministry of Agriculture, Livestock and Food Supply (MAPA) traditionally has had a strong focus on large-scale agribusiness production systems (www.agricultura.gov.br), but since 2003–2004, MAPA has included policies toward small-scale rural producers. Thus, both MAPA and MDA now have programs to support small-scale agriculture, such as credit lines, capacity building, and commercialization aid programs (see Secretaria da Agricultura Familiar http://portal.mda.gov.br/portal/saf/). To meet all these demands together is a challenge to many people involved in this process, and the decisions made by different Brazilian official institutions come into conflict.

11.6.1 Policies Toward Rural Development

For more than 5 years, Ipanema farmers have been participating in the MDA-Brazil Zero Hunger Program (created in July 2003, see www.fomezero.gov.br) in order to commercialize their farming products. They sell all kinds of products with this government support, from vegetables to fruit jellies, milk and cheese, and chicken eggs, once a week. A municipality's truck goes to settlements and stops at various fixed points in order to collect them to be distributed to social care institutions (childcare centers, orphanages, nursing homes, among others). Farmers were very pleased to participate in this program because one of the main constraints to small-scale farmers has been transportation and commercialization of agricultural products, and because they are proud to help those less-privileged citizens.

Despite cattle being considered one of the main causes of deforestation in Brazil (Hetch 1993; Walker et al. 2000), small-scale farmers are very much attracted to raise these animals. Small-scale rural producers have been blamed for deforestation by some researchers and by some parts of the government (Scinocca 2008). In the Ipanema settlements, local farmers declared that cattle were important to any rural producer, including family-based farming, even if a farmer has only one or two heads of cattle. Scholars studying why small-scale farmers are so attracted to cattle came up with a set of motives that might drive them to invest in this activity (Porro et al. 2004; Veiga et al. 2004), as previously mentioned (see Sect. 11.4.2). All these benefits were also pointed out by families in the Ipanema settlements. A group of 16 families in Ipanema purchased a refrigerator to store milk collected from both settlements.

Families in Ipanema Flona reported that because the Zero Hunger Program was helping them to commercialize their products and guaranteeing a minimum income by selling their products, cultivation of plants had become as attractive as cattle for family-based farmers. But, if the government stops these incentives and support, cattle might become more attractive again. These families were also trying to invest more in chemical-free farming. The former Ipanema Flona director (2004–2007) encouraged this organic production, and the local farmers demonstrated willingness to do so. They perceived organic production to be less damaging to human health and to natural systems, particularly to natural vegetation in Ipanema Flona. Overall, the combination of cropping and cattle raising was the main source of income of most families in the two rural settlements in the State of São Paulo and the Tapajós communities in the State of Pará.

Tourism, as previously mentioned, was another economic activity adopted by some families in both regions, as they had incentives from the government to invest in this activity. MDA (2005) provided farmers with a program that promoted rural tourism (see http://comunidades.mda.gov.br/dotlrn/clubs/redenacionaldeturismo/one-community). This program was created in 2003 in partnership with the Ministry of Tourism (MTur), the Network of Rural Tourism and the Family-Based Agriculture, and several other technicians and institutions. Tourism is considered by many—scholars, officials, and civilians—as a reduced-environmental-impact activity and a source of supplemental household income. Another benefit related to rural tourism refers to appraisal of the value of traditional culture and practices behind the family-based agriculture and rural lifestyle.

11.6.2 Policies Toward Environmental Conservation

What are the environmental outcomes with regard to public policy incentives and patterns of land use? Since the 1990s, MAPA and MDA have tried to include in their agenda environmental goals to be achieved within programs related to sustainable development. Simultaneously, MMA has tried to find legal instruments to force rural producers and others ministries to follow environmental rules (www.mma.gov.br). Two main rules that have been causing disagreements among official institutions at federal and state levels are the Forest Reserve (FR) and APP.

The FR, based on the Forest Law of 1965, refers to a continuous chunk or patch of forest that a property owner must preserve. The minimum size of an FR varies according to the region in Brazil: 80% of a property for the Amazon region, 50% of a property for the midwest and northeast regions, and 20% for the south and southeast regions (Machado 1989). According to APP rules, farmers should not cultivate or raise animals at the top of a hill or on a hillside with a slope greater than or equal to 45°, and a minimum width of 30 m of vegetation (gallery forest) must be left standing along rivers or any other type of water stream (Machado 1989). In those areas, a rural producer can extract some non-timber products, such as fruits, nuts, and oils, but she/he may not cut down trees.

A large number of small-scale farmers have cultivated for decades or even centuries annual and perennial crops in areas considered APPs by MMA. Some smallholders in the south region of Brazil cultivated grapefruits and apples, and in the State of Minas Gerais, farmers planted coffee on the hillsides

(ABC Digital 2009). In the Ipanema settlements, by 2009, the majority of farmers sampled did not follow the gallery forest preservation rules (Venturino and Futemma 2009; see Table 11.3).

Farmers in the Ipanema settlements were interested in investing in the agroforestry system (AFS) for dual purposes: income and reforestation of gallery forests. Researchers and extensionists of two public universities of the State of São Paulo tried to work with these farmers, particularly involving young people, to implement an AFS project by providing them with training courses, seeds of annual and perennial plants, and exchange experiences with some AFS producers. The team of researchers intended to work with families from both Ipanema 1 and Ipanema 2.

In 2009, the project was interrupted in Ipanema 1 because the settlement was located within Ipanema Flona boundaries and the current national forest director (2008–present) did not give permission to carry out the project, even though reforestation of gallery forests was one of the main goals of this endeavor. The director argued that the settlement was not formally regularized by INCRA; thus, he could not allow any activity in this area. It is worth mentioning that this AFS project had financial support from the Brazilian National Council for Scientific and Technological Development. This director also prohibits free entrance of settlers into Ipanema Flona boundaries and reported that he does not negotiate any collaboration while they are illegally in the area. The problem is that legal recognition of these properties may take years, even decades, to happen if it ever happens. Thus, the director's decision to interact with local people only after the settlements' regularization might indicate that he is not willing to include locals in the process of governing the area.

MAPA argued that if smallholders reduced the cultivating area to follow the environmental rules regarding APP or FR, they would suffer a reduction in their already low income, and their existence as small-scale farmers would be threatened (Scolese 2009). MAPA suggested that MMA make the APP rules more flexible in order to allow traditional coffee and grapefruit growers to keep their production systems. In 2008, the State of Santa Catarina tried to create a state forest code, and one of the rules reduced APP from 30 to 5 m. However, this new forest code went against a federal rule, and attorneys warned the state legislators about breaking this rule (Agência Estado 2009; Agência O Globo 2009). As far as small-scale farmers or smallholders are concerned, is the size of a property a constraint or limiting factor to achieve both conservation and economic goals? To answer this question, some aspects should be analyzed, which are not fully addressed in this study: the roles of property regimes and biophysical conditions (e.g., types of soils, water availability, variety of natural resources) in reconciling economic and conservation targets.

The data show that farmers in the Tapajós communities kept their FR in that they presented a minimum of 80% of continuous vegetation within community boundaries (Fig. 11.1) as required for the Amazon region. It is worth mentioning that in the Tapajós case, the forested land was held and managed on a communal basis in contrast to the Ipanema settlements, in which land was held by individual private owners. Because those Tapajós communities were located inside Tapajós Flona, farmers had to follow the rules for PAs. These rules might have contributed to keeping the 80% of forest but also led to local conflicts. For instance, by the mid-1990s, an official manager made the extreme decision to prohibit all clearings for farming and wood extraction. Manioc flour has always been the main staple food and source of calories among native Amazonian people and has been always produced locally. Thus, such prohibition caused a great revolt among local residents. Afterward, some adjustments of these rules occurred but it was not enough, because discontent continued. Local farmers had to ask permission to clear an area, and there was a size limit for cropping. Similar to the Ipanema case, the relationship between the Tapajós farmers and the manager of Flona-Tapajós had become tense.

Overall, Ipanema private properties had no forest reserve. Historical facts indicate that previous to the INCRA settlements, the area was already covered by abandoned pasture and this was one of reasons that drove the landless families to occupy this area. The former landless and now smallholders reported willingness to reconcile their household income with activities more appropriate to the conservation of natural resources, such as tourism, agroecology, and reforestation of gallery forests.

Farmers from both national forests—Tapajós and Ipanema—faced similar opportunities and constraints. They invested mostly in activities directly related to use of land: cropping and livestock. Regarding source of income, on one hand, the Tapajós farmers had the advantage of exploiting a dense, native Amazonian forest to extract either non-timber (e.g., vegetal oils) or timber (e.g., wood debris to make furniture) products. By having a larger variety of natural resources to extract and generate income might also have contributed to Tapajós families keeping forest stands in comparison to their Ipanema counterparts, who had no forest products to exploit. On the other hand, Ipanema's farmers were close to main urban centers, which gave them easier access to markets to sell and buy products and roads that were in much better condition for transit than in the Amazon region. However, as Forsyth (2003) argues, some policies, particularly environmental policies, unfairly penalize land users, mainly in developing countries, which may even increase environmental degradation and poverty by threatening their livelihoods.

The MDA and MAPA policies created incentives to farmers to participate actively in local markets to help improve their household income and thus promote rural development. The Zero Hunger Program, besides generating larger incomes, also helped food security through production of vegetables, fruits, roots, and milk and their by-products. Moreover, the Zero Hunger Program, by attracting a farmer to invest more into cropping than cattle, might contribute to a lower environmental impact. In addition, incentives to sustainable tourism might contribute to a more environmentally friendly economy as well as to create opportunities, especially for youth (farmers' children) by keeping them on rural properties. Moreover, ProManejo in collaboration with IBAMA provided support to Tapajós families to conduct projects, such as extraction of vegetal oils.

Political ecology allowed us to uncover political factors underlying changes in access to natural resource as well as government incentives to enhance some economic activities. Political ecology contributed to bring into this analysis conflicting public policies, mainly within the federal (ministry) level, which caused uncertainties among small-scale farmers. In Brazil, history is full of government interventions that enhance both environmental degradation and social disruption and inequalities. In the 1970s and 1980s, government incentives, such as tax exemptions and other advantages, contributed to clearing millions of hectares in the Amazon region for pasture, mining, and construction of hydroelectric power plants and roads (Moran 1996). Likewise, in Latin American countries, government policies throughout the twentieth century have caused deforestation, social exclusion, and rural poverty (Durham 1988; Schmink and Wood 1992; Sheridan 1988; Stonich 1993).

However, it is important to recognize that other factors rather than political might play important roles in explaining environmental changes or decisions regarding patterns of land use among farmers (Vayda and Walters 1999). In Ipanema rural properties, type of soil was a biophysical factor that defined a more or lesser investment in cattle, as mentioned earlier. In fact, multiple factors are behind attraction to cattle as far as small-scale farmers are concerned (Porro et al. 2004; Veiga et al. 2004). As previously questioned, issues on size of property and type of property regime might be a constraint to small-scale rural producers' adequately conciliating economic goals with conservation; thus, institutional analysis might shed some light on this puzzle (Gibson et al. 2000; Hanna et al. 1996; Ostrom 2005).

11.7 Final Considerations

Public policies directly and indirectly affect decision making among family-based farmers with regard to governance of a rural property, including forms of land use and ways of conserving the natural environment. Family-based farmers from both national forests wished to invest more into cropping and livestock to fulfill both subsistence needs and household income, but they have been confronting barriers due to environmental regulations. Because they were living in a PA, data indicate that Tapajós and Ipanema small-scale farmers were trying to change some practices toward more sustainable activities by adopting chemical-free agriculture and agroforestry systems, exploiting non-timber products, and introducing rural and ecological tourism into their household economies. Thus, some public policies opened new opportunities through which these rural families could participate more in the market while other public policies created some constraints to do so. Therefore, there is a long way to go to find a reconciliation among social justice, rural development, and environmental conservation.

References

ABC Digital. (2009). *Projeto que institui Código Ambiental estará no Congresso em maio*. http://floripamanha.org/2009/03/projeto-que-institui-codigo-ambiental-estara-no-congresso-em-maio/. Accessed 10 Feb 2010.

Agência Estado. (2009). *Lei ambiental causa conflito entre Santa Catarina e Ministro Carlos Minc*. http://www.agroin.com.br/novo/?p=view&id=203. Accessed 10 Feb 2010.

Agência O Globo. (2009). *Minc critica lei de Santa Catarina que reduziu mata ciliar*. 03/04/2009. http://www.gazetadopovo.com.br/vidaecidadania/conteudo.phtml?id=873731. Accessed 10 Feb 2010.

Almeida, M., & Cunha, M. (2000). Populações Tradicionais e Conservação Ambiental. In A. Veríssimo, A. Moreira, D. Sawyer, I. dos Santos, e L.P. Pinto (Orgs.), *Biodiversidade na Amazônia Brasileira: Avaliação e Ações Prioritárias para a Conservação, Uso Sustentável e Repartição de Benefícios* (pp. 184–193). São Paulo: ISA e Estação Liberdade.

Alston, L. J., Libecap, G. D., & Mueller, B. (1999). *Title, conflict, and land use: The development of property rights and land reform on the Brazilian frontiers*. Ann Arbor: University of Michigan Press.

Altieri, M. A. (1999). The ecological role of biodiversity in agroecosystems. *Agriculture, Ecosystems & Environment, 74*(1–3), 19–31.

Altieri, M. A., Anderson, M. K., & Merrick, L. C. (1987). Peasant agriculture and the conservation of crop and wild plant resources. *Conservation Biology, 1*(1), 49–58.

Benatti, J. H. (2001). Presença humana em unidade de conservação: um impasse científico, jurídico ou político? In J. P. Capobianco, A. Veríssimo, A. Moreira, D. Sawyer, I. Santos, & L. P. Pinto (Orgs.), *Biodiversidade na Amazônia* (pp. 299–305). São Paulo: Editora Estação Liberdade/Instituto Socioambiental-ISA.

Beni, M. C. (2002). Conceituando turismo rural, agroturismo, turismo ecológico e ecoturismo. In M. Barreto & E. Tamanini (Eds.), *Redescobrindo a ecologia do turismo* (pp. 31–34). Caxias do Sul: EDUCS.

Brondízio, E., Cak, A., Caldas, M. M., Mena, C., Bilsborrow, R., Futemma, C., Moran, E., Ludewigs, T., & Batistella, M. (2009). Small farmers and deforestation in Amazonia. In M. Keller, M. Bustamante, J. Gash, & P. Silva Dias (Eds.), *Amazonia and global change* (1st ed., Vol. 186, pp. 117–147). Washington, DC: American Geophysical Union.

Doolittle, A. A. (2010). Stories and maps, images and archives: Multimethod approach to political ecology of native property rights and natural resource management in Sabah, Malaysia. *Environmental Management, 45*, 67–81.

Dubois, J. (1976). *Silviculture preliminary forest management guidelines for the national Forest of the Tapajós*. FAO/FO/BRA-71-545, Projeto de Desenvolvimento e Pesquisa Florestal. Belém: Região Norte Region.

Durham, W. H. (1988). Political ecology and environmental destruction in Latin America. In M. Painter & W. H. Durham (Eds.), *The social causes of environmental destruction in Latin America* (pp. 249–264). Ann Arbor: University of Michigan Press.

Faria, A. A. C. (2004). *Uma visão da Floresta Nacional do Tapajós: diagnóstico rural participativo (DRP) – junho a setembro de 2003*. Belterra: IBAMA-Promanejo.

Fávero, O. A., Nucci, J. C., & Biasi, M. (2004). Vegetação natural potencial e mapeamento da vegetação e usos atuais das terras da Floresta Nacional de Ipanema, Iperó: conservação e gestão ambiental. *Revista RA´E GA, 8*, 55–68.

FLONA/IBAMA. (2006). *Census 2006*. www.icmbio.gov.br/tapajs. Accessed 9 Feb 2010.

Flona-Tapajós Report. (1996). *Plano diretor e zoneamento comunitário da margem direita do Tapajós*, versão 3. Piracicaba: MMA/IBAMA-IMAFLORA, Programa Piloto para Proteção das Florestas Tropicais do Brasil.

Forsyth, T. (2003). *Critical political ecology: The politics of environmental science*. London: Routledge.

Fox, J., Fujita, Y., Ngidang, D., Peluso, N., Potter, L., Sakuntaladewi, N., Sturgeson, J., & Thomas, D. (2009). Policies, political-economy, and swidden in Southeast Asia. *Human Ecology, 37*, 305–322.

Futemma, C., & Brondízio, E. (2003). Land reform and land-use changes in the lower Amazon: Implications for agricultural intensification. *Human Ecology, 31*(3), 369–402.

Futemma, C., & Brondízio, E. (2004, July 27–29). *Institutional dimensions and land-cover change: social groups, time of settlement, property size, and location and access affecting intra-regional variation in the Lower Amazon*. Presented at the II LBA conference, Brasília, DF.

Geist, H. J., & Lambin, E. F. (2002). Proximate causes and underlying driving forces of tropical deforestation. *BioScience, 52*(2), 143–150.

Gerber, J. F., Veuthey, S., & Martínez-Alier, J. (2009). Linking political ecology with ecological economics in tree plantation conflicts in Cameroon and Ecuador. *Ecological Economics, 68*, 2885–2889.

Gibson, C. C., McKean, M. A., & Ostrom, E. (Eds.). (2000). *People and forests: Communities, institutions, and governance*. Cambridge, MA: The MIT Press.

Hanna, S. S., Folke, C., & Maler, K.-G. (1996). *Rights to nature: Ecological, economic, cultural, and political principles of institutions for the environment*. Washington, DC: Island Press.

Hetch, S. (1993). The logic of livestock and deforestation in Amazonia. *BioScience, 43*(10), 687–695.

Lage, B. H. G., & Milone, P. C. (1991). *Economia do turismo* (3rd ed.). Campinas: Papirus Editora.

Luchiari, M. T. (2002). Turismo e território: sustentabilidade para quem? In M. Barreto & E. Tamanini (Eds.), *Redescobrindo a ecologia do turismo* (pp. 111–125). Caxias do Sul: EDUCS.

Machado, P. A. L. (1989). *Direito Ambiental Brasileiro* (2nd ed.). São Paulo: Editora Revista dos Tribunais Ltda.

Machado, A. M. B. (1998). *A produção do saber sobre a floresta pelos assentados na Fazenda Ipanema, Iperó (SP)*. Master thesis. University of São Paulo, Brazil.

Martins, M. R., & Futemma, C. (2009, October 5–9). *O jovem e o turismo rural: o caso do assentamento Ipanema, Iperó (SP)*. Presented at XVII Congresso de Iniciação Científica, Universidade Federal de São Carlos, São Carlos, SP, Brazil.

MDA (Ministério do Desenvolvimento Agrário). (2005). www.mda.gov.br/saf. Accessed 10 Feb 2010.

MMA (Ministério do Meio Ambiente). (2003). *Plano de manejo: Floresta Nacional de Ipanema*. Brasília: Ministério do Meio Ambiente/IBAMA/Via Oeste Concessionária de Rodovias do Oeste De São Paulo.

Moran, E. (1996). Deforestation in the Brazilian Amazon. In E. S. Leslie, T. N. Headland, & R. C. Bailey (Eds.), *Tropical deforestation: The human dimension* (pp. 149–164). New York: Columbia University Press.

Moran, E., & Ostrom, E. (Eds.). (2005). *Seeing the forest and the trees: Human-environment interactions in forest ecosystems*. Cambridge, MA: The MIT Press.

Myers, N. (1993). Tropical forests: The main deforestation fronts. *Environmental Conservation, 20*, 9–16.

Ostrom, E. (2005). *Understanding institutional diversity*. Princeton: Princeton University Press.

Painter, M. (1988). Introduction: Anthropological perspectives on environmental destruction. In M. Painter & W. H. Durham (Eds.), *The social causes of environmental destruction in Latin America* (pp. 1–21). Ann Arbor: The University of Michigan Press.

Porro, R., Mesquita, B. A., & Santos, I. J. P. (2004). *Expansão e trajetórias da pecuária na Amazônia: Maranhão, Brasil*. Brasília: Editora UNB.

Ranjan, R., & Upadhyay, V. P. (1999). Ecological problems due to shifting cultivation. *Current Science, 77*(10), 1246–1250.

Ruschmann, D. V. M. (2000). Gestão da capacidade de carga turístico recreativa como fator de sustentabilidade ambiental: o caso da ilha João da Cunha. In B. H. Lage, & P. C. Milone (Orgs.), *Turismo: teoria e prática* (pp. 71–79). São Paulo: Atlas.

Schellas, J., & Sánchez-Azofeifa, G. A. (2006). Post-frontier forest change adjacent to Braulio Carrillo National Park, Costa Rica. *Human Ecology, 34*(3), 407–431.

Schmink, M. (1987). The "political ecology" of Amazonia. In P. D. Little & M. M. Horowitz (Eds.), *Lands at risk in the Third World* (IDA Monographs in Development Anthropology, pp. 38–57). Boulder: Westview Press.

Schmink, M., & Wood, C. (1992). *Contested frontiers in Amazônia*. New York: Columbia University Press.

Scinocca, A. P. (2008). *TCU diz que pequenos produtores respondem por 18% do desmatamento*. São Paulo: Jornal o Estado de São Paulo.

Scolese, E. (2009). *Agricultura quer aliviar regra para fazendeiro*. São Paulo: Jornal Folha de São Paulo. http://www1.folha.uol.com.br/folha/brasil/ult96u644869.shtml. Accessed 29 Oct 2009.

Sheridan, T. (1988). *Where the dove calls: The political ecology of a peasant corporate community in Northwestern Mexico*. Tucson: University of Arizona Press.

Shriar, A. (2007). In search of sustainable land use and food security in the arid hillside regions of Central America: Putting the horse before the cart. *Human Ecology, 35*, 275–287.

Soares, E. S. (2004). *Desafios, resultados, ameaças e oportunidades em uma unidade de conservação na Amazônia: a Floresta Nacional do Tapajós*. Manaus: Brazilian Institute of Environment and Renewable Natural Resources–ProManejo.

Stonich, S. C. (1993). *"I am destroying the land!" The political ecology of poverty and environmental destruction in Honduras*. Boulder: Westview Press.

Terborgh, J., & Peres, C. A. (2002). O problema das pessoas nos parques. In J. Terborgh, C. van Schaik, L. Daveport, & M. Rao (Orgs.), *Tornando os parques mais eficientes: estratégias para a conservação da natureza nos trópicos* (pp. 334–346). Curitiba: Editora UFPR.

Vayda, A. P., & Walters, B. B. (1999). Against political ecology. *Human Ecology, 27*, 167–179.

Veiga, J. B., Tourrand, J. F., Piketty, M. G., Poccard-Chapuis, R., Alves, A. M., & Thales, M. C. (2004). *Expansão e trajetórias da pecuária na Amazônia*. Brasília: Editora UNB.

Venturino, M. C. C., & Futemma, C. (2009, October 5–9). *Análise da conservação ambiental nos assentamentos rurais da Floresta Nacional de Ipanema, Iperó, SP: áreas de preservação permanente*. Presented at XVII Congresso de Iniciação Científica. Universidade Federal de São Carlos, São Carlos, SP, Brazil.

Veríssimo, A., Cochrane, M. A., & Souza, C., Jr. (2002). National forests in the Amazon. *Science, 297*, 1478.

Walker, R. T., Moran, E., & Anselin, L. (2000). Deforestation and cattle ranching in the Brazilian Amazon: External capital and household processes. *World Development, 28*, 683–699.

Chapter 12
Institutional Evolution, Forest Conservation, and Rapid Change in Rural Honduras

Catherine M. Tucker

Abstract Research increasingly suggests that effective institutional arrangements constitute a critical component for environmental conservation and development efforts, but many questions remain as to how institutions form and evolve, particularly in contexts of rapid change. This study examines the evolution of institutional arrangements for forest conservation in a rural municipality of western Honduras during a period of dramatic social and economic transformation. In recent years, the municipality has experienced expansion of export coffee production, infrastructural improvements, and increasing ties to national and international markets. The municipality has created a cloud forest reserve and gained national recognition for forest conservation. The analysis examines the transformation of traditional institutions, the emergence of new ones, and factors that influence apparent successes, failures, and ongoing challenges. Building on recent scholarship on institutional diversity and complex social-ecological systems, the discussion considers the implications of ongoing transformations for local forests, land-use and land-cover change, and vulnerability (as well as adaptation) to rapid change.

12.1 Introduction

Many of the world's surviving forests are managed by de facto or de jure local institutions, and local groups' participation in forest management institutions is associated with better forest conditions when compared to forests managed without local participation (Gibson et al. 2005; Hayes 2006; White and Martin 2005). Currently, many places with surviving forests have been experiencing rapid

C.M. Tucker (✉)
Department of Anthropology, Indiana University, Student Building 130,
Bloomington, IN 47405, USA
e-mail: tuckerc@indiana.edu

E.S. Brondízio and E.F. Moran (eds.), *Human-Environment Interactions: Current and Future Directions*, Human-Environment Interactions 1,
DOI 10.1007/978-94-007-4780-7_12,

Fig. 12.1 View of La Campa landscape

changes linked to globalization, market integration, and higher-level government interventions. These processes often have been associated with a breakdown of local institutional arrangements and concomitant degradation of natural resources (Brouwer 1995; Gadgil and Iyer 1989; Jodha 1992; Tucker 2004). When conditions change rapidly, people may not have time to develop effective institutional responses. Conscious efforts to create or adjust institutions typically move slowly (Young 2010). Even so, a few studies indicate certain communities or groups have created or adapted effective institutions despite rapid change or external interventions (e.g., Agrawal 2005; Ashenafi and Leader-Williams 2005; Ghate 2004).

This study examines how institutions for communal forest management have been transforming in La Campa, a municipio (similar to a county) in western Honduras. In recent decades, important changes have occurred in national policies, market linkages, and transportation and communication infrastructures. Although such changes often have been linked to dissipation of local institutions and deforestation, La Campa generally has maintained forest cover, with some areas experiencing regeneration, and others subject to limited deforestation (Fig. 12.1). This study addresses the circumstances in which La Campa has mitigated some of the destabilizing impacts of rapid change on community members and their forest resources. It also considers the new challenges posed by evolving socioeconomic, political, and environmental conditions.

The discussion opens with a review of key theoretical perspectives, followed by an introduction to the research site and data collection methods. Then I examine forest management institutions and how they have evolved through national policy transformations, processes of market integration, and developmental processes that have impacted forests and the community. Examination of satellite images allows me to evaluate changes in forest cover and what this indicates about the effects of institutions on forest conservation and land-cover change. I point to apparent successes and shortcomings and the implications for social-ecological vulnerability and resilience through transformations in socioeconomic conditions, municipal governance, and institutions for forest management.

12.2 Theoretical Perspectives on Institutional Change and Stressors

This research draws on emerging perspectives for studying complex social and ecological systems, and theoretical approaches to institutional change. Institutions are rules that define what individuals may, may not, or must do (Ostrom et al. 2002). Key concepts for understanding complex social and ecological systems' responses to change include vulnerability, robustness, and resilience. *Vulnerability* refers to the inherent characteristics of a system that create susceptibility to harm from social and environmental stressors, combined with lack of capacity to adapt (Adger 2006; Cutter et al. 2008). *Robustness* refers to the capacity of a system to absorb disturbances without needing to make adaptive changes. This concept has been applied particularly to social systems. Robust social systems use and reinforce existing mechanisms to maintain themselves without making fundamental changes when undergoing stressors or shocks (Young 2010). *Resilience*, a concept emergent from ecology and mathematical models, refers to the capacity of a system to make adaptive changes in response to destabilizing impacts (Janssen et al. 2006). In resilient systems, functions and key characteristics are renewed through adaptive processes when disturbances occur; thus, they exhibit dynamic equilibrium under stress. Resilience scientists are interested in how much disturbance a system can experience before it becomes incapable of returning to its characteristic conditions. A tipping point occurs when stressors exceed the adaptive capacity of a system, and a fundamentally different system emerges. When the concept of resilience is applied to institutions, it involves appropriate adaptation of institutions to changing circumstances. Instead of maintaining or returning to equilibrium, resilient institutions evolve to remain relevant and useful as their contexts change (Young 2010).

Theoretical considerations of complex social-ecological systems suggest that the capacity to respond flexibly to internal or external disruptions and disjunctures is crucial for resilience, but uncertainty exists as to what contexts or factors contribute to resilience given the diverse pressures and disruptions that may impact a system. Researchers have not been able to predict confidently when existing institutions will endure despite transformative pressures, undergo adaptive change, or suffer abrupt dissolution. Institutional arrangements may endure even when they are shown to be unsustainable or inefficient. North (1990) pointed out that organizations and institutions can intertwine to create a "lock in" that resists fundamental transformation, even as change occurs. In other cases, institutions may change dramatically or dissolve unexpectedly.

Types of stressors that affect institutional arrangements include acute (single, short-term) stressors, cumulative stressors, and interactive stressors. Social-ecological systems and institutions may be able to recover fairly quickly from acute stressors (such as a period of drought) in the absence of additional stressors. Cumulative and interactive stressors present greater challenges for institutional arrangements. Cumulative stressors gradually increase pressures on existing institutions, and may be internal, external, or multilevel in origin (Young 2010). A growing population, changes in climatic patterns, or the influx of new ideas that undermine existing institutions are examples of cumulative stressors. Interactive stressors result from a combination of different pressures that interact synergistically on existing institutions, such that the overall pressure is much greater than any specific stressor. For example, the imposition of top-down policies, construction of new roads, and new communications technology may interact to change incentives and perceptions of a local population toward their natural resources, and compel institutional dissolution or transformation. In the research site, cumulative and interactive stressors have impacted institutions and forest management, but the effects have sometimes have been contradictory, or multifaceted.

12.3 Methods

The data and information were collected between 1992 and 2010. I lived in La Campa for 11 months in 1993–1994 conducting dissertation research and have returned 12 times to continue fieldwork during briefer visits of 2 weeks to 3 months. Data collection methods included household surveys, participant observations, archival research, meetings with local groups and cooperatives, and interviews with residents, local authorities, and representatives of government programs and nongovernment organizations (NGOs). In collaboration with colleagues, I collected measurements of forest conditions on the ground (see Tucker 1999) and examined forest-cover change through analysis of satellite images. An over-time satellite image analysis (1996, 1999, 2006) was conducted with Landsat TM (1996, 1999) and ETM+ (2006) images acquired during the dry season (November–March) to control for seasonal variability in vegetative cover and because cloud cover is less. Images were geometrically rectified and calibrated for sensor drift, sun angle, and atmospheric conditions. An overlay function was used to verify that the image alignment matched precisely. Each image was classified into forest (forest canopy cover above 25%), non-forest (agricultural land, settlements, and rocky outcroppings), and missing information (areas under cloud). The classifications were verified with ground-truthing observations (training samples) collected in 1996, 2000, and 2008–2009. The sensor gaps in the March 2006 ETM+ image were filled with data from the next available ETM+ image from April 2006.

12.4 Site Description

The municipio of La Campa is located in western Honduras in the Department of Lempira, which ranks as the least developed department (similar to a state) in the nation according to the United Nations Human Development Index. The department has difficulties providing basic health care and primary education compared to other departments. Nationally, Lempira has the highest illiteracy rate (37.2% of adults aged 15 and older cannot read or write), and the highest level of malnutrition (affecting 71.8% of children under the age of 5 years). Within the department, municipios' specific conditions vary.

La Campa shares many characteristics with other municipios in the department. Subsistence production of maize and beans provides a foundation for livelihoods, and many people work seasonally as agricultural laborers and coffee pickers. La Campa and neighboring municipios are known as strongholds of Lenca Indian heritage; they hold special celebrations and practice syncretic customs that express Lenca identity even though the language disappeared in the early twentieth century. La Campa, similar to its neighbors, has a strong tradition of participatory governance. An elected alcalde (mayor) and council members meet regularly in consultation with elected representatives from the municipio's villages. Together, they deliberate problems, answer petitions from residents, respond to national mandates, and make decisions regarding natural resource management and funding for municipal projects. Most meetings are open to any Campeño (native of La Campa) who wishes to observe, offer an opinion, or make a request or complaint. Local institutional arrangements and enforcement emerge through council deliberations and decisions, which are recorded in municipal archives and become the basis for precedent.

Despite these similarities with neighboring municipios, La Campa presents some unusual dimensions. Its geographic setting is unique in the area; the municipal seat (Centro Urbano) is located at the base of a sheer escarpment that dwarfs the cluster of houses on the narrow valley floor (Fig. 12.2). The escarpment divides the lowlands from the highlands, and until 1994, when the first road was completed, only a few precipitous paths linked the distinct zones. The range in elevation—from below 800 to 1,850 msl—gives Campeños access to a range of microclimatic zones, and they take advantage of them by planting crops at different altitudes to increase their chances of a harvest. This provides an advantage, because western Honduras experiences periodic droughts and highly variable rainfall. Of Lempira's municipios, La Campa has the highest rank in human development except for Gracias, the department's capital, and the highest percentage of households with potable water (82.2%) (PNUD 2006). Its artisanal pottery, produced mainly by

Fig. 12.2 Escarpment above the Centro Urbano of La Campa

women, has gained national fame after visits from national journalists, researchers, and documentary filmmakers (e.g., Ardón Mejía 1989; Castegnaro de Foletti 1989). Moreover, the municipio has established a reputation for forest conservation. In 1994, it received second place in a national competition for municipal natural resource conservation, and when the national government created the "Green Municipio" award in 2000, La Campa was among the first named (Tucker 2008). The recognition reflects a continuing predominance of forest cover, which has fluctuated between 60 and 70% over the past 20 years.

Changes in transportation, communications, property rights, and agricultural activities have occurred in tandem with improved access to electricity, education, health care, information, and external funding for development. When I first visited La Campa in 1992, the only vehicle access was a rough dirt road, which was impassable during rain due to lack of bridges over major streams. One decrepit bus made a daily round trip to Gracias when it was in working order, and if the weather allowed. Only three individuals owned a vehicle: the priest, the mayor, and the telegraph operator. By 2010, the rough dirt road had been widened and graded, and bridges had been built over the major streams. New roads had been built to reach all of the municipio's villages, and the road through La Campa had been extended southward through neighboring municipios to the border with El Salvador. Five buses from villages in La Campa and neighboring municipios made daily round trips to Gracias. Motorcycles and pick-up trucks had become too common to count accurately, especially because the improved roads allowed constant traffic of merchants, visitors, representatives of development programs, and residents of other municipios through La Campa. Electric lines reached the municipal seat in 2003. Soon after, La Campa was connected to the national phone lines and obtained national funding to establish a small Internet facility with computers for educational and general public use (but national support for rural Internet service ended with the coup d'état that deposed President Manuel Zelaya on June 28, 2009). In 2008, a cell phone tower further revolutionized communications. Within months, many Campeños had purchased cell phones, which are priced within reach of most Hondurans. Perhaps the most dramatic change since 1992, however, has been the transition from subsistence production of maize, beans, and fruit to increasing dependence on export coffee production. The expansion of coffee production, the privatization of land, as well as improvements in transportation and communications have contributed to changes in La Campa's institutions for forest management, to which the discussion now turns.

12.5 Early Forest Management Institutions (1921–1973)

Formal institutions for forest management in La Campa can be documented back to 1921, when the municipio was founded and the municipal council began keeping records of council meetings and decisions. Forest management norms probably existed earlier, but they elude identification without written records. Between 1921 and 1972, the municipal council endeavored to limit forest clearing around the Centro Urbano, because they hoped to maintain nearby sources of firewood. Nevertheless, the forests retreated to steep hillsides and ravines as people built houses and cleared land for fields, orchards, and gardens. With greater success, the council supported local customs and national laws that prohibited tree cutting and planting near water sources (springs, streams, rivers). They also enforced communal rights to forest products and water.

Examination of records kept by the municipal law enforcement judge (Juez de Policía) from 1921 to 1970 shows cases in which judges fined residents for cutting trees down next to water (where erosion could pollute the water source). Judges also enforced communal access to forests and water sources; anyone who built a fence across a path to a spring or a communal forest had to remove it, or build a gate for public access. Additional laws required farmers to seek municipal permits to clear land for temporary slash-and-burn fields and to claim lots for house building, pasture, or a permanently owned plot for agriculture. Farmers had to specify whether they wanted land for permanent use or temporary agricultural fields; the former required the construction of sturdy fences within a limited time frame. Temporary fields were fenced as well, but usually with makeshift arrangements of branches piled between posts from trees felled to create the field. After several harvests, slash-and-burn fields returned to the commons. Makeshift fences had to be partially destroyed to allow livestock in to graze on crop remnants, and trees grew back. Many Campeño families planted fields in the lowlands as well as the mountains, where the higher humidity and cooler temperatures provided a contrasting environment for crop production. Through most of the twentieth century, competition for land was greatest around the Centro Urbano because it had a school, a church, municipal offices, and a river that provided water year round. Therefore, rules for permanent and temporary fields were strictly enforced in the area adjacent to the Centro Urbano. In the mountains, the population density was low, and rules were enforced less consistently because few people were likely to observe violations such as fencing or clearing land without permission. But if complaints were brought to the council, those who lacked permits risked losing their land rights.

Nearly all conflicts over natural resources, as documented in municipal archives and records of the Juez de Policía, related to competing claims to parcels of choice land. Through most of the twentieth century, forest management aimed to maintain residents' access to shifting agricultural fields, water sources, and forest products. Although the population grew slowly, land scarcity did not become an issue in this period. The households appear to have been relatively similar in access to land, livestock ownership, educational attainment, and participation in municipal governance. Importantly, Campeño institutions limited the amount of land that an individual could claim to what was seen as reasonable for household subsistence. Moreover, local rules prohibited outsiders from claiming land. In several instances, outsiders who tried to purchase land were turned away. If a Campeño tried to sell land, the council reversed the sale and forced any payment to be returned to the prospective buyer. This rule, perhaps more than any other, assured residents' access to land and forests, and forestalled the emergence of landless and land-poor households. It also supported egalitarian social relationships and prevented nonresidents from insinuating themselves in community affairs. Enforcement did not prove difficult, because the arrival of any outsider was so unusual at this time that it drew residents' avid curiosity, and every movement was followed closely.

Through the 1960s, the majority of the landscape evidently cycled through forest, agriculture, and long fallows. Although the municipal council allowed timber harvesting, requests were few. Examination of municipal records revealed several small timber sales, mainly for construction or repair of buildings in Gracias and a few limited extractions by regional sawmills. Then in 1973, a major sawmill requested permission to harvest a large quantity of timber from forests along La Campa's access road. After a great deal of debate among residents and extensive negotiations over price and compensation for La Campa, the sawmill received permission. The municipio hired several young men to serve as forest guards to monitor timber extraction, and they received formal training to fulfill their duties (Tucker 2008).

From the 1920s through 1973, residents used forests for subsistence. Residents did not design institutions to conserve forest resources; forests seemed abundant. Because institutions prevented incursions by outsiders, and because Campeños had limited means or rights to claim land permanently, no one became a large landowner. Under existing technology, and without chemical fertilizers, permanent planting of agricultural fields would have resulted in declines in productivity as soils became exhausted. Forest institutions thus protected communal rights and maintained everyone's rights to forest goods and agricultural clearings for subsistence. This period set a precedent for participatory forest management and institutional design. Effective government and institutions that fit the local contexts were supporting resilience. All Campeños had rights to use land, but the community owned it under communal titles. There were no landless families. Institutional arrangements reinforced egalitarian social relationships and impeded the concentration of land and limited the influence of special interest groups. Although Campeños did not lack for land or forest resources, they were vulnerable to drought, famine, epidemics, and national civil conflicts, which at times disrupted social life, caused hardships, and increased the death rate. Lack of roads, distance from markets, and scarcity of medical care throughout the region meant that Campeños had to be self-sufficient.

12.6 Forest Management under COHDEFOR (1974–1987)

In 1974, the Honduran government nationalized the nation's trees, including those on private property. The government simultaneously created the Honduran Forestry Development Corporation (COHDEFOR) and charged it with conserving forests and capturing forestry income for the national government. It was like hiring a fox to guard the chicken coop. COHDEFOR had to make a profit for the nation by selling timber, yet also protect forests. As might have been predicted, these contradictory mandates confounded COHDEFOR's capacity to manage forests and allowed COHDEFOR to prioritize income generation over forest conservation.

COHDEFOR had immediate and profound impacts on La Campa's forest management. The nationalization of forest management terminated municipal rights to manage communal forests, monitor forest appropriators, or impose fines for violations. La Campa's forest guards lost their authority and their jobs. COHDEFOR granted timber harvesting leases to a series of sawmills to extract timber from La Campa, but COHDEFOR had too few employees to consistently monitor sawmill operations and timber extraction activities. Sawmill companies ignored timber harvesting regulations; the risk of being fined was negligible. They extracted timber with heavy equipment that disturbed the fragile soils, exacerbated erosion, and caused lingering damage to the forests. Although COHDEFOR foresters delimited areas for harvesting and marked trees to be left for regeneration, loggers usually took the best trees. The decrepit trees that remained had little potential for fostering healthy regrowth.

Sawmills not only had free rein to ignore forest laws, they also exceeded logging permits by sending out loaded trucks after 5 pm and on weekends. COHDEFOR employees and the police only had authority to check logging trucks during their working hours. In interviews that I conducted in 1993–1995 with former and current COHDEFOR foresters, I learned that they felt powerless to enforce the rules against sawmills during the period of nationalized forests (1974–1992). COHDEFOR's administration failed to suspend logging permits or fine companies, even when foresters documented egregious violations (Tucker 2008).

Slash-and-burn agriculture began to disappear under COHDEFOR. Not only did COHDEFOR oppose slash-and-burn agriculture, it prohibited conversion of forest to fields. Thus, land available for agriculture became scarcer under COHDEFOR rules. With La Campa's conditions of population growth and limited availability of new land for agriculture, fallow length began to fall, particularly near the Centro Urbano. This process fits Boserup's hypothesis, which predicts that the transition from extensive to intensive agriculture occurs as population demand exceeds the productive capacity of a system and there is no additional land suitable for agriculture (Boserup 1965). Other signs of intensification emerged as farmers began to use agricultural fields for successive years by applying fertilizers and began to adopt soil conservation methods, such as contour plowing and planting living barriers (especially pineapple) to reduce erosion.

Thus, the Honduran government's forest policy undermined La Campa's traditional forest management institutions and also pushed Campeños toward more intensive agriculture that reduced dependence on clearing forests for fields. Meanwhile, sawmills harvested lowland forests near the road with minimal monitoring. In La Campa, public outrage grew over sawmill companies' clear-cutting and damage that logging trucks were causing to their road and bridges, which residents had built with their voluntary labor. While sawmills hauled away truckloads of timber, residents found it difficult to obtain permits from COHDEFOR to cut a few trees to temper pottery, build a house, or make furniture. Conflicts erupted between residents and COHDEFOR foresters when residents tried to establish houses and fields in areas cleared by sawmills. People felt that their survival and their children's futures were at stake and formed a grassroots movement to end logging. In 1987, after several years of concerted effort, the movement ousted COHDEFOR and outlawed commercial logging in La Campa (see Tucker 2008). COHDEFOR agreed to depart, with the understanding that it would return if the municipio attempted to allow commercial logging. La Campa regained its rights to manage its communal forests 5 years before the national government passed the Agricultural Modernization Law of 1992, which ended nationalization of trees and returned management rights to forest owners.

Interviews indicate that people had as much desire to end COHDEFOR's meddling with local land-use decisions as to end logging. By the time Campeños ousted COHDEFOR, they had realized that careless practices could destroy forests. Witnessing forest degradation at the hands of outside companies made the risks apparent. Traditional practices had been transforming forests gradually, but under COHDEFOR, people saw loggers transform green pine forests to eroded hillsides scattered with discarded branches, ragged stumps, and scraggly trees deemed too small for a sawmill. These sights elicited concerns for the future of their forests. While COHDEFOR weakened local institutions and allowed degradation of lowland forests, it also compelled changes in people's perceptions of forest resources, encouraged the adoption of more intensive agriculture, and through its blatant failure to prevent degradation, sparked concern for forest conservation. The events of this period also increased vulnerability among families whose livelihoods depended on forest resources.

12.7 Institutional Evolution and Forest Governance Following COHDEFOR's Expulsion (1987–1994)

By the time La Campa expelled COHDEFOR, few of the traditional institutions that had governed forest and land use had survived intact. The rule against clearing land near water sources held because it conformed to national law, and at first the prohibition against outsiders buying land survived. The council quickly instated a rule against commercial sales of La Campa's timber, which reflected local sentiments as well as COHDEFOR's condition on departing. The council also forbade the use of chainsaws to remove the temptation to cut trees carelessly.

Certain traditional institutions had also fallen out of favor as local conditions changed. During the 1980s, roads had improved, more schools had been built, and the population had grown. The long fallow cycles had become less viable as good agricultural land became scarcer. Residents stopped requesting council permission for temporary agricultural clearings in forest commons and instead requested perpetual usufruct. Although the council continued to grant common land to residents at no charge, it became harder for residents to find unclaimed land suitable for agriculture. Land scarcity was exacerbated because the municipal council ceased to enforce limits on the amount of land that an individual could claim. While petitioners generally received permission to claim only 1 or 2 ha, those with resources proceeded to buy barbed wire and fence in as much land as possible. The municipal council looked the other way. Some Campeños argued openly that economic development depended on increasing productivity; therefore, more land should be given to those who had the means and ambition to manage larger tracts. Even residents of modest means requested private rights to forested land, often with plans to plant coffee, pasture livestock, or reserve land for their children. Barbed wire became much more common and available through the 1990s, when Care International and other development programs offered it with hope that better land demarcation might support staple crop production and food security. Mainly, barbed wire accelerated the conversion of communal forests to private land. Villages requested formal council designation of communal woodlots, because residents feared that private claims to forest commons would keep the general public from accessing firewood, timber, and non-timber forest products.

At the same time that La Campa expelled COHDEFOR, the national government passed laws to expand coffee production and build roads in coffee-producing areas. The municipal council looked favorably on the expansion of coffee plantations because it was a taxable land use, and growth in coffee production would allow the municipio to obtain a subsidy for road building. Campeños have produced coffee for household consumption for generations, but in the late 1980s, several coffee producers began to plant improved coffee varieties with an eye to external markets. Technical advisors visited La Campa from Honduras' National Coffee Institute and encouraged producers to plant coffee at elevations above 1,200 m, which qualifies for the coveted "strictly high-grown" ranking on international markets. A national road-building subsidy for coffee-producing municipios further motivated Campeños to plant coffee. In less than a decade, La Campa's communally owned highland mountain forests had been largely subdivided into privately claimed parcels by farmers hoping to improve their lives through coffee production. Mountain forests became pockmarked with coffee plantations and then transformed to complex mosaics of pine-oak trees and shade-grown coffee as farmers planted useful leguminous and fruit trees as shade.

12.8 Privatization and Land-Use Change (1995–2000)

The process of privatization was supported by a government land-titling program, which reached Lempira in 1995. This change in property-rights regimes brought major changes to La Campa by awarding formally recognized private titles with greater ease than had previously existed. Campeños had the choice of purchasing a private title to existing land claims, or letting their land claims become part of village land titles. Many Campeños lacked adequate funds to purchase titles to all of their private claims, and most preferred at any rate to hedge their bets by letting some of their land fall under a village land title even as they bought titles to some parcels. Although the top-down imposition of new institutional arrangements can destroy common-property regimes, La Campa's common-property regimes had already been transforming due to demand for land. The national land-titling program accelerated Campeños' slide toward private tenure on land appropriate for agriculture and coffee but also left space for village titles and communal ownership of some municipal forests.

The effects of the titling program were mixed for forests and related institutions. Forest governance decisions under the municipal council and public oversight were diminished as private owners gained more rights to make land management decisions. The area in common-property forests was reduced, but not eliminated because nearly every village maintained a communal woodlot within its titled area. The municipio also retained several large patches of forested land where no one had made land claims. The private titles helped better-off residents disproportionately, because less advantaged Campeños could not afford the discounted private titles. Thus, it exacerbated social inequity, as better-off families acquired more land and private titles. It favored coffee producers over subsistence farmers, because coffee provided cash resources that subsistence producers lacked.

12.9 Breaking with the Past: Outsiders Gain Legal Entry (2000–Present)

In 2000, a new and ambitious mayor convinced the council to allow outsiders to buy land. Steeped in the rationale of free markets and neoliberal economics, the mayor believed that Campeños' best chance to make economic progress would be to sell land to wealthy outsiders, because La Campa had little else of worth. From the perspective of municipal coffers, outsiders would pay more for land, pay more taxes, and bring land prices in line with the broader region. If Campeños had not already titled land officially, outsiders might have found it even easier to acquire land by dealing directly with municipal authorities. Indeed, a subsequent mayor subdivided a large remaining parcel of common forest among well-off friends and acquaintances.

In evaluating this break with the past, discussions with Campeños reveal interesting perspectives. The mayor's decision appears to have given legal recognition to the surreptitious acquisition of land by outsiders that began to occur during the early 1990s when coffee emerged as a profitable crop. A few Campeños had begun to serve as fronts for outsiders with whom they shared confidence. In this arrangement, Campeños claimed the land and managed it, but the outsider covered the costs and gained the profit. These arrangements required quite a bit of trust on both sides, and the outsider risked losing the investment if the arrangement was revealed to the council. Through the mid-1990s, coffee prices stayed relatively high, and outsiders seeking land for coffee visited in hopes of obtaining land. The neighboring municipio of Caiquín did not have any restrictions on land sale, and by the end of the 1990s, Caiquín's best land had been acquired by outside landowners. The municipal council of 2000 was well aware of the generous land prices offered by outsiders (only later would they realize that La Campa's land was sold for a pittance compared to other municipios).

As these incidents indicate, the mayors' power increased during the 1990s and the first decade of the twenty-first century, to the detriment of community participation in decision making. Mayors appear to have expanded their influence with the increased funds made available from national and international sources to the mayor's office for municipal development projects. Development programs, promoted by the Food and Agriculture Organization of the United Nations, Care International, Spain's Solidaridad, and the Honduran government (with international donor funding), brought funds to La Campa to support improved housing, agricultural modernization, small monthly payments for elders, and school supplies for children. Many of these funds were funneled through the mayor's office, giving the mayor unprecedented power to accomplish favorite projects, fulfill promises to faithful constituencies, and deny requests from those who questioned mayoral decisions. The emphasis on consensus building and participatory processes weakened as mayors gained influence.

The appropriation of commons by better-off residents and outsiders added to social inequality, which in turn increased vulnerability among marginalized or disadvantaged members of the population (Tucker 2008). The most vulnerable group in La Campa is composed of landless and land-poor households, which have emerged since 2000 with the influx of outside landholders and the expansion of better-off coffee producers. Households in La Campa typically sell their land when confronting family emergencies, such as a severely ill family member, or a death in the family. Once they sell land, they may not be able to obtain more, because land prices have risen above the capability of most Campeños to pay and nearly all the remaining land in commons is regarded as too steep, rocky, or inaccessible to be worth planting. Typically, landless farmers have ended up working for coffee producers, in some cases living as hired hands on land that they once owned.

12.10 Evolving Principles of Land Use, Forest Transformation, and Conservation

While mayors and council members supported privatization of remaining commons and the advent of absentee landowners, they also made decisions that supported forest conservation. In some cases, the rules evolved with the participation and input of residents during council meetings. The rules constrained residents' rights to fell trees and clear land without permission by imposing fees for harvesting timber and forbidding the clearing of land with mature pine trees. The municipal council and many residents cooperated with COHDEFOR, and its successor, National Forest Conservation and Development Institute (ICF), to prevent unauthorized felling of healthy, mature pine trees. The council hired a full-time forest guard, paid in part by COHDEFOR/ICF, to patrol municipal forests, prevent forest fires, and assure that residents did not cut trees without permits or permission to clear land. When I asked the forest guard about how Campeños reacted to his remonstrations, he replied that he had few problems. Most violations of tree cutting are reported by angry neighbors, and residents volunteer to fight forest fires.

A curious tension developed between concern for forest conservation and the desire for export coffee production. Coffee can grow well on slopes, but forests had to be cleared to plant it. Residents wanted to produce more coffee and gain subsidies for the road improvements associated with coffee production. For many households, coffee became the major source of income. Mayors could not have been elected by opposing coffee expansion and road improvements. At the same time, public sentiment remained set against outsiders' commercial logging. As a result, Campeño councils and residents moved toward norms that assessed land on its relative merits for coffee, annual crops, forest, or water sources. Honduran law aims to prevent clearing on any "land suited for forest," which means most of the country, given historical forest coverage. Campeños have interpreted the law to mean that forest belongs anyplace where agriculture and coffee are unlikely to do well, or where clearing would threaten a water source or important forest area. New fields for annual crops can be planted on flat land as long as erosion is unlikely and runoff will not affect rivers, streams, or springs. Oaks and scrubby trees are considered "unimportant" because they have little market value, and therefore can be cleared. Pine trees, however, are considered a prime resource, and ideally should not be felled without a productive justification or well-defined need.

Several cases of serious transgressions occur every year. In 2009, one farmer cleared a hillside of mature pine trees. If the clearing had been on land suited for coffee, the action would have been tolerated as long as it did not affect a water source. But this man offered no legitimate excuse. Through negotiation with the forest guard, the municipal judge, and disgruntled neighbors, the man agreed to replant the entire area with pine saplings at his own expense. The forest guard then checked on the man every few months. In 2010, another man bought a large parcel of private forest land and began to clear it for coffee. The land contained a spring that 20 families had been using for their water source. The people who depended on the water turned out en masse and stopped the man from cutting trees. The prospective coffee planter and the village residents requested mediation. The forest guard and the head of the municipal natural resources office came to examine the area and concurred that the trees should not have been cut. They told the man that he was at risk of being prosecuted by ICF for breaking national forest laws, and that it would be better for everyone to find an alternative solution. After several meetings, the man agreed to turn the land over to the village. The villagers agreed to repay the man for the land, minus a deduction for the illegally cut trees that had reduced the land's value—a remarkable reversal of Hondurans' standard conceptualization that land increases in value when cleared. Thereafter, the village declared it a protected area and collectively built a fence around the perimeter (J. Pérez and E. García, personal communications, 14–15 April 2010). These examples represent the procedures for addressing violations of forest laws in La Campa, and they manifest several principles originally identified by Ostrom's analyses of enduring and effective common-property regimes (Ostrom 1990, 2005). Specifically, the rules have been designed with participation of the stakeholders, the rules fit the local contexts, and stakeholders have recourse to low-cost conflict resolution mechanisms. Although Ostrom focused on cases where people shared communal rights, La Campa's understanding of property rights blurs the line between private and communal rights where forests and water sources are concerned.

The institutions evolved for forest management have included a set of norms to classify proper land uses. When Campeños assess the suitability of land for a given use, they consider topography, direction of prevailing winds, soil appearance, and type of vegetation present. In general, they view flat areas as suitable for traditional subsistence crops (corn, beans, squash) and house lots. Sloped land in the higher elevations falls into the category of land appropriate for coffee, as long as the soil is not shallow or rocky, winds do not hit the slope directly, and it is neither damp nor cold. On a number of occasions, I have been walking through a forest and a Campeño companion would stop and gesture to a patch of flat land and comment that it would be good for planting. Two examples stand out, because they occurred with authorities with reputations for protecting forests. In the first one, I had tagged along with the alcalde and several other council members to meet with disputants in a land conflict. As we walked through a protected forest, they pointed out several patches of forest and debated the chances that maize would grow well there. I inquired as to whether they would permit agricultural clearings in an area set aside for forest, and they told me that it was fine to clear fields in a forest as long as the land was fairly flat. The alcalde added that the council could not oppose productive activities when many people suffered from poverty and needed land to grow crops. He explained that his duty was to approve productive activities in areas likely to produce well, and unlikely to cause erosion or extensive degradation. On another occasion, I accompanied the forest guard as he patrolled municipal borders. His dedication to preventing forest fires and stopping unnecessary tree felling was well known in the municipio. Suddenly the guard stopped and gazed admiringly at a flat patch of land with a few pines and smaller trees. "Now, that would be a good place to plant corn," he noted reflectively. I stared in surprise and then realized that it was clearly suitable for a small field. Few trees would need to be cut, it did not appear rocky, and risks of runoff or soil erosion seemed minimal.

Most Campeño land falls into their "appropriate for forest" category. It seems that their criteria have become stricter under the influence of COHDEFOR, national forest laws, and workshops presented by representatives of environmental NGOs. In 1993 and 1994, I saw farmers planting maize in slash-and-burn fields on slopes so steep that one could stand up, hold an arm out straight, and touch the side of the hill. Today maize fields can still be found on steep slopes, but they are rarely new clearings, which would draw the ire of the forest guard and neighbors. Instead, people have been managing sloping fields for a long time with rotational planting and soil conservation measures. The use of fertilizer is a critical factor in the intensification of agricultural fields and the apparent reduction of staple production on steep slopes. With judicious use of fertilizer, weeding, and maintenance, farmers can plant more maize in smaller fields and harvest more. The labor investment and costs rise with the intensification of land use, but it has allowed them to meet or approach subsistence demands on scarce agricultural lands. Forest cover continues to transform as agricultural fields and coffee plantations create landscape mosaics with secondary successional forests. Nearly all of the coffee grows under diverse shade trees, including a variety of domesticated and native fruit and leguminous trees.

While institutions continue to evolve, they have supported a flexible and practical approach to combining livelihood needs, crop production, and forest management. The institutions have been enforced by forest guards, and by concerned residents. A number of studies indicate that enforcement of rules is key to forest conservation. In La Campa, the resulting secondary successions and shaded coffee plantations have helped to maintain forest cover as a dynamic mosaic. In addition, Campeños have taken steps to protect their cloud forest, Montaña Camapara, from clearing. The mountain has a number of important springs, and by the end of the 1990s, 17 communities had built water transport systems and depended on them for their water supply (Navarro 2002). In the same time frame, farmers had begun to clear Camapara's slopes for maize and coffee. Village water committees, whose members had expended great effort to build and maintain the potable water systems, worried that the clearings would contaminate their water with runoff and agricultural chemicals. They lobbied for a reserve by approaching the councils of the municipios that shared the mountain (La Campa, Caiquín, and Santa Cruz). Beneficiaries of water projects included every member of the municipal councils as well as the mayors and their families, and they shared concerns for the future availability of clean water. Two mayors expended effort to create the reserve on the mountain. After painful negotiations with the farmers who had cleared land on the mountain, Campeño farmers agreed to leave Camapara. In 2002, La Campa declared the Montaña Camapara Reserve and demarcated it with fences built by the beneficiaries of the water projects—representing nearly all the households in La Campa's largest villages.

12.11 Forest Change and Dynamic Stability

While fieldwork has shown a transformation in forest composition, time-series analyses of La Campa's forest cover indicate that it has been largely stable across the past 20 years (Fig. 12.3). Analyses of land-cover change from 1987 to 2000 revealed a slight increase in forest cover for the boundaries that La Campa recognizes as its own, including a disputed area (Tucker 2008). A subsequent 1996–1999–2006 analysis of the officially recognized area, which excluded the disputed area, found that forest cover had been maintained. Over the 10-year interval, the total changes summed to 120 ha that gained forest cover and 122 ha that lost forest cover.

The dynamic equilibrium in La Campa's forest cover indicates that the evolution of local institutions has been able to mitigate the potential for deforestation that market integration and export coffee production can bring. The creation of the Montaña Camapara Reserve represents the greatest success in recent forest management efforts and has been associated with new institutions that forbid incursions and require joint monitoring by forest guards and members of water committees. For the most part, these institutions have been enforced and the vast majority of the

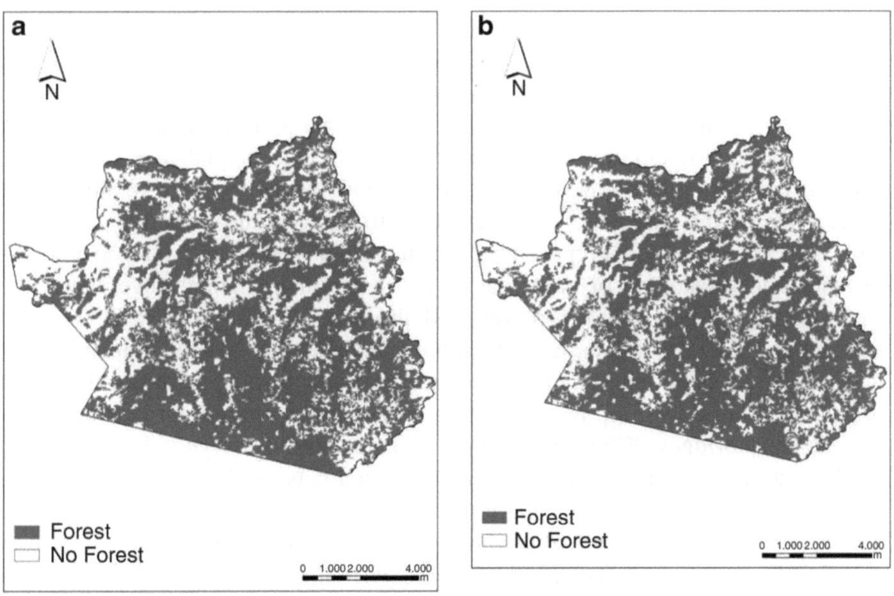

Fig. 12.3 La Campa's forest cover in 1996 (**a**) and 2006 (**b**)

population respects the rules. The rest of the forests in La Campa present a more confounded history. Most forest cover is composed of younger trees, and a different species composition from that of 40 years ago. Mature pine-oak forests have become rare due to COHDEFOR-led logging, export coffee expansion, and predominance of secondary successions in fallowed fields and woodlots. Communal woodlots and grazing areas are used for firewood collection and limited harvesting for local construction; they regenerate naturally but slowly. In the mountains, clearings for coffee plantations gradually become coffee forests under diverse leguminous and fruit trees. In studies of coffee forests similar to La Campa's, researchers have found that shade-grown coffee provides habitat for insects, songbirds, and small wildlife and can serve as a refuge for biodiversity (Perfecto et al. 1996; Vandermeer and Perfecto 2005). Shade coffee also inhibits the soil erosion that accompanies the major alternative land uses: annual crops and pasture. Coffee forests, however, contrast with the pine-oak forests that preceded them. Thus, stable forest cover obscures changes in vegetation composition. Even so, the maintenance of forest cover in La Campa indicates that the institutions for assessing appropriate land uses and managing land-cover transformation have been achieving what Campeños intend: Their land has not been degraded, areas important for forest and watersheds have been conserved, and most private and communal forest owners act as stewards of the forest and water while carrying out productive activities.

12.12 Concluding Thoughts

In summary, the institutional and social transformations in La Campa present mixed outcomes for forest management, social vulnerability, and resilience. While forest cover and forest-related institutions have thus far survived and been adapted to changed economic and political conditions, other aspects of La Campa's land-cover and socioeconomic circumstances have changed. Social inequality has increased, and it is evident in the skewed distribution of land and the gains in wealth among a minority of the population. Vulnerability to hunger and privation appear to have increased for the poorer members of the population as they have lost access to land. The weakening of participatory processes in municipal governance, the strengthening of the mayor's power, and the lack of transparency in financial allocations have also increased vulnerability by allowing wealthier members of the population and council members to channel more resources toward their own priorities. Among their top priorities have been infrastructural improvements such as roads and bridges. While these improvements benefit everyone by facilitating transportation to urban areas with hospitals, markets, and colleges, they have occurred in lieu of addressing more critical needs. Most Campeños would gain greater benefit and a higher quality of life with local access to health care, improved nutrition for malnourished children, and reapportionment of communal land from private interlopers to land-poor and landless farming households. La Campa's recent municipal administrations have paid inadequate attention to pressing health issues. Over two-thirds of children under the age of 5 years suffer from malnutrition, and one of every five children born is unlikely to live past age 40 (PNUD 2006). Institutional evolution and development have fallen short in meeting critical human needs. Even so, participation in small cooperative enterprises, school committees, and water project maintenance activities represent collective action and opportunities to build understanding and trust that may help to enhance resilience and commitment to push for solutions to these problems.

La Campa, similar to many places in the world, struggles to adapt to change. Where people have found incentives to work collectively to meet shared needs, they have created institutions that work fairly well. In this regard, La Campa appears to have developed some institutions that may mitigate vulnerability and augment resilience. Water committees, and the highest provision of potable water in Lempira, represent successes for collective action to meet shared needs. At the same time, the instances in which people have neglected to participate, or been forcibly excluded from decision making, have led to undermining or elimination of institutions that support equity and access to resources. The clearest examples of failures to support equity are found in the rule change that permitted outsiders to buy land in La Campa and the councils' failure to monitor and enforce restrictions on the amount of land that an individual could fence in the commons.

Instead of success or failure, La Campa's experience points to mixed and sometimes contradictory outcomes of institutional change during rapid socioeconomic and environmental transformations. Mixed outcomes and varied trends may be more typical than complete successes or utter failures, particularly given nonlinearity and unpredictability of change processes in complex social and ecological systems. While La Campa has retained forest cover to a greater extent than most of its neighbors, it has not avoided the social inequality that in the rest of Honduras has become associated with rising rates of delinquency and crime. But compared to surrounding municipios, La Campa has less crime. It has retained niches for collective action and community involvement in natural resource management decisions. The situation in La Campa continues to evolve. It has not escaped the dangers of rapid change, but traditions of participatory governance have shaped the direction of institutional change and moderated the pace of social and environmental change. It seems institutional adaptation to cumulative and interactive stressors becomes more feasible where people have social and organizational capacities for collective action and decision making, but preventing inequitable accumulations of wealth and resources associated with capitalist markets appears to be a nearly insurmountable challenge for La Campa and similar communities.

References

Adger, W. N. (2006). Vulnerability. *Global Environmental Change, 16*, 268–281.

Agrawal, A. (2005). *Environmentality: Technologies of government and the making of subjects*. Durham: Duke University Press.

Ardón Mejía, M. (1989). Panorama de la alfarería tradicional de La Campa, Honduras. *Folklore Americano, 48*, 69–80.

Ashenafi, Z. T., & Leader-Williams, N. (2005). Indigenous common property resource management in the central highlands of Ethiopia. *Human Ecology, 33*, 539–563.
Boserup, E. (1965). *The conditions of agricultural growth: The economics of agrarian change under population pressure*. Chicago: Aldine.
Brouwer, R. (1995). Baldíos and common property resource management in Portugal. *Unasylva, 46*, 37–43.
Castegnaro de Foletti, A. (1989). *Alfarería Lenca contemporanea de Honduras*. Tegucigalpa: Editorial Guaymuras, SA.
Cutter, S. L., Barnes, L., Berry, M., Burton, C., Evans, E., Tate, E., et al. (2008). A place-based model for understanding community resilience to natural disasters. *Global Environmental Change, 18*, 598–606.
Gadgil, M., & Iyer, P. (1989). On the diversification of common-property resource use by Indian society. In F. Berkes (Ed.), *Common property resources: Ecology and community-based sustainable development* (pp. 240–255). London: Belhaven.
Ghate, R. (2004). *Uncommons in the commons: Community initiated forest resource management*. New Delhi: Concept.
Gibson, C. C., Williams, J. T., & Ostrom, E. (2005). Local enforcement and better forests. *World Development, 33*, 273–284.
Hayes, T. (2006). Parks, people and forest protection: An institutional assessment of effectiveness of protected areas. *World Development, 34*, 2064–2075.
Janssen, M. A., Schoon, M. L., Ke, W., & Börner, K. (2006). Scholarly networks on resilience, vulnerability and adaptation within the human dimensions of global environmental change. *Global Environmental Change, 16*, 240–252.
Jodha, N. S. (1992). *Common property resources: A missing dimension of development strategies*. Washington, DC: The World Bank.
Navarro, E. (2002). *Agua para más de 500 años*. In the Socializando la Esperanza series, a publication of the Regional Office for Latin America and the Caribbean of the Food and Agriculture Organization of the United Nations. http://www.rlc.fao.org/es/desarrollo/particip/doctos/libro3/camapara.pdf
North, D. C. (1990). *Institutions, institutional change and economic performance*. New York: Cambridge University Press.
Ostrom, E. (1990). *Governing the commons*. Cambridge: Cambridge University Press.
Ostrom, E. (2005). *Understanding institutional diversity*. Princeton: Princeton University Press.
Ostrom, E., Dietz, T., Dolšak, N., Stern, P. C., Stonich, S., & Weber, E. U. (Eds.). (2002). *The drama of the commons*. Washington, DC: National Academies Press.
Perfecto, I., Rice, R. A., Greenberg, R., & van der Voort, M. E. (1996). Shade coffee: A disappearing refuge for biodiversity. *BioScience, 46*, 598–608.
PNUD (Programa de las Naciones Unidas para el Desarrollo). (2006). *Informe sobre desarrollo humano Honduras 2006: Hacia la expansión de la ciudadanía*. San José: Litografía e Imprenta Lil, S.A.
Tucker, C. M. (1999). Private versus common property: Forest conditions and tenure in a Honduran community. *Human Ecology, 27*, 201–230.
Tucker, C. M. (2004). Community institutions and forest management in Mexico's Monarch Butterfly Reserve. *Society and Natural Resources, 17*, 569–587.
Tucker, C. M. (2008). *Changing forests: Collective action, common property and coffee in Honduras*. Dordrecht: Springer Academic Press.
Vandermeer, J., & Perfecto, I. (2005). *Breakfast of biodiversity: The political ecology of rainforest destruction*. Oakland: Food First Books.
White, A., & Martin, A. (2005). Who owns the world's forests? Forest tenure and public forests in transition. In J. Sayer (Ed.), *The Earthscan reader in forestry and development* (pp. 72–103). London: Earthscan.
Young, O. R. (2010). Institutional dynamics: Resilience, vulnerability and adaptation in environmental and resource regimes. *Global Environmental Change, 20*, 378–385.

Chapter 13
Land-Use Institutions and Natural Resources in Fast-Growing Communities at the Urban-Rural Fringe

Abigail M. York and Darla K. Munroe

Abstract In the last several decades, urban decentralization and the conversion of formerly natural or agricultural areas have become the norm in much of the United States. Effective policies to constrain or mediate such growth and its effect on rural landscapes are a major priority at local and regional levels. Past research on land-use policies' ability to protect natural resources has not paid sufficient attention to the effects of the land market; rising land values, particularly when spatially differentiated, complicate policy efforts to stave off development in environmentally valuable areas. In this chapter, we review key lessons from the literature and provide examples from empirical work in Ohio, Indiana, and Arizona. Better understanding of spatial impacts of land-use institutions across a wide range of contexts will enable planners and policy makers to craft more effective policies balancing costs and benefits of development.

The Indiana work was supported by the Biocomplexity Project at the Center for the Study of Institutions, Population, and Environmental Change, Indiana University, through National Science Foundation grant SES0083511. The Arizona work was supported by the National Science Foundation under grant number DEB-0423704, Central Arizona-Phoenix Long-Term Ecological Research.

A.M. York (✉)
School of Human Evolution and Social Change, Arizona State University, P.O. Box 872402, Tempe, AZ 85287, USA
e-mail: abigail.york@asu.edu

D.K. Munroe
Department of Geography, Ohio State University, 1036 Derby Hall, 154 N. Oval Mall, Columbus, OH 43210, USA
e-mail: munroe.9@osu.edu

E.S. Brondízio and E.F. Moran (eds.), *Human-Environment Interactions: Current and Future Directions*, Human-Environment Interactions 1,
DOI 10.1007/978-94-007-4780-7_13,

13.1 Land-Use Institutions and Natural Resources in Fast-Growing Communities at the Urban-Rural Fringe

The USA is an urban nation with approximately 239 million people in urban areas (82.9%) and 19 million in rural areas (6.9%) in 2003. The remaining 10.1% (or 29 million) live in micropolitan areas (Slifkin et al. 2006: 1–6). While some urban growth is contiguous, "leapfrog" developments, edge cities, and exurban enclaves have created discontinuous patterns of settlement across the American landscape, fragmenting agricultural and natural areas (Berube et al. 2006; Clark et al. 2009). This new urban spatial structure has a host of implications for rural communities and landscapes. Socially, exurban landscapes may be characterized by a diverse group of residents, with a variety of demographic groups, livelihood strategies, and visions for the landscape, potentially leading to conflicts (Rudel 1989; Walker and Fortmann 2003). Because low-density development often requires more extensive road and utility infrastructure (Burchell and Downs 2005), agricultural and natural areas may suffer from land fragmentation and other environmental effects such as soil erosion and altered water runoff patterns (Hasse and Lathrop 2003).

Rural communities are concerned about managing the social and environmental effects of new urban development on the urban-rural fringe. Land-use policy tools utilized by local communities include zoning and development impact fees or exactions, while state and federal policies, such as public land sales, also fuel growth, especially in the western USA context. In this chapter, we review the impacts and merits of these policies and evaluate their effects on the local political economy, land markets, and exurbanization processes. We use institutional analysis to understand the dynamic spatial and temporal impacts of local and state policy on exurbanization in three cases: Ohio, Indiana, and Arizona. Institutions are defined as the rules-in-use, norms, and shared strategies (Ostrom 2005) that affect individual decision making. We focus our attention on land-use institutions relevant in our empirical cases but also commonly used in the USA and investigate how these institutions factor into the development process at the urban-rural fringe.

The three cases provide a diverse array of experiences with exurbanization due to varied population pressure, politics, and development legacies. Although the state of Ohio experienced only modest population growth between 1970 and 2000, there were significant increases in the built environment, predominantly associated with the loss of farmland at the urban-rural fringe (Clark et al. 2003). In Indiana, exurbanization processes are largely driven by land rent, although farmland preservation objectives and policies within county comprehensive plans may slow the rate of farmland conversion. Arizona has experienced exponential growth since the post-World War II period, as the pattern of development in Phoenix, has been influenced by local and state land-use policies, regional population shifts, water institutions, and changing consumer preferences.

Understanding the spatial and temporal dynamics of the land market is critical for growth management in exurban landscapes. These regions are characterized by rapidly changing land values, particularly as land is converted from agriculture or other primary uses into commercial or residential use. The rising land values, particularly when spatially differentiated, complicate the development of effective land-conservation policies (Newburn et al. 2006). At each of the three sites, we evaluate and discuss the spatialization of land-use policies through property markets. The study areas depict varying levels of growth intensity and different types of urban-rural conflict. Nevertheless, we summarize some of the more salient lessons to be drawn across these case studies. In later sections, we review relevant literature concerning the urban-growth process and the efficacy of various policy responses, with attention to how policy affects and is affected by changes in land value. Then, we highlight interesting findings in each study region. Finally, we conclude with a discussion about cross-cutting issues, especially the role of institutions in speeding or slowing land conversion, how the local political economy affects implementation of policy, and the importance of spatialization of institutional research.

13.2 Driving Forces Underlying Exurbanization

One hotly debated question in the literature is the degree to which current patterns of exurbanization are qualitatively different from past periods of urban growth. Some argue that fragmentation is temporary and will eventually be offset by infill (Burchfield et al. 2006), while other studies have shown that development is of significantly lower density than in past periods and does not fill in over time (Travis 2007). What underlies low-density, fragmented urban growth? New urban growth in North America is generally attributed to some combination of the following factors: rising average incomes, new transportation infrastructure (particularly the interstate system and continuing road improvements), flight from perceived urban ills, greater natural amenities in rural areas, and lower taxes and land values in outlying areas (Berry 1980; Berube et al. 2006; Clark et al. 2009; Downs 1999). Between 2000 and 2008 (peaking in 2005) there was a large increase in the financing of new single-family homes in outlying areas (R. W. Burchell, personal communication, May 6, 2010), no doubt also reflecting new home financing options in the mortgage market.

From expansion of suburbs to clearing of forests, transportation is a key land conversion factor. The creation of the interstate system and continued road expansions and improvements are a critical component of exurbanization in the postwar period (Ewing 1994). One of the best ways to predict land change is the development of new transportation corridors. This is especially the case when combined with an understanding of existing land uses (Iacono and Levinson 2009; Yang et al. 2008). Transportation affects land-use decision making through location decisions and through technological innovation such as the invention and adoption of the railroad, automobile, or light rail and mass public transit technologies. The ability to build in certain locations nevertheless is limited by regulatory institutions, especially zoning and master plans (Dow 2000; Lambin and Geist 2006), as well as individual location preferences. Tiebout (1956) introduced the idea that individuals "vote with their feet" through their residential choices. Because communities in the USA are reliant on property tax dollars for a large portion of their municipal budgets, politicians seek to attract the highest income residents possible (Peterson 1981). Of course, not all cities are able to attract the wealthy, so cities differentiate themselves offering public goods preferred by their residents resulting in residential sorting (V. Ostrom et al. 1999). Based on underlying citizen preferences, bureaucratic or policy-maker views, state mandates, enabling acts, and home rules, counties and cities select from a wide range of land-use institutions to shape urbanization and exurbanization patterns and processes. It has been noted that urban land conversion at the urban-rural fringe is often detrimental to the fiscal health of the central jurisdiction, especially if it occurs outside its boundaries (Burchell and Downs 2005; Travis 2007). In order to prevent losing tax revenue to surrounding suburbs, cities in the US Midwest and West have pursued aggressive annexation policies (Esparza and Carruthers 2000). Municipalities and other authorities often face a balancing act in trying to facilitate the "right" type of growth: weighing the increasing cost of service provision to low-density subdivisions against the possibility of losing the growth (and residents) to a proximate jurisdiction.

13.3 Land-Use Institutions

Current urban spatial structure exhibits the impact of past land-use institutions, such as the zoning history of a region. At the same time, communities are often in the position of facilitating institutional changes to grapple with current challenges. Thus, we run through several types of land institutions and analyze the role they have traditionally played in the development process. Land-use institutions constrain or encourage developers' and property owners' decision making, fueling, or braking the exurbanization process; institutions spatially distribute costs and benefits and spur growth in particular directions. Institutional constraints such as growth-management policies or urban-growth boundaries are a means to protect agricultural and forestland, yet traditional zoning remains the most prevalent land-use control in the USA, especially in rural regions (Diamond and Noonan 1996).

13.3.1 Zoning

Zoning enables the government to separate incompatible uses and may minimize the volume of nuisance litigation through creation of limited property rights for particular parcels (Nelson 1980). Zoning occurs at a local level, typically with no or limited coordination with neighboring jurisdictions or resolution of differences between the authorities of counties and cities. Carruthers (2003) identifies inconsistent local land-use regulation as a contributing factor to urban sprawl because counties, in most states, are unable to regulate land use in the same manner as incorporated municipalities. Frequently urban/suburban municipalities have tighter growth policies to limit development (Rudel 1989), whereas areas beyond the municipality have limited land-use regulations or pro-growth policies, again creating the potential for the leapfrog phenomenon (Carruthers 2003). But zoning is not the only land-use institution at the local level; development impact fees are increasingly utilized throughout the country to manage residential growth.

13.3.2 Development Impact Fees

Development impact fees spread throughout high-growth regions during the 1980s–2000s (Burge and Ihlanfeldt 2009; Jeong and Feiock 2006) based on the idea that developers and homebuyers do not bear the full cost of new development, that is, the increased congestion of roads, utility infrastructure costs, and service costs. Impact fees are one means to reduce "excessive growth" or sprawl (Brueckner 2000; Wassmer 2000), as the fees require developers to internalize some of the external costs associated with infrastructure development, which theoretically reduces city size (Brueckner 1997). These fees are exactions that communities assess during the plan approval process (Altshuler et al. 1993) whereby developers make "payments for permissions that can be withheld" (Fischel 2001: 6). Historically, most developers provided in-kind exactions through infrastructure, roads, and sidewalks that were built at the same time as a housing or commercial development (Burge and Ihlanfeldt 2009; Jeong and Feiock 2006), but increasingly fees are used instead. Turnbull (2004) demonstrated analytically that development impact fees are a more efficient means to reduce sprawl than urban-growth boundaries when considering the policy effect over time on urban-growth boundaries led to faster development and more inflated land prices than development impact fees.

13.3.3 Public Land Institutions

In much of the West, the majority of land is held by state and federal governments, including military bases, Bureau of Land Management (BLM) lands, US Forest Service lands, and state trust lands. Growth patterns in the West are partially determined by sales of public land holdings. For example, in Las Cruces, New Mexico, BLM sales contributed to extensive suburbanization (Nash 1994; York et al. 2011); in Phoenix, sale and conversion of state trust land is a major factor in development of the valley (Gammage 1999). Public land sales and public landholdings provide constraints and opportunities for development. The goals of the holding agency help to determine the growth potential of particular peri-urban areas surrounding western cities. Some agencies are more likely (BLM), unlikely (National Park Service), or mandated (State Land Department) to sell land.

13.4 Institutions' Impacts on Property Markets

Zoning, development impact fees, and public land policies affect the local political economy and property markets in different ways. We explore the effects of these institutions on various actors, as well as the effect on the property market and housing markets as a whole. Zoning is local government's monopoly power through the assertion of property owners' collective property rights to protect land market value through the prevention of nuisances and undesirable uses (Hamilton 1978). In many suburban communities, this monopoly power of the local government is controlled by residential interests seeking to restrict nonresidential designations, that is, mixed-use or multifamily structures in single-family-home neighborhoods (Babcock 1966; Babcock and Siemon 1985) or limit growth of new residential developments (Rudel 1989). Landowners who are able to attain a rezoning are able to reap substantial benefits via increased land prices (Babcock and Siemon 1985). Local governments have great control over who receives the economic benefits of rezoning.

The stabilization of property values is seen as a huge benefit for certain groups, such as homeowners, but also provides a stable base for property taxes that provide the largest portion of most local governments' revenue stream. Through zoning ordinances, homeowners may fail to realize higher benefits from alternative uses, but their properties will not lose value from neighbors using their land for factories, for example. Some researchers have argued that zoning won acceptance throughout the USA because it strengthened property rights, especially for homeowners, through the prevention of nuisances (Babcock 1966). Zoning has given jurisdictions the power to prevent some nuisance suits through the establishment of usage zones. This generally held up the property values of certain areas, protecting the rights of some established owners and leading to spatially distributed rights to develop less exposure to environmental hazards and nuisances and access to amenities such as open space.

Likewise, zoning has been a controversial source of class and racial conflict, as was demonstrated in the famous *Mount Laurel I* and *II* cases. In 1975, *Southern Burlington County NAACP v. Township of Mount Laurel*, 67 N.J. 151, established that zoning could not prevent affordable housing and therefore access for minorities to move into communities, through excessive restrictions. Because of problems with the implementation of the court decision, the Southern Burlington County NAACP brought a second suit against the township alleging continued discriminatory land-use regulatory practices, which excluded low-income and minority families (Hughes and Vandoren 1990). These famous cases highlight exclusionary practices that sometimes are hidden behind land-use institutions, particularly zoning. Similarly, one concern about development fees is the resulting lack of supply of affordable housing (Downs 2005). Most communities assess development fees based on per unit costs instead of square footage or lot size; thus, the already slim margin for affordable housing is reduced further, decreasing the supply of these units. Because zoning and development impact fees may increase the costs associated with provision of affordable housing, they sometimes are considered exclusionary.

Altshuler et al. (1993) argue that development fees allow communities to reduce political conflicts between pro- and antigrowth constituencies by improving infrastructure concurrency and reducing equity and efficiency issues. Growth regulations, such as zoning, may slow the development process and reduce housing starts (Mayer and Somerville 2000), while development impact fees, as Altshuler et al. (1993) observed, may in fact reduce political friction and have no effect on the time associated with approval, resulting in no effect on housing starts. One concern about impact fees has been their effect on economic growth, yet Jeong and Feiock (2006) actually find a positive relationship. They argue that increased infrastructure development and associated property value increases drive local economic growth. There is great debate about who bears the cost of the development impact fees, although Anderson (2005) suggests that the burden relates to the competitiveness of the market; in regions where homebuyers are able to purchase homes in different jurisdictions, it is likely that the developer bears most of the burden. Yinger (1998), Evans-Cowley and Lawhon (2003), and Evans-Crowley et al. (2005) argue that impact fee burdens shift back, reducing the value of land, especially land that is not yet subdivided. Anderson (2005) and Ihlanfeldt and Shaughnessy (2004) and Evans-Cowley and Lawhon (2003) find that current homeowners capitalize on impact fees as there is an indirect, positive effect on existing home prices. Fischel (2001) argues that the competitiveness of the market affects whether purchasers or developers bear the cost of the impact fees, but the effect on existing homeowners is always positive or neutral through reductions in property taxes or capitalization in home value. Therefore, the costs and benefits of development impact fees affect political economic actors differently; likewise, the benefits are largely accrued in single-family-home areas, while the costs are borne by developers or owners of undeveloped land, particularly the land on the urban-rural fringe.

In contrast, public land sales largely have negative effects on property owners as they decrease land values through increased supply and often reduce provision of local public goods, such as open space. Public land sales provide some of the primary locations for large-scale new home construction in western metropolitan areas and are supported by homebuilders and many local governments with pro-growth agendas (Gammage 1999; Logan 1994). These policies affect property markets through the increase in land supply in local, neighborhood markets directing growth within a metropolitan area (York et al. 2011). Preservation of open space increases the value of nearby land (Abbott and Klaiber 2010). The decisions to protect land from development and conserve recreational, aesthetic, and cultural values, or other ecosystem services typically increase nearby land values, while decisions to sell may decrease local land prices through expansion of supply and loss of valuable open space. Thus, the decisions by the state and federal governments to sell off public land affect communities through expansion of local land supply in a spatially distributed manner. These decisions result in benefits to the development community, via increased land supply and decreased land prices, and potentially to homebuyers seeking affordable homes, while current property and homeowners, especially those adjacent to the public lands, face reduced property values and decreased access to open space.

In the above discussion of land-use institutions, we explored how zoning, development impact fees, and public land sales affect particular parties within the local political economy and land markets. These costs and benefits are spatially distributed through a region, as evidenced by politically contentious public hearings during land conversion processes or institutional changes, that is, rezonings or land sales; these debates highlight the impacts on neighboring landowners and communities. Because of the spatial aspect of this institutional change and land conversion, we introduce classic theories of urbanization and development from the economics and economic geography literature.

13.5 Spatial Differentiation of the Land Market

Land-use institutions affect actors' decision making within the land market. These institutions primarily affect exurbanization through the development process and land conversion. Land values and land-rent returns via conversion vary over space. Many spatial factors influence the relative profitability of a parcel of land and its urban or development potential. These include accessibility to the centers of commerce and employment, the characteristics of a local neighborhood (e.g., parks, crime rate, school quality), and the various landscape features such as tree cover, topography, viewsheds, flood risks, microclimates, and other environmental features. Therefore, as land is converted at the urban-rural fringe, landowners attempt to maximize their returns over space and time. Because this process is dynamic, both an increasing speed in development and more fragmented development may occur (Irwin and Bockstael 2007).

Classic models of land development dating back several decades call our attention to the temporal nature of this process (Capozza and Helsley 1989; Capozza and Li 1994). For example, the value of urban land conversion in a rural area may rise over time. The flipside of this conversion is the rural landowner looking to sell off his or her property and make a profit. Research has shown that rural landowners (e.g., farmers) will delay the sale of their parcels until they believe they can earn the most profit. Therefore, decisions farmers are making about their farm, their production systems, and plans for the future depend strongly on their expectations of future land value (Irwin and Bockstael 2002). These individual decisions on rural landscapes affect natural resources, ecosystems, and ecological services by collectively resulting in the loss of significant land areas or invaluable open space corridors to residential development. Natural resources are particularly vulnerable to urban growth at the urban-rural fringe in two ways. First, ecosystems are often threatened by fragmentation. Research has shown that a variety of plant and animal species respond in a nonlinear way to changes and discontinuities in their habitats (Ellis et al. 2010; Vitousek et al. 1997). More interestingly, the symbiotic relationship between natural amenities and urban land conversion implies that the most scenic or otherwise desirable landscape features may attract land conversion.

If these scenic areas are also abundant in fragile resources, however, protecting these areas from further development may be especially challenging (Johnson and Klemens 2005).

A government's decisions to protect open space, retain public landholdings, enact zoning, or increase development impact fees affect the land conversion process at the urban-rural fringe. In the next section, we investigate how zoning and planning via minimum lot sizes (Ohio), forest conversion zones (Indiana), and comprehensive county planning (Indiana) affect land conversion. We also investigate the adoption of smart growth principles (Talen and Knaap 2001) with Indiana as our example and citizen support for increasing development impact fees in Arizona. Through spatially explicit analyses at the regional and jurisdictional levels, we evaluate the impacts of jurisdictional policy difference and public land sales on fragmentation, land conversion, and exurbanization at the urban-rural fringe, enabling us to connect policy to process on particular parcels, cities, and counties.

13.6 Spatially Explicit Analysis of Land-Use Institutions

The exurbanization process is affected by a variety of institutions, including zoning, development fees, taxes, and public land sales. Institutions influence developers' and landowners' decisions about land conversion. The study sites present variation in population growth, land-use histories, and institutions utilized. In Table 13.1, we lay out each of these characteristics.

The three cases are at the household, parcel, municipality, and county levels of analysis covering a relatively small Midwestern area—Monroe County, Indiana—to the large western Maricopa County, Arizona. Southern Indiana faced slow population growth of 0.7% per year throughout the 1970s, 1980s, and 1990s, while in Ohio and Arizona, annual growth was 6 and 4.7% per year, respectively, during the 1990s. These case studies also survey important land-use institutions across the country and specific forms of implementation, such as conversation zones and minimum lot sizes.

13.6.1 Ohio

Delaware County, Ohio, is part of the Columbus metropolitan area and is just to the north of Franklin County, occupied by Columbus. This county was one of the fastest-growing regions in the country between 1995 and 2005. In an analysis by Munroe (2010), statistical models were developed to explain the observed development patterns between 1989 and 2003. Complementary log-log hazard models were used to identify the average effect of a particular factor given the likelihood that development varied over time (McCullagh 1980). Then, using estimated parameters,

Table 13.1 Comparison of three study sites

	Ohio	Indiana		Arizona	
Cited works	Munroe (2010)	Munroe and York (2003), York (2005), York and Munroe (2010)	Munroe et al. (2005)		York et al. (2009, 2011)
Study extent	Delaware County	40 Southern counties	Monroe county	Maricopa county	Maricopa county
Level of analysis	Parcel	County	Parcel	Household	Municipality
Population of study area	151,569 (2005)	1,507,243	120,563	4,039,182	3,199,440
Annual population growth rate	6% (1995–2000)	0.7% (1970–2000)	0.7% (1996–1997)	4.2%	4.6%
Area, sq kilometers	1,181	20,916	1,065	23,890	23,890
Institutions	Minimum lot sizes	County level zoning & planning	Forest conservation zones	Development impact fees	State trust land and zoning
Years of study	1991–2003	1970–2000	1997	2006	1992–2001
Study focus	Impact of variable lot sizes on probability for conversion	Effect of zoning on land use	Effect of forest zone on fragmentation	Citizen support for increasing fees	Effect of land use policies on fragmentation

simulations were run holding one factor constant at a time. In this way, observed development patterns were compared to stimulated development patterns if that one factor did not matter. Landscape metrics were calculated to quantify the patterns of development, with observed data and for each of the simulations.

In Delaware County, one of the major policies set at the local level, is a maximum density requirement. In general, local jurisdictions within the county are most worried about property price stabilization, and therefore, multifamily housing, as one example, is nearly nonexistent. The way in which these density requirements are enforced is through minimum lot-size codes, defined as the allowable number of units per area. Within Delaware County, there are 25 separate townships and municipalities, and the maximum density ranges from as low as 1.0 unit per 2 acres to 9.09 units per acre. When holding this minimum lot-size constant, the resulting landscape was significantly less fragmented than observed. This finding implies that developers may be avoiding areas with stricter policies and conversely seeking those tax districts with less stringent restrictions. To the extent that this greater fragmentation is undesirable for the rural landscape (e.g., due to conflicts between agricultural land use and new urban residents), townships may want to work together to harmonize their policies on lot-size density. There may also be undesirable environmental effects of these locally inconsistent policies. When overall land fragmentation is higher, there may be greater average effects due to increased runoff or soil erosion. However, interestingly, there is some evidence that suggests that having a mixed landscape (e.g., interspersion of agricultural and residential land uses) may have positive benefits, all things being equal, in containing the runoff of pollutants (Tong and Chen 2002). In any case, research suggests that agricultural land uses may be more negatively impacted by sharing common borders with urban neighbors than vice versa (Clark et al. 2009; Parker and Meretsky 2004).

Another interesting process studied in this research was the effect of natural amenities on urban land conversion. Delaware County has a significant number of recreational water bodies, notably Hoover Reservoir. How did these water bodies play into the speed and pattern of urban land conversion? What was interesting was that access to these water bodies was associated with a decrease in the speed of urban conversion. In other words, all things being equal, parcels proximate to water bodies were converted later. This finding is surprising because one would expect high-amenity areas to be particularly attractive for new development. However, because the goal of agricultural landowners seeking to sell their parcels and developers looking for prime real estate is to maximize the returns from land conversion, they may have an incentive to delay the conversion process (Bunce 1985). If returns to urban development rise over time, there may be observable "upgrading" in the housing stock. In this case, the most desirable parcels will be sold only when the demand for new urban land is at its highest point. If policy is required to manage and shape the development that occurs around fragile environmental resources, policy makers must have very good knowledge about not only current trends but also more speculative trends in the land market.

13.6.2 Indiana

Southern Indiana is a region that has experienced profound shifts in economic structure over the last several decades. These shifts include a decline in average real commodity prices for both agriculture and forestry. Land-use systems have adjusted to the shifts in that agricultural land use has been replaced by urban and forest uses, and forestland has been cleared for urban expansion. In response to a perceived threat of urban encroachment, land-use policies, primarily zoning, have been enacted at the county level unevenly across the region.

In a series of analyses (Munroe and York 2003; Munroe et al. 2005; York 2005; York and Munroe 2010), we have investigated how land use has changed across this landscape, and intensively within Monroe County, as a result of changing economic conditions, land-use transitions, and land-use policies. In particular, we wanted to know whether policies such as zoning are effective in a context where land use is changing temporally as land-use profitability changes and spatially as various physical features (e.g., topography) and infrastructure (e.g., access to urban centers) come to matter more or less depending on how higher-level changes filter down to the heterogeneous landscape.

13.6.2.1 Economic Structure

Land-use patterns are hypothesized to relate to land-use profitability. This profitability is determined by the collective impact of prices (determined in national or global markets, varying temporally), land rent (physical suitability and market access, varying spatially), and land-use policy (which can shape allowable land uses in a given location). To estimate how land-use patterns relate to land-use profitability, we developed a model of land-use shares at the county level from 1970 to 2000. We compared the ratios of agriculture-to-urban land use and forest to urban land use and as a function of topography, urban accessibility, the profitability of forest and agriculture, average land values, and the concentration of employment in service sectors.

Land use is changing against the backdrop of trends in regional labor markets. Labor markets effectively shape employment patterns, in various extractive sectors as well as residential land use. Increased urbanization and rural residential land conversion in southern Indiana are linked to changes in the regional land market (tertiary/quaternary), sector job growth increases, and the effect of relatively wealthy individuals moving into rural areas. We found that the share of urban land relative to agricultural land increased with steeper terrain and lower agricultural profitability, indicating that urban encroaches on agricultural land use in areas more marginally suited to agriculture. Interestingly, the profitability of forest did not matter to the share of observed forestland, but given the large proportion of nonindustrial private forestry in the region, it is not surprising that profits from forestry are not the prime determinant of forest area. Higher service-sector employment was associated with a lower share of forest than urban area, indicating that urban encroachment is occurring in those areas where the regional economy is experiencing growth in higher-ordered sectors. Therefore, land-use policy, which is enacted at the county level, is likely to be most effective when it is not at odds with the adjustments in economic structure that happen as the region's traditional economic base shifts from extractive industry to services.

13.6.2.2 Land-Use Policy

In order to understand communities' adoption of zoning and planning, we first evaluated the smart growth principles incorporated by southern Indiana counties in comprehensive plans. Then, we investigated the impact of this county-level zoning using a land-use shares model addressing two main questions: (1) Does zoning restrict urban encroachment and the conversion of open space, both agricultural and forest? (2) Are the impacts of urban conversion zoning distinct from forest or agriculture?

The planning community identified principles that enable communities to grow but in a manner that is "smart," typically referred to as smart growth principles (Talen and Knaap 2001). These principles include compact urban form, maximizing existing infrastructure, walkable communities, and quality of life (Talen and Knaap 2001) principles that are supposed to allow a community to develop in a way that limits negative environmental impacts and promotes a community's sense of place, its individuality. The planning community's support for smart growth is not always mirrored within the community. Planners may find it difficult to realize smart growth principles in typical Indiana communities that prefer single-family homes to multifamily housing (Clark et al. 2009). A community's political orientation and preferences are important components in the effectiveness of zoning to achieve particular principles, such as smart growth.

We analyzed 25 county plans in southern Indiana using a modified version of smart growth principles in Talen and Knaap (2001); as can be seen in Table 13.2, planning and dealing with growth is the most common smart growth principle that these plans addressed. The majority of plans incorporated a livable built environment and responsible regionalism concepts. Responsible regionalism usually was discussed in the plan as positioning the community for economic development, whereas the livable built environment was typically addressed through discussion of zoning for compatible uses. Harmony with nature and planning and funding infrastructure were also discussed in the majority of plans. Harmony with nature usually meant a focus on agricultural preservation programs or goals. Communities often discussed planning for infrastructure, but only two counties addressed funding for the infrastructure. The weakest principles for these counties were implementation and assessment. The majority of counties were concerned about the protection of their rural and agricultural resources but were also concerned about economic development.

Table 13.2 Comprehensive plan principles

Principles	Identifier used	*n*	% age
Planning for growth	References to population growth/loss	16	64
Livable built environment	Pedestrian uses, compatibility of uses	14	56
Responsible regionalism	Competitive advantage, economic development, promotion of region	14	56
Harmony with nature	Preservation natural resources and/or agricultural land	13	52
Plan and fund infrastructure	Use of a capital improvements plan	13	52
Compact urban form	Infill development in areas with infrastructure in place	10	40
Variety of choice in housing	Affordable housing, advocating different types of housing	10	40
Quality of life	References to parks, open space, sense of identity, sense of community	9	36
Maximizing infrastructure	Using infrastructure that already exists	9	36
Equity	Affordable housing, impact fees	8	32
Balance of multimodel transportation system	Public transit and/or alternative forms of transportation like greenways, bike paths, and pedestrian walkways or sidewalks	8	32
Government collaboration	Reference to need for current of future government collaboration	8	32
Infrastructure development	Impact fees or requirements that developers pay	7	28
Place-based environment	Preservation of natural resources or agriculture, energy conservation	6	24
Historic preservation	References to preserving historic structures or districts	6	24
Citizen participation	Means to encourage citizen participation	6	24
Costs of implementation	Identification of how costs of plan recommendations will be paid for, strategies for securing funds	5	20
Improving development review processes	Allowing more ease for implementation and permitting flexibility	4	16
Reasonable, predictable, fair plan review processes	More predictable review process for both developers and public interests	4	16
Zoning enforcement	Mention of need to enforce existing or new codes	4	16
Mixed-use, walkable neighborhoods	Sidewalks and development with different types of uses, receptiveness to mixed use	3	12
Implementation responsibility	Identification of person or agency responsible for plan implementation	2	8
Implementation schedule	Timeline for implementation of plan recommendations	1	4
Assessment of progress	Identification of methods for measurement of plan's success	1	4

To evaluate the influence of these comprehensive plans on land-use conversion, we created a land-use shares model and included county zoning as one of the variables. We find that zoning, although only weakly significant in the agriculture-to-urban model, has different impacts on rural land uses positively associated with agriculture and negatively with forest land use. Our land-use shares model indicates that the impact of zoning on forestland decision making is not significant, but population pressure and changing economic structure are strong predictors of forest use. Zoning is weakly significant in the protection of farmland and has the second-largest effect in the agriculture-to-urban submodel, after slope. Our results at the regional scale do not refute the claim that zoning, through policies like minimum lot requirements, accelerates exurban development, but neither do we find conclusive evidence that zoning causes sprawl. Rather, we find evidence of the importance of land rents and land characteristics in land-use decision making, as well as evidence of structural changes in urbanization processes. Thus, county zoning might protect some farmland in urbanizing areas but most likely does not slow forest loss, failing to protect culturally, ecologically, and economically valuable forestland. In addition, because the economic value of forestland actually corresponds with greater urbanization in the region, additional policy instruments to protect forestland may be warranted.

Monroe County has adopted specific zoning to conserve culturally, economically, and ecologically valuable forestland. In our study of the relationship between these zones and forest fragmentation, we found that the estimated impact of zoning was weakened once we accounted for variations in topography and accessibility across the county, that is, zoning was a significant factor but secondary to the other influences underlying land-use change (Munroe et al. 2005). We found that areas that were both flatter (lower development costs) and close to major arterial roads exhibited the highest landscape fragmentation. Not surprisingly, the areas with the least restrictive regulations had the most fragmentation (though it might fill in over time, resulting in a less fragmented but more urbanized landscape).

Our studies illustrate that the greatest protection for forestland in this region is simply slopes that are too difficult to convert to residential uses, as well as shifting economic structure from extractive industries to the service sector. Zoning, as many scholars argue, is clearly not a panacea for protection of open space, although it may be somewhat more effective than no land-use regulation in reducing farmland conversion in the Midwest. Even in the context of communities that include agricultural preservation within their comprehensive plans, the impact of county zoning is limited. Zoning slows the agricultural conversion process but provides little protection for the preservation of forestland at the regional level; land rents and economic structure are more important for our understanding of urban encroachment on rural lands.

13.6.3 Arizona

Arizona provides a very different context, that of the American West, to explore the influence of institutions on exurbanization processes. Between 1990 and 2000, total population of the American West region surged by 19.7%, the fastest among all four regions in the country (Perry and Mackun 2001). Western cities provided burgeoning economic opportunities for people in the region and for those seeking to retire in a place with better "quality of life" and amenities—especially a warmer climate, year-round sunshine, and wilderness (Duncombe et al. 2003; Frey 2003).

Booming job markets and aggressive economic growth also changed the regional migration pattern, sending migrants to the Southwest (Johnson and Klemens 2005; Mueser and Graves 1995). Metropolitan Phoenix experienced exponential population growth between 1992 and 2000, mainly from an influx of new migrants attracted by booming economic opportunities in the valley (Gober and Burns 2002). Economic growth via residential homebuilding is the major driver of the regional economy (Gober 2006), providing a substantial hurdle for Phoenix area policy makers striving to balance growth with other resource demands.

In a study of 12 major jurisdictions, we found differences in fragmentations (York et al. 2009), which are associated with institutional differences (zoning policies and state land sales), legacies of land-use decision making, and decisions regarding infrastructure development. Our research results indicate that rapid urbanization resulted in increased fragmentation in peri-urban areas, while fragmentation near the urban core decreased via in-fill processes. Fragmentation is not only related to urbanization, increased change to the developed classification, but is also associated with jurisdiction, reflecting the impact of land-use institutions on development patterns and water extension decisions (York et al. 2011). Differences between fragmentation and conversion processes are due to the legacies of development, as well as the local political economy and institutional implementation.

To understand the public's support for changing land-use institutions, namely, increasing development impact fees, to manage fragmentation and conversion on the urban-rural fringe, we analyzed data from the Phoenix Area Social Survey, a survey of over 800 individuals' attitudes and behaviors surrounding environmental issues in the Phoenix Valley (Harlan et al. 2007). Impact fees are one means for jurisdictions in this housing growth economy to offset costs, raise revenue, and keep up with the demand for new infrastructure (Been 2005). Chapman (2008) has noted that Arizonan cities, similar to Californian cities, are limited in their ability to increase property taxes: in Arizona, through the referendum requirements for tax increases, while in California, through proposition 13. Because of the rapid growth and inability to raise monies for capital improvements through more traditional tools, Arizona cities utilize fees extensively, although unlike most western states, municipalities cannot assess impact fees for school development, so fees are generally lower than those in neighboring states (Planning Research 2002). Legal mandates, such as in Arizona, restrict assessment of fees, so governments are unable to set fees equal to marginal social costs, limiting their effectiveness in reducing sprawl (Bluffstone et al. 2008); however, our results indicate that there is political will within the valley to increase the fees. The majority of respondents support increasing fees, yet using a cumulative logit model, we find that their individual characteristics, especially gender, neighborhood type, and the length of time they have lived in the valley, shape their policy preferences. We reject the idea that citizens' exclusionary attitudes based on race or class are driving development impact fee adoption, a concern raised by Been (2005). We have some evidence that long-time Phoenicians and city dwellers may want to maintain the status quo. Women are very supportive of increasing development impact fees, which may be associated with women's concern about community in general (Mohai 1997). Political attitudes also are not driving citizen support for development impact fees, which was unanticipated. Although development impact fees differentially impact homeowners and non-homeowners, homeownership is not a major determinant of policy preference in Phoenix. There is political support for changing impact fees in order to better contain growth and conserve rural, open spaces in Arizona, but this support varies across the local population and is shaped by individuals' characteristics.

In the West, unlike the Midwestern examples of Indiana and Ohio, another political actor influences the local political economy and urbanization: state and federal public land managers and agencies. The bureaucrats and policy makers face a challenging and intense political environment where competing public land values, uses, and interests collide on huge parcels of land held by the federal government. BLM, US Forest Service, National Park Service, and Fish and Wildlife Service hold approximately 626 million acres, with the vast majority in the West and Alaska (Clarke and Angersbach 2001). Additionally, states manage a significant portion of public land: Congress granted at statehood over 97 million acres of state trust land to the lower 48 states with approximately 47 million acres currently retained (Culp et al. 2006; Souder and Fairfax 1996). In our study of state ballot initiatives on trust land, we found that Arizonans frequently rejected attempts to alter state trust land sales and management, although typically with a narrow margin, which may illustrate a slim majority's support for the status quo. These public land sales were a major determinant of growth and observed fragmentation in Phoenix (York et al. 2011). Although the public supports changing development impact fees to slow conversion at the urban-rural fringe, they have not supported institutional changes for public land sales at the ballot box. Phoenix illustrates the complexity of land-use institutions that is typical in the American West where federal, state, and local policies collide with intense population growth. We find that citizen preferences, differences in jurisdictional zoning and planning policies, and legacies of land-use decisions are important drivers in the observed fragmentation patterns. Institutions spatially distribute costs and benefits associated with urbanization across the Phoenix valley.

13.7 Land-Use Institutions' Impact on Urbanization

Overall, the empirical evidence of the effectiveness of zoning in the mitigation of urban sprawl and protection of rural lands is mixed (Bengston et al. 2004; Erickson 1995), perhaps due to the fact that zoning is implemented to reduce conflicts between land uses, maintain or increase property values for preferred land uses, and promote orderly development not to slow urbanization. What is particularly interesting in Indiana is the interplay between the replacement of agriculture by urban land versus urban encroachment into forest. Variations in land rent are a better predictor of the loss of agricultural land, as this use is more directly tied to agricultural profits (although transfer payments to farmers are significant). On the other hand, smaller plots of forested land may be subjected to periodic selective harvesting, depending on the value of tree species. However, there is a literature that suggests that the non-timber value of forest, particularly in regions with a significant presence of private, small-scale landowners, may be very significant (Hardie and Parks 1997). Interestingly in Ohio, high-value properties near a water body are slower to convert, most likely due to property owners' attempts to maximize sales revenue by holding onto the land and maintaining agricultural production in the interim.

Because land-use changes, such as urban conversion of natural lands occurs at a regional level, scholars characterize local-level zoning as ineffective in the absence of a comprehensive regional approach (Carruthers 2003; Diamond and Noonan 1996). Carruthers (2003) has clearly articulated the issue of leapfrogging onto unzoned or inconsistently zoned land in rural regions. Most studies assessing zoning's impact on forest and farmland conversion have focused on isolated plots, parcels, or individual cities (Brabec and Smith 2002) but do not evaluate the broader impact of zoning on land-use conversion in a regional context. In our southern Indiana and Phoenix regional studies, we evaluate the effect of different policies on land-use conversion and fragmentation, while in Ohio, we evaluate the effects across a region at a parcel level.

Differences in the types of institutions adopted, and the implementation of these institutions affects land conversion processes and observed fragmentation patterns. Our cases illustrate the importance of the local political economic actors, whether in the language of comprehensive plans or support for increasing development impact fees.

In each study, we focus on different types of institutions:

- In Ohio, lower minimum lot-size zoning attracts development, fueling the exurbanization patterns observed.
- Indiana shows the dominance of land rents in exurbanization, although zoning does appear to reduce forest fragmentation and slow agricultural land conversion.
- In Arizona, the differences in public landholding policies, diversity of land-use institutions across jurisdictions, citizen support for growth management, and legacies of development create spatially variable exurbanization patterns at the urban-rural fringe.

It is important to understand the historical legacies in these communities in order to understand the observed patterns, but it is critical to explore the different and complex institutions that affect the development process, differentially impacting political economic actors and shifting property markets. Assessing the effect of land-use institutions in a spatially explicit manner allows us to disentangle the urbanization processes and patterns.

13.8 Conclusion

Economic changes, such as shifts from extractive activity to manufacturing and services, lead to land-use readjustments as residential and commercial land uses replace traditional rural economies. Institutions affect the landscape by directing this exurbanization and growth, whether these policies are variability of minimum lot-size requirements across a region (Ohio), zoning and planning policy that only weakly protects farmland (Indiana), or variation in city development impact fees and zoning coupled with state public land sales (Arizona). The variability of policies within a region appears to be related to the political will of the constituency, which fits with traditional models of local political economy (see, e.g., Peterson 1981; Ostrom et al. 1999). Economists' models of development decision making are also supported by our results whereby institutions increase or decrease the costs associated with development (see, e.g., Fischel 1985). Understanding the impacts of institutions on these dynamic landscapes requires multiple methods and approaches, some of which we have utilized.

Our results indicate that different institutions affect development across the USA—in the West, the federal and state governments have a direct role in urbanization through public land management, while in the Midwest, zoning remains the primary land-use institution. Development impact fees vary across the Phoenix metropolitan area but are not utilized in either the Indiana or Ohio study sites. Because of the rapid growth in Delaware County, Ohio, we might expect citizens to push for policies that shift the costs and burdens of new development onto developers and homeowners, although the current economic downturn likely has slowed growth and its negative externalities.

Institutions slow or speed the process of land-use conversion on the urban-rural fringe. These institutions affect property owners' decisions, expand or shrink the developable land supply, and increase or decrease the costs of development. Our evidence indicates that some institutions affect land conversion (minimum lots in Ohio) and fragmentation (zoning policies in Arizona and Indiana and impact fees and public land sales policy in Arizona), while in other cases, uneven implementation or limited political will reduces the impact of institutions on land-use decisions (county zoning in Indiana). The political economy leads to different impacts on stakeholders and differences in the willingness to implement impact fees (Arizona) or smart growth principles (Indiana). Through our cross-site examination, we illustrate the usefulness of spatial analyses of land-use institutions and the importance of greater understanding of the spatialization of institutions.

Land-use institutions in the USA primarily are at the local level, although within the western context state and federal land agencies, play a significant role in management, land-use decisions, and public land sales. Development impact fees are growing in popularity but are more frequently used in high-growth areas such as Arizona. Zoning on the other hand is utilized throughout the country, but the implementation of zoning and its impacts vary, as demonstrated by our cases. Land-use decision making occurs within the context of a land market that is shaped by a local political economy and political actors. Institutions developed and implemented within this political economy spatially distribute costs and benefits among actors and influence the rate of conversion on the urban-rural fringe and patterns of development. Better understanding of spatial impacts of land-use institutions across a wide range of contexts will enable planners and policy makers to craft more effective policies, balancing costs and benefits of development. This chapter represents a first attempt to move toward a comparative, empirical understanding of spatialized land-use institutions.

References

Abbott, J. K., & Klaiber, H. A. (2010). Is all space created equal? Uncovering the relationship between competing land uses in subdivisions. *Ecological Economics, 70*(2), 296–307.

Altshuler, A. A., Gomez-Ibanez, J. A., & Howitt, A. M. (1993). *Regulation for revenue: The political economy of land use exactions*. Washington, DC: Brookings Institution Press.

Anderson, J. E. (2005). Taxes and fees as forms of land use regulation. *Journal of Real Estate, Finance and Economics, 31*(4), 413–427.

Babcock, R. F. (1966). *The zoning game*. Madison: The University of Wisconsin Press.

Babcock, R. F., & Siemon, C. L. (1985). *The zoning game revisited*. Boston: Oelgeschlager, Gunn & Hain.

Been, V. (2005). Impact fees and housing affordability. *Cityscape, 8*(1), 139–185.

Bengston, D. N., Fletcher, J. O., & Nelson, K. C. (2004). Public policies for managing urban growth and protecting open space: Policy instruments and lessons learned in the United States. *Landscape and Urban Planning, 69*(2–3), 271–286.

Berry, B. J. L. (1980). Urbanization and counterurbanization in the United States. *The Annals of the American Academy of Political and Social Science, 451*(1), 13–20.

Berube, A., Singer, A., Wilson, J. H., & Frey, W. H. (2006). *Finding exurbia: America's fast-growing communities at the metropolitan fringe*. Washington, DC: The Brookings Institution.

Bluffstone, R., Braman, M., Fernandez, L., Scott, T., & Lee, P. Y. (2008). Housing, sprawl, and the use of development impact fees: The case of the Inland Empire. *Contemporary Economic Policy, 26*(3), 433–447.

Brabec, E., & Smith, C. (2002). Agricultural land fragmentation: The spatial effects of three land protection strategies in the eastern United States. *Landscape and Urban Planning, 58*, 255–268.

Brueckner, J. K. (1997). Infrastructure financing and urban development: The economics of impact fees. *Journal of Public Economics, 66*(3), 383–407.

Brueckner, J. K. (2000). Urban sprawl: Diagnosis and remedies. *International Regional Science Review, 23*(2), 160–171.

Bunce, M. F. (1985). Agricultural land as a real estate commodity: Implications for farmland preservation in the North American urban fringe. *Landscape Planning, 12*(2), 177–192.

Burchell, R. W., & Downs, A. (2005). *Sprawl costs: Economic impacts of unchecked development*. Washington, DC: Island Press.

Burchfield, M., Overman, H. G., Puga, D., & Turner, M. A. (2006). Causes of sprawl: A portrait from space. *Quarterly Journal of Economics, 121*(2), 587–633.

Burge, G., & Ihlanfeldt, K. (2009). Development impact fees and employment. *Regional Science and Urban Economics, 39*(1), 54–62.

Capozza, D. R., & Helsley, R. W. (1989). The fundamentals of land prices and urban growth. *Journal of Urban Economics, 26*(3), 295–306.

Capozza, D. R., & Li, Y. (1994). The intensity and timing of investment: The case of land. *American Economic Review, 84*(4), 889–904.

Carruthers, J. L. (2003). Growth at the fringe: The influence of political fragmentation in United States metropolitan areas. *Papers in Regional Science, 82*(4), 475–499.

Chapman, J. I. (2008). State and local fiscal sustainability: The challenges. *Public Administration Review, 68*, 115–131.

Clark, J. K., Sharp, J. S., Irwin, E., & Libby, L. (2003). *Growth and change at the rural-urban interface: An overview of Ohio's changing population and land use*. Columbus: Exurban Change Project, Department of Agricultural, Environmental, and Development Economics, The Ohio State University.

Clark, J. K., McChesney, R., Munroe, D. K., & Irwin, E. G. (2009). Spatial characteristics of exurban settlement pattern in the United States. *Landscape and Urban Planning, 90*(3–4), 178–188.

Clarke, J. N., & Angersbach, K. (2001). The federal four: Change and continuity in the Bureau of Land Management, Fish and Wildlife Service, Forest Service, and National Park Service, 1970–2000. In C. Davis (Ed.), *Western public lands and environmental politics* (pp. 35–41). Boulder: Westview Press.

Culp, P. W., Laurenzi, A., & Tuell, C. C. (2006). *State trust lands in the West: Fiduciary duty in a changing landscape* (Policy Focus Report). Cambridge, MA: The Lincoln Institute of Land Policy.

Diamond, H. L., & Noonan, P. F. (1996). *Land use in America*. Washington, DC: Island Press.

Dow, K. (2000). Social dimensions of gradients in urban ecosystems. *Urban Ecosystems, 4*(4), 255–275.

Downs, A. (1999). Some realities about sprawl and urban decline. *Housing Policy Debate, 10*(4), 955–974.

Downs, A. (2005). Smart growth: Why we discuss it more than we do it. *Journal of the American Planning Association, 71*(4), 367–380.

Duncombe, W., Robbins, M., & Wolf, D. A. (2003). Place characteristics and residential location choice among the retirement-age population. *Journal of Gerontology B Psychological Social Science, 58*, S244–S252.

Ellis, E. C., Goldewijk, K. K., Siebert, S., Lightman, D., & Ramankutty, N. (2010). Anthropogenic transformation of the biomes, 1700 to 2000. *Global Ecology and Biogeography, 19*(5), 589–606.

Erickson, D. L. (1995). Rural land use and land cover change: Implications for local planning in the River Raisin watershed. *Land Use Policy, 12*(3), 223–236.

Esparza, A. X., & Carruthers, J. I. (2000). Land use planning and exurbanization in the rural mountain West. *Journal of Planning Education and Research, 20*(1), 23–36.

Evans-Cowley, J. S., & Lawhon, L. L. (2003). The effects of impact fees on the price of housing and land: A literature review. *Journal of Planning Literature, 17*(3), 351–359.

Evans-Crowley, J. S., Forgey, F. A., & Rutherford, R. C. (2005). The effect of development impact fees on land values. *Growth and Change, 36*(1), 100–112.

Ewing, R. H. (1994). Characteristics, causes, and effects of sprawl: A literature review. *Environmental and Urban Issues, 21*(2), 1–15.

Fischel, W. A. (1985). *The economics of zoning laws: A property rights approach to American land use controls*. Baltimore: The Johns Hopkins University Press.

Fischel, W. A. (2001). *The homevoter hypothesis: How home values influence local government taxation, school finance, and land-use policies*. Cambridge, MA: Harvard University Press.

Frey, W. H. (2003). *Boomers and seniors in the suburbs: Aging patterns in census 2000* (The Living City Census Series). Washington, DC: The Brookings Institution Center on Urban and Metropolitan Policy.

Gammage, G., Jr. (1999). *Phoenix in perspective: Reflections on developing the desert*. Tempe: Herberger Center for Design Excellence, College of Architecture and Environmental Design, Arizona State University.

Gober, P. (2006). *Metropolitan Phoenix: Place making and community building in the desert*. Philadelphia: University of Pennsylvania Press.

Gober, P., & Burns, E. K. (2002). The shape and size of Phoenix's urban fringe. *Journal of Planning Education and Research, 21*, 379–390.

Hamilton, B. W. (1978). Zoning and the exercise of monopoly power. *Journal of Urban Economics, 5*(1), 116–130.

Hardie, I. W., & Parks, P. J. (1997). Land use with heterogeneous land quality: An application of an area base model. *American Journal of Agricultural Economics, 79*, 299–310.

Harlan, S., Budruk, M., Gustafson, A., Larson, K., Ruddell, D., Smith, V., et al. (2007). *Highlights of the 2006 Phoenix area social survey: Community and environment in a desert metropolis. Contribution No. 4, Central Arizona – Long-Term Ecological Research Project*. Tempe: Global Institute of Sustainability, Arizona State University.

Hasse, J. E., & Lathrop, R. G. (2003). Land resource impact indicators of urban sprawl. *Applied Geography, 23*(2–3), 159–175.

Hughes, M. A., & Vandoren, P. M. (1990). Social policy through land reform: New Jersey's Mount Laurel controversy. *Political Science Quarterly, 105*(1), 97–111.

Iacono, M. J., & Levinson, D. M. (2009). Predicting land use change: How much does transportation matter? *Transportation Research Record, 2119*, 130–136.

Ihlanfeldt, K. R., & Shaughnessy, T. M. (2004). An empirical investigation of the effects of impact fees on housing and land markets. *Regional Science and Urban Economics, 34*(6), 639–661.

Irwin, E. G., & Bockstael, N. E. (2002). Interacting agents, spatial externalities and the evolution of residential land-use patterns. *Journal of Economic Geography, 2*(1), 31–54.

Irwin, E. G., & Bockstael, N. E. (2007). The evolution of urban sprawl: Evidence of spatial heterogeneity and increasing land fragmentation. *Proceedings of the National Academy of Sciences of the United States of America, 104*(52), 20672–20677.

Jeong, M. G., & Feiock, R. C. (2006). Impact fees, growth management, and development: A contractual approach to local policy and governance. *Urban Affairs Review, 41*(6), 749–768.

Johnson, E. A., & Klemens, M. W. (2005). *Nature in fragments: The legacy of sprawl*. New York: Columbia University Press.

Lambin, E. F., & Geist, H. J. (Eds.). (2006). *Land use and land cover change: Local processes, global impacts*. New York: Springer.

Logan, M. F. (1994). *Fighting sprawl and city hall: Resistance to the urban growth in the Southwest*. Tucson: The University of Arizona Press.

Mayer, C. J., & Somerville, C. T. (2000). Land use regulation and new construction. *Regional Science and Urban Economics, 30*(6), 639–662.

McCullagh, P. (1980). Regression models for ordinal data. *Journal of the Royal Statistical Society B, 42*(2), 109–142.

Mohai, P. (1997). Gender differences in the perception of most important environmental problems. *Race, Gender and Class, 5*(1), 153–169.

Mueser, P., & Graves, P. (1995). Examining the role of economic opportunity and amenities in explaining population redistribution. *Journal Urban Economics, 37*, 176–200.

Munroe, D. K. (2010). Pattern-based evaluation of peri-urban development in Delaware County, Ohio, USA: Roads, zoning and spatial externalities. In A. Páez, J. Gallo, R. N. Buliung, & S. Dall'erba (Eds.), *Progress in spatial analysis* (pp. 149–169). Berlin: Springer.

Munroe, D. K., & York, A. M. (2003). Jobs, houses, and trees: Changing regional structure, local land-use patterns, and forest cover in southern Indiana. *Growth & Change, 34*(3), 299–320.

Munroe, D. K., Croissant, C., & York, A. M. (2005). Land use policy and landscape fragmentation in an urbanizing region: Assessing the impact of zoning. *Applied Geography, 25*(2), 121–141.

Nash, G. D. (1994). New Mexico since 1940: An overview. In R. W. Etulain (Ed.), *Contemporary New Mexico* (pp. 1940–1990). Albuquerque: University of New Mexico Press.

Nelson, R. H. (1980). *Zoning and property rights: An analysis of the American system of land use regulation*. Cambridge, MA: The MIT Press.

Newburn, D. A., Berck, P., & Merenlender, A. M. (2006). Habitat and open space at risk of land-use conversion: Targeting strategies for land conservation. *American Journal of Agricultural Economics, 88*(1), 28–42.

Ostrom, E. (2005). *Understanding institutional diversity*. Princeton: Princeton University Press.

Ostrom, V., Tiebout, C. M., & Warren, R. (1999). The organization of government in metropolitan areas: A theoretical inquiry. In M. D. McGinnis (Ed.), *Polycentricity and local public economies: Readings from the workshop in political theory and policy analysis* (pp. 31–51). Ann Arbor: The University of Michigan Press.

Parker, D. C., & Meretsky, V. (2004). Measuring pattern outcomes in an agent-based model of edge-effect externalities using spatial metrics. *Agriculture, Ecosystems & Environment, 101*(2–3), 233–250.

Perry, M. J., & Mackun, P. J. (2001). *Population change and distribution: 1990 to 2000* (2000 Census Brief C2KBR/01-2). Washington, DC: U.S. Census Bureau.

Peterson, P. E. (1981). *City limits*. Chicago: University of Chicago Press.

Planning Research, Inc. (2002). *Development impact fees*. Best Practices Paper: Growing Smarter Implementation Project. Phoenix: Maricopa Association of Governments.

Rudel, T. K. (1989). *Situations and strategies in American land-use planning*. Cambridge: Cambridge University Press.

Slifkin, R. T., Randolph, R., & Ricketts, T. C. (2006). The changing metropolitan designation process and rural America. *The Journal of Rural Health, 20*(1), 1–6.

Souder, J. A., & Fairfax, S. K. (1996). *State trust lands: History, management, and sustainable use*. Lawrence: University Press of Kansas.

Talen, E., & Knaap, G. (2001). *Legalizing smart growth: An empirical study of land use regulation in Illinois*. Champaign: Department of Urban and Regional Planning, University of Illinois at Urbana-Champaign. Published in *Journal of Planning and Education Research, 22*, 345–359 in 2003.

Tiebout, C. (1956). A pure theory of public expenditures. *Journal of Political Economy, 64*, 416–424.

Tong, S. T. Y., & Chen, W. (2002). Modeling the relationship between land use and surface water quality. *Journal of Environmental Management, 66*(4), 377–393.

Travis, W. R. (2007). *New geographies of the American West: Land use and the changing patterns of place*. Washington, DC: Island Press.

Turnbull, G. K. (2004). Urban growth controls: Transitional dynamics of development fees and growth boundaries. *Journal of Urban Economics, 55*(2), 215–237.

Vitousek, P. M., Mooney, H. A., Lubchenco, J., & Melillo, J. M. (1997). Human domination of earth's ecosystems. *Science, 277*(5325), 494–499.

Walker, P., & Fortmann, L. (2003). Whose landscape? A political ecology of the 'exurban' Sierra. *Cultural Geographies, 10*(4), 469–491.

Wassmer, R. W. (2000). Fiscal equalization for state and local government finance. *Economics of Education Review, 19*(4), 451–452.

Yang, Q. S., Li, X., & Shi, X. (2008). Cellular automata for simulating land use changes based on support vector machines. *Computers & Geosciences, 34*(6), 592–602.

Yinger, J. (1998). The incidence of development fees and special assessments. *National Tax Journal, 51*(1), 23–41.

York, A. M. (2005). *Land use institutions in an urbanizing environment*. Ph.D. dissertation, Indiana University, Bloomington.

York, A. M., & Munroe, D. K. (2010). Urban encroachment, forest regrowth and land-use institutions: Does zoning matter? *Land Use Policy, 27*(2), 471–479.

York, A. M., Zhang, S., Boone, C. G., & Shrestha, M. (2009). Landscape fragmentation under rapid urbanization. *UGEC Viewpoints, 20*, 20–24.

York, A. M., Shrestha, M., Boone, C. G., Zhang, S., Harrington, J. A., Jr., Prebyl, T., et al. (2011). Land fragmentation under rapid urbanization: A cross-site analysis of southwestern cities. *Urban Ecosystems, 14*(3), 429–455.

Part IV
Historical and Archeological Approaches

The three chapters in this part, ranging from decadal to millennial time scales, demonstrate the importance of long-term perspectives to different questions concerning human-environment interactions. Different historical, environmental, and sociocultural contexts provide the backdrop for the chapters to examine conditions underlying long-term trajectories of social change and land-use intensification. The chapters illustrate the integration of various methodologies and forms of evidence to interpret the temporal intensity and the spatial pattern of changes and their social and environmental consequences. The reader will see examples of archival research, institutional analysis, meta-analysis of published studies, remote sensing of various time depths, vegetation ecology, and archeological field investigation.

In Chap. 14, Batistella, Bolfe, and Moran provide us a window to the fascinating history of Japanese colonization in the Amazon and their institutional and land-use trajectories, particularly since WWII when local farmers became global leaders of black pepper production. The crash of this commodity due to *Fusarium* infestation is iconic in how vulnerable a monocultural system is, based on a limited genetic diversity and dependent on a competitive global market. Their economic rebound has been equally impressive, this time through the development of intensive agroforestry systems using local and exotic fruit crops. They link research on historical land-use analysis to changes in land cover by coupling remote sensing analysis and detailed field inventories. Property maps reconstructed from the early phase of settlement allow the analysis of land-cover change and carbon stocks at the farm level. The study goes further in examining the consequences of agroforestry intensification for land-cover change, carbon sequestration, and other ecosystem services vis-à-vis a growing tendency toward conversion to pasture in the region. The authors suggest that intensive agroforestry systems contribute to carbon sequestration and land-cover diversity while supporting a good economic return for local farmers.

In Chap. 15, Vogt extends the time scale to the late nineteenth century using archival and historical records, aerial photography, satellite remote sensing, institutional analysis, and ethnohistorical interviews. Vogt reconstructs institutions affecting land use and land tenure in the west Mengo region of Uganda that borders the rapidly growing capital Kamapala, in the Baganda region of the Lake Victoria basin. The chapter reconstructs the landscape of the region since 1880 in four main periods: pre-colonial (late 1800s), early colonial (1900–WWII), late colonial to early independence (WWII–1995), and the era of structural adjustments (1990s–2002). He contrasts narratives of degradation of forest resources by smallholders with data demonstrating stable forest cover in reserves created during colonial times and smallholder areas where intensive agroforestry has developed. Tree cover in village areas has increased since WWII as a response to local needs and to the increasing market demand for natural resources (fuelwood, construction timber, and staple foods) in growing regional urban areas. Vogt analyzes these processes as a result of the interplay among customary and formal land-tenure arrangements, market fluctuations, and sociopolitical shifts shaping each historical period.

Neves offers a fascinating discussion in the closing chapter of Part IV, where the Amazon serves as a window to take us back to the fundamental question of the relationship between environmental conditions and the rise of social and cultural complexity in human history. He starts by reviewing evidence that places the Amazon as an early center of independent technological innovation, i.e., of pottery and plant domestication. Reviewing the archeological record on the expansion and retraction of cultural and political integration of the region, Neves proposes that linguistic diversity in the Amazon (where language is not segmented by major physical barriers) can be used as a proxy to understand the process of plant and landscape domestication that happened in the early and mid-Holocene. This argument is then linked to the process of political fragmentation and decentralization in the region. Particular attention is paid to the farming-language dispersal hypothesis. The chapter offers detailed discussions of the social and political landscapes of the Amazon while exploring theoretical parallels to other regions of the world. Within this context, Neves flips the environmental determinism argument (iconic to the scholarship of the Amazon region) about the limitations of soil, climate, protein, and staple crops to the formation of complex polities in the Amazon to one of resource abundance. Abundance, however, has not resulted in marked patterns of social hierarchy and related material culture (as typical of the Andes and Mesoamerica). That is, he argues that the abundance of resources and their distribution and the dissemination of knowledge and technology to manage aquatic and terrestrial resources may have set the conditions to prevent the emergence of institutionalized social hierarchies in pre-Columbian Amazon.

Chapter 14
Agroforestry in Tomé-Açu: An Alternative to Pasture in the Amazon

Mateus Batistella, Édson Luis Bolfe, and Emilio F. Moran

Abstract This chapter focuses on historical perspectives of land use in Tomé-Açu using remote sensing as a tool to assess land use/land cover and ecosystem goods and services such as carbon sequestration. The results show that agroforestry systems in Tomé-Açu play an important role at the property and landscape scales, trends that should be monitored as they appear to offer an alternative to conversion of forest to pasture. The Amazon-wide dominant trajectory toward conversion to pasture, based on reduced labor needs, if compared to intensive agroforestry systems, requires much more land to yield the same returns to a household at current management levels. If one considers the potential of agroforestry as a source of carbon sequestration, the returns are even higher and may contribute to the mitigation of greenhouse effects.

We thank the National Aeronautics and Space Administration (NASA) for funding the research project entitled "Human and Physical Dimensions of Land Use/Cover Change in Amazonia: Towards Sustainability" (NASA LBA-ECO project no. NNG06GD86A). This project is part of the Large-Scale Biosphere-Atmosphere Experiment in Amazônia Programme, which examined the human and physical dimensions of land-cover change. We also thank Bibiana Teixeira de Almeida and Vera Viana for reviewing the manuscript.

M. Batistella (✉) • É.L. Bolfe
Embrapa Monitoramento por Satélite, Av. Soldado Passarinho, 303, Fazenda Chapadão, CEP 13070-115 Campinas, SP, Brazil
e-mail: mb@cnpm.embrapa.br; bolfe@cnpm.embrapa.br

E.F. Moran
Department of Anthropology, Anthropological Center for Training and Research on Global Environmental Change, School of Public and Environmental Affairs, Indiana University, Bloomington, IN, USA
e-mail: moran@indiana.edu

E.S. Brondízio and E.F. Moran (eds.), *Human-Environment Interactions: Current and Future Directions*, Human-Environment Interactions 1,
DOI 10.1007/978-94-007-4780-7_14, © Springer Science+Business Media Dordrecht 2013

14.1 Introduction

Land-use/land-cover change due to deforestation has been interpreted as a major process occurring in ecological systems (Arrow et al. 1995; Kates 1994; Nobre et al. 2007; Stern et al. 1992; Turner et al. 1990, 1995). The Brazilian Amazon is one of the most dramatic examples of these dynamics (Becker 2001; Moran 1993; Wood and Skole 1998). Many communities have used a range of strategies to cope with the needs of production and subsistence, but several case studies indicate that smallholders have failed to thrive in the Amazon due to lack of credit and inferior marketing infrastructure (Fearnside 1986; Moran 1981; Smith et al. 1992). One of the best examples of relative economic success of colonization in the region occurred in Tomé-Açu[1] (Anderson 1992; Subler 1993; Yamada 2009). The process began over 80 years ago, when a group of Japanese families immigrated from Japan to an area located in Pará State as part of a binational effort to resettle Japanese families in Brazil, most going to São Paulo. More immigrants came later, with the challenge of being productive under conditions of high rainfall in a forested ecosystem. They tried annual crops, jute, cocoa, rubber, black pepper, fruits, pasture, and other production systems. They dealt with several difficulties, such as forest clearing, lack of knowledge about tropical agriculture, World War II, poor transportation conditions, pests, and diseases that infected plants.

Even though the first years of colonization were difficult, the hard work of these people, their willingness to try alternatives, and their household cohesiveness, based on the nuclear family, were central to their success (Tsunoda 1988). Parents always worried about the possibility of their sons leaving the land in search of more attractive economic activities (Yamada 1999; Yamada and Gholz 2002), which actually happened as younger Japanese moved to Japan during the economic crisis of the 1990s. This loss of labor led to a decline in agricultural activities that contributed to pasture conversion as a way to stay on the land with less labor. But newer trends toward the expansion and intensification of agroforestry systems have created new opportunities. This chapter shows how Japanese immigrants have achieved the status of successful settlers in the Amazon, through an analysis of the Tomé-Açu community, land use, land cover, and production arrangements.

[1] In 1952, Tomé-Açu inhabitants initiated an emancipation movement at the municipality of Acará. The state government of Pará, by means of Law No. 1,127 in 1955, authorized the establishment of the municipality of Tomé-Açu. However, in October 1955, the Brazilian Supreme Federal Court ruled the law unconstitutional. This decision caused the recommencement of several political movements with the aim of reestablishing Tomé-Açu, and on March 17, 1959, the state government promulgated a new law (No. 1,725) which reinstated it as a municipality. On September 1, 1959, the municipality of Tomé-Açu was officially established (Tomé-Açu 2007).

14.2 Historical Background

The process of colonization in the Amazon Basin has attracted a great deal of attention. Research has focused on the amount and rate of deforestation (FAO 1993; INPE 2002, 2006–2009; Skole and Tucker 1993) and related social, economic, and ecological processes involved (Becker 2006; Brondizio et al. 1994; Conant et al. 1983; Dale and Pearson 1997; Gerber 1997; Moran et al. 1994b; Nepstad et al. 1999; Porto-Gonçalves 2001; Turner et al. 1993). Large-scale deforestation in the Brazilian Amazon began after the construction of Brasília, the national capital, and its connection to the north through the Belém-Brasília Highway (Moran 1996). After the announcement of the Program of National Integration (PIN) in 1971, other roads were constructed: BR-364 (north-south) and Trans-Amazonian (BR-230, east-west) highways connecting sites in the Amazon Basin with other parts of the country (Moran 1993).

Rates of deforestation in the Brazilian Amazon resulted in national and international pressure that led to major forest policy shifts (Batistella and Moran 2005; INPE 1996; Moran et al. 1996). Other transitions have transformed deforested areas into productive agricultural land or secondary successional vegetation cover (Brondizio et al. 1996; Homma 2003; Sader and Joyce 1985). Understanding land-use changes to achieve a sustainable strategy is a matter of growing interest to communities in the Amazon, to regional and national policy makers, and to everyone concerned with the consequences of deforestation (Goodland et al. 1993; Laurance et al. 2001; McCracken et al. 1998; NRC 1998).

Tomé-Açu is an example of dynamics following deforestation in the Brazilian Amazon. Early Japanese migrants to the region cleared significant areas of forest and established production systems based on polyculture and cattle ranching. The spatial patterns in that landscape are a result of ecological and socioeconomic processes, such as secondary succession, land-use changes, and alterations in social organization. To better understand landscape changes, it is central to analyze how human action occurred in time, space, and culture (Crumley 1994; Yamada 2009).

The first migration trip occurred in 1929, when 43 Japanese families (189 people) arrived in Tomé-Açu. This was a consequence of previous agreements between the state of Pará and the Japanese government. With the availability of land given by the state, several trips brought other migrants to start a new life in the Brazilian Amazon. After clearing the forest, they started growing cocoa and rice without immediate success (Tsunoda 1988). In 1931, Cooperativa de Hortaliças (the Vegetables Cooperative) was founded with the intention of stimulating agriculture production and reaching the market in Belém. In 1935, this cooperative became stronger, organizing all settlement production and changing its name to Cooperativa Agrícola do Acará. The disappointment was clear during the first 10 years. Several people left the community due to economic failure or diseases such as yellow fever and malaria. World War II was another challenge for those families, as Brazil declared war against the Axis. Tensions occurred in the community when Japanese became isolated by the Brazilian population and had problems getting access to the market and producing just enough for subsistence (CAMTA 1967).

In 1947, the cooperative had its first indication of success, based on the production of black pepper. Production increased year after year, impelling the farmers to reorganize the cooperative and legalize it with the name that remains today: Cooperativa Agrícola Mista de Tomé-Açu (CAMTA). From 1947 to 1957, Tomé-Açu was in a period of prosperity, becoming one of the largest black pepper–producing communities in the world (Uhl and Subler 1988). In 1959, Tomé-Açu became a new county in the state of Pará. In 1963, the community elected a Japanese descendant as its mayor, the first one in the Amazon. The increase in production and the cooperative system were followed by improvements in infrastructure, investments in health and education, and expansion of black pepper agriculture to adjacent lands (CAMTA 1975).

During the late 1960s and early 1970s, fungal disease (*Fusarium* sp.) and international market oscillations forced the community to try other agricultural products, specifically perennials like cocoa, rubber, and cultures of intermediate cycles, such as passion fruit and papaya (Yared and Veiga 1985). Lately, agroforestry characterized by overlapping crops and cattle ranching have become the dominant farming systems in Tomé-Açu (CAMTA 1994; Stolberg-Wernigerode and Flohrschutz 1982).

According to Homma (1998), the immigrants' way out of their "ecological crisis" was the diversification of activities from the 1970s on, with highlights in fruticulture, especially Hawaii papaya, melon, acerola, orange, oil palm tree, *cupuaçu*, and passion fruit, among other native and exotic fruit trees, which generated a new economic cycle in the region. Crop diversification was associated with a new production system—agroforestry—which developed from several local "experiments," generated different production arrangements with different species, and promoted the insertion of rural producers in new markets. Homma (2004) approaches agroforestry systems' dynamics at the Tomé-Açu agricultural colony in different cycles, which can be summed up in eight phases, as described in Table 14.1.

Tomé-Açu is still known as "pepper land" due to the fact that this crop drove the region and the country, for the first time, to the status of a large world producer of black pepper. Several areas with agroforestry systems are currently present in the municipality. This is the alternative found by producers to resist the decrease in the market price of black pepper and the excessive incidence of diseases in this crop, thus keeping up the region's development.

14.3 The Place: A Challenge for the Migrants

The study area is part of Tomé-Açu County, Pará State, in eastern Brazilian Amazon (Fig. 14.1). It is currently connected by paved roads in the north and the Belém-Brasília Highway to the south. The original settlement is located along the left bank of Acará River, approximately at 2°40′ S and 48°20′ W. The Landsat TM 2005 color composition in Fig. 14.1 gives a general idea about the spatial organization of the

Table 14.1 Phases and characteristics of agroforestry systems' dynamics in Tomé-Açu, PA, Brazil

Period	Main characteristics
1933–1950	Introduction of black pepper crops
	Exclusive Japanese-Brazilian monocultural crops
	Economic climax in the postwar international market
1957–1970	Beginning of agricultural mechanization and use of NPK
	Dissemination of black pepper cultivation in other regions
	Fusarium uprising
	Reduction in black pepper crops' lifetime
1970–1978	Spatial shift of crop areas due to *Fusarium*
	Interest in activity diversification
	Introduction of Hawaii papaya, melon, cocoa, passion fruit, and oil palm tree
	Agricultural policy with intense rural credit
	Fusarium persistence and increasing pepper production
1978–1982	Brazil is the greatest world producer/exporter for pepper (1982)
	Pepper prices decrease
	Increase in supply costs and decrease in credit
	Introduction of animal husbandry and food crops
1982–1987	Maintenance of low prices for pepper in the market
	High inflationary rates stagnating large producers
	Decrease in pepper production and export
1987–1991	High cocoa and Hawaii papaya production
	National recession situation
	Pepper production stagnation
	Expansion of pepper crops in small properties
1991–1999	Despite the contrasts, Brazil has record production and export (1991)
	New crisis causes reduction of pepper prices internationally
	Environmental severity diminishes new areas for pepper
From 2000 on	National and international visibility of fruit produced in Amazonia
	Record local productions of passion fruit and acerola
	New pepper-plant areas incorporating other crops
	Introduction of several AFSs (cocoa, cupuaçu, acai berry)
	Increase in fruit pulp exports (cupuaçu and acai berry)
	New alternatives with the expansion of other fruit crops

Source: Based on text in Homma (2004)

study area. The farmlands of Tomé-Açu are located at the center of the image, where brighter tones reveal deforested areas, mainly converted to pasture or agroforestry systems. The total resident population of Tomé-Açu is 47,081, of which 27,161 are urban inhabitants and 19,920 are rural inhabitants. In 2006, the region's population density was approximately 10 people km^{-2}, the human development index or HDI was 0.676, the total GDP was R$ 169,226,321 (US$ 77,855,953), and the GDP per capita was R$ 3,410.31 (US$ 1,568.98) (IBGE 2007).

The geological Boa Vista and Pará formations, belonging to the Holocene, constitute the parent material of the soils (Falesi et al. 1964). The terrain is

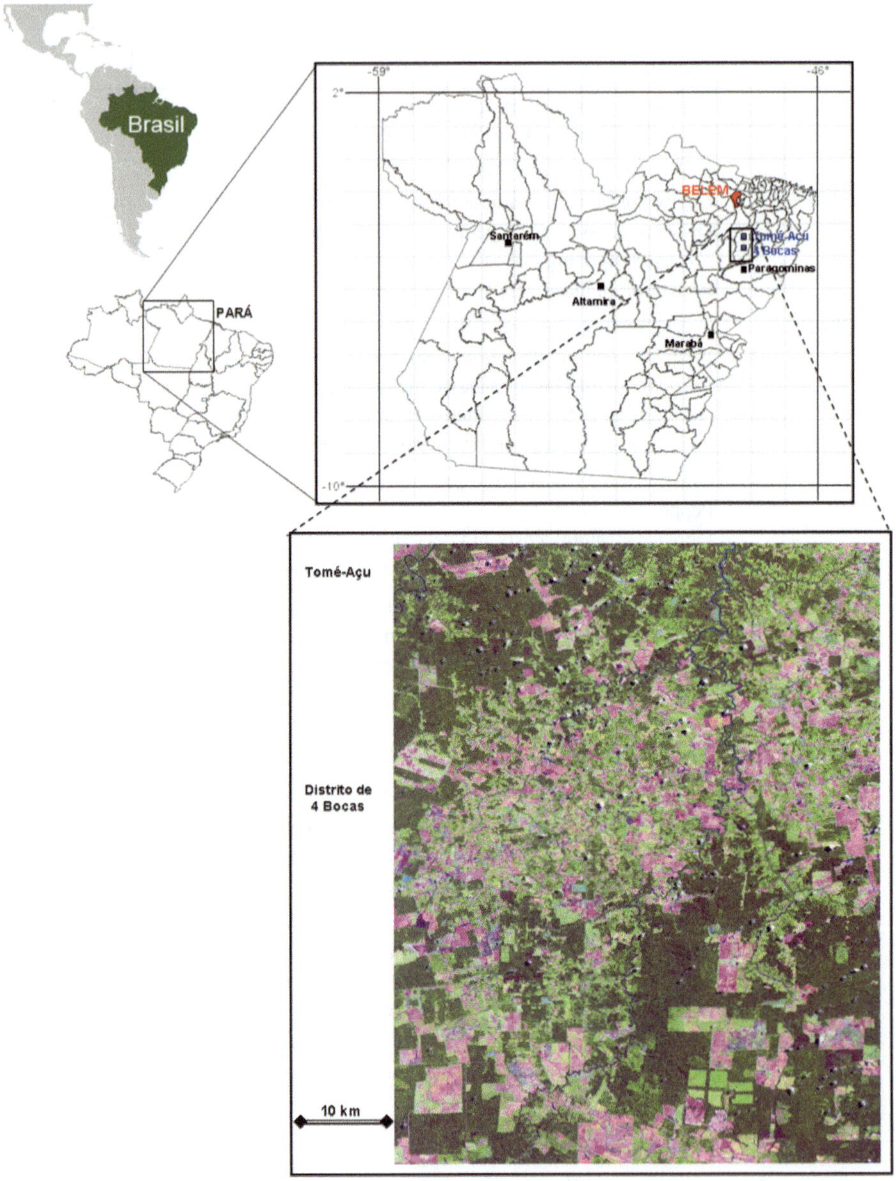

Fig. 14.1 Study area in Tomé-Açu, Pará State, eastern Brazilian Amazon

primarily upland, with highly weathered acid Oxisols characterized by low nutrient supply and retention capabilities. Topography is generally flat, with red-yellow concretionary latosol in more undulating areas (Falesi et al. 1964; Rodrigues et al. 2001; Sanchez 1976).

The hot and humid climate corresponds to the type *Af*, according to Koppen's (1948) classification. The annual average temperature is 27.9°C. The area receives an annual average of 2,600 mm of precipitation with a drier season between July and October (Bastos 1972; SECTAM 2009).

Falesi et al. (1964) found four main soil types, generally showing good drainage potential: yellow latosol of heavy clay content, ranging from 50% to 70% at the B horizon; yellow podzolic latosol of rather heavy texture, clay content from 35% to 50% at the B horizon; yellow latosol of medium texture, clay content from 15% to 35%; and red-yellow concretionary latosol.

Land-use/land-cover characteristics are defined by different stages of land occupation and secondary succession. Vegetation is primarily moist forest, including several species with a high timber value (Rodrigues and Baena 1974; Subler and Uhl 1990). According to Flohrschutz (1983), there are two main types of rural properties in Tomé-Açu. Small areas cultivated with manioc or rice, generally owned by native Brazilian smallholders, characterize the first type. The other type is predominantly used for perennial agriculture and owned by Japanese descendants. Besides social origin and organization, size of cultivated area, invested capital, fertilization, production, and productivity also differentiate these properties.

The population oscillated during the first 50 years of colonization, depending on the periods of success achieved by the community. Between 1985 and 1991, approximately 280 Japanese families lived there, with a local population of about 3,000 people (Yamada 1999). Each family holds the tenure of one or more lots of 50 ha each, being responsible for their own production and outcomes.

As the characteristics of goods generally affect institutions (Crawford and Ostrom 1995), it is important to understand how the access to natural resources takes place. The previously forested land was divided into private lots. After the settlement was established, each plot owner acquired control of his or her property. Using the classification of goods given in McGinnis and Ostrom (1996), subtractability is high for all the goods produced in Tomé-Açu, but exclusion varies, depending on the type of land. Croplands are usually monitored, with a high level of exclusion, but forested lands are not, although access is restricted.

The tiers of decision-making units proposed in Moran et al. (1998) have had very clear impacts over forest and property management conditions during colonization in Tomé-Açu. Some critical examples of these impacts are listed below:

- International tier: World War II was a tough period for the Japanese, as they were seen as enemies and faced many problems in making their products available to local and regional markets.
- National tier: a federal law required 80% of each plot to be maintained under forest cover, limiting the expansion of production systems.
- Regional tier: Pará State government gave the land to Japanese settlers as an incentive to stimulate production and development.
- Settlement tier: the settlers organized themselves in a cooperative system as an adaptation to severe conditions of production and commercialization.

- Individuals and households tier: plot owners made their decisions about fertility, planting, harvesting, and other relevant activities, always giving importance to education.

This mosaic of situations affected community outcomes. But on a daily basis, the settlement and individual tiers show more dynamic relations, adopting different rules to maximize payoffs. The most important institution in Tomé-Açu is the cooperative system. All members of the settlement contribute to CAMTA by giving a percentage of their production. This input is used for several purposes, including storehouse maintenance, agricultural experiments, equipment, and savings for future strategies (CAMTA 1975). However, making CAMTA strong did not keep individuals from making their own experiences, both positive and negative, creating a synergism between the settlement organization and the individuals' decision-making processes at the private plot level.

14.4 Adaptation and Patterns of Interaction

When studying adaptation and patterns of interaction at the community level, it is not easy to distinguish where one situation ends and the next begins (Ostrom et al. 1994). However, for analytical purposes, it is essential to define boundaries in space and time in order to focus on a single event. For this study, we selected a portion of the Japanese settlement in Tomé-Açu, using satellite images and field data collected in 2008.

The actions related to the community production systems were basically elaborated after individual farmers carried out different experiments. The outcomes of these production systems have at least one characteristic in common. They are based in polycultural activities where crops are overlapped, sometimes to the point of making it difficult to differentiate between a forest and an agroforestry system, without some knowledge of botany or agriculture. Production and productivity outcomes are determined by several factors, such as household structure, property characteristics, investments, and management (Flohrschutz 1983; Yamada 1999).

Following cooperative patterns of interactions (CAMTA 1967, 1975, 1994), the flux of information is a very important element of the community's success (Uhl and Subler 1988). Challenges, such as the fungal disease in black pepper fields, have been faced taking into account the experiences of each settler through the process of sharing information.

Besides the experience achieved by the settlers with their production systems, one of the most important elements to maximize payoffs during the development of the settlement was related to the transportation problem. Using the cooperative structure and building storehouses in Tomé-Açu and Belém, they became independent in taking their products to the market, overcoming one of the major problems in Amazonian colonization projects.

Self-governance and cooperation are significant patterns of interaction in Tomé-Açu. Since the first years of colonization, Japanese immigrants have worked together to clear the forest, to adapt agricultural techniques, and to determine strategies to be taken. Their motivations based on hope, determination, and willingness to take risks represented driving forces for their success (Uhl and Subler 1988).

Some particularities of the community of Tomé-Açu partially explain the success of the region's development, among them are the strong association of producers engaged not only in production but also in the commercialization of their products and a higher level of education compared to the region's average (Smith et al. 1998). In the surroundings of Tomé-Açu, several other producers are currently adopting diversified agroforestry systems similar to those of the Japanese community, which shows that the system can be successfully reproduced by non-Japanese farmers.

The social capital associated with Japanese culture was central for the achievement of better outcomes, compared to other settlers in the Brazilian Amazon. Even the characteristic of their farming systems, based on intensive practices, may be related to the farming systems present in Japan, where having 50 ha of land to cultivate is hardly common.

The ability to negotiate and look for market opportunities and accessibility is another characteristic of Tomé-Açu settlers. Using evaluative criteria based on periodic meetings at CAMTA, they produce a financial report every year. Discussing the results obtained with each product, they are flexible enough to go in new directions but always take into account the experiences of the past (CAMTA 1994). CAMTA (2009) has 243 associated producers, and 100% of them work with agroforestry systems.

After the success with black pepper and the subsequent challenges of dealing with international market demand fluctuations and fungal diseases, they realized that monocultural systems could represent a risky activity. As commented before, they had had this perception with other products as well (e.g., annual crops, cocoa, and jute). Choosing polycultural systems was a fundamental strategy to get higher payoffs, even when specific products have low market demands. At the same time, CAMTA solved temporarily the problem of storage and transportation. However, archival reports indicate efforts were made to develop a better road network in order for products to reach the market more promptly (Yamada 1999). Otherwise, activities such as logging and grazing were favored, which happened during a period when the road to Belém was in particularly poor condition in the early 1990s.

Menezes et al. (2004) studied the influence of the Japanese-Brazilian colony in the formation of the small family farmers' agroforestry systems (AFSs) in Tomé-Açu. They concluded that even though their model cannot be used generally in Amazonia, its replication and adaptation by small producers show the Japanese immigrants' influence on the adopted AFSs.

The community uses evaluative criteria based on market behavior for their products. For instance, a CAMTA meeting reported difficulties in selling the fruit. To aggregate value to passion fruit, *cupuaçu* (an Amazonian fruit), and *acerola* (a berry rich in vitamin C), they established the factory in order to send the frozen pulp to national and international markets (CAMTA 1994).

Fig. 14.2 Property based on agroforestry systems in Tomé-Açu in 2005 (Photo: M. Batistella, region of Tomé-Açu, Pará, Brazil, 2005)

Other challenges are related to pasture and cropland conversion, together with the opportunity of the younger descendants to increase their incomes by going to work in Japan. Fieldwork has demonstrated, however, that a significant portion of the younger generation learned from experience and is now investing in agroforestry systems. To better understand and monitor land-use and land-cover dynamics in the region, we propose a method to map different stages of such productive arrangements.

14.5 Agroforestry Systems in Tomé-Açu

Agriculture is still the basis of Tomé-Açu's economy. Besides black pepper, which is beginning to regain strength again after a long period of decline, fruticulture has currently gained greater economic relevance in rural properties, as illustrated in Fig. 14.2, and its production is distributed to several Brazilian states, North America, and the European Community. Several native and exotic tropical fruits are cultivated and industrialized through CAMTA (especially in the form of pulp), e.g., *açaí* berry (*Euterpe oleracea*), *cupuaçu* (*Theobroma grandiflorum*), *corossol* (*Annona muricata*), passion fruit (*Passiflora edulis*), cashew (*Anacardium occidentale*), pineapple (*Ananas comosus*), *taperebá* (*Spondias mombin* L.), guava (*Psidium guajava*), acerola (*Malpighia glabra*), *murici* (*Byrsonima crassifolia*), cocoa (*Theobroma cacao*), and carambola (*Averrhoa carambola*).

Yamada and Gholz (2002) highlight that the AFSs found in Tomé-Açu are a possibly sustainable alternative to converting forests to pastures in Amazonia. They also point out that, besides the high rate of employment generation, the income obtained by Japanese-Brazilian farmers from their properties, which vary between 10 and 20 ha, is comparable to the income of farmers who use between 400 and 1,200 ha of land as pastures, thus maintaining the population in rural areas and promoting the conservation of forest remnants.

The AFSs at rural properties in Tomé-Açu produce a mosaic of ecosystems and landscapes, configuring a space of heterogeneous patterns and of great diversity in rural production (Bolfe 2010). In this sense, there is a relevant need to map and monitor the expansion trend of the region's agroforests using remote sensing, thus supporting analyses of land use and land cover as well as aboveground biomass and carbon—an important component of ecosystem goods and services.

To estimate biomass data and to correlate them to vegetation's spectral responses, a field sampling system was initially developed, enabling the analysis of different agroforestry stages. Agroforestry producers associated with CAMTA were interviewed, and information on land-use history and production arrangements was obtained. Forty sampling plots composed of three units each, in a total of 120 sampling units, formed by 10×10, 3×3, and 1×1 m squares, were inventoried (Fig. 14.3).

The basal area (G) and the percentage of canopy cover (C) were calculated based on the analysis of vegetation structure parameters (diameter at breast height, DBH; total height, H) and enabled the definition of a conceptual framework of four agroforestry classes (Fig. 14.4). This classification was tested using data obtained in the field, and descriptive statistical analyses were performed to validate the proposed classification system.

Data obtained showed that AFS 1 has the smallest numbers of families (8) and species (9), a fact related to the initial phase of establishment of these areas (up to 2 years). The absolute abundance (N) was 747 individuals ha^{-1}, and C was 29.5%. The average DBH was 5.2 cm, G was 2.3 m^2 ha^{-1}, and the average height was 2.9 m. AFS 2 had 19 different species from 15 families. The abundance of 1,770 individuals ha^{-1} was the greatest observed. C was 45%; the average DBH measured was 6.4 cm, with G of 7.2 m^2 ha^{-1} and an average height of 3.7 m. The richness at AFS 3 was the same as that of AFS 2, with 19 species from 15 families. N was 1.723 individuals ha^{-1}, with C of 63%. The average DBH was 10.7 cm, G was 20.4 m^2 ha^{-1}, and the average height observed was of 5.9 m. AFS 4 registered the greatest number of species, 40, from 25 families. Its N was 1.457 individuals ha^{-1} with C of 78%. The average DBH measured was 13.7 cm, with G of 36.3 m^2 ha^{-1} and an average height of 7.4 m. Figure 14.5a, b, c, and d illustrate AFSs 1, 2, 3, and 4, respectively.

At the next stage, six allometric equations developed for the Amazonian region were used with the aim of estimating the individual aboveground biomass for vegetation species, as shown in Table 14.2.

To estimate the carbon stock (CS) in the dry biomass, a multiplication factor of 0.45 was used (Higuchi et al. 1998; Silva 2007). Thus, the aboveground biomass

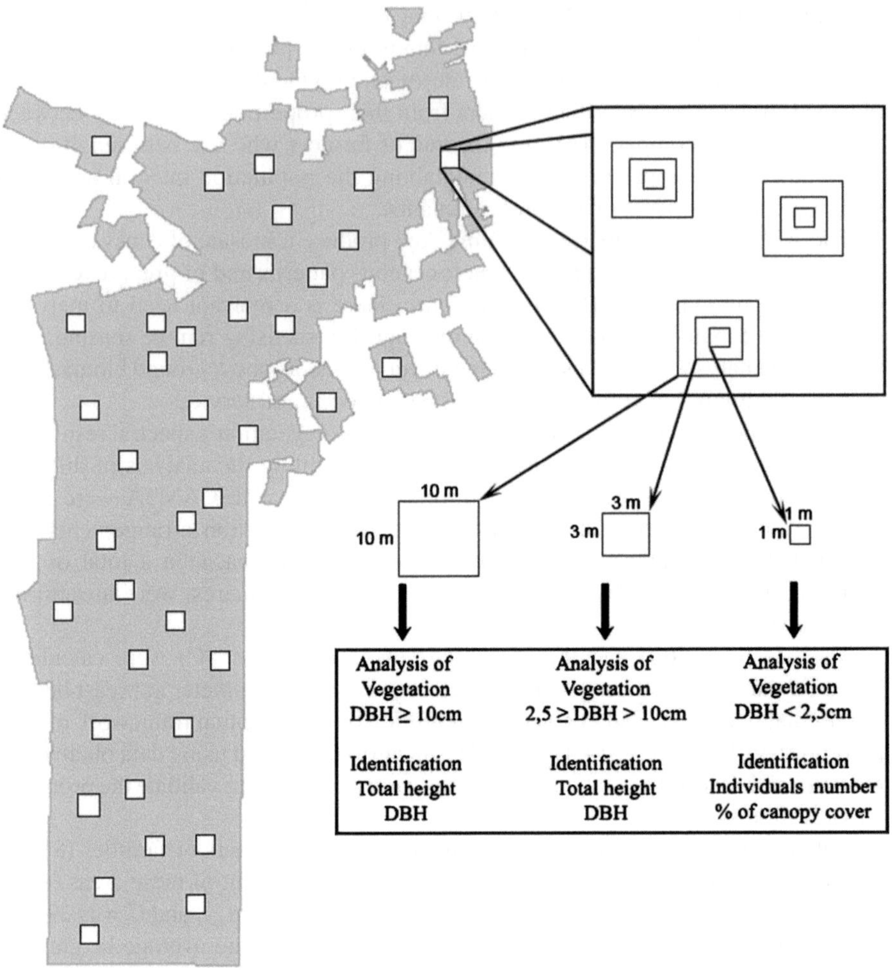

Fig. 14.3 Sampling strategy for agroforests in Tomé-Açu

and the carbon content per hectare of vegetation were estimated for each sampling unit, as well as their average for the sample plots, represented in Fig. 14.6.

Digital maps were generated through data integration in a geographic information system (Fig. 14.7). Carbon values and different vegetation indices were analyzed (i.e., simple ratio indices, normalized indices, and complex indices).

Table 14.3 presents some of the best adjusted regression models relating carbon data obtained in the field and spectral data extracted from satellite imagery (TM 2005 image), which can predict carbon stock in the vegetation of different AFSs for the region of Tomé-Açu. The greatest correlations were obtained for the Normalized Difference Vegetation Index (AFS 1 and AFS 3), Ratio Vegetation Index (AFS 2), and Transformed Soil-Adjusted Vegetation Index 1 (AFS 4).

Classes	Average parameters	Canopy cover estimation	Profile estimation
ASF 1	Age: 0 to 2 years Basal area: 0 to 5m²/ha Canopy cover: 0 to 35% Total height: 0 to 3m	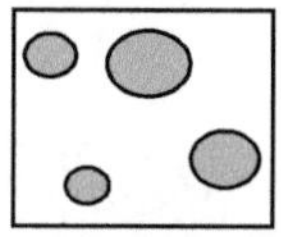	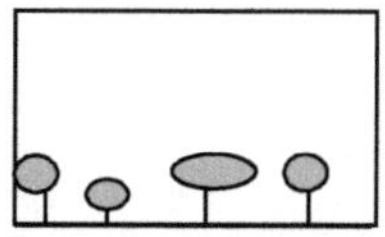
ASF 2	Age: 2 to 5 years Basal area: 5 to 10m²/ha Canopy cover: 35 to 50% Total height: 3 to 5m	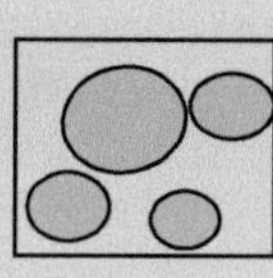	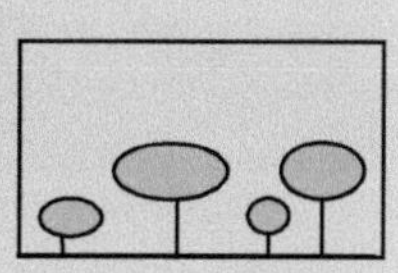
ASF 3	Age: 5 to 12 years Basal area: 10 to 30m²/ha Canopy cover: 50 to 70% Total height: 4.5 to 6m	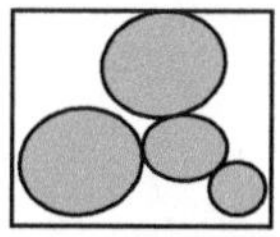	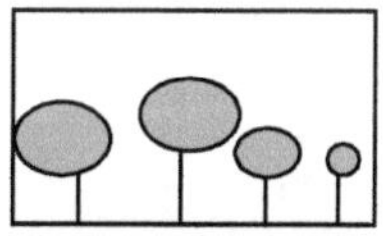
ASF 4	Age: more than 12 years Basal area: more than 30m²/ha Canopy cover: more than 70% Total height: more than 7m	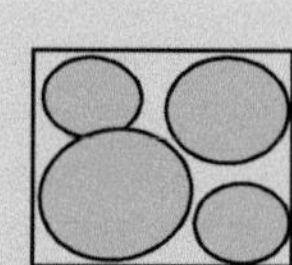	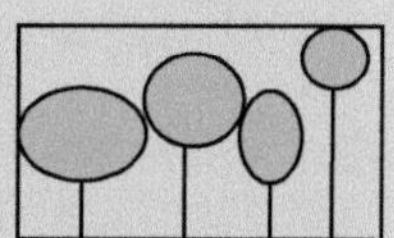

Fig. 14.4 Classes, parameters, and schematic representation of agroforestry systems in Tomé-Açu

The generated models were used to map aboveground biomass and carbon content (MgC.ha^{-1}) for the study area and for each agroforestry class (AFS 1, AFS 2, AFS 3, and AFS 4). Figure 14.8 illustrates the carbon content of AFS 3 within property lots associated with CAMTA.

The generation of these maps provided a synthetic view for the spatial distribution of carbon contents in the AFSs of Tomé-Açu. Such maps are essential when planning land use and land cover.

14.6 Conclusions

Beyond individual ambitions, the primary characteristics of the Tomé-Açu community were organization and cooperation, a "gainful strategy" when risk is high (Pruitt and Kimmel 1977). The challenge of being successful in the Amazon substantially affected the decision-making process, overcoming the role of beliefs and motivations as driving forces for cooperation (Hayashi et al. 1999). But the colonists' social capital was also central for the achievement of success. For example, the role of language in institutional analysis has been discussed as a way to express

Fig. 14.5 Different classes of agroforestry systems in Tomé-Açu (*Source*: Bolfe 2010: Figs. 5a and 5b on p. 137, Fig. 5c on p. 139, and Fig. 5d on p. 142)

ideas, develop common understanding, share learning, and explain the foundation of social order (Ostrom 1997). The fact that the Japanese immigrants kept their language also made Tomé-Açu an "island" of information flow, distinguished from other settlements in the region.

This chapter focused on historical perspectives of Tomé-Açu settlement, using remote sensing as a complementary tool to assess land-use/land-cover characteristics. The procedure was interactive and still may be improved through additional techniques. Further research in the area will be done to integrate previous experiences on spectral identification of land use and land cover in the Amazon (Mausel et al. 1993; Moran et al. 1994a, b).

The results show that AFSs in Tomé-Açu play an important role at the property and landscape scales, a trend that should be monitored. The trajectory toward conversion to pasture, related to its lower labor requirement, if compared to the intensive agroforestry systems, is now concurrent to the implementation of high carbon content land covers. Bolfe et al. (2009) highlight that the AFSs in Tomé-Açu are important carbon accumulators and may contribute significantly to the mitigation of the greenhouse effect.

Table 14.2 Allometric equations used to estimate aboveground biomass

Species	Equation	R^2	Sampling level	Source	Area
Euterpe oleracea Mart.	$BE1 = Exp(-0.0550 + 0.0451*DBH)$ *(acai berry tree leaves)*	0.92	DBH > 5 cm	Bartelt et al. (2000)	Cametá (PA)
	$BE2 = Exp(-0.0470 + 0.0750*DBH)$ *(acai berry tree stipes)*	0.99			
Theobroma cacao L.	$BE3 = Exp(-1.3200 + 0.0566*DBH)$ *(cocoa and cupuacu tree leaves)*	0.96	DBH > 5 cm	Bartelt et al. (2000)	Cametá (PA)
Theobroma grandiflorum Schum.	$BE4 = Exp(0.0320 + 0.0810*DBH)$ *(cocoa and cupuacu tree wood)*	0.99			
Other species	$BE5 = 0.0336*DBH^{2.171}*H^{1.038}*0.6$	0.94	DBH < 20 cm	Higuchi et al. (1998)	Tomé-Açu (PA)
	$BE6 = 0.0009*DBH^{1.585}*H^{2.651}*0.6$	0.92	DBH ≥ 20 cm		

Notes: *BE* is the dry aboveground biomass in $Mg.ha^{-1}$, *Exp* is the natural logarithm, *DBH* is the diameter at breast height (cm), *H* is the total height (m)

Fig. 14.6 Schematic view and carbon content estimated for agroforestry systems in Tomé-Açu (*Source*: Bolfe 2010: 149)

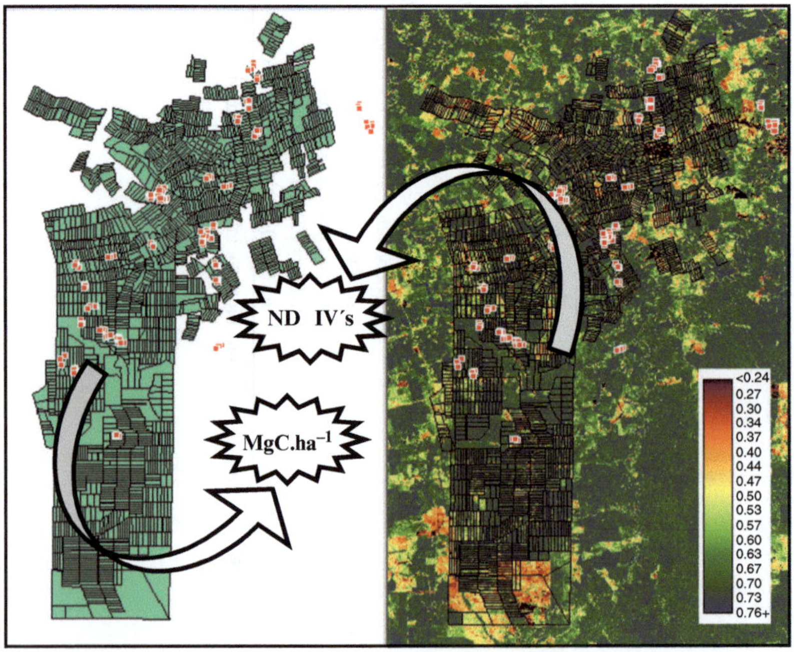

Fig. 14.7 Extraction of values for carbon content and vegetation indices in Tomé-Açu

Table 14.3 Best adjusted models for carbon estimation in Tomé-Açu agroforestry systems

AFS class	Regression model	r^2
AFS 1	Carbon = −15.368 + 81.976 * NDVI[a]	0.84
AFS 2	Carbon = 41.527–55.370 * RVI[a]	0.80
AFS 3	Carbon = −34.349 + 134.145 * NDVI[a]	0.63
AFS 4	Carbon = −59.502 + 6.2956 * TSAVI 1[a]	0.59

r^2 coefficient of determination

[a]Significant according to student's *t* test at the 1% probability error level

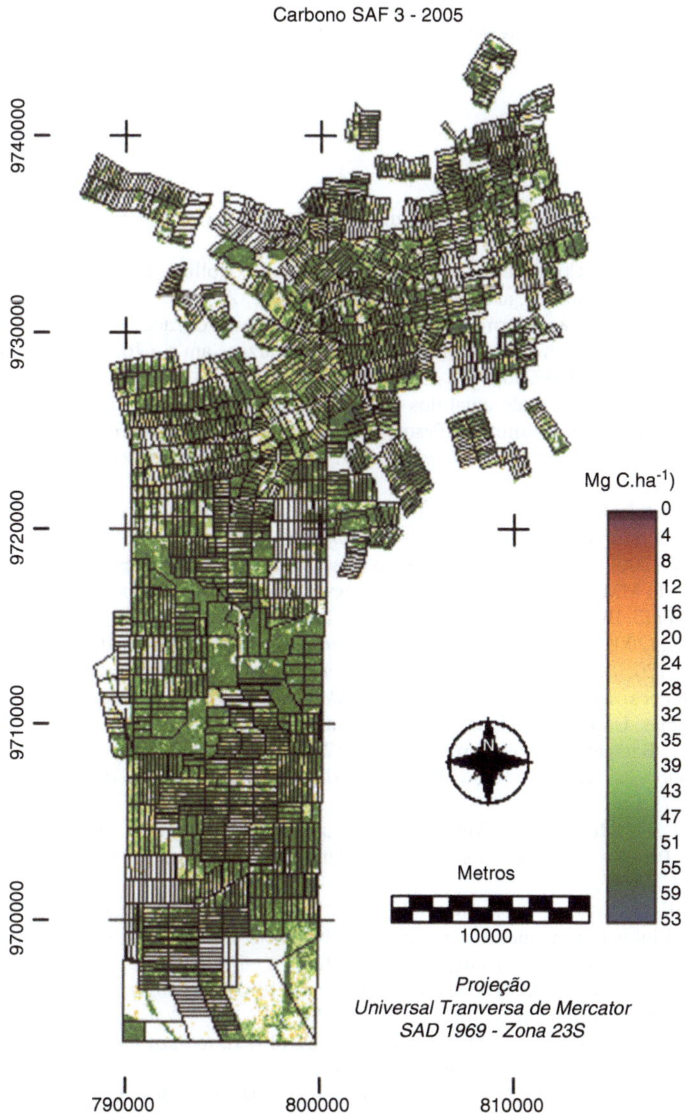

Fig. 14.8 Carbon content estimates for agroforestry class AFS 3 in 2005

The community of Tomé-Açu is aware of the challenges they face. The older farmers want the younger generations to continue their effort to develop an "Amazonian agriculture," based on agroforestry, local species, and technology. Their goal is to focus on products that avoid market competition, such as exotic fruits and valuable timber species. Recent initiatives have attempted to enrich the fallow areas with timber, seeking to produce better extractive results and increase the value of their properties. The younger farmers are also recognizing the potential of such strategies as an alternative to cattle ranching. Further research in Tomé-Açu will provide a better understanding about the ecological results in terms of spatial organization of production systems and management of natural resources.

References

Anderson, A. B. (1992). Land-use strategies for successful extractive economies. *Advances in Economic Botany, 9*, 667–678.

Arrow, K., Bolin, B., Costanza, R., Dasgupta, P., Folke, C., Holling, C., et al. (1995). Economic growth, carrying capacity, and the environment. *Science, 268*, 520–521.

Bartelt, D., Koch, J., & Tourinho, M. M. (2000). Anbau von acai (*Euterpe oleracea*) und kakao (*Theobroma sylvestre*) in primärwäldern der varzeas am rio Tocantins (Brasilien/Para). *Forstarchiv, 71*(6), 250–256.

Bastos, T. X. (1972). O estado atual dos conhecimentos das condições climáticas da Amazônia brasileira. In IPEAN (Instituto de Pesquisas e Experimentação Agropecuárias do Norte) (Ed.), *Zoneamento agrícola da Amazônia: 1a aproximação* (Boletim técnico, no. 54, pp. 68–122). Belém: IPEAN.

Batistella, M., & Moran, E. (2005). Dimensões humanas do uso e cobertura das terras na Amazônia: uma contribuição do LBA. *Acta Amazonica, 35*, 251–259.

Becker, B. K. (2001). Síntese do processo de ocupação da Amazônia: lições do passado e desafio do presente. In MMA (Ministério do Meio Ambiente) (Ed.), *Causas e dinâmica no desmatamento da Amazônia*. Brasília: MMA.

Becker, B. K. (2006). *Amazônia: geopolítica na virada do III milênio*. Rio de Janeiro: Garamond.

Bolfe, E. L. (2010). *Desenvolvimento de uma metodologia para a estimativa de biomassa e de carbono em sistemas agroflorestais por meio de imagens orbitais*. Doctoral thesis, Instituto de Geociências, Universidade Estadual de Campinas, Brazil.

Bolfe, E. L., Batistella, M., & Ferreira, M. C. (2009). Biomassa epígea e estoque de carbono de agroflorestas em Tomé-Açu, PA. Cruz Alta. *Revista Brasileira de Agroecologia, 4*(2), 2171–2175.

Brondizio, E. S., Moran, E. F., Mausel, P., & Wu, Y. (1994). Land-use change in the Amazon estuary: Patterns of caboclo settlement and landscape management. *Human Ecology, 22*(3), 249–278.

Brondizio, E. S., Moran, E. F., Mausel, P., & Wu, Y. (1996). Land cover in the Amazon estuary: Linking of the thematic mapper with botanical and historical data. *Photogrammetric Engineering and Remote Sensing, 62*(8), 921–929.

CAMTA (Cooperativa Agrícola Mista de Tomé-Açu). (1967). *Relatos históricos da Cooperativa Agrícola Mista de Tomé Açu*. Tomé-Açu: CAMTA.

CAMTA (Cooperativa Agrícola Mista de Tomé-Açu). (1975). *Relatos históricos da Cooperativa Agrícola Mista de Tomé Açu*. Tomé-Açu: CAMTA.

CAMTA (Cooperativa Agrícola Mista de Tomé-Açu). (1994). *Relatório da diretoria da Cooperativa Agrícola Mista de Tomé Açu*. Tomé-Açu: CAMTA.

CAMTA (Cooperativa Agrícola Mista de Tomé-Açu). (2009). *Associados da CAMTA: Informações disponibilizadas verbalmente pela diretoria da cooperativa*. Tomé-Açu: CAMTA.

Conant, F., Rogers, P., Baumgardner, M., Mckell, C., Dasmann, R., & Reining, P. (1983). *Resource inventory and baseline study methods for developing countries*. Washington, DC: American Association for the Advancement of Science.

Crawford, S., & Ostrom, E. (1995). A grammar of institutions. *American Political Science Review, 89*(3), 582–600.

Crumley, C. L. (1994). Historical ecology: A multidimensional ecological orientation. In C. L. Crumley (Ed.), *Historical ecology* (pp. 1–16). Santa Fe: School for Advanced Research Press.

Dale, V. H., & Pearson, S. M. (1997). Quantifying habitat fragmentation due to land-use change in Amazonia. In W. F. Laurance & R. Bierregaard Jr. (Eds.), *Tropical forest remnants: Ecology, management, and conservation of fragmented communities* (pp. 400–409). Chicago: University of Chicago Press.

Falesi, I. C., Santos, W. H. dos, & Vieira, L. S. (1964). *Os solos da colônia agrícola de Tomé Açu* (Boletim técnico, no. 44). Belém: Instituto de Pesquisas e Experimentação Agropecuárias do Norte.

FAO (Food and Agriculture Organization of the United Nations). (1993). *Forest resources assessment 1990: Tropical countries* (Forestry Paper 112). Rome: FAO.

Fearnside, P. M. (1986). *Human carrying capacity of the Brazilian rainforest*. New York: Columbia University Press.

Flohrschutz, G. H. H. (1983). *Análise econômica de estabelecimentos rurais no município de Tomé Açu, Pará: um estudo de caso* (Documentos, 19). Belém: Centro de Pesquisa Agropecuária do Trópico Úmido, Empresa Brasileira de Pesquisa Agropecuária.

Gerber, J. (1997). Beyond dualism: The social construction of nature and the natural *and* social construction of human beings. *Progress in Human Geography, 2*(1), 1–77.

Goodland, R. J. A., Daly, H. E., & Serafy, S. (1993). The urgent need for rapid transition to global environmental sustainability. *Environmental Conservation, 20*(4), 297–309.

Hayashi, N., Ostrom, E., Walker, J., & Yamagishi, T. (1999). Reciprocity, trust, and the sense of control: A cross-societal study. *Rationality and Society, 11*, 27–47.

Higuchi, N., dos Santos, J., Ribeiro, R. J., Minette, L., & Biot, Y. (1998). Biomassa da parte aérea da vegetação da floresta tropical úmida de terra-firme da Amazônia Brasileira. *Acta Amazonica, 28*(2), 153–165.

Homma, A. K. O. (1998). *Amazônia: meio ambiente e desenvolvimento agrícola*. Brasília and Belém: Informação Tecnológica and Centro de Pesquisa Agropecuária do Trópico Úmido, Empresa Brasileira de Pesquisa Agropecuária.

Homma, A. K. O. (2003). *História da agricultura na Amazônia: da era pré-colombiana ao terceiro milênio*. Brasília: Informação Tecnológica, Empresa Brasileira de Pesquisa Agropecuária.

Homma, A. K. O. (2004). Dinâmica dos sistemas agroflorestais: o caso da Colônia Agrícola de Tomé-Açu, Pará. *Revista Instituto de Estudo Superiores da Amazônia, 2*(1/2), 57–65.

IBGE (Instituto Brasileiro de Geografia e Estatística). (2007). *Censo Agropecuário 2006: Resultados Preliminares*. http://www.ibge.gov.br/home/estatistica/economia/agropecuaria/censoagro/2006/agropecuario.pdf. Accessed Mar 2010.

INPE (Instituto Nacional de Pesquisas Espaciais). (1996). *Projeto PRODES*. São José dos Campos: INPE, demonstration CD-ROM.

INPE (Instituto Nacional de Pesquisas Espaciais). (2002). *Monitoring of the Brazilian Amazon forest by satellite 2000–2001*. São José dos Campos: INPE.

INPE (Instituto Nacional de Pesquisas Espaciais). (2006–2009). *PRODES*. www.dpi.inpe.br/prodesdigital. Accessed 15 July 2006, 15 Aug 2007, 11 Sept 2008, 17 Aug 2009.

Kates, R. W. (1994). Sustaining life on the Earth. *Scientific American, 271*(4), 114–122.

Koppen, W. (1948). *Climatologia* (3rd ed.). Mexico: Fondo Cultura Económica.

Laurance, W. F., Cochrane, M. A., Bergen, S., Fearnside, P. M., Delamônica, P., Barber, C., et al. (2001). The future of the Brazilian Amazon. *Science, 291*(5503), 438–439.

Mausel, P., Wu, Y., Li, Y., Moran, E. F., & Brondizio, E. S. (1993). Spectral identification of successional stages following deforestation in the Amazon. *Geocarto International, 8*(4), 61–81.

McCracken, S., Brondízio, E., Moran, E. F., Nelson, D., Siqueira, A., & Rodriguez-Pedraza, C. (1998). The use of remote sensing and GIS in the collection of survey data on households and land use: Example from the agricultural frontier of the Brazilian Amazon. In *Proceedings of the IX Brazilian remote sensing symposium*. CD-ROM.

McGinnis, M., & Ostrom, E. (1996). Design principles for local and global commons. In O. R. Young (Ed.), *The international political economy and international institutions* (Vol. II, pp. 465–493). Cheltenham: Edward Elgar.

Menezes, A. J. E. A. de, Frazão, D. A. C., Homma, A. K. O., Matos, G. B. de, Ishisuka, Y., Rocha, A. C. P. N. da, et al. (2004). Influência da colônia nipo brasileira na formação dos sistemas agroflorestais dos pequenos agricultores familiares de Tomé Açú, Pará. In *Anais V Congresso Brasileiro de sistemas agroflorestais: SAFs: desenvolvimento com proteção ambiental, Curitiba* (pp. 105–107). Document 98. Colombo: Florestas, Empresa Brasileira de Pesquisa Agropecuária.

Moran, E. F. (1981). *Developing the Amazon*. Bloomington: Indiana University Press.

Moran, E. F. (1993). Deforestation and land-use in the Brazilian Amazon. *Human Ecology, 21*, 1–21.

Moran, E. F. (1996). Deforestation in the Brazilian Amazon. In L. E. Sponsel, T. N. Headland, & R. C. Bailey (Eds.), *Tropical deforestation: The human dimension* (pp. 149–164). New York: Columbia University Press.

Moran, E. F., Brondizio, E. F., Li, Y., Mausel, P., & Wu, Y. (1994a). Discrimination between advanced secondary succession and mature moist forest near Altamira, Brazil, using Landsat TM data. In *Proceedings, 60th annual meeting for American society of photogrammetry and remote sensing, Reno, Nevada* (pp. 350–364). Bethesda: American Society of Photogrammetry and Remote Sensing/American Congress on Surveying and Mapping.

Moran, E. F., Brondizio, E. S., Mausel, P., & Wu, Y. (1994b). Integrating Amazonian vegetation, land-use, and satellite data. *BioScience, 44*(5), 329–338.

Moran, E. F., Packer, A., Brondizio, E. S., & Tucker, J. (1996). Restoration of vegetation cover in the eastern Amazon. *Ecological Economics, 18*, 41–54.

Moran, E., Ostrom, E., & Randolph, J. C. (1998). *A multilevel approach to studying global environmental change in forest ecosystems* (CIPEC Working Paper CWP-98-07). Bloomington: Center for the Study of Institutions, Population, and Environmental Change (CIPEC), Indiana University.

Nepstad, D. C., Verssimo, A., Alencar, A., Nobre, C., Lima, E., Lefebvre, P., et al. (1999). Large-scale impoverishment of Amazonian forests by logging and fire. *Nature, 398*, 505–508.

Nobre, C. A., Sampaio, G., & Salazar, L. (2007). Mudanças climáticas e Amazônia. *Ciência e Cultura, 59*(3), 22–27. http://cienciaecultura.bvs.br. Accessed 27 Aug 2007.

NRC (National Research Council). (1998). *Global environmental change: Research pathways for the next decade*. Washington, DC: National Academies Press.

Ostrom, V. (1997). *The meaning of democracy and the vulnerability of democracies: A response to Tocqueville's challenge*. Ann Arbor: University of Michigan Press.

Ostrom, E., Gardner, R., & Walker, J. (1994). *Rules, games, and common-pool resources*. Ann Arbor: University of Michigan Press.

Porto-Gonçalves, C. W. (2001). *Amazônia, Amazônias* (2nd ed.). São Paulo: Contexto.

Pruitt, D. G., & Kimmel, M. J. (1977). Twenty years of experimental gaming: Critique, synthesis, and suggestions for the future. *Annual Review of Psychology, 28*, 363–392.

Rodrigues, T. E., & Baena, A. R. C. (1974). *Solos da Rodovia PA-02; Trecho km 65 ao km 111 (Tomé Açu – Paragominas)* (Boletim técnico, no. 55). Belém: Instituto de Pesquisas e Experimentação Agropecuárias do Norte.

Rodrigues, T. E., Santos, P. L. dos, Rolim, P. A. M., Santos, E., Rego, R. S., Silva, J. M. L. da, et al. (2001). *Caracterização e classificação dos solos do município de Tomé-Açu, Pará* (Embrapa Amazônia Oriental Documentos, 117). Belém: Amazônia Oriental, Empresa Brasileira de Pesquisa Agropecuária.

Sader, S. A., & Joyce, A. T. (1985). Relationship between forest clearing and bio-physical factors in a tropical environment. *Applied Geography, 17*, 143–162.

Sanchez, P. A. (1976). *Properties and management of soils in the tropics*. New York: Wiley.

SECTAM (Secretaria Executiva de Ciência, Tecnologia e Meio Ambiente do Pará). (2009). *Meteorologia 2009*. http://www.para30graus.pa.gov.br/meteorologia.htm. Accessed 15 Aug 2009.

Silva, R. P. (2007). *Alometria, estoque e dinâmica da biomassa de florestas primárias e secundárias na região de Manaus (AM)*. Doctoral thesis, Ciências Florestais, Instituto Nacional de Pesquisas da Amazônia, Brasil.

Skole, D., & Tucker, C. J. (1993). Tropical deforestation and habitat fragmentation in the Amazon: Satellite data from 1987 to 1988. *Science, 260*, 1905–1910.

Smith, N., Williams, J., Plucknett, D., & Talbot, J. (1992). *Tropical forests and their crops*. Ithaca: Cornell University Press.

Smith, N., Dubois, J., Current, D., Lutz, E., & Clement, C. (1998). *Experiências agroflorestais na Amazônia Brasileira: restrições e oportunidades*. Brasília: Programa Piloto para a Proteção das Florestas Tropicais do Brasil.

Stern, P. C., Young, O. R., & Druckman, D. (Eds.). (1992). *Global environmental change: Understanding the human dimensions*. National Research Council Committee on the Human Dimensions of Global Change. Washington, DC: National Academies Press.

Stolberg-Wernigerode, A. G. Zu, & Flohrschutz, G. H. H. (1982). *Levantamento de plantios mistos na colônia agrícola de Tomé Açu, Pará* (Document 6). Belém: Centro de Pesquisa Agropecuária do Trópico Úmido, Empresa Brasileira de Pesquisa Agropecuária.

Subler, S. (1993). *Mechanisms of nutrient retention and recycling in a chronosequence of Amazonian agroforestry systems: Comparisons with natural forest ecosystems*. Ph.D. dissertation, The Pennsylvania State University.

Subler, S., & Uhl, C. (1990). Japanese agroforestry in Amazônia: A case study in Tomé Açu, Brazil. In A. B. Anderson (Ed.), *Alternatives to deforestation: Steps toward sustainable use of the Amazon rainforest* (pp. 152–166). New York: Columbia University Press.

Tomé-Açu (Prefeitura Municipal de Tomé-Açu). (2007) *História de Tomé-Açu*. http://www.pmtomeacu.com.br. Accessed Sep 2007.

Tsunoda, F. (1988). *Canção da Amazônia: uma saga na selva*. Rio de Janeiro: Francisco Alves.

Turner, B. L., II, Clark, W. C., Kates, R. W., Richards, J. F., Mathews, J. T., & Meyer, W. B. (Eds.). (1990). *The Earth as transformed by human action*. New York: Cambridge University Press.

Turner, B. L. II, Moss, R. H., & Skole, D. L. (1993). *Relating land-use and global land cover change: A proposal for an IGB-HDP core project*. Joint publication of International Geosphere-Biosphere Programme (Report No. 5) and International Human Dimensions Programme (Report No. 24). Stockholm: Royal Swedish Academy of Sciences.

Turner, B. L. II, Skole, D., Sanderson, S., Fischer, G., Fresco, L., & Leemans, R. (Eds.). (1995). *Land-use and land-cover change: Science/research plan*. Joint publication of International Geosphere-Biosphere Programme (Report No. 35) and International Human Dimensions Programme (Report No. 7). Stockholm: Royal Swedish Academy of Sciences.

Uhl, C., & Subler, S. (1988). Asian farmers: Stewards of Amazônia. *Garden, 12*(5), 16–20.

Wood, C. H., & Skole, D. (1998). Linking satellite, census, and survey data to study deforestation in the Brazilian Amazon. In D. Liverman, E. Moran, R. Rindfuss, & P. Stern (Eds.), *People and pixels: Applications of remote sensing in the social sciences* (pp. 70–93). Washington, DC: National Academies Press.

Yamada, M. (1999). *Japanese immigrant agroforestry in the Brazilian Amazon: A case study of sustainable rural development in the tropics*. Ph.D. thesis, University of Florida.

Yamada, M. (2009). Uma breve história de desenvolvimento agroflorestal Nikkei na Amazônia: o caso da colônia de Tomé-Açu. In R. Porro (Ed.), *Alternativa agroflorestal na Amazônia em transformação* (pp. 691–704). Brasília: Empresa Brasileira de Pesquisa Agropecuária.
Yamada, M., & Gholz, H. L. (2002). An evaluation of agroforestry systems as a rural development option for the Brazilian Amazon. *Agroforestry Systems, 55*(2), 81–87.
Yared, J. A. G., & Veiga, J. B. (1985). Sistemas agroflorestais na colônia agrícola de Tomé Açu, Pará, Brasil. In *Informe del taller sobre investigacion agroforestal en la region Amazonica* (pp. 128–164). Nairobi: International Center for Research in Agroforestry/Inter-American Institute for Cooperation on Agriculture/United States Agency for International Development.

Chapter 15
Changing Driving Forces, Imposed Tenure Regimes, and Tree-Cover Change on Village Landscapes in the West Mengo Region of Uganda, 1890–2002

Nathan D. Vogt

Abstract Rapid growth in local and urban populations in Africa over the last decades of the twentieth century greatly increased demand for fuelwood, construction timber, and staple foods and drew concerns over the fate of forest and tree cover across the continent. Few places have greater pressures on land and forest resources in Africa than the densely populated West Mengo district of Uganda that borders the rapidly growing capital Kampala, in the Buganda region of the Lake Victoria basin. Recent studies have shown and clarified why forest cover in West Mengo remained largely stable through much of the colonial era, under increasing market and population pressures. There is evidence, but little understanding, that tree cover on lands used for perennial cropping (coffee and banana) has gradually increased over the same period, despite the extreme pressures. This chapter argues that tree-cover growth resulting from increased intensity of agroforestry is explained by the interplay of markets and land tenure and important shifts in those factors over the study period.

15.1 Introduction

Human-environment interactions have been ongoing in East Africa, the cradle of humanity, longer than anywhere else on earth. Imposition of Western institutional arrangements and land-use zones by colonial governments since the late 1800s makes it difficult to reconstruct the precolonial forms of land use and understand contemporary human-environmental interactions in East Africa. Contemporary relationships of smallholders in Africa to their environment are highly impacted by the growth in external demand for fuelwood, construction timber, and staple foods

N.D. Vogt (✉)
Instituto de Pesquisa Ambiental da Amazônia, Belém, Brazil
e-mail: ndvogt@gmail.com

E.S. Brondízio and E.F. Moran (eds.), *Human-Environment Interactions: Current and Future Directions*, Human-Environment Interactions 1,
DOI 10.1007/978-94-007-4780-7_15, © Springer Science+Business Media Dordrecht 2013

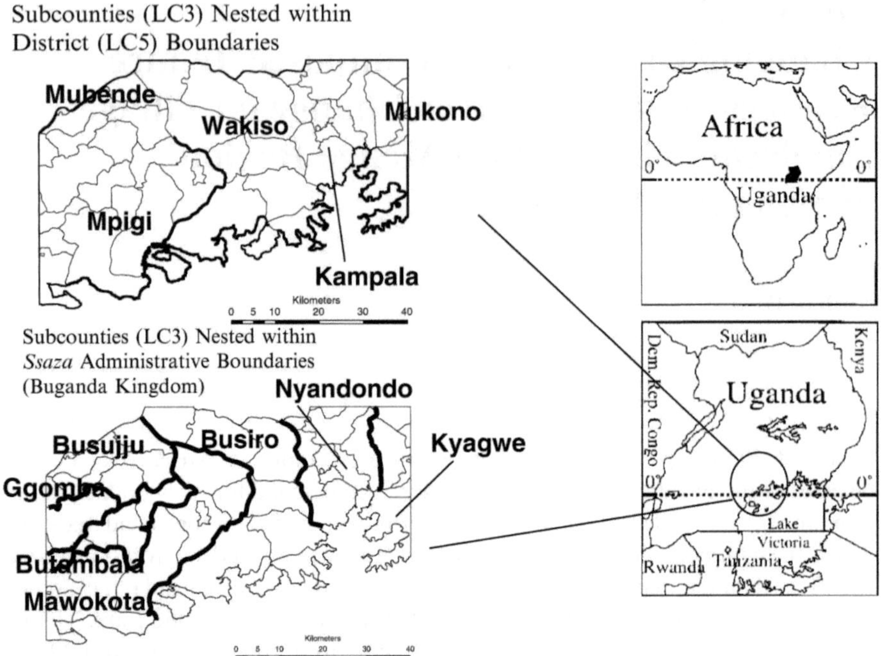

Fig. 15.1 Maps of the study area, including those showing the subcounty (*gombolola*) units nested within both district and traditional county (*ssaza*) units. These maps illustrate clear territoriality, fixing boundaries between groups and defining rights to and control over specific sets of natural resources that evolved long before British colonization

over recent decades, drawing concerns over the fate of forest and tree cover across the continent. Nowhere are these pressures greater than in the West Mengo region of Uganda (Fig. 15.1). Prevailing Western assessment is that land and forests are degrading and smallholders, whose resource use and tenure regimes are deemed anachronistic and maladapted to these pressures, are to blame (Fairhead and Leach 1996). These narratives describe a mechanism whereby the growing population and market pressures drive smallholders to shorten fallows on agricultural lands, which degrades agricultural soils. Degraded soils provide incentives to heavily harvest forests and convert them to agricultural purposes in short fallow cycles, producing a "downward spiral" (Cleaver and Schreiber 1994; Otsuka and Place 2001). Conversion of forests to agriculture by smallholders in the region is often identified as the cause of increased siltation and eutrophication in Lake Victoria (Reid 2000). This chapter argues that smallholders in West Mengo, after a dramatic decline in forest and tree cover in the early colonial era, have increased aggregate tree cover within traditional Buganda village units since World War II (WWII) despite extreme pressures for staple and woody products. This is contrary to common results from ahistorical, coarse-scale analyses from visions that continue to blame smallholders for deforestation in the Lake Victoria basin.

Western sectoral visions based on individuals' maximizing production on distinct resource units (land, forest, wetlands, and grasslands) often interpret smallholder land-use systems as resource degrading. These constructs perpetuate an oversimplified dichotomy between nature and culture (Brondízio 2006; Fairhead and Leach 1996). Other approaches stress more linked social-ecological approaches less based on maintaining ecological equilibrium within individual resource units (Berkes and Folke 1998; Berkes et al. 2003; Folke 2006; Ostrom 2005, 2009; Scoones 1999), as does historical ecology that highlights the spatially and temporally complex configurations of land-use/cover systems that smallholders often maintain across all resource units in traditional territorial units and their efficiency to adapt to external forces (Balée and Erickson 2006; Fairhead and Leach 1998; Hornborg et al. 2007; Scoones 1999).

Despite high growth in market and population pressures in West Mengo, forest cover within forest reserves created in the colonial era remained stable, due mostly to the strength of the Bugandan chiefs and their ability to negotiate locally recognized reserve boundaries and institutional arrangements governing them. When control over the forest reserves was decentralized to local administrative units, harvesting and conversion to alternative land uses increased (Banana et al. 2007), with results suggesting smallholders continuing the "downward spiral" when resources are placed in their care. However, in this chapter, a more holistic study of landscape change across traditional village units of West Mengo, at greater historical depth, is presented to bring new insights. This chapter argues that smallholders have increased tree cover on the village *mitala*—arable land in villages where agroforestry is practiced—while managing the less arable village valleys in more spatially and temporally complex ways. These adaptations are facilitated by the flexibility in traditional rules and norms of resource use across the village unit, despite perverse incentives created by British-imposed tenure regimes.

To better understand how extreme growth in resource pressures and imposed tenure regimes are impacting forest/tree cover in the West Mengo region of Uganda, an analysis of change in smallholder land-use/cover patterns at the village landscape level from 1880 to 2002 was conducted. Fixed traditional village boundaries and land-use systems in them reflect ecological patterns characteristic of the West Mengo region. These patterns are shaped by the regular undulation of the soil catena across the region (Fig. 15.2). Patterns of land use/cover on each of the major soil types found in a typical West Mengo village unit are reconstructed for four distinct periods between 1880 and 2002. Major socioeconomic pressures for each of these periods, and changes in formal tenure regimes, are also reconstructed to explain village-level change in smallholder land-use/cover patterns.

In this chapter, I argue that the highly dynamic change in pattern of forest and tree cover across West Mengo village units and its long-term increase are due to the diversified and ecosystem-based forms of institutions and resource use in Buganda villages, which include use and management of seasonally inundated forests (both young and mature), upland forests (both young and mature), agroforests, agricultural fields, and natural pasture across the soil types found in a typical village unit. This diversity of land-use/cover systems at the village level increases their flexibility

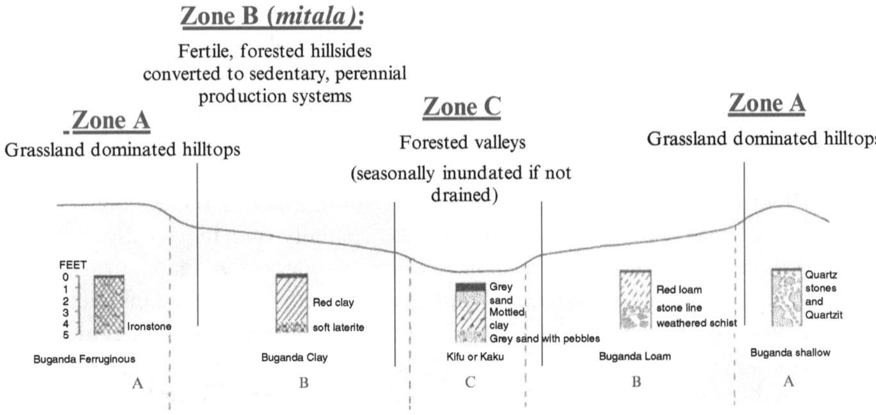

Fig. 15.2 Soil catena of typical, traditional village unit in West Mengo, Uganda

and facilitates more efficient adaptation to changing external forces. However, the flexibility of village-level land-use/cover systems continues to be constrained by the colonial imposition of formal tenure over a century ago, when village chiefs were given individual titles over their village units, and management of the land in distinct agricultural and forestry units, rather than more integrated agroecological systems reflecting traditional, village-level management.

This chapter begins with a description of the study area and methods of data collection and analysis. It then develops a description of the soil catena defining traditional village units of the West Mengo region. A reconstruction of village landscapes in West Mengo in four distinct periods—circa 1880–1900, 1900–1945, 1945–1995, and 1995–2002—follows. In each period, there are four sections: (1) introduction to the period and major socioeconomic and land tenure changes affecting land-use decisions; (2) incentives created by socioeconomic and land tenure changes affecting private land (*mailo*) owners, the landed class created during British rule, and village resource-use decisions; (3) incentives created by socioeconomic and land tenure changes affecting smallholder tenants' village resource-use decisions; and (4) village landscape changes. A final discussion of the importance of historical context for understanding contemporary smallholder land-use problems concludes the chapter.

15.2 Study Area

West Mengo is part of the Lake Victoria agroecological zone and lies in the northwest quadrant of the Lake Victoria basin of East Africa between 0.06′S and 0.30′N and between 32°00′E and 32°45′E (see Fig. 15.1). The research team chose West Mengo to establish a biophysical and cultural context for the driving forces, tenure regimes, and land-cover patterns under investigation. West Mengo receives between 1,500 and 1,800 mm of rainfall per year. The topography is characterized by regularly spaced, flat-topped hills. Altitudes range from approximately 1,100 m above sea level in the valleys to 1,200 m above sea level on the hilltops. Vegetation is further differentiated within this agroecological zone by topographic relief, i.e., soil drainage differentiates soil types (see Fig. 15.2).

This investigation was conducted in a portion of the forest biome of the Lake Victoria basin, occupied by the Buganda Kingdom since as early as the thirteenth century (Reid 2002). It covers portions of Mpigi, Mukono, Mubende, and Wakiso districts (administrative units defined by the colonial government). It also covers portions of Kyadondo, Busiro, Busujju, Gomba, Butambala, and Mawokota *ssaza*s (traditional administrative units of the Buganda Kingdom, which are now formally called counties). West Mengo experienced extreme growth in demand for food and forest products during the study period. Rural population densities in West Mengo for 1980, 1991, and 2002 were 110, 150, and 220 persons/km^2, respectively (100% growth), and total urban population for the same dates was 496,781, 854,272, and 1,334,106 (169% growth) (Vogt et al. 2006a, b). Over 90% of all Ugandans rely on biomass fuels for energy (Arnold et al. 2003), and over 80% of urban populations rely on charcoal (MWLE 2000).

15.3 Data Collection and Analysis

Land-use/cover patterns for each of the four study periods were reconstructed from various combinations of remotely sensed products, vegetation plot data, historical forestry department data, and interviews with elder key informants. To reconstruct driving forces and tenure systems affecting village landscape change in each period and land-use/cover systems in use in each period of interest, we draw on archival data, census data, information from past research of Buganda Kingdom, research of the International Forestry Resources and Institutions program, and personal interviews with key informants, who aided in interpreting change observed in remotely sensed products.

In 2003, the team interviewed district forest and agriculture officers from the study area, along with a sampling of their subcounty agents. Four interviews were conducted, and 11 key informants participated. Two ssaza chiefs and two subcounty administrators who were also mailo landlords from the Mpigi district were interviewed. These informants aided in explaining the impacts of factors at subcounty, district, and state levels on broader trends in land-cover change observed across the West Mengo region in remotely sensed products, including long-term boundary stability and tree-cover change on cultivated lands. We used the aid of both 2002 Landsat images and a 1955–1995 time series of aerial photographs. To protect the anonymity of subjects, we created a confidential list of persons interviewed and identified them by letters and numbers for referencing purposes: FO = forest officer or ranger, AO = agriculture officer or extension agent, LC = local council member,

LL = mailo owner and subcounty councilor, CE = village elder, and SC = ssaza chief. Letters were added to the base codes to denote districts, and numbers were added to denote individuals.

15.4 Soil Catena Defines and Shapes Village Landscapes

Administrative limits in West Mengo are adapted to the regularly patterned soil and vegetation types of the region, so almost every village contains the entire profile of the soil catena characteristic of the region (see Fig. 15.2). Land uses at the village level and institutions regulating them vary with the physical characteristics of the soil catena. Within each village unit, some soils are conducive to more permanent agricultural uses and division (fertile hillsides traditionally known as mitala, zone B in Fig. 15.2), and other soils are more conducive to extractive and communal uses (grassland hilltops and seasonally inundated valley forests, zones A and B in Fig. 15.2).

Fertile hillsides (zone B) were once covered with forests but are now dominated by a banana-coffee-fruit tree matrix. Seasonally inundated valleys (zone C) also support productive forests. Valley soils are less conducive to the more continuous, perennial systems of the mitala, but productivity can be increased through more spatially and temporally complex land management patterns. The summits are short-grassed savannas less conducive to crop production and primarily used for communal grazing (zone A).

15.5 Reconstructions of Past Village Landscapes in West Mengo, Uganda

15.5.1 Precolonial (Late 1800s): Bugandan Village Landscapes upon British Arrival

When Europeans arrived in the Buganda Kingdom in the late 1800s, they found a clearly divided region, administered by a nested, five-level hierarchy of chiefs defining control over access and use of resources across the region. Administrative boundaries persisting until today mostly follow ecological transitions such as streams in valleys or hilltops. Village boundaries, the smallest administrative unit in West Mengo, contain a portion of the primary ecotypes found in West Mengo, including arable forested land on the mitala conducive to sedentary agriculture, hilltop grasslands for grazing, and seasonally inundated valley forests for hunting, fuelwood, construction timber, and a water supply (see Fig. 15.2).

Traditional Buganda land boundaries are associated with rights to control the land. In Buganda, there are four categories of rights to land (Gombya-Ssembajwe 1985; Mukwaya 1953): (1) clan rights, (2) rights of kings and chiefs, (3) individual hereditary rights, and (4) smallholder tenant rights of occupation. During this time, no administrator or clan elder had permanent individual rights over land, only the king (*kabaka*). As contact with Arabs and Europeans increased in the second half of the nineteenth century, the kabaka had more incentives to grant permanent holdings to administrators and clan elders to gain goods (e.g., ivory) for trade with these foreigners (Richards et al. 1973).

Key commercial products traded within the kingdom in this period included *matooke* (plantains), beer brewed from matooke, and bark cloth (*Ficus natalensis*), used widely in Buganda ceremonies. Ivory and cattle were traded to, or captured from, markets external to the kingdom. Stanley (1878) described the crop and tree cover on a typical holding in a West Mengo village, giving additional insights into period socioeconomics:

> In it grow large sweet potatoes, yams, green peas, kidney beans, some crawling over the ground, others clinging to supporters, field beans, vetches and tomatoes. The garden is bordered by castor-oil, manioc, coffee, and tobacco plants. On either side are small patches of millets, sesamum, and sugar-cane. Behind the house and courts, and enfolding them, are the more extensive banana and plantain plantations and grain crops, which furnish his principal food, and from one of which he manufactures his wine and from the other his potent pombe [beer]. Interspersed among the bananas are the umbrageous fig-trees, from the bark of which he manufactures his cloth. (Stanley 1878: 300)

The fig tree (*Ficus* spp.) is indigenous, but many of the other crops mentioned by Stanley are not, revealing the impact of trade with Arabs in this period. Many fruit trees found pervasively today were introduced through trade with Arabs at this time (Reid 2002; Tucker 1908). Much speculation exists as to when matooke, characteristic of Buganda, was introduced, but that discussion is beyond the scope of this work.

15.5.1.1 Village Chiefs

Village chiefs (lowest administrative unit) made most decisions regarding resource access, settlement pattern, and allocation of land use including grazing on hilltop grassland, individual cultivation in forested areas, and hunting and extraction of products from forests. Administrators at each successive level could intervene, and ultimate authority was with the territorial governor (ssaza chief) to make decisions about land allocation on behalf of the kabaka. But, day-to-day allocation and resource conflict resolution was the responsibility of the village chief and local clan elders. When decisions of the village chief were not satisfactory to smallholders, appeals could be made to the higher courts or they could migrate to another village (Mukwaya 1953).

Administrative duties of chiefs included collecting taxes for the kabaka, organizing men for military campaigns and big-game hunts, holding councils with constituents over kingdom matters, and assembling labor for constructing royal structures and for clearing and maintaining roads through forests. The number of individuals or households within a village unit was highly variable due to frequent need of village members to temporarily leave to carry out obligations to higher-level chiefs and clan members outside the village and due to smallholder migrations between villages arising from conflicts with village chiefs. Chiefs were very familiar with organizing members of their administrative units for various purposes within the village unit or across the broader kingdom (Richards et al. 1973), including spatial reorganization of holdings and land-use patterns across the village to adapt to the unpredictable shifts in village population that occurred.

Village chiefs may be either clan elders (and thus hold permanent positions) or appointed by higher administrators in the kingdom. They could exact labor and tribute from tenants for their own needs. Exacted tributes included labor to construct housing or fences, food (maize, plantains) to sustain the chief conducting official business, beer brewed from bananas, and bark cloth for official clothing (from *Ficus natalensis* and *Antiaris toxicaria*) (Richards et al. 1973). Any members of the kingdom, smallholders and clan elders alike, could be appointed to these positions, creating considerable potential for vertical and horizontal social mobility for Bugandan subjects. In addition to subsistence activities, smallholders put considerable time into gaining political position and favor among chiefs.

15.5.1.2 Smallholders

Tenure security for smallholders (*bibanja*), or village subjects, was greater within a village whose chief was of the same clan. However, mobility was high in the kingdom, and subjects could relocate to administrative units with more favorable or more powerful clan chiefs (Mukwaya 1953). This may have served to reduce excessive exploitative behavior toward tenants. Smallholders enjoyed relative security over holdings by paying tribute to village chiefs (Mukwaya 1953).

Subjects made decisions over disposal of land and trees on individual holdings while they occupied it. They could leave the holding to a relative, who would assume authority, or if they simply abandoned it, authority returned to the relevant chief. Decisions of use and disposal of grazing areas on hilltops and forests within the village were made by the administrative hierarchy (primarily the village chief). However, at this time, resources were abundant and smallholders within a given administrative unit could typically graze animals and harvest freely. Subjects' rights included (1) the right to undisturbed occupation dependent on correct social and political behavior; (2) grazing rights, water rights, and rights to trees and firewood; and (3) the right to remain in possession of land upon succession (Mukwaya 1953: 14).

Subjects had rights and duties to the village in addition to the obligations of tribute and labor to the village administrator. Members had to join with other members in activities of welfare. Communal activities, coordinated by the village administrator or clan elders, included clearing village paths and kingdom paths that passed through their village and improvement of wells. A well-maintained network of paths was essential for the kingdom's administration of taxation, trade, and mobilizing fighters during times of war, as the kingdom did not have a permanent standing army (Mukwaya 1953).

15.5.1.3 Impacts of Driving Forces and Imposed Tenure Regimes on Village Landscapes

Forest cover (patches of late succession and mature forest) on both the mitala and valley soils of a typical West Mengo village was likely more extensive and more stable in the late nineteenth century than today, as interpreted from writings of early foresters and viewing isolated areas with low populations today. Villages contained a high diversity of land uses, including natural grasslands, agricultural field, home gardens, matooke plantations, and forests in different stages of succession, which persist today.

Land use/cover on the village mitala was likely highly dynamic, as home gardens and plots of annual and perennial crops were almost entirely found in forest clearings on the mitala in this period (Richards et al. 1973). Two key factors likely contributed to high dynamism on the mitalas: smallholders practicing a mix of both shifting and sedentary perennial agriculture across the mitala (Stanley 1878) and frequent movements of households into and out of villages for kingdom obligations, resulting in frequent clearing for site establishment and abandonment for service. Smallholdings were adequate to grow both perennial and annual crops. Trees, especially *Ficus natalensis*, were also encouraged or intentionally planted to provide bark cloth (a tradable commodity), shade, boundary demarcation, windbreaks, or to produce fruit.

Land use/cover on valley soils was likely stable relative to hillsides, with extensive forest patches maintained in advanced to mature stages of succession. Valley forests were used for extractive purposes of timber, fuelwood, hunting, and water but not for charcoal or temporary agriculture at this time. Extensive patches of stable valley forests, likely in different stages of advanced succession, are still present in later study periods, as observed in the1955 aerial photographs (Fig. 15.3) and again in the 2002 satellite image.

15.5.2 Early Colonial Period (1900 to WWII): Village Power Changes, Regional Zoning into Agriculture and Forestry Spaces, and Forest Decline

Colonial planning of West Mengo was driven by a vision of a landed class, similar to England, producing commercial cotton and coffee for European markets. These estates would be surrounded by a patchwork of forests on less arable soils that would provide high-value timber to European markets and ensure adequate rainfall for farming on estates (common belief at this time) (Richards et al. 1973). To achieve these economic goals, all land in the former Buganda Kingdom was formally titled as either mailo land or crown land under the 1900 Buganda agreement between the regents of Buganda Kingdom and the British colonial government. Lands designated

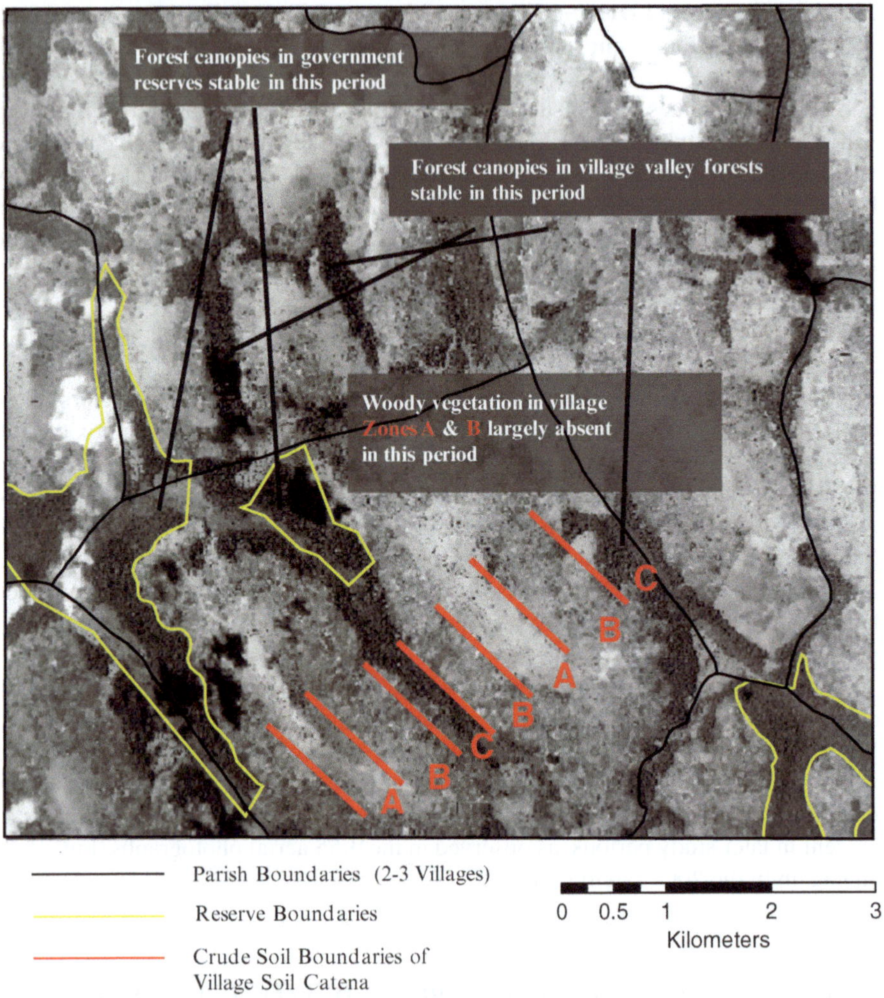

Fig. 15.3 1955 aerial photograph showing village *mitala* (zone B) void of forest and trees, cleared in the early colonial era, and stable forest cover in village valleys

to agriculture, and cotton and coffee production, were overseen by the colonial agricultural department, and lands designated to forestry, and timber production, overseen by the colonial forestry department.

The purpose of the ensuing 1908 Land Law was to emphasize the rights of mailo owners, clearly shifting rights over these holdings from the kabaka to the individual holder. It was an attempt to shift incentives of landlordship from a political to an economic realm (Mukwaya 1953). The immediate effect of granting private ownership, however, was the solidification of the political positions of mailo holders in the kingdom. Land ownership was perceived more as a means to gain political power (e.g., an appointment in the administrative hierarchy) in the kingdom rather than a means for economic gains through cash crop production on newly granted private holdings (coffee and cotton markets introduced by the colonial government) (Richards et al. 1973).

15.5.2.1 Village Chiefs

Limits of formal private holdings created by the colonial government were not conterminous with limits of traditional villages, and village lands in the early colonial era were often owned by more than one mailo owner. Villages still had chiefs, with overlapping duties and privileges with new mailo owners. Chiefs, who were often also village mailo owners, had privileges, as landowners, to exact tributes and labor as chiefs had had in precolonial times. These privileges, as in the precolonial era, freed them from subsistence activities and allowed them to focus energy on matters of the kingdom.

Administrators were still appointed by the administrator of the next level up in the hierarchy and approved by the kabaka and *lukiiko* (parliament). The decision for appointment as an administrative chief, judge, or central kingdom government official was often based on landownership, size of holding, and number of tenants on the holding. These were positive incentives for chiefs to increase tenants on their new private holdings.

Day-to-day decisions regarding access and use of resources—grazing on hilltop grassland, allocation of forests (now private) to individuals for cultivation, and extraction of products from private forests—within the lowest administrative levels were made by mailo owners at this time, rather than the village administrator (on behalf of the kabaka). However, members of village administrative units (mailo owners and bibanja) still had rights and duties to the village administration. For example, appeals to extract forest products from within the now private village forests would be channeled through the mailo owner of the forest, clan elders, or kingdom hierarchy depending on the customary role of the forests (e.g., sacred or productive). That is, there was clear resistance to absolute control of these resources.

Higher-level administrators and clan elders did not have authority to intervene in mailo decisions about use and disposal of forest resources on their holdings. But upper echelons in the kingdom (e.g., ssaza chiefs) maintained some influence over resource uses within lower-level administrative units. Likewise, clan elders maintained similar influence over mailo owners who were not clan elders on land sales, allocations to tenants, and inheritance. They would influence management and use of forest resources practiced by mailo owners, particularly if that forest had traditional significance (e.g., grave sites). And if mailo owners' use of the forest jeopardized the supply of subsistence resources to other members in the village (e.g., high levels of harvesting or conversion to other uses), then clan elders, the village chief, and tenants could organize and intervene, including making appeals to higher authorities (discussed more in Sect. 15.5.2.2), to find solutions.

Since the precolonial era, village members have not used lands outside those allotted to them without permission from authorities. Bibanja would not convert a forest patch to create a personal agriculture field or banana plantation unless permitted by a village administrator. Now, with all forests owned either privately or by the state, the tenants ask permission from either the mailo owner or administrator, although the village administrator no longer has de jure authority to grant such permission in relation to state reserves that may fall within their jurisdiction.

15.5.2.2 Mailo Owners

Mailo owners could still exact labor and food tributes. With the introduction of cash crop production (cotton and coffee), the mailo owners could also exact cash tributes (rent) from tenants. Mailo owners drew profits from their lands in ways other than granting to tenants. Some rented portions of holdings seasonally to immigrants from regions of Uganda outside Buganda Kingdom on a shareholder basis. After harvesting cotton, these laborers would return to their home regions with a share of the profits of cotton sales (Richards et al. 1973). Mailo owners with larger holding sizes, and larger numbers of tenants on their holdings, were often appointed to higher-level kingdom positions. Thus, by increasing the number of tenants and seasonal sharecroppers on private holdings, a mailo owner, like chiefs, could improve political position in the kingdom, which was highly desirable in the early part of the twentieth century.

Over time, however, as sizes of mailo holdings were diminished through division among heirs and upon sales of parcels to tenants, the number of mailo owners increased, and there were not enough administrative or kingdom positions for all of them. Mailo owners with diminishing estate sizes and tenants often had no more political status or control over village resources than many other smallholders in the village at the end of this period.

15.5.2.3 Smallholders

Under the 1900 Buganda Agreement, tenant rights under mailo owners were recognized in a similar, feudal form as had previously existed between tenants and clan elders or administrative chiefs within the local governance units (Mukwaya 1953). Mailo owners could still demand tributes, but payment of rent and tribute also provided tenants with relative security of occupancy. And, if conditions were unfavorable, the tenant could still relocate and obtain land from another mailo owner, similar to the past movements between administrative units. Administrators could no longer evict members of their administrative units using customary law, and mailo owners could evict only tenants on their own holdings. Evictions could be challenged, but adjudication occurred through the existing judicial system that was staffed by mailo owners appointed by the lukiiko, which sometimes disadvantaged the tenants (Mukwaya 1953).

In the first years after the 1908 Land Law, smallholders were mostly prevented from making permanent improvements on their holdings located on village mitala, such as planting coffee plantations, permanent trees, or permanent structures, so the landlords could readily evict or relocate them if they chose to put that portion of the holding to other uses or farm it personally. Also, any mitala land allotted to a tenant that was not converted from forest to annual and banana crop production could be granted to another tenant (MLL5 and SCM7). This was also the case in the precolonial era, but rarely practiced, as land was abundant and administrators had less incentive to actively increase the number of tenants within the administrative unit. Now, as mailo owners increased the number of tenants per holding, the tenants had incentive to keep all parts of their holdings clear of forests in fear of losing them to one of the increasing number of neighboring tenants invited by the landlord.

Cotton production by tenants expanded rapidly when global prices increased following World War I (WWI) (Mukwaya 1953). Entrepreneurial tenants used profits from cotton production to purchase mailo lands and enjoy social and political advancement associated with ownership, including increasing tenants on new holdings (Richards et al. 1973).

Shortly after implementation of the 1900 Buganda Agreement and 1908 Land Law, it became apparent to the colonial administration that the tenants were not legally protected and were being exploited by mailo owners. Rents and labor exacted were ever increasing. In 1928, the law for *busulu* (payment in lieu of free labor) and *envujjo* (tribute on produce) was enacted. Tenants could no longer be evicted unless the land was unutilized for a period of time and formalized through a court order. Owners could not prevent tenants from building permanent houses but could still grant unutilized portions of the holding to another tenant. Upon death, the holding would go to an heir, not the mailo owner. Mailo owners could no longer exact tribute and rent from tenants. Busulu was permanently fixed and legally replaced obligations of cash crop tributes and labor, although they were still obliged to provide food. Customary obligations to the village remained and, again, were coordinated through customary institutions of the clans and administrative hierarchy (Mukwaya 1953).

After 1928, tenants had formal rights to trees other than timber species. Trees with commercial value could be harvested if required for constructing a house on the holding. Tenants could plant trees on the holding, but timber species (with a diameter at breast height [dbh] >20 cm) belonged to the mailo owner, and all trees, and other improvements, would revert to the mailo owner if land was surrendered (Mukwaya 1953). Tenants had great incentive to clear trees before they reached 20 cm dbh in the early part of the twentieth century.

15.5.2.4 Impacts of Driving Forces and Imposed Tenure Regimes on Village Landscapes

A dramatic decline in tree and forest cover occurred on the mitalas of West Mengo villages from the early decades of the colonial era to WWII. Extent of valley forests in villages remained stable, but levels of timber extraction of stems in these stands increased (Fig. 15.3).

The rapid conversion of forests (primarily on mitala portions of villages) in West Mengo at this time is alluded to in a statement by the colonial governor in 1949:

> I have noticed in my travels through Buganda that widespread felling of forest on African-owned land continues. Many small but beautiful patches of forest have already disappeared completely; others are well on the way to vanishing. In addition to the many forests which are being cut out completely and replaced by banana gardens, others are being exploited for timber (Forest Department of Uganda 1949: 9).

Key drivers of rapid clearing of forest/tree cover on the mitala in the early decades of the twentieth century were the perverse incentives created by the specific private tenure regimes the colonial government imposed on West Mengo villages. These regimes created great incentives for the mailo landowners to rapidly increase the number of tenants on the village mitala and require them to clear forests for cotton production, thus increasing rents and political status. Owners also had incentive to prevent tenants from pursuing coffee production or planting trees, as either could be deemed an improvement, and the owner could not evict the tenant, further decreasing tree cover. They did allow banana plantations to fulfill the need for food and beer at kingdom functions but limited tree planting among them. Smallholder tenants had great incentives to convert all forests on their holdings to fields or home gardens, as owners could grant forest patches on a tenants' holding to other tenants.

Between WWI and WWII, forests on the village mitala declined even further, with owners demanding smallholder tenants to further expand cotton production on their holdings in response to increased world demand in these decades. By WWII, only a few small patches of forest remained on mitalas, while others had already been cleared for cotton fields and banana plantations. Trees were not as important in cotton production as they were in banana or coffee production, so even fewer were maintained on village mitalas. However, changes in land and tree rights in 1928 and restrictions on tributes owners could extract limited owners' ability to grant patches of forests on tenant holdings to other tenants or prevent tenants from planting trees. Tenants were able to expand banana plantations and to plant trees, and utilize those up to 20 cm dbh, after which, they were property of the owners. Tree cover on the mitala, among banana plantations and home gardens, was just beginning to recover at the onset of WWII (see Fig. 15.3).

Dramatic forest and tree decline on the village mitala suggests that there was a decline in availability of timber, fuelwood, and non-timber forest products in villages outside of valley forest patches. Dependence on valley forest patches for the supply of these products was likely increasing. At the onset of WWII, the entire demand for forest products from village populations, and the still small urban populations, was supplied through more selective extractivism from the village valley forest patches found on mailo holdings. With total population still low and only selective stem harvesting, these stands remained largely intact with few gaps, though pressure was on the increase (see Fig. 15.3).

15.5.3 Late Colonial to Early Independence (WWII to 1995): Population Growth, Global Markets, Dictators, War, and the Intensification of Agroforestry Systems on the Village Mitala

Domestic and international prices for plantains and coffee, respectively, grew and persisted between 1945 and the mid-1990s, and cotton prices declined, driving increased production of banana and coffee in West Mengo villages and a decline in cotton production. The characteristics of tenure arrangements continued to create incentives to owners to increase the numbers of smallholder tenants in villages. Maintenance of banana-coffee plantations over several decades, and change in tree rights giving smallholders the ability to prevent mailo owners from harvesting stems within their holdings, resulted in a steady increase in tree cover on the village mitalas between 1945 and 1995. In addition to shade for perennial crops, trees on the mitala provided several goods and services on which village smallholders depended for daily subsistence as construction, fruit, fences, and erosion prevention, among others. Once again, village demands for forest products were also supplied from agroforestry systems on the mitala and were no longer exclusively supplied from valley forests in the village as in the early colonial era.

Political changes affecting land use/cover in West Mengo villages in this era were vast and tumultuous. Uganda gained independence in 1962 and started a period as a highly centralized and abusive state. The Buganda Kingdom was abolished in 1967 by a powerful central government that feared the kabakas' vast political influence, to the chagrin of Bugandans throughout West Mengo. A coup in 1971 brought dictator Idi Amin to power, a regime based on terror until 1979 that nationalized all land and industries, including coffee trade. Amin was replaced by an elected government but with similarly poor human rights and disdain for Bugandans and other opposing tribes. Yoweri Museveni staged an insurgency from West Mengo against the government aimed at ameliorating human rights in Uganda among all tribes. In return for support from chiefs in West Mengo for his insurgency, he promised, if he won, the Bugandan Kingdom would be formally recognized again. Thus, many skirmishes and battles occurred in West Mengo through the mid-1980s. Although smallholder casualties were small, there were several periods of different lengths, some permanent, when internal households were displaced as in the precolonial era when tribal conflicts occurred.

15.5.3.1 Village Chiefs

How village chiefs gained their positions, their duties to the government, and how they were compensated changed several times in this tumultuous period. However, their traditional role and responsibilities in day-to-day activities in the village remained largely the same. Resource conflict resolution within a village is still the responsibility of the village administrator (village chief and later resistance councilor), clan elders, and relevant mailo owner. When decisions of the village authorities were not satisfactory to village members (owners or tenants), they could still appeal to the higher courts or authorities as before. Functions of these village leaders still include adjudicating inheritance claims, conflicts over land and forest resources, and walking and demarcating boundaries.

15.5.3.2 Mailo Owners

After WWII, prices of cash crops (primarily coffee) increased, making production very profitable and fertile lands in village mitalas of West Mengo highly desirable (Richards et al. 1973). Size of landholdings for mailo owners had greatly declined as they allotted much of their land to tenants (who could no longer be easily evicted) over the decades prior to WWII. And those holdings not occupied by tenants were further divided among heirs of mailo owners. There was great demand for land but few holdings available of substantial size (Richards et al. 1973). As sums of rent and tributes were fixed, the real value of tenant holdings to mailo owners declined over time, and by the end of the war, there was little incentive for mailo owners to invite more tenants, so they turned to pursuing production themselves on personal holdings unoccupied by tenants. The extent of a mailo holding under personal production of the owner is often no larger than the holding of a tenant.

15.5.3.3 Smallholders

The busulu and envujjo laws of 1928 provided tenants with greater tenure security and enabled production of perennial crops (e.g., coffee and banana). Coffee production remained profitable for both bibanja and mailo owners until the mid-1990s when global prices declined and coffee blight spread through West Mengo (all CEs). Plantain production remains profitable due to steady increases in urban demand. Numbers of smallholders in villages grew substantially due to invitations by mailo owners and increased population growth rates from introduced medicines. However, growth in the number of holders in a village was tempered and in some cases declined, during the insurgency of the 1980s.

More recent land laws have given tenants the right to alienate land with the mailo owner's approval. The tenant can only be evicted if the mailo owner sells the land to another owner who compensates each tenant evicted for the amount they invested in developing the property in addition to the value of the land. Mailo owners still own trees greater than 20 cm dbh on tenant holdings but cannot destroy property (structures or crops) to extract them. In general, land and tree tenure improved for smallholders.

In this period, both tenants and mailo owners had rights to bequeath holdings. Holdings of both types of tenure continue to be subdivided through a process of inheritance, resulting in an increased density of smallholdings practicing the banana-coffee-fruit tree production system. Trees are an important component of this production system, and with improved tree security, a greater number and diversity of trees were planted among coffee and banana plantations over this time period. An increasing variety of trees are also being planted by smallholders in the home gardens of their more secure, well-demarcated holdings (types of trees emerging on the village mitala and their uses in this period are described in Sect. 15.5.3.4). Mailo owners cannot remove these stems among perennial crop plantations.

15.5.3.4 Impacts of Driving Forces and Imposed Tenure Regimes on Village Landscapes

Smallholders expanded tree and perennial crop cover on the village mitalas between 1955 and 1995 (compare Figs. 15.3 and 15.4). Production systems on the mitala changed markedly from a largely treeless cotton-banana system in the early colonial era to a banana-coffee-fruit tree system during this period. This change was a result of increased security of tenants to produce perennial crops and plant trees, increased external demands for coffee and bananas, and increased internal demands for tree products and services. Regeneration also occurred on the mitalas of some villages in the 1980s due to village abandonment during the civil war.

Land use/cover in village valley forests began to change at the end of this period. Valley forest use changed from more extractive uses that maintained the extent of the patch without gap formation to uses including charcoaling, brick making, and sand extraction that would temporarily open gaps in valley forests (see Fig. 15.4). Even though village members began to increase charcoaling and timber extraction from village valley forests for external markets, the increase by smallholders of tree cover on the village mitalas increased availability of and maintained the flow of key forest products within the village unit. Although national law makes these activities illegal, traditional controls are in place. Activities need to be discussed and approved by appropriate mailo owner, smallholders, elders, and chiefs within the village, i.e., not engaged in outside of local rules and norms.

15.5.4 Era of Structural Adjustments (1990s to 2002): Extreme Urban Demand for Forest Products, Coffee Blight, and Increased Spatial and Temporal Land-Use Complexity in Village Valley Forests

One result of structural adjustments in West Mengo was the reduction in funding from the central government to district-level forest and agriculture offices, which limited forest monitoring by the former and extension activities by the latter, giving smallholders, chiefs, and mailo owners even greater de facto control over land-use decisions concerning village landscapes. Global coffee prices declined in the

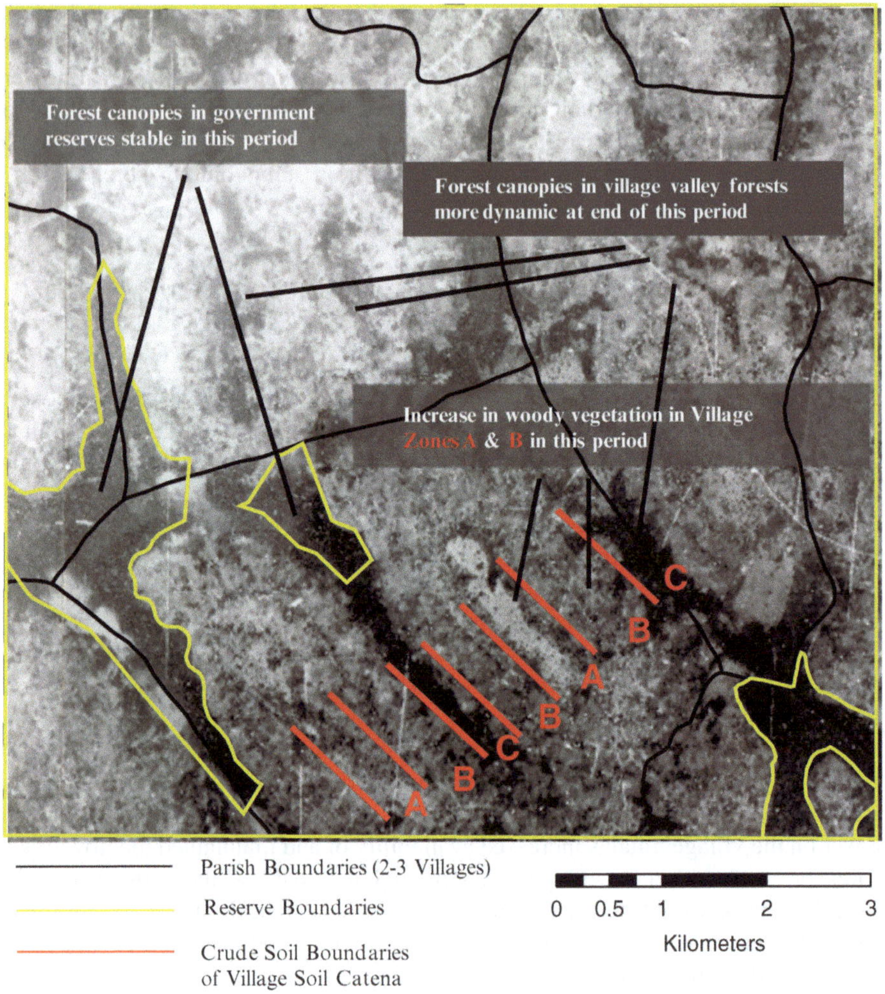

Fig. 15.4 1995 aerial photograph showing increased tree cover and valley forest dynamism on village landscapes compared to 1955 aerial photo (Fig. 15.3)

mid-1990s, and at the same time, coffee blight struck West Mengo and swiftly spread, destroying an important source of cash income on which village members had relied for the previous half century (all CEs). Since then, village members have sought other ways to supplement cash income. Hyper urban growth in this era increased demands from villages for horticultural crops, construction materials (poles, lumber, bricks, and sand), and energy (charcoal and fuelwood), which villagers began to extract more heavily from valley forests.

Extraction of timber for poles, lumber, charcoal, and fuelwood aimed at urban markets produces openings in forest cover. Horticultural crops may be grown both on the mitala during normal growing seasons and in openings in valley forests during the dry season. Sand and bricks are also extracted from soils in openings in village valley forests, but horticultural production and forest regeneration is retarded after this type of extraction. Commercial use of valley forests, previously conducted primarily on village grasslands and mitala, expanded substantially in this period and included both extraction and swidden production in diverse spatial and temporal patterns.

This recent heavy utilization of valley forests in West Mengo is interpreted as predatory. Once again, a new generation of forestry officials and conservationists is lamenting the smallholders of West Mengo and their inability to conserve resources due to their anachronistic land-use systems and traditional institutions, preventing efficient adaptation to the extreme energy demands for fuel.

15.5.4.1 Village Chiefs

The Local Government Act of 1997 clarified the roles of the local councilors and village chiefs. Village-level local councilors (LC1) have the same roles and responsibilities as formerly held by village chiefs, working together with clan elders and mailo owners for day-to-day village administrative activities. However, these lower-level administrators are voluntary, and they do not have as many incentives (tributes, labor, resource control, land grants, etc.) for carrying out responsibilities requested of them by officials in higher-level sectors of government. Requests include managing valley forests exclusively for maintenance of wildlife habitat and the production of high-value timber on long rotation cycles, which are of interest to wildlife and forestry officials still envisioning a West Mengo landscape of distinct agriculture and forestry/wild spaces.

15.5.4.2 Mailo Owners and Tenants

Under the 1998 Land Act, village tenants gained the option to become registered owners of their holdings and receive certificates of occupancy. Mailo owners still may sell unregistered tenant holdings. However, if owners decide to evict unregistered tenants, they must compensate them for the value of land and developments, but this is increasingly rare. Costs of compensation are high and resistance is high, as occupation of land has now passed through generations, resulting in greater ties to a particular village (e.g., gravesites of more ancestors are now often within a single village). Few differences exist between mailo owners and other smallholders in the village, but mailo owners still control larger extents of the ever important village valley forests than non-mailo smallholders.

Even though most lands in each village are formally private, individual smallholders within the enduring clan and administrative structure may still organize to regulate the use or management strategies of subsistence timber and non-timber forest resources on private lands. This mostly occurs if the supply of these products is perceived to be threatened or if they anticipate that it will be threatened after major shifts in land use within the village (e.g., temporary clearance for cash income or more permanent conversion to cash crop production). For example, members of village CE6 obtained subsistence timber and non-timber forest products from both individual holdings and a mailo forest in the administrative unit until the late 1990s. Then, the owner decided to harvest all of the trees for charcoal and commercial wood. Legally, owners can dispose of their forests as they wish. However, social and kinship ties make disposing of resources in a manner that jeopardizes the flow of forest products to village members challenging, as disputes will likely ensue and the threatening (perceived or real) change in use checked. In the above case, forest clearance was discussed with tenants, and they were employed to cut and sell so as to also profit. In this case, members of the village anticipated that they could not rely on substitutes for these products found on individual holdings alone. So, measures were taken to harvest in a way that left some products and some areas untouched after harvesting to regenerate more quickly. Also, villagers needed to monitor the area and prevent any members of neighboring villages to extract products during this period when they might be more vulnerable. The forest owner allowed members to plant seasonal crops in the valley after charcoal production and plant *Maesopsis* sp. on tenant holdings. So they all profited and maintained the flow of subsistence timber and non-timber products. These institutions successfully maintained the flow of timber and non-timber forest products but did so by clearing most of the forest temporarily and relying during that time on the substitutes found on the village mitala.

Despite the coffee blight of this period, both mailo owners and other smallholders in the village continued to intensify agroforestry systems on the village mitala. Key informant interviews were conducted to determine what trees were common on the mitala, how they were managed, and their uses. In these interviews, several tree-cover species were commonly cited as encouraged or propagated on the village mitala for reasons including cultural importance, wind breaks for plantations, improvement of soils, boundary demarcation, fruit, and timber and poles, among others.

Ficus natalensis and *Albizia coriaria* are perceived to enhance soil fertility. *Ficus natalensis* was identified as a historically important species in Buganda, used for administrative boundary demarcation and production of bark cloth while also providing shade for banana and coffee plants, fixing nitrogen, and providing fuel wood. *Maesopsis eminii* and *Markhamia platycalyx* are hardwood species commonly found in smallholdings on the village mitala. These species may be used for timber and firewood and provide shade and wind blocks for banana and coffee. They are not "grade 1" timber species as identified by the forest department but do have desirable attributes for local construction (e.g., tin roof supports), furniture, and tools (e.g., hoe handles). *Eucalyptus* spp. production is also increasing in this study area. It is produced primarily for commercial purposes but also for local uses. Production is not as widespread because they are considered incompatible with coffee and banana production. Thus, they are found as monocrop plantations, typically near the toe-slopes where more moisture is available.

Jackfruit and mango trees were cited as the most common fruit trees on the village mitala together with papaya, avocado, and guava. Fruit trees readily germinate from seeds disposed after human consumption. Thus, with increasing density of households in villages, we observe an increase in fruit tree cover across fields and home gardens on the village mitala. They are compatible with the banana-coffee production system, on rare occasions are used for timber production, and sometimes dead branches are used as fuelwood.

Use of village valley forests by both mailo owners and other smallholders increased and diversified in this period. Smallholders are replacing profits lost from coffee production with profits from the production of charcoal, commercial fuel wood, bricks, and sand from forests and year-round production of banana and horticultural crops. Specific cash crops (e.g., *Amaranthus* sp.) may be grown on valley forest floors in the dry season after timber or charcoal harvesting. There are two growing seasons (January to March and June to August) in West Mengo as the rainfall is bimodal. After one or two dry seasons of production, the forest is left to regenerate.

15.5.4.3 Impacts of Driving Forces and Imposed Tenure Regimes on Village Landscapes

Woody cover on hilltop grasslands continued to expand in this period through a combination of bush invasion and attempts to expand agroforestry practiced on the mitala to the hilltops (Fig. 15.5). Communal cattle grazing remains the most common use of village hilltop grasslands.

Woody cover also expanded on village mitala from growth in the numbers of smallholdings and further intensification of traditional agroforestry systems on those holdings, continuing the long-term expansion of woody cover in villages after the dramatic clearing of forest and tree cover that had occurred in the early colonial era. Trees encouraged and propagated in the agroforestry system not only provide subsistence needs and services to village smallholders but in later stages of tree life cycle are cut and sold in urban markets. Several tree species emerging in the black lumber markets of Kampala, meeting demands of recent urban immigrants, are among those with highest importance values on the village mitalas (see Table 15.1). These species include jackfruit (*Artocarpus heterophyllus*), *Ficus natalensis*, *Markhamia platycalyx*, and *Eucalyptus* spp.

An analysis of tree-cover structure and composition on the mitala lands of West Mengo was conducted based on plot data collected systematically by the NBS across this study area in 2000. A tree of 10 cm dbh or greater was found in 96 of the 114 plots of the settled lands. Average basal area was 1.9 m^2/ha and the average total aboveground woody biomass was 9.5 Mg/ha. Basal area ranged from 0 to 14 m^2/ha and aboveground woody biomass ranged from 0 to 77 Mg/ha. The most important species on the settled lands are primarily fruit trees (#1 jackfruit, #2 mango, #5 avocado), which supports results from the key informant interviews. Rankings of

Fig. 15.5 Increased tree cover in agroforestry systems and home gardens on village *mitala* since the early colonial era (see Fig. 15.3) (Photos by Nathan Vogt, 2002)

the other species identified as important by key informants are #4 *Ficus natalensis*, #12 *Albizia coriaria*, #6 *Maesopsis eminii*, #13 *Markhamia platycalyx*, and #23 *Eucalyptus saligna*. Specific uses of the species found on village mitala, ranked by importance value, are presented in Table 15.1.

A method of using home digital video to analyze percent tree cover on the mitala was also tested in 2002 to complement the NBS plot data (Vogt 2005). Percents of tree cover on the mitala derived from two video mosaics were 23% and 12%. Land that is used primarily for agriculture and contains over 10% tree cover is classified as agroforestry in the FAO land-use classification. Both estimates are considerably lower than the 60% tree-cover estimates found for West Mengo in coarse East Africa (McNeely and Scherr 2003) and global (DeFries et al. 2000) studies of tree cover. However, when we sum both the woody and herbaceous classes in the video dataset, we get estimates of 51% and 60% for the cover of all photosynthetically active material, more in line with tree-cover estimates from lower-spatial-resolution studies. This suggests that East Africa and global studies of tree cover are overestimating tree cover by capturing both perennial crops (e.g., banana and coffee) and tree cover.

Use of village valley forests increased dramatically in this period, and clearings in the forests were widespread across West Mengo (Fig. 15.6). Smallholder land uses after deforestation in valley forests became spatially and temporally complex and little understood. One land-use sequence to meet diverse urban demands for

Table 15.1 Tree species on village *mitala* ranked by importance value

Species	Local name	Uses	Importance value[a]
Artocarpus heterophyllus	*Ffene*	Fruit (food), firewood, shade	0.1278
Mangifera indica	*Muyembe*	Fruit (food), firewood, shade	0.1187
Sapium ellipticum	*Musasa*	Fuelwood, charcoals	0.1038
Ficus natalensis	*Mutuba*	Bark cloth, agroforestry (shade), fodder	0.0883
Persea americana	*Ovakedo*	Fruit (food), firewood, shade	0.0367
Maesopsis eminii	*Musizi*	Timber, fuelwood	0.0354
Bridelia micrantha	*Katazamiti*	Poles, firewood	0.0291
Ficus ovata	*Kokowe*	Firewood	0.0245
Albizia zygia	*Nongo*	Timber, fuelwood	0.0217
Ficus exasperata	*Luwawu*	Fuelwood	0.0200
Aleurites moluccana	*Kabakanjagala*	Ornamental, firewood	0.0200
Albizia coriaria	*Mugavu*	Timber, agroforestry (nitrogen fixing), firewood	0.0191
Pseudospondias microcarpa	*Muziru*	Fuelwood, charcoals	0.0168
Erythrina abyssinica	*Jjirikiti*	Fencing posts, medicine	0.0161
Markhamia platycalyx	*Musambya*	Building pole, timber, firewood	0.0150
Pittosporum mannii	*Nabuluka*	Fuelwood	0.0146
Spathodea campanulata	*Kifabakazi*	Medicinal, firewood	0.0139
Sterculia dawei	*Mutumbwe*	Timber, firewood charcoal	0.0111
Entada abyssinica	*Mwolola*	Medicinal, firewood	0.0108
Margaritaria discoideus	*Kamenyambazi*	Timber, commercial firewood, charcoal	0.0107

[a]Calculations based on Randolph (2000)

valley forest products is identified here (Fig. 15.6). Land use changed from forestry to charcoaling, to *Amaranthus* sp. cash crop production (claimed to provide a cash crop and to improve soils after charcoaling to speed regeneration), to stages of secondary succession (depending on products desired). One land-use pathway is to select only some commercially viable (though lower value and quality) secondary species and allow the remainder to advance to later stages of succession. Another land-use pathway is to clear all stems in the early stages of succession for timber, poles, fuelwood, or charcoaling. Using this pathway, soils are exposed again and cash crops planted, driving very rapid temporal change in land uses and forest-cover dynamism (Fig. 15.6).

15.6 Discussion and Conclusions

This study finds that smallholders in West Mengo increased tree cover, and the spatial and temporal land-use/cover complexity, on village landscapes as demands for a range of food and forest products from both village and urban populations grew

Fig. 15.6 Increasing spatial and temporal land-use complexity in village valley forests: one land-use sequence following deforestation (Photos by Nathan Vogt, 2002)

between WWII and 2002. This runs counter to most regional-level studies of human-environment interactions in the Lake Victoria basin, which often do not account for historical or political-economic context and produce results that smallholders are converting forests to agricultural uses, reducing livelihood security, and increasing siltation in Lake Victoria (Reid 2000).

Long-term tree recovery followed the dramatic decline in forest/tree cover that occurred on village landscapes in the early colonial era. The dramatic forest/tree cover decline was caused by imposition of colonial tenure regimes that constrained traditional, village-level land-use systems and institutions regulating them. The imposed tenure regimes were designed by agricultural department officials to create incentives for chiefs to produce economic activities for commercial cotton and coffee production. However, they created incentives for chiefs to increase the number of smallholder tenants and to prohibit them from planting trees or perennial crops. They created incentives for the new class of smallholder tenants to clear forests on individual holdings and to plant only cotton. This process of forest/tree decline slowed, and reversed, only when the tenure regimes were altered to bring the checks and balances in power of control over natural resources between chiefs, clan elders, and smallholders of the village to arrangements more similar to the precolonial era. The enduring, ecosystem-based land-use systems in West Mengo villages, and traditional institutions regulating them, are diverse and flexible, facilitating efficiency to adapt land-use/cover configurations and settlement patterns to changing and increasing demands for food and forest products at both household and village landscape levels.

Forest clearings created by smallholders likely contribute little to siltation observed in Lake Victoria. Smallholder land uses after deforestation are diverse, including crop production and encouraging forest regeneration to maintain flow of benefits and likely maintain a more closed nutrient cycle than other land uses after deforestation in the region. For example, many valley forests in West Mengo have been permanently converted to urban areas, with impermeable surfaces increasing runoff. Foreign direct investments and land purchases have resulted in the construction of resorts, hotels, and factories on formerly forested lands, also producing greater runoff and unknown impacts on smallholder livelihoods in neighboring villages. Finally, although rare, entire holdings in some villages have been purchased by wealthy and politically powerful (still possible under current law) individuals and converted to pasture, forcing smallholder emigration to cities or to negotiate space in distant villages and creating a landscape more conducive to erosion. More spatially and temporally explicit studies are needed to link the siltation and eutrophication of Lake Victoria to specific land uses causing them, including studies clarifying the diverse land-use/cover changes that occur after deforestation across the region and the impact on smallholder livelihoods in villages adjacent to those landscape simplifications.

Regional models of environmental change in East Africa, of importance to policy makers, can be greatly improved by integrating social and ecological assessments of forest/tree-cover change at the level of traditional territorial units and at greater historical depth. In East Africa, this approach can help clarify enduring legacies of complex, precolonial institutions and land-use/cover systems that persist to help village populations remain flexible to uncertain social (tribal conflicts, cycles of household immigration and emigration, etc.) and biophysical (ancient soils, weather patterns, etc.) forces and important variations in these contexts across the region. Integrated, historical approaches also can provide insights to how modern institutions and technologies may enhance these diverse, ecosystem-based systems that persist across the region to better adapt to emerging external forces of globalization and improve livelihoods. Three countries, and several tribes, with very different histories occupy the Lake Victoria basin, suggesting great variation across the basin in the flexibility (opposite of vulnerability) of land-use systems, constraints to them, and ability to efficiently adapt to changing external pressures. Further comparative historical studies are needed to improve policy prescriptions for sustainable smallholder-environment relationships and livelihood improvements in the Lake Victoria basin.

References

Arnold, M., Kohlin, G., Persson, R., & Shepherd, G. (2003). *Fuelwood revisited: What has changed in the last decade?* (CIFOR Occasional Paper, No. 39). Bogor: Center for International Forestry Research.

Balée, W., & Erickson, C. L. (2006). *Time and complexity in historical ecology*. New York: Columbia University Press.

Banana, A., Vogt, N., Bahati, J., & Gombya-Ssembajjwe, W. (2007). Decentralized governance and ecological health: Why local institutions fail to moderate deforestation in Mpigi District of Uganda. *Scientific Research and Essay, 2*(10), 434–445. http://www.academicjournals.org/SRE. Accessed 10 Oct 2010.

Berkes, F., & Folke, C. (Eds.). (1998). *Linking social and ecological systems: Management practices and social mechanisms for building resilience*. Cambridge: Cambridge University Press.

Berkes, F., Colding, J., & Folke, C. (Eds.). (2003). *Navigating social-ecological systems: Building resilience for complexity and change*. Cambridge: Cambridge University Press.

Brondízio, E. S. (2006). Landscapes of the past, footprints of the future: Historical ecology and the study of contemporary land use change in the Amazon. In W. Balée & C. Erickson (Eds.), *Time and complexity in the neotropical lowlands: Studies in historical ecology*. New York: Colombia University Press.

Cleaver, K., & Schreiber, G. (1994). *Reversing the spiral: The population, agriculture, and environment nexus in Sub-Saharan Africa*. Washington, DC: World Bank.

DeFries, R., Hansen, M., & Townshend, J. (2000). Global continuous fields of vegetation characteristics: A linear mixture model applied to multi-year 8 km AVHRR data. *International Journal of Remote Sensing, 21*, 1389–1414.

Fairhead, J., & Leach, M. (1996). *Misreading the African landscape: Society and ecology in a forest-savanna mosaic*. Cambridge: Cambridge University Press.

Fairhead, J., & Leach, M. (1998). *Reframing deforestation*. New York: Routledge.

Folke, C. (2006). Resilience: The emergence of a perspective for social-ecological systems analyses. *Global Environmental Change, 16*(3), 253–267.

Forest Department of Uganda. (1949). *Annual report*. Entebbe: Author.

Gombya-Ssembajwe, W. (1985). *Proposal for the development and organizational strategy of forestry for basic needs in Uganda*. Master of Science thesis, Australia National University, Canberra, Australia.

Hornborg, A., McNeill, J. R., & Martinez-Alier, J. (Eds.). (2007). *Rethinking environmental history: World-system history and global environmental change*. Lanham: AltaMira Press.

McNeely, J., & Scherr, S. (2003). *Ecoagriculture: Strategies to feed the world and save biodiversity*. Washington, DC: Island Press.

Mukwaya, A. B. (1953). *Land tenure in Buganda: Present day tendencies*. Kampala: Eagle Press.

MWLE (Ministry of Water, Lands, and Environment). (2000). *National forest plan*. Kampala: Author.

Ostrom, E. (2005). *Understanding institutional diversity*. Princeton: Princeton University Press.

Ostrom, E. (2009). A general framework for analyzing sustainability of social-ecological systems. *Science, 325*(5939), 419–422.

Otsuka, K., & Place, F. (2001). *Land tenure and natural resource management*. Washington, DC: International Food Policy Research Institute.

Randolph, J. C. (2000). *Forest ecology and management*. Bloomington: School of Public and Environmental Affairs, Indiana University.

Reid, W. (2000). Ecosystem data to guide hard choices. *Issues in Science and Technology, Spring, 16*(3), 37–44.

Reid, R. (2002). *Political power in pre-colonial Buganda: Economy, society & warfare in the nineteenth century*. Oxford: James Currey.

Richards, A. I., Sturrock, F., & Fortt, J. M. (Eds.). (1973). *Subsistence to commercial farming in present-day Buganda: An economic and anthropological survey*. Cambridge: Cambridge University Press.

Scoones, I. (1999). New ecology and the social sciences: What prospects for a fruitful engagement? *Annual Review of Anthropology, 28*, 479–507.

Stanley, H. M. (1878). *Through the dark continent*. London: Sampson Low, Marston & Company.

Tucker, A. R. (1908). *Eighteen years in Uganda and East Africa*. London: Edward Arnold.

Vogt, N. (2005). *Mechanisms of land-cover change in Uganda: Longer-term analyses of the role of institutional arrangements*. PhD dissertation, School of Public and Environmental Affairs, Indiana University, Bloomington.

Vogt, N., Bahati, J., Unruh, J., Green, G., Banana, A., Gombya-Ssembajjwe, W., et al. (2006a). Integrating remote sensing data and rapid appraisals for land-cover change analyses in Uganda. *Land Degradation & Development, 17*, 31–43.

Vogt, N., Banana, A., Gombya-Ssembajjwe, W., & Bahati, J. (2006b). Understanding the stability of forest reserve boundaries in the West Mengo forest region of Uganda. *Ecology and Society, 11*(1), 38 [online]. http://www.ecologyandsociety.org/vol11/iss1/art38

Chapter 16
Was Agriculture a Key Productive Activity in Pre-Colonial Amazonia? The Stable Productive Basis for Social Equality in the Central Amazon

Eduardo Góes Neves

Abstract Despite a long-standing debate on its intensity and economic role, it has been generally assumed that swidden or extensive agriculture was an important economic component for Amazonian pre-colonial tropical forest societies. However, the available data do not back such a claim. In the central Amazon, the archaeological record shows that the establishment of sedentary societies in the area is quite recent, going back to no longer than 500 BC. Despite good conditions of preservation, so far, no evidence of manioc cultivation has been found during a record of 2,000 years of human occupation. In other areas such as Marajó Island, at the mouth of the Amazon, no evidence of agriculture whatsoever has been found so far, despite the presence of artificial earth mounds and elaborated pottery. Such evidence, when put together, suggests that although plant domestication may have been very ancient in the tropical lowlands, the advent of predominantly agricultural-based economies was much more recent. It is proposed here that agriculture was much more an opportunistic activity based on the intense and sophisticated management with stone axes of gardens and forest under different stages of ecological succession than the pattern of extensive cultivation with metal axes or chain saws of the large manioc gardens known today. If true, the hypotheses have as a corollary that the emergence of the pattern of extensive manioc-based agriculture typical of the tropical forest results from demographic and technological changes brought by the European conquest.

E.G. Neves (✉)
Museu de Arqueologia e Etnologia, Av. Professor Almeida Prado, 1466,
Cidade Universitária, São Paulo, SP, Brazil 05508-070
e-mail: edgneves@usp.br

E.S. Brondízio and E.F. Moran (eds.), *Human-Environment Interactions: Current and Future Directions*, Human-Environment Interactions 1,
DOI 10.1007/978-94-007-4780-7_16,

16.1 Introduction

It is widely accepted today that the occupation of the Amazon Basin is as old as the occupation of other parts of South America (Barse 2003; Meggers and Miller 2003; Roosevelt et al. 2002). This evidence contradicts hypotheses that proposed the occupation of tropical forests would not be possible by populations with economies not based on intensive agricultural food production (Headland and Bailey 1991; Roosevelt et al. 2002). It is also clear that the Amazon was also a stage for early cultural innovations in South America. Among these, there is independent production of pottery, found in shell mounds spread east of the mouth of the Amazon river, with dates going back to ca. 3,500 BC (Roosevelt 1995; Simões 1981). Other sites, located in the lower Amazon, may have even earlier pottery, dating back to 6,000 BC (Roosevelt 1995; Roosevelt et al. 1991), although their chronological context is not totally clear, and it is unclear whether they were an independent innovation that developed earlier in the Amazon than in the rest of the continent (Meggers 1997).

At any rate, there is good ground to affirm that the Amazon Basin was a stage for early cultural development in South America, in some cases preceding other areas. However, by the time of the arrival of the first European colonizers, in the early sixteenth century, highly diversified patterns of social and political organization were described across the continent. Among such plethora, there were centralized states, powerful chiefdoms, and bands of highly mobile foragers. In the central Amazon, early chroniclers describe densely occupied settlements under the leadership of paramount chiefs who could mobilize individuals dispersed over different villages in a multitiered hierarchy resembling chiefdoms. The archaeological data of some areas in the Amazon support such early colonial reports (Heckenberger et al. 2003, 2008). The explanation of the history of the development of the different social formations prevailing in South America in the sixteenth century has been the major task of the archaeology done in the continent for the last century. Why did comparatively more centralized and hierarchical social formations develop in areas such as the central Andes, the Peruvian coast, and parts of northern South America and apparently did not develop elsewhere in the continent in places such as the Amazon? This question is almost as old as Amazonian archaeology itself, and several kinds of explanations, ranging from materialistic- to ideological-based hypotheses, have been presented to answer it. Other authors, on the other hand, propose that Amazonian pre-colonial social formations were indeed centralized and hierarchical, being comparable to other areas of the world where states or empires developed.

In the long history of occupation of the Andes and adjacent areas, one sees the alternation of horizons of cultural or political integration by intermediate horizons of dispersed, localized, or autonomous political groups. In the Amazon, a similar pattern can be visualized albeit not with such widespread distribution. Different from the Andes, however, one sees in the Amazon an amazingly wide distribution of distinct languages and linguistic families without major physical barriers such as deserts or high mountain ranges. In this chapter, I will suggest that, in the Amazon, great linguistic diversity may be used as a proxy to understand the process of plant and landscape domestication that happened in the early and mid-Holocene. The argument will be further expanded, as I will also propose that great linguistic diversity be used as an indicator of political fragmentation and decentralization across the Amazon.

16.2 Language Dispersals and the Expansion of Agriculture in Tropical Lowland South America

In lowland South America, the association between patterns of language distribution and the expansion of ethnic groups in the past was initially proposed, almost 100 years ago by Max Schmidt. Erland Nordesnkiöld, in his brief but insightful synthesis of Amazonian archaeology (1930), took that correlation further, proposing an association between the wide distribution of incised-modeled ceramics decorated with bird heads, found from Trinidad, in the Antilles, to the delta of the Paraná River. For him, such wide distribution resulted from the fact that, in lowland South America, the three major river basins—Orinoco, Amazon, and Paraná—are all geographically integrated.

It was Donald Lathrap, however, who more elegantly formulated a series of hypotheses proposing the association of the patterns of distribution of languages from the Arawak and Tupi families to agricultural expansions in the past. Together with his former graduate students José Brochado and José Oliver, Lathrap, influenced by Schmidt, Nordenskiold, and Carl Sauer, suggested that the central Amazon would have been an early center of population growth and dispersal in the American continent (Brochado 1984; Lathrap 1970, 1977; Oliver 1989). For these authors, such dispersals resulted from a process of population growth and agricultural colonization of the fertile floodplains of the Amazon Basin leading eventually to the occupation of other alluvial and non-alluvial settings in South America. Perhaps, the great merit of this so-called cardiac hypothesis has been the fact that it proposed an actual mechanism for diffusion. In this case, diffusionism was not employed as an explanatory mechanism whose causes were unknown. Rather, the expansion of people, languages, and ceramics was seen as a result of population growth of well-adapted agricultural groups colonizing contiguous areas in alluvial settings. Lathrap's brand of diffusionism was in many ways similar to the "demic diffusion hypothesis" proposed by Ammerman and Cavalli-Sforza (1984) to explain the distributions of languages and genetic frequencies in the European Neolithic.

Lathrap's work focused more on the dispersal of peoples who spoke languages from the Tupi-Guarani, Arawak, and Pano linguistic families. For him, such population and language dispersals would be correlated to the expansion of ceramics of the polychrome tradition, in the case of Tupi-Guarani and Arawak, and of the

Cumancaya tradition in the case of the Pano family. Later studies in the central Amazon, after the publication of Lathrap's original hypothesis, verified that the archaeological record of the area did not match his expectations (Heckenberger et al. 1998; Neves 2006). Moreover, the correlation of ceramic complexes, or material culture in general, and language groups has been severely criticized by the processual approach of the 1960s to the 1980s, rendering such kind of archaeology almost obsolete for some.

The problem with this criticism, however, is that such perspective was retaken in a series of works done in different parts of the world, most notably the Pacific (Kirch 2000) and the European Neolithic (Anthony 2007; Renfrew 2000). Such approach, denominated in the "farming-language dispersal hypothesis" (FLDH), proposes that the distribution of some of the most widespread language families today result from demographic dispersals resulting from the adoption of farming by different populations of the world. Thus, the dispersal of the Lapita complex of objects, including ceramics, in Melanesia and eastern Polynesia would correlate to the early expansion of Austronesian speakers in the area. In the same way, the expansion of Linearbandkeramik pottery in Europe would correlate to the expansion of early farmers coming from Anatolia and similarly with Bantu languages in Sub-Sahara or the Arawakan colonization of the insular Caribbean. All these cases demonstrate that the FLDH is a powerful paradigm in archaeology today, recycling some of the cherished themes of cultural-historical archaeology—such as the use of diffusionism as an explanatory mechanism or the correlation between the distribution of languages and groups of artifacts—that were almost abandoned by archaeology.

For several reasons, the archaeology of lowland South America could provide a good testing ground for FLDH. Such an attempt, however, was never taken. Among the reasons calling for this testing is that the area has one of the widest distributions of languages and linguistic families in the world. For instance, while most contemporary European languages belong to a single language family, the Indo-European, there are in lowland South America at least four large families with continental-scale distributions—Arawak, Tupi-Guarani, Carib, and Gê—together with several other families with large regional distributions, such as Pano and Tukanoan, and several isolated languages apparently with no established connection with other language families in the area. Another reason for testing the strength of FLDH in lowland South America stems from the fact there has never been large state-like social formations in the area. It is known that these social formations can have a skewing effect in the distribution of languages on a continental scale, such as what happened with Quechua in Andean South America or Latin in Europe. So, whichever were the means for language dispersal in lowland South America, the development of the state was not one of them.

In order to test the FLDH, a series of assumptions have to be accepted. First, one needs to be willing to accept that there is, to some measure, a positive correlation between language variability and the variability in the archaeological record. In order words, since languages cannot be excavated and since there were no writing systems known in pre-colonial Amazonia, one could use the variability in the archaeological record as a proxy for language variability in the past. Such assumption, although necessary to address the questions raised here, is extremely complex. Since at least the 1960s, there have been many case studies demonstrating that there is no universal correlation between language and material culture. Such studies have indeed generated some of the major theoretical changes undergone by Anglo-American archaeology in the period: the development of the processual and post-processual approaches.

Moreover, in the particular case of Amazonia and northern South America, the ethnographic and ethnohistorical literature is full of evidence that in the sixteenth century, and in some areas until the present, local indigenous groups were regionally integrated in multiethnic networks including specialized production and exchange of goods, the mobilization for warfare, and the periodic condensation into hierarchical chiefdom-like social formations. These social formations were multilinguistic, sometimes with the development of lingua francas or pidgins, despite patterning in material culture generated by exchange networks. It is likely that many of the Amazonian social formations in the 500 years that preceded the European conquest had this general structural pattern (Neves 2008).

How, then, can archaeologists working in the lowland South American tropics, where ceramic artifacts and their distribution patterns are the primary archaeological record, establish a long-term history of indigenous peoples before the arrival of the Europeans? The answer to this question may rest in the identification of the historical contexts where positive correlations between the variability in the archaeological record and languages could be stronger. Which contexts could these be? First, there are the cases of colonization of previously empty areas (Renfrew 2000). This was, for instance, what happened in western Polynesia, where an association between the Lapita complex, identified by patterns in the archaeological record including rock-stamping pottery, and a branch of the Austronesian language family was established (Kirch 2000). Another potential context for this correlation could be the initial decades or centuries of occupation of a previously settled area by external populations arriving with a new technology or a different political, religious, or ideological system (Renfrew 2000). This is what happened in the insular Caribbean with the early colonization of Arawak-speaking groups bringing with them Saladoid pottery and settling in ring-shaped villages dating back to ca. 500 BC (Petersen 1996; Rouse 1992). This was also the case of the colonization of the Atlantic shore of eastern Brazil by the Tupinambá Indians who spoke a language of the Tupi-Guarani family and are associated with sites bearing a distinctive pottery with polychrome decoration. These groups, who arrived in the area around the beginning of the Christian era or even earlier, completely replaced the shell-mound builders who had lived there for many millennia, in some cases since the beginning of the Holocene. In both cases, the replacement can be explained by the fact that the newcomers brought with them a different technology—agriculture. In the particular context of the Tupinambá, there was also a political system strongly based on warfare, captive taking, and cannibalism that was firmly associated with the expansion of these groups.

Turning back to the Amazon, it could, in light of the previous discussion, be possible to identify a historical context where a stronger correlation between ancient languages and patterns in the archaeological record could be established. In most of the Amazon, such context developed in the period ranging from ca. 1,000 BC to AD 500. This was the time when a true cultural explosion happened in the area, marked by the replacement in some areas of long-established lifestyles from the early Holocene by a basic general pattern of economic and social organization that prevailed until the arrival of the Europeans and in some cases until today. In accordance with FLDH, these changes were probably brought by the expansion of agricultural-based societies over areas previously occupied by societies with economies based on a wide range of resources, including the cultivation of domesticated plants, fishing, collecting, and agroforestry. Contrary, however, to what was verified in Europe, Polynesia, and Sub-Saharan Africa, one does not see in tropical lowland South America the prevalence of the expansion of one language family over wide expanses. Rather, what one sees is a mosaic-like pattern with several language families and many small families or isolated languages distributed on a continental scale.

16.3 Domestication, Agriculture, and Climatic Changes in the Late Holocene

The explanation for this pattern can be sought in two factors: first, in an important distinction that has to be made between domestication and agriculture in the archaeology of tropical lowland South America and second, in the potential correlation between climatic and social changes starting around AD 1,000 in this area. I suggest that plant domestication and forest management started quite early, at the beginning of the Holocene, in tropical lowland South America. Such practices involved the replanting of seedlings and the cultivation in house gardens of economically useful plants, including medicines and other kinds of drugs. Overall landscape domestication can be defined as the "conscious process by which human manipulation of the landscape results in changes in landscape ecology and the demographics of its plant and animal populations, resulting in a landscape more productive and congenial for humans" (Clement 1999: 190). During this period, however, plant cultivation never quite became the major source of resources for the populations settled in this area. This is inferred from the small frequency, and in some cases even the absence, in the archaeological record of evidence of large and sedentary settlements across the Amazon dating prior to ca. 1,000 BC. The exceptions come from Marajó Island at the mouth of the Amazon, where large Ananatuba-phase settlements date back to ca. 1,200 BC (Schaan 2004; Simões 1969); the lower Tapajós River at the lower Amazon, where ring-shaped villages dating from 1,400 BC were identified (Gomes 2008); and the upper Madeira Basin close to the current borders of Bolívia and Brazil, where there is evidence of continuous occupation throughout the Holocene, in what seems to be the longest continuous sequence known for the Amazon (Miller 1992, 1999).

Elsewhere in the Amazon, one sees, from ca. 1,000 BC onward and in many cases already in the Christian era, what appears to be the sudden evidence of human occupation after long intervals in the mid-Holocene with no or very scant signs of human occupation (Neves 2008). These changes may be correlated to more mobile societies with economies focused on, and opportunist exploitation of, natural resources being replaced by societies with fully sedentary lifestyles based on economies more dependent on agriculture or the intensive management of natural resources such as fish banks or aquatic corrals adjacent to villages. A major consequence of such replacement was a considerable increase in archaeological visibility. All of the currently known signs of landscape transformation in the Amazon date from this period (Neves and Petersen 2006). Among them are the artificial earth mounds of Acre and Marajó Island (Meggers and Evans 1957; Roosevelt 1991; Schaan 2004, 2008), the anthropic and fertile dark soils known as *terras pretas* (Neves et al. 2003; Petersen et al. 2001), the large sites at the mouth of Tapajós River (Gomes 2002), the cluster of large villages connected by roads in the upper Xingu Basin (Heckenberger et al. 2003, 2008), the ring villages of central Brazil (Wüst and Barreto 1999), and so forth.

The fact that these distinct events happened in sequence, and almost simultaneously, at least from a long-term perspective, in the first millennium AD, calls for a common causal explanation for them. The point to be made here is that these different places or regions were so directly or indirectly connected to each other that any major changes in terms of social and political organization in one area could have had implications in terms of political organization in the other ones. The historical record of the Amazon has some examples of it, provided that one is comparing contexts of state-based societies in the early colonial era with non-state societies in the pre-colonial area. It is known, for instance, that in the eighteenth century, the establishment of the English, Dutch, and French at the Guianese Coast and the Portuguese in the central Amazon generated a series of conflicts among indigenous groups that were located far away from the settlements of these European powers (Dreyfus 1993). It is likely that these conflicts did not happen randomly but that they profited from previous networks of trade and even warfare among the same groups.

One should, however, turn to other sources of evidence to try to explain the sudden changes in the archaeological record of the Amazon visible from ca. 1,000 BC onward. The figure of 1,000 BC is somewhat arbitrary, but it may help to provide a benchmark against which one could establish a comparison with previous periods. One such source could be the paleoclimatic record. It may be useful to verify whether there were distinctive climatic or ecological changes happening prior to this period that could be associated to the social transformations verified in the archaeological record. The point here is not to search for causal relations between climatic changes and social changes but to check whether there could be a correlation between these phenomena. The search for these correlations is, however, problematic. The paleoecological record of the Amazon is still poorly known: The traditional methods of paleopalinology, for instance, which privilege the coring of fossil lakes, isolated from the main stream of rivers, left the main floodplain of the

Table 16.1 Climate changes in the Amazon Basin

Area	Evidence	Chronology (cal years BP)	Source
Middle Caquetá River (Western Amazon)	Drying	11,500–4,700	Behling et al. (1999)
	Increase in rainfall	3,000	Berrío (2002)
Loma Linda (Eastern Colombia)	Increase in rainfall	3,600	Behling and Hooghiemstra (2000)
Porto Velho-Humaitá Transect (SW Amazon)	Drying	9,000–3,000	Freitas et al. (2001)
Rio Beni, Bolivia (SW Amazon)	Forest expansion	2,000	Burbridge et al. (2004)
Titicaca Lake	Increase in humidity	2,000	Baker et al. (2001)
BR-174, Manaus (Central Amazon)	Drying	7,700–3,000	Piperno and Becker (1996)
Caxiuanã (Eastern Amazon)	Increase in rainfall	2,700	Behling and Lima da Costa (2000)
Caeté Estuary (Eastern Amazon)	No mangrove vegetation	5,900–2,800	Souza-Filho et al. (2008)
Estuary (Eastern Amazon)	General decrease in mangrove species	5,600–3,600	Behling (2002)
Estuary (Eastern Amazon)	Reduced discharge of the Amazon River	8,000–5,000	Toledo and Bush (2008)
	Resume of sedimentation in Amazonian floodplain	1,710	

Amazon and its adjacent areas unsampled. This situation is beginning to change now, and new methods, such as the search for carbon isotopes in stable organic matter of soils, are applied. Accordingly, sampling for fossil pollens is also being done directly from soil or alluvial lakes as well. These limitations notwithstanding, it is possible to compile the available literature to verify whether climatic changes are visible in the transition from the middle to the late Holocene. There was a trend toward a general increase in humidity and of forest expansion in different parts of the Amazon starting ca. 1,500 BC. The nature of these changes is not well known, but it is likely that one is dealing here with drastic transitions from, say, parkland savanna open areas to close, tall-canopy equatorial forests (Table 16.1).

What is important for the argument presented here is that these ecological changes may have created conditions for the spread of the kind of economic strategies later denominated as "tropical forest pattern" across the whole of the Amazon. Such pattern, initially described in *Handbook of South American Indians* (Lowie 1948), would have as one of its major characteristics the slash-and-burn cultivation of manioc in itinerant gardens. Indeed, manioc cultivation is so widespread today in the lowland tropics that it is hard to imagine agriculture without it in those areas. The paleobotanical data seem to back that claim, since there is evidence that manioc was already under cultivation along Rio Porce in the tropical northern Andes of Colombia, ca. 7,000 years ago.

Interestingly enough, however, there is so far little, if any, direct evidence of pre-colonial manioc cultivation in the Amazon. The study of chipped stones from griddles of the upper Orinoco area of Venezuela has shown that these artifacts were used for the grating and processing of a number of roots and tubers, including *Dioscorea* (Perry 2005). In the central Amazon, despite good conditions of preservation, so far, no evidence of manioc cultivation has been found during a record of 2,000 years of human occupation. Moreover, in areas such as Marajó Island, no evidence of agriculture has been found so far, despite the presence of artificial earth mounds and elaborated pottery (Schaan 2008). Such evidence, when put together, suggests that although plant domestication may have been very ancient in the tropical lowlands, the advent of predominantly agricultural-based economies was much more recent. The data also show that even in these latter cases, it was likely that agriculture was much more an opportunistic activity based on the intense and sophisticated management of gardens and forest under different stages of ecological succession than the pattern of extensive cultivation with metal axes or chain saws of large manioc gardens known today (Denevan 1996).

If this is true, alternations from more agricultural to more hunter-gathering lifestyles were quite common in the past, and, more important, the emergence of the pattern of reliance on extensive manioc-based agriculture consolidated in the literature as "the" tropical forest pattern resulted from the demographic and technological changes brought by the European conquest. Based on these considerations, it is probably better to denominate such tropical forest pattern more as "agroforestry" than as "agriculture." Agroforestry can be described as the combination of mixed strategies that include the cultivation of domesticated plants in gardens and the management, in areas of fallow or in other places dispersed through the forest. Agroforestry is, in a way, opportunistic, since it profits from areas already opened in the forest, connected by paths, old gardens, or large trees felled by the wind to establish new cultivation areas (Clement 1999). Agroforestry is also intensive in the sense that it invests a lot of time and energy in the tending or cultivation of the gardens or individual trees over longer periods. The technology involved in these agroforestry systems demanded few instruments, among them stone axes and digging sticks. As important as these tools was the knowledge of the forest and its plants by the Indians. The combination of opportunism and intensive care and use of stone axes probably resulted in a redundancy of cultivation and occupation in the same spots in single regions, creating "anthropic islands" in a sea of forests. This was the case of the central Amazon, where there is evidence of long-term occupation of single sites with spans of several centuries, generating terras pretas in the process (Arroyo-Kalin 2008; Neves and Petersen 2006).

If agroforestry was opportunistic, the structure and composition of gardens and managed areas also change a lot across the Amazon. The scant evidence available so far tends to support this hypothesis: In the central Amazon, there is so far no evidence of cultivation of large gardens, not even of manioc, in the lower Tapajós area; on the other side, the dispersal of large areas of the so-called *terras mulatas*—dark brown soils of human origin but not associated with artifacts such as pottery—

suggests that these areas were used as gardens. In the upper Xingu, Heckenberger (2008) has demonstrated through ethnographic analogy that manioc was the staple that sustained the dense settlements of the area at the transition from the first to the second millennia AD.

The upper Madeira Basin, next to the current border between Brazil and Bolivia, provides an interesting case study where a long-term trend can be identified. Botanical evidence suggests that this area was the initial center of domestication of manioc and the peach palm (*Bactris gasipaes*), the only fully domesticated palm in Amazonia. This is also the area where the oldest-known evidence for sedentary occupations in the Amazon, going back to ca. 2,500 BC, has been identified. This evidence is supported by the presence of strata of anthropically created terras pretas with that age (Miller 1992, 1999). Finally, this is the place where one of the longest-known uninterrupted sequences has been identified in Amazonia, spanning almost the whole Holocene (Miller 1992). It is thus likely that the upper Madeira was a center of cultural innovation and the development of a particular agroforestry system based on the cultivation of manioc and peach palms, among other plants, throughout the Holocene. Interestingly, the linguistic data also suggest that this area was the center of origin of the Tupi stock. As research progresses in the Amazon, it is likely that other such core areas generating particular agroforestry systems with long-term sequences in the Holocene also will be identified. Interestingly, the expansion of the upper Madeira system, attested by the expansion of sites with ceramics from the polychrome tradition, dates from the middle to the end of the first millennium AD.

As already mentioned, lowland South America has a remarkable linguistic diversity. There is no single linguistic family that dominates the area on a large scale in the same way as the Indo-European language in Europe or Bantu in Sub-Saharan Africa. Such diversity probably resulted from a conjunction of these factors:

- Agroforestry systems of opportunistic and variable nature developed in the area, without the prevalence of one system over the other.
- The climate changed in the transition from mid- to late Holocene, which probably triggered a stronger reliance on these diverse agroforestry systems and the establishment of large sedentary settlements across the area.
- No social formation associated with a particular language was strong enough to politically spread itself on a large scale.

The remainder of this chapter will present data from the central Amazon that builds on the above argument. I will suggest that, in the central Amazon, the alternation between political centralization and fragmentation mirrors, at a local scale, the same overall pattern of linguistic and cultural diversity, or fragmentation, verified across the Amazon. The underlying explanation for these distinct phenomena rests in the opportunistic use of an ample resource base, in the case of the central Amazon in a setting where natural resources are abundant and relatively predictable, coupled with an easily available technological complex. Such factors prevented the emergence and reproduction of institutionalized political centralization.

16.4 Complexity and Abundance in Pre-colonial Amazonia: A View from the Central Amazon Basin

Research done during the last decade in the central Amazon may help to illuminate long-term dynamics of political centralization and fragmentation that could be used as reference for other cases in the Amazon. Human occupation of the central Amazon has been characterized by the alternation of long-term episodes of stability interrupted by what seems like abrupt events of rapid change. This pattern can be associated with different sources: Gaps in the early part of the chronology may be associated with mid-Holocene climatic change, whereas the changes seen around the early second millennium AD are associated with the replacement of ethnic groups in the area. A regional survey allowed for the identification of around 100 archaeological sites in an area located between the cities of Manaus and Coari. This number, by no means an underestimation of the total number of sites occurring in the area, is a sample resulting from three different regional projects: the Central Amazon Project, the Archaeological Survey of the Coari-Manaus Pipeline, and the PIATAM environmental assessment project. The historical reconstruction presented here results from the excavation and mapping of 14 sites: Dona Stella, Açutuba, Laguinho, Hatahara, Lago Grande, Osvaldo, Pilão, Antonio Galo, Lago do Limão, Jacuruxi, Nova Esperança, Lauro Sodré, São Paulo II, and Perpétuo Socorro. It shows a trend of continuous and long-term occupation of the area from ca. 500 BC to ca. AD 1,500. During this period, there were noticeable changes in settlement size and shape, as well as in patterns of ceramic decoration, allowing for the establishment of a chronology comprising four distinct components: Açutuba phase from ca. 400 BC to AD 400, Manacapuru phase from ca. AD 400 to AD 900, Paredão phase from ca. AD 700 to ca. AD 1,250, and Guarita phase from ca. AD 900 to AD 1,500.

The regional chronology shows that there are interpolations among the occupations: at least of 200 years between Manacapuru and Paredão and more than 300 years between Paredão and Guarita. Such interpolations in the sequence show that people making different pottery and occupying villages of different shapes lived not far from each other, in some cases during several decades, in the first millennium AD in the central Amazon. Such was, for instance, the case of Osvaldo and Lago Grande sites, located less than 10 km from each other, adjacent to the complex of lakes that compose the floodplain of the Solimões (Amazon) River. Osvaldo is a single-component Manacapuru-phase site occupied during the seventh century (Chirinos 2007; Neves et al. 2004), whereas Lago Grande is a multicomponent site with a long Paredão-phase occupation dating from the end of the seventh century to the beginning of the eleventh century (Neves and Petersen 2006). The small but constant amount of Manacapuru shards on Lago Grande and vice versa on Osvaldo indicates at least two different possibilities: first, that there was trade in pottery associated with these contemporary occupations and second, that these local groups were regionally integrated in patrilocal exogamous regional systems not unlike those verified today in areas such as the northwest Amazon (Jackson 1983).

Together with the evidence of trade or exogamy, or both, Manacapuru- and Paredão-phase sites also share another distinctive feature: a general ring or horseshoe shape (Moraes 2007). Ring sites are quite common in the archaeology and ethnography of central Brazil, where they appear around the eighth century AD (Wüst and Barreto 1999). Such sites or villages are normally associated with Gê-speaking groups such as the Kayapó and Bororo, described in the first half of the twentieth century by the likes of Curt Nimuendajú and Claude Lévi-Strauss, although the introduction of this pattern in the area probably resulted from the cultural influence of Arawak groups arriving from the west (Heckenberger 2002). Regardless of their history in central Brazil, ring villages were not generally known archaeologically or ethnographically in the Amazon. The archaeological data now being uncovered show that such patterns prevailed in the central Amazon in the second half of the first millennium AD.

Manacapuru- and Paredão-phase sites were sedentary and quite large. They are associated with terras pretas, also known as anthropic dark earths (ADEs). These soils, which are quite fertile and sought after by farmers today, were formed through the deposition of slow-burning charcoal, food residues, and other organic remains in long-term, stable occupations. ADE sites have deposits more than 200 cm deep, densely packed with ceramic remains. They also have features such as artificial mounds made by the piling of soil and pot shard strata. Some of these mounds reach more than 3 m in height and 20 m in length. The functions of these artificial mounds are not yet clear, but it is plausible that they were built as platforms for houses or longhouses. Their construction required in some cases the mobilization of labor, involved in digging, carrying, and piling of soil, and the accumulation of pot shards. In this sense, they can be considered monumental, but there is nothing in them that indicates large-scale planning or engineering. A potential explanation is that these structures were built in events of feasting with the collective participation of individuals of different communities who would contribute labor, beer, and other products or services.

For the argument of this chapter, the evidence presented so far can be hypothetically interpreted in the following way: Manacapuru- and Paredão-phase societies were regionally integrated into exchange networks that included trade and exogamy. Despite differences in pottery, the sharing of a basic village layout shows that such a regional pattern could probably include other commonly shared traits in material culture in, for instance, featherwork or basketry. Regional systems such as this were described historically or ethnographically in areas such as the northwest Amazon, the upper Xingu Basin, the middle Orinoco Basin, and the lower Antilles. Such systems are normally multiethnic, although they tend to be associated with Arawak-speaking groups (Heckenberger 2002; Hornborg 2005). One of their defining trends is that such regional systems did not allow for the emergence of permanent or stable regional hierarchies.

Around the early first millennium AD, one notices visible changes in the occupation history of the central Amazon. The most visible of these changes correspond with a rupture in the ceramic sequence, where Paredão ceramics disappear to have been replaced by occupations with Guarita ceramics. Guarita pottery is remarkably different from both Paredão and Manacapuru ceramics: They display black and red on white polychrome decoration, which places them among the so-called Amazon polychrome tradition (APT). APT sites are found throughout the Amazon Basin, from Marajó Island to the upper Amazon Basin in Peru, Ecuador, and Colombia. Initially, their center of origin was supposed to be located at the foothills of the Andes (Evans and Meggers 1968), then in the central Amazon (Brochado 1984; Lathrap 1970), but at the moment, the oldest evidence of polychrome tradition ceramics comes from the upper Madeira Basin, near the current border of Brazil and Bolivia. Polychrome occupations in the central Amazon are superficial and normally placed on top of previous occupations of the Paredão or Manacapuru phase. The distribution of diagnostic shards shows that their area of dispersion within sites is smaller than that of preceding occupations, indicating a decrease in settlement size. Conversely, APT sites tend to be more dispersed and more regionally spread than with the preceding occupations (Lima 2003). Their occupation span is also smaller, never reaching the many decades or even centuries of continuous occupation found among preceding occupations.

Given the continental scale of the distribution of APT sites, virtually covering an area from the foothills of the Andes all the way to the mouth of the Amazon, albeit with localized regional gaps, one should expect a large measure of variability in terms of site size, density, and length of occupation. It is therefore possible that the pattern seen in the central Amazon could not apply to what is verified upstream in the Amazon River. Indeed, this is what was verified by the first Europeans who traveled down the Amazon in the early fifteenth century. These early chroniclers reported the presence of large sites regionally integrated into hierarchical social formations led by paramount chiefs.

At any rate, in the central Amazon, the social and cultural changes brought about the local replacement of Paredão-phase occupations by Guarita-phase (polychrome tradition) occupations did not bring any visible changes in terms of increase in political centralization or social hierarchy. To the contrary, if any changes happened, they were toward political decentralization and decrease in settlement size.

16.5 Discussion

The long-term patterns emerging from the data of the central Amazon allow for the discussion of a series of topics pertinent to South American archaeology. Among those topics are the development of the formative and the emergence of political centralization.

In what relates to the formative, one is faced with what may be the most important question of South American archaeology: Why did populations, derived from probably small initial groups of colonizers sometimes toward the end of the Pleistocene, take such different political and social paths during the Holocene? The available data show that some of the most important cultural innovations on the continent, such as pottery manufacture and plant domestication, started earlier in the tropical lowlands than in other areas. However, the development of settled life with a stronger reliance on agriculture seems to have developed much later in the Amazon than elsewhere in South America. How do we account for this apparent paradox?

The other question is the emergence of political centralization: The data from the central Amazon show a clear trend toward population growth and visible landscape modifications starting around 500 BC. Associated with such trends are clear signs of sedentary life, sometimes in large settlements reaching areas of many hectares. There is nothing, however, in the archaeological record that shows the development of marked regional hierarchies, wealth accumulation, or other traits that could be linked with the long-term production and reproduction of institutionalized social inequalities.

Traditionally, those processes have been interpreted by arguments of environmental limitations: lack of soil nutrients, lack of animal protein, unfavorable climatic conditions, or the right staple to support complex polities. Maybe the time has arrived to flip those arguments upside down and look at such processes in a different way. Instead of limitation, it could well be that important factors may have provided the material setting for the political processes discussed here. There is an abundance of resources along the large Amazonian rivers and their main tributaries. Even the present major upstream Amazonian cities, such as Manaus with almost 2,000,000 people, are daily provisioned by fish caught in the central Amazon. It is clear that current levels of exploitation are not sustainable and that new technologies, unavailable prior the twentieth century, have been introduced. Still, anyone familiar with the main Amazonian floodplain is well aware of the wealth of resources, mostly animal protein, to be found there. It may well be that such wealth of resources accounts for one peculiar fact related to the long-term history of the interplay of plants and people in the Amazon: Many species of palms have been managed since the beginning of the Holocene, but only one palm species, *Bactris gasipaes* (peach palm, chonta, pejibaye, or pupunha), has been fully domesticated. It can be argued that the lack of domestication means that there was no selective pressure for domestication, since resources were abundant enough to be managed in the wild or in domestic settings.

16.6 Concluding Remarks

This chapter started with the premise that early cultural developments in South America, such as plant domestication and pottery making, started as early, if not earlier, in the tropical lowlands as in the Andean area, traditionally seen as the major single center of cultural development in the continent. However, after this initial burst, developments in the lowlands seemed to stall, whereas in the central Andes, one sees, already in the late Archaic, the construction of monumental structures indicating the emergence of institutionalized social hierarchies. In the tropical lowlands, on the other hand, the archaeological record of the middle Holocene is characterized by gaps, followed by a cultural and demographic explosion around the beginning of the Christian era.

It has been suggested here that such gaps probably stem from poor archaeological visibility resulting from climatic change toward drier conditions in the middle Holocene. It is likely as well, however, that mid-Holocene populations of the Amazon had mixed economies marked by the cultivation of domesticated plants but also by the management of wild resources. Such strategies may have contributed to increased mobility and hence diminished archaeological visibility.

The noticeable changes in demography and settlement patterns visible after the beginning of the Christian era were probably associated with the establishment of climatic conditions similar to the ones verified in the Amazon today. From this time on, in the central Amazon and elsewhere, occupations became sedentary and permanent, urban clusters developed in some areas, and visible signs of landscape change became abundant. Interestingly though, such changes are not associated with the rise of intensive agriculture or, in some cases, with the development of institutionalized social hierarchies. For instance, in the central Amazon, the long-term archaeological record shows signs of political equality and not the development of centralized chiefdoms. Traditionally, such factors would have been interpreted as the result of adaptations to poor or limiting environmental conditions. It has been proposed here that the opposite may have happened: The abundance of resources and the technology to exploit and manage them, readily available and difficult to control at the institutional level, may have set the material conditions that prevented institutionalized social hierarchies from emerging and reproducing themselves.

If these arguments are correct, the Amazonian formative can be seen as a long-term process that lasted through a large part of the Holocene, only to be interrupted by the arrival of the Europeans in the sixteenth century. Does this mean that social evolutionary categories have no heuristic value for Amazonian archaeology? The answer is probably no, but it could be that the archaeological record of the Amazon may help us to use other benchmarks to evaluate social development or complexity. After all, what is preferable, to work compulsorily building monumental structures and maintaining irrigation channels or to roam freely in the semiwilderness where an abundance of resources is available?

References

Ammerman, A., & Cavalli-Sforza, L. (1984). *The Neolithic transition and the genetics of populations in Europe*. Princeton: Princeton University Press.

Anthony, D. (2007). *The horse, the wheel, and language: How Bronze Age riders from the Eurasian steppes shaped the modern world*. Princeton: Princeton University Press.

Arroyo-Kalin, M. (2008). *Steps towards an ecology of landscape: A geoarchaeological approach to the study of anthropogenic dark earths in the central Amazon region, Brazil*. PhD dissertation, Department of Archaeology, University of Cambridge.

Baker, P. A., Seltzer, G. O., Fritz, S. C., Dunbar, R. B., Grove, M. J., Tapia, P. M., et al. (2001). The history of South American tropical precipitation for the last 25,000 years. *Science, 291*, 640–643.

Barse, W. (2003). Holocene climate and human occupation in the Orinoco. In J. Mercader (Ed.), *Under the canopy: The archaeology of tropical rain forests* (pp. 249–270). New Brunswick: Rutgers University Press.

Behling, H. (2002). Impact of the Holocene sea-level changes in coastal, eastern and central Amazonia. *Amazoniana, XVII*, 41–52.

Behling, H., & Hooghiemstra, H. (2000). Holocene Amazon rainforest-savanna dynamics and climatic implications: High-resolution pollen record from Laguna Loma Linda in eastern Colombia. *Journal of Quaternary Science, 15*(7), 687–695.

Behling, H., & Lima da Costa, M. (2000). Holocene environmental changes from the Rio Curuá records in the Caxiuanã region, Eastern Amazon basin. *Quaternary Research, 53*(3), 369–377.

Behling, H., Berrío, J. C., & Hooghiemstra, H. (1999). Late Quaternary pollen records from the middle Caquetá River Basin in Central Colombian Amazon. *Palaeogeography, Palaeoclimatology, Palaeoecology, 146*, 193–213.

Berrío, J. C. (2002). Synthesis of the paleoenvironmental changes in the Colombian lowlands. In J. C. Berrío (Org.), *Lateglacial and Holocene vegetation and climatic change in lowland Colombia* (pp. 219–231). Amsterdam: University of Amsterdam.

Brochado, J. (1984). *An ecological model of the spread of pottery and agriculture into eastern South America*. PhD dissertation, Department of Anthropology, University of Illinois.

Burbridge, R., Mayle, F., & Killeen, T. (2004). Fifty-thousand-year vegetation and climate history of Noel Kempff Mercado National Park, Bolivian Amazon. *Quaternary Research, 61*, 215–230.

Chirinos, P. R. (2007). *Padrões de assentamento no sítio Osvaldo, Amazonas*. Dissertação de Mestrado, Programa de Pós-Graduação em Arqueologia, Museu de Arqueologia e Etnologia, Universidade de São Paulo.

Clement, C. (1999). 1492 and the loss of Amazonian crop genetic resources I: The relation between domestication and human population decline. *Economic Botany, 53*(2), 188–202.

Denevan, W. (1996). A bluff model of riverine settlement in prehistoric Amazonia. *Annals of the Association of American Geographers, 86*(4), 654–681.

Dreyfus, S. (1993). Os empreendimentos coloniais e os espaços políticos indígenas no interior da Guiana ocidental (entre o Orenoco e o Corentino) de 1613 a 1796. In E. Viveiros de Castro & M. Carneiro da Cunha (Eds.), *Amazônia: Etnologia e História Indígena* (pp. 19–41). São Paulo: Núcleo de História Indígena e Indi-genismo – USP.

Evans, C., & Meggers, B. (1968). *Archaeological investigations on the Rio Napo, Eastern Ecuador* (Smithsonian Contributions to Anthropology, Vol. 6). Washington, DC: Smithsonian Institution Press.

Freitas, H., Pessenda, L., Aravena, R., Gouveia, S., Ribeiro, A., & Boulet, R. (2001). Late Quaternary vegetation dynamics in the southern Amazon Basin inferred from carbon isotopes in soil organic matter. *Quaternary Research, 55*, 39–42.

Gomes, D. (2002). *A cerâmica arqueológica da Amazônia: vasilhas da coleção Tapajônica MAE-USP*. São Paulo: Imprensa Oficial SP/Editora da Universidade de São Paulo.

Gomes, D. (2008). O uso social da cerâmica de Parauá, Santarém, baixo Amazonas: uma análise funcional. *Arqueología Sul-Americana, 4*, 4–33.

Headland, T., & Bailey, R. (1991). Introduction: Have hunter-gatherers ever lived in tropical rain forest independently of agriculture? *Human Ecology, 19*(2), 115–122.

Heckenberger, M. (2002). Rethinking the Arawakan diaspora: Hierarchy, regionality, and the Amazonian formative. In J. Hill & F. Santos-Granero (Eds.), *Comparative Arawakan histories: Rethinking language and culture areas in the Amazon* (pp. 99–122). Urbana: University of Illinois Press.

Heckenberger, M. (2008). Amazonian mosaics: Identity, interaction, and integration. In H. Silverman & W. Isbell (Eds.), *Handbook of South American archaeology* (pp. 941–962). New York: Springer.

Heckenberger, M., Neves, E., & Petersen, J. (1998). De onde surgem os modelos? As origens e expansões Tupi na Amazônia central. *Revista de Antropologia, 41*(1), 69–96.

Heckenberger, M. J., Kuikuro, A., Kuikuro, U. T., Russell, J. C., Schmidt, M., Fausto, C., et al. (2003). Amazonia 1492: Pristine forest or cultural parkland? *Science, 301*, 1710–1714.

Heckenberger, M. J., Russell, J. C., Fausto, C., Toney, J. R., Schmidt, M. J., Pereira, E., et al. (2008). Pre-Columbian urbanism, anthropogenic landscapes, and the future of the Amazon. *Science, 321*, 1214–1217.

Hornborg, A. (2005). Ethnogenesis, regional integration, and ecology in prehistoric Amazonia: Towards a systemic perspective. *Current Anthropology, 46*(4), 589–620.

Jackson, J. (1983). *The fish people: Linguistic and exogamy and Tukanoan identity in northwest Amazonia*. Cambridge: Cambridge University Press.

Kirch, P. V. (2000). *On the road of the winds: An archaeological history of the Pacific Islands before European contact*. Berkeley: University of California Press.

Lathrap, D. (1970). *The upper Amazon*. London: Thames & Hudson.

Lathrap, D. (1977). Our father the cayman, our mother the gourd: Spinden revisited or a unitary model for the emergence of agriculture in the new world. In C. Reed (Ed.), *Origins of agriculture* (pp. 713–751). The Hague: Mouton.

Lima, L. F. (2003). *Levantamento arqueológico nas áreas de interflúvio da área de conflueência dos rios Negro e Solimões*. Dissertação de Mestrado, Programa de Pós-Graduação em Arqueologia, Museu de Arqueologia e Etnologia, Universidade de São Paulo, Brasil.

Lowie, R. (1948). The tropical forests: An introduction. In J. Steward (Ed.), *Handbook of South American Indians* (Bulletin 143, Vol. 3, pp. 1–56). Washington, DC: Bureau of American Ethnology, Smithsonian Institution.

Meggers, B. J. (1997). La cerámica temprana en América del Sur: invención independiente o difusión? *Revista de Arqueología Americana, 13*, 7–40.

Meggers, B. J., & Evans, C. (1957). *Archaeological investigations at the mouth of the Amazon* (Bulletin No. 167). Washington, DC: Bureau of American Ethnology.

Meggers, B. J., & Miller, E. T. (2003). Hunter-gatherers in Amazonia during the Pleistocene-Holocene transition. In J. Mercader (Ed.), *Under the canopy: The archaeology of tropical rain forests* (pp. 291–316). New Brunswick: Rutgers University Press.

Miller, E. T. (1999). A limitação ambiental como barreira à transposição do period formativo no Brasil: tecnologia, produção de alimentos e formação de aldeias no sudoeste da Amazônia. In P. Ledergerber-Crespo (Ed.), *Formativo Sudamericano: Homenaje a Albert Rex González y Betty J. Meggers* (pp. 331–339). Quito: Ediciones Abya-Yala.

Miller, E. T. (Coord.). (1992). *Arqueologia nos empreendimentos hidrelétricos da Eletronorte; resultados preliminares*. Brasília: Eletronorte.

Moraes, C. P. (2007). *Levantamento arqueológico da região do Lago do Limão, Iranduba, Amazonas*. Dissertação de Mestrado, Programa de Pós-Graduação em Arqueologia, Museu de Arqueologia e Etnologia, Universidade de São Paulo, Brasil.

Neves, E. G. (2006). *Arqueologia da Amazônia*. Rio de Janeiro: Jorge Zahar.

Neves, E. G. (2008). Ecology, ceramic chronology and distribution, long-term history and political change in the Amazonian floodplain. In H. Silvermann & W. Isbell (Eds.), *Handbook of South American archaeology* (pp. 359–379). New York: Springer.

Neves, E. G., & Petersen, J. B. (2006). Political economy and pre-Columbian landscape transformations in central Amazonia. In W. Balée & C. Erickson (Eds.), *Time and complexity in historical ecology: Studies in the neotropical lowlands* (pp. 279–309). New York: Columbia University Press.

Neves, E. G., Petersen, J. B., Bartone, R. N., & da Silva, C. A. (2003). Historical and socio-cultural origins of Amazonian dark earths. In J. Lehmann, D. Kern, B. Glaser, & W. I. Woods (Eds.), *Amazonian dark earths: Origins, properties, management* (pp. 29–50). Dordrecht: Kluwer Academic Publishers.

Neves, E. G., Petersen, J. B., Bartone, R. N., & Heckenberger, M. J. (2004). The timing of terra preta formation in the central Amazon: Archaeological data from three sites. In B. Glaser & W. I. Woods (Eds.), *Amazonian dark earths: Explorations in space and time* (pp. 125–134). Berlin: Springer.

Nordenskiöld, E. (1930). *Ars Americana: L'archéologie du bassin de l'Amazone*. Paris: Les Éditions G. Van Oest.

Oliver, J. (1989). *The Archaeological, linguistic and ethnohistorical evidence for the expansion of Arawakan into northwestern Venezuela and northeastern Colombia.* PhD dissertation, Department of Anthropology, University of Illinois.
Perry, L. (2005). Reassessing the traditional interpretation of "manioc" artifacts in the Orinoco valley of Venezuela. *Latin American Antiquity, 16*(4), 409–426.
Petersen, J. B. (1996). The archaeology of Trants, Montserrat, part 3. Chronological and settlement data. *Annals of the Carnegie Museum, 65*, 323–361.
Petersen, J. B., Neves, E. G., & Heckenberger, M. J. (2001). Gift from the past: *Terra Preta* and prehistoric Amerindian occupation in Amazonia. In C. McEwan, C. Barreto, & E. Neves (Eds.), *Unknown Amazon: Culture in nature in ancient Brazil* (pp. 86–105). London: British Museum Press.
Piperno, D., & Becker, P. (1996). Vegetational history of a site in the central Amazon basin derived from phytolith and charcoal records from natural soils. *Quaternary Research, 45*, 202–209.
Renfrew, C. (2000). At the edge of knowability: Towards a prehistory of languages. *Cambridge Archaeological Journal, 10*(1), 7–34.
Roosevelt, A. (1991). *Moundbuilders of the Amazon.* New York: Academic.
Roosevelt, A. (1995). Early pottery in the Amazon: Twenty years of scholarly obscurity. In W. K. Barnett & J. Hoopes (Eds.), *The emergence of pottery: Technology and innovation in ancient societies* (pp. 115–131). Washington, DC: Smithsonian Institution Press.
Roosevelt, A., Housley, R. A., Imazio Da Silveira, M., Maranca, S., & Johnson, R. (1991). Eighth millennium pottery from a prehistoric shell midden in the Brazilian Amazon. *Science, 254*, 1621–1624.
Roosevelt, A., Douglas, J., & Brown, L. (2002). The migrations and adaptations of the first Americans: Clovis and pré-clovis viewed from South America. In N. Jablonski (Ed.), *The first Americans, the Pleistocene colonization of the new world* (Memoirs of the California Academy of Sciences, No. 27, pp. 159–235). San Francisco: University of California Press.
Rouse, I. (1992). *The Tainos: Rise and decline of the people who greeted Columbus.* New Haven: Yale University Press.
Schaan, D. (2004). *The Camutins Chiefdom.* PhD dissertation, Department of Anthropology, University of Pittsburgh.
Schaan, D. (2008). The non-agricultural chiefdoms of Marajo Island. In H. Silverman & W. Isbell (Eds.), *Handbook of South American archaeology* (pp. 339–357). New York: Springer.
Simões, M. (1969). The Castanheira site: New evidence on the antiquity and history of the Ananatuba phase. *American Antiquity, 34*(4), 402–410.
Simões, M. (1981). Coletores-pescadores ceramistas do litoral do Salgodo (Pard). *Boletim do Museu Paraense Emilo Goeldi* (NS) 78.
Souza-Filho, P. W. M., Lessa, G. C., Cohen, M. C. L., Costa, F. R., & Lara, R. J. (2008). The subsiding macrotidal barrier estuarine system of the eastern Amazon Coast, Northern Brazil. In S. F. Dillenburg & P. A. Hesp (Eds.), *Geology of Brazilian coastal barriers* (Lecture Notes in Earth Sciences, Vol. 107, pp. 347–375). New York: Springer.
Toledo, M., & Bush, M. (2008). Vegetation and hydrology changes in Eastern Amazonia inferred from a pollen record. *Anais da Academia Brasileira de Ciências, 80*(1), 191–203.
Wüst, I., & Barreto, C. (1999). The ring villages of central Brazil: A challenge for Amazonian. *Latin American Antiquity, 10*(1), 3–23.

Part V
Future Directions

Chapter 17
Human-Environment Research: Past Trends, Current Challenges, and Future Directions

Eduardo S. Brondízio and Rinku Roy Chowdhury

Abstract This chapter reflects on issues and questions underscoring the preceding chapters while providing a brief overview of current and future directions in the study of human-environment interactions. A discussion of future directions begins with recognizing that the challenges of studying complex coupled human-environment systems are not new, although current processes of globalization and climate change render such challenges pressing, and of wider societal relevance. In this brief overview, we highlight questions related to institutions, ecosystem services, health, adaptation to climate change, urbanization, and methodological challenges.

This chapter reflects on issues and questions underscoring the preceding chapters while providing a brief overview of current and future directions in the study of human-environment interactions (HEI). A discussion of future directions begins with recognizing that the challenges of studying complex coupled human-environment systems are not new, although current processes of globalization and climate change render such challenges pressing, and of wider societal relevance.

As we look into the present and future of HEI research, we are challenged with many of the same long-standing questions confronted from a variety of disciplinary angles as in past decades. These challenges pertain to complexity, scale, heterogeneity, governance, and barriers to interdisciplinary bridging, among

E.S. Brondízio (✉)
Department of Anthropology, Anthropological Center for Training and Research on Global Environmental Change, School of Public and Environmental Affairs, Indiana University, Bloomington, IN, USA
e-mail: ebrondiz@indiana.edu

R.R. Chowdhury
Geography Department, Indiana University, Student Building 120, Bloomington, IN 47405, USA
e-mail: rroychow@indiana.edu

E.S. Brondízio and E.F. Moran (eds.), *Human-Environment Interactions: Current and Future Directions*, Human-Environment Interactions 1,
DOI 10.1007/978-94-007-4780-7_17,

other themes. Although we are constantly improving upon methodological tools and analytical concepts in human-environment research, a key challenge remains to bridge disciplinary knowledge to the demands of (interdisciplinary) problems we face today. This has been a central concern for the contributors to this volume. Different disciplinary traditions have confronted the question of how local populations interact with larger social, ecological, and economic processes and how we understand complex pathways of change and transitions in HEI systems and their variance across scales. These questions have required the research community to consider the tradeoffs of scaling up (and down) human-environment analysis and confront common yet questionable assumptions about different degrees of social and environmental homogeneity. We have learned that context, units of analysis, and the choices of temporal and spatial scales regarding the questions asked can strongly influence the understanding obtained. Yet, the many challenges of scaling (e.g., how to gain insights into lower-level phenomena from broader-scale behavior or vice versa) continue to define an important conceptual and analytical frontier for the HEI research community and the social sciences in general.

It is important to recognize the evolution of such efforts, as discussed in the introduction to this volume and illustrated in different chapters, as part of the effort put toward interdisciplinary approaches to HEI research. We have come a long way from opportunistic methodological and theoretical borrowing to increasingly interdisciplinary and collaborative research frameworks (Moran 2010; Ostrom 2009). The emergence of social-ecological systems frameworks, as discussed below, illustrates relevant advancements on this front. Progressively, these developments are influencing the training of a new generation of scholars and practitioners equipped with language and skills to address old, new, and pressing questions confronting us today. In these and other frameworks, there is a critical need to detail the historical trajectories of HEI systems in order to generate accurate and robust understanding of their dynamics.

17.1 Global and Regional System: HEI Research for a Changing World

Acknowledging the long history of challenges for HEI research as referred to above does not preclude us from recognizing that we are confronting social and environmental problems at a scale that is unprecedented in human history. As a result of social-economic-political changes intensified during the past 50 years, there are indications that we are beginning a period of global and regional shifts (Lenton et al. 2008; Steffen et al. 2004). The (tele)coupling of climate change, heretofore uncoordinated (or only loosely linked) economic systems, demographic transitions, and sociocultural and political transformations are altering human-environment systems, introducing new patterns of complexity, higher levels of unpredictability, and uncertain outcomes (Rockström et al. 2009). Social and environmental systems in the Global North as well as South are facing critical thresholds and tipping points, and these social-environmental interactions present further surprises. Agreement is growing in the diverse HEI research community that understanding these questions requires more attention to longer temporal scales and to their underlying historical processes (Balée and Erickson 2006; Dearing et al. 2010). These questions raise opportunities for HEI research and numerous methodological challenges.

No matter which continent, local and regional changes are related to larger national and global historical contexts that pose key methodological and conceptual challenges to HEI research. How can we account for historical differences while also learning from comparative analysis? How can HEI research assess or adjudicate among the path dependencies of human-environment outcomes in different societies? How do problems of scale and governance express themselves in different places? Are there common patterns or "syndromes" (Geist and Lambin 2004; Petschel-Held et al. 1999) of changes in HEI through time and across regions?

Research needs to consider the various historical bases for different social-political-economic processes shaping HEI. Since WWII, development policies and demographic trends have set in motion major regional transformations around the world. More recently, a variety of structural adjustment and economic development programs, trade liberalization policies, and networks of commodity exports have spread and intensified. The social and environmental impacts of structural adjustment programs have been diverse. In some cases, they have been marked by the exponential expansion of large-scale agriculture export, deforestation, and pressure on small-scale producers. In other cases, dependence on imported food has been growing, in parallel to retracting rural economies and aging populations, followed by reforestation. Throughout the world, however, the expansion of urban areas has been striking and marked by significant infrastructural deficiencies, high poverty rates, and various types of links to rural areas. The scope of cultural changes has been equally remarkable. Many regions of the world have seen parallel processes of "de-agrarization" (i.e., diversification in economic activities, income sources, and social identity of rural producers with rural households moving to nonagricultural activities), "de-peasantization" (i.e., relocation of rural families to urban areas with shifts in livelihood basis and social identity), as well as new kinds of ethnogenesis (i.e., reclaiming of indigenous identities frequently associated with land rights, ecotourism, and access to incentives). In parallel to infrastructure and urban expansion, the number and expanse of protected areas for environmental conservation have increased exponentially, which have offered both opportunities and conflicts for different social groups. As illustrated by the preceding chapters, today, infrastructure expansion, commodity production and deforestation, social change and urbanization, and a myriad of new institutional arrangements underlie an evolving matrix of social-territorial complexity worldwide. This is the context defining current and future directions for HEI research.

17.2 Current and Future Directions in HEI Research

As the preceding chapters illustrated, a suite of new questions, themes, and methodological approaches are defining the current and future landscape of HEI research. In this brief overview, we highlight questions related to institutions, ecosystem services, health, adaptation to climate change, urbanization, and methodological challenges.

The diffusion of **new institutional arrangements** and forms of defining territoriality for the purpose of resource governance is altering social and ecological environments and introducing new configurations of HEI worldwide. The spread of new institutional arrangements extends to resource governance as evidenced in the increase in protected areas and community-based conservation. Also of note is the overlap among government, private, and common-pool arrangements and the risks of compounding institutional failure (e.g., Acheson 2006). Significantly, current shifts in markets and governance globally may increase the connectivity and functional interdependence among different institutional arrangements on one hand and reducing or severing such functional connectivity on the other. For instance, the emergence of new commodity markets has linked local livelihoods as well as ecologies to global economic boom-bust cycles. At the same time, traditional dependencies in local places may be disrupted. Thus, the rise of capital-intensive commodity farming or land tenure "reform" is altering the decision logics of sedentary agriculturalists (such as in East and West Africa) who previously allowed herds owned by itinerant pastoralists to graze their fallow fields while benefitting from the resultant "natural" subsidy to soil fertility. As a result, local forms of use and regulation of resources, although potentially effective at one level, are affected and in some cases overwhelmed by resource use or a shift in a different part of the larger system (Brondizio et al. 2009). Such changes raise important questions for HEI research, such as how different types of management arrangements may help facilitate solutions to intricate cross-level problems.

During the past decade, the concept of **ecosystem services** has emerged as a way to call attention to human dependence on natural resources and ecosystem functioning. Frameworks for the analysis of ecosystem services, such as that proposed by the Millennium Ecosystem Assessment (2005), have focused on the dependence of human well-being on provisioning, regulating, supporting, and cultural ecosystem services. While the focus on human well-being represents a major advance for HEI analysis, the concept of ecosystem services has been used in particular to highlight the economic value of biodiversity and ecosystem functioning. New research and policy agendas for payment for different types of ecosystem services have emerged and are rapidly creating new forms of institutional arrangements and economic incentives of importance to local populations as well as to national governments. For instance, programs such as Reduced Emissions through Deforestation and Degradation (REDD) are quickly developing, particularly in tropical regions. The effectiveness and social impact of these programs remain to be seen and represent an important dimension of HEI in coming years.

The new United Nations Intergovernmental Platform for Biodiversity and Ecosystem Services (IPBES) calls for interdisciplinary research very much along the lines of HEI. The research community of HEI is uniquely positioned to contribute to IPBES's mission of assessing interactions among ecosystem services, biodiversity, and human well-being at multiple levels of analysis, and in ways that include local and indigenous knowledge systems and that are relevant to policy and to different social groups. In this context, **land use and cover change** continues to be an important theme in HEI, both in terms of understanding trajectories of landscape change and its implication for biogeochemical cycles and biodiversity, as well as in terms of food production systems and local economies. Global climate change scenarios point to environmental shifts in food production areas representing risks and opportunities for local populations. The role of small-scale, decentralized production around urban areas is growing in importance and deserves close attention in terms of their contribution to food security and to trajectories of land-use change around urban areas. Similarly, the impact of so-called land grabs on local populations and on land-cover change in different parts of the world, particularly in the tropics, deserves attention from the HEI community. **Consumption** and changing consumption patterns represent another important linkage to, among other issues, land-use and land-cover change. Changing consumption patterns in rural and urban areas link local to global economic systems through feedback loops with far-reaching environmental and social consequences. Understanding the drivers and outcomes of changing consumption patterns in different parts of the world require new research approaches linking local-level studies to commodity chain and social networks that embed local decisions within wider social and economic networks.

Human adaptation to global climate change represents another major area of research in HEI. Studies of human adaptation to the environment have a long history in anthropology and cultural ecology. New research approaches are developing to study how local adaptive processes are incorporating new types of information and developing an understanding about new climate patterns and uncertainties (Brondizio and Moran 2008). Global environmental changes are affecting the geographic distribution of many vector-borne diseases, leading to increased human health risk (Confalonieri et al. 2007; Gubler 1998). A recent report by the National Research Council (2010) notes that *where* people live now affects their health more strongly than ever, necessitating spatially explicit approaches to modeling and predicting health outcomes. Malaria and dengue, respectively the world's deadliest and fastest-growing vector-borne diseases, continue to claim lives in many regions including Africa, Asia, and Latin America. These diseases are borne by mosquito vectors that have complex life cycles and exhibit clear links to temperature, rainfall, and habitat conditions (DeFries et al. 2002; Halstead 2008; Hopp and Foley 2003), which are further linked to land/resource use by managers of different socioeconomic backgrounds and decision strategies (Molyneux 2002; Patz et al. 2004; Vanwambeke et al. 2007). As the ultimate local agents of land management, households make land-use and related decisions (e.g., migration-related farmland abandonment, extended outdoor field labor, water storage practices) that may strongly affect disease vector exposures and infection risk (e.g., Somboon et al.

1998; Suwonkerd et al. 2002; Van Benthem et al. 2005). For instance, rural-to-urban migration is expected to change the worldwide distribution of disease and mortality (Harpham 1997). Of great concern is how rapid urbanization, social inequality, and climate variability in the developing tropics are jointly driving shifts in land cover and regional disease burdens.

Worldwide, urbanizing regions represent Earth's most rapidly changing landscapes, with strong impacts on global biodiversity and biogeochemistry. Despite increasing social and ecological research in **urbanizing systems**, much remains unknown about their complex dynamics or about their resultant resource demand implications. Given the increasing concentration of the world's population in urban areas, there is also a pressing need to better understand the connectivity and vulnerability of urbanizing coastal regions and their hinterlands. Patterns of urban development have multiple and significant ecological impacts. Urban sprawl may reflect variable urbanization densities; rates of loss in forests, farmland, and natural wetlands; and increases in impervious surfaces (Hasse and Lathrop 2003; Jantz et al. 2005; Medley et al. 1995). Landscape patterns and edge effects affect urban biodiversity and the connectivity of social processes; increase vulnerability to further landscape fragmentation, fire, drought, and other stressors; and affect regional and global biogeochemistry (Cochrane et al. 1999; Laurance and Williamson 2001; Laurance et al. 2001; Melles et al. 2003; Stein et al. 2005; Stohlgren et al. 1998). Urban design drives the extent of and access to urban green space and infrastructure, with implications for environmental justice and human health (Jackson 2003). The determinants and dynamics of urban development and density are a significant focus of urban economics, geography, and planning (Alonso 1964; Alperovich 1982; Craig and Haskey 1978; Kau and Lee 1976; Mills 1970; Muth 1969). At the regional scale, urban expansion has been related to "pull" factors, such as employment and retail clustering, infrastructure, services, amenities, and cheap land, and "push" factors such as crime, taxes, land costs, and pollution (Mieszkowski and Mills 1993). Building densities in North American and European cities are a result of these processes operating over multiple time scales, including the clustering of industries and transportation infrastructure (Antrop 2004; Geyer and Kontuly 1993).

The **heterogeneity and dynamics of suburban growth and decline** in regions under variable stages of economic development remain arenas for further research. Formal and informal structures (institutions) of landownership/management drive urban land-use change (Bryant et al. 1982). In particular, rural-urban land conversion is a political process: it is embedded in broader sets of power relations mediating land change and entails policy choices designating land values for various uses, e.g., through zoning (Kelly 1998; see also York and Munroe in this volume). Social structures, local decisions, and ecological factors interact in complex ways to shape all urban environments at multiple scales (Roy Chowdhury et al. 2011). Developing urban areas worldwide reflect state, private, communal, and other property ownership structures; however, many urban and suburban landscapes are increasingly under private landownership and residential land use. Understanding these landscapes depends on adequately integrating their multiscalar, social-ecological dynamics, e.g., linking dynamics in social neighborhoods to municipal and state governance, and on integrating both social and environmental aspects. A deeper, integrated understanding is especially critical given the spatial pervasiveness of residential landscapes. The rapid expansion of residential land covers is driven by complex interactions among socioeconomic, political, and environmental factors. Within the United States, as in other parts of the world, losses of agricultural and forested land to urban and suburban uses are among the greatest sources of anthropogenic landscape change in the twentieth and twenty-first centuries, a concern intensified by the large ecological footprints of urban areas as they mobilize resource, commodity, and material flows affecting agricultural, forested, and other non-urban landscapes (Munroe et al. 2005). While expanding residential spaces are embedded with personal and social values (e.g., recreational, aesthetic, property), they also affect ecological structure (e.g., habitat fragmentation) and function (e.g., nutrient flows), creating a "self-imposed burden and hazard" (Fissore et al. 2011; Robbins 2007).

Methodological challenges in HEI research will continue to push the boundaries of disciplinary methodologies. The human-environment relationship at the local level, for instance, is in itself a multilevel process. Local resource users do not operate in a vacuum; they are embedded in systems of resource access and tenure, social-political relations, and external forces and create multilevel alliances and social movements that influence larger political processes. They also adopt and adapt multiscalar influences to foster their own livelihoods, including the way knowledge is developed. The complexity of these social-ecological systems also means that conventional models of causality are limited in capturing the direct and indirect "teleconnections" created by the growing globalization of resource use systems, the interactions created by distant demands for local resources (and vice versa), and the responses of different sectors of society to such demands. These interactions are mediated by social, political, and biophysical conditions that may function across levels (e.g., constitutional arrangements, major infrastructure, macroeconomic processes), while others are level specific (e.g., land tenure systems, resource use technologies). In order to address these challenges, it is important for the HEI research community to build upon proposed frameworks, such as the social-ecological systems framework proposed by Ostrom (2009), which have mechanisms to integrate local-level processes and institutional arrangements into cross-scale assessments. Many ongoing efforts are developing common language, protocols, and frameworks to collect and analyze and compare knowledge about resource management systems in a cumulative fashion.

Reconciling our analytical paradigms and disciplinary contributions with the complexity required to understand societal and environmental change today and in coming decades remains a stimulating challenge for the broader HEI research community. The chapters in this book illustrate a range of productive, collaborative, and inspiring approaches to examine past, present, and future human-environment interactions in different parts of the world. We hope this volume will support and encourage readers to reflect on and to engage with new frontiers of research and practice needed to inform society of more sustainable and equitable pathways for a changing world.

References

Acheson, J. M. (2006). Institutional failure in resource management. *Annual Review of Anthropology, 37*, 117–134.
Alonso, W. (1964). *Location and land use*. Cambridge, MA: Harvard University Press.
Alperovich, G. (1982). Density gradients and the identification of the central business district. *Urban Studies, 19*(3), 313–320.
Antrop, M. (2004). Landscape change and the urbanization process in Europe. *Landscape and Urban Planning, 67*, 9–26.
Balée, W., & Erickson, C. (Eds.) (2006). *Time and complexity in historical ecology: Studies in the neotropical lowlands*. New York: Columbia University Press.
Brondizio, E. S., & Moran, E. F. (2008). Human dimensions of climate change: The vulnerability of small farmers in the Amazon. *Philosophical Transactions of the Royal Society, 363*, 1803–1809.
Brondizio, E., Ostrom, E., & Young, O. (2009). Connectivity and the governance of multilevel social-ecological systems: The role of social capital. *Annual Review of Environment and Resources, 34*(253–2), 78.
Bryant, C. R., Russwurm, L. H., & McLellan, A. G. (1982). *The city's countryside: Land and its management in the rural–urban fringe*. London: Longman.
Cochrane, M. A., Alencar, A., Schulze, M. D., Souza, C. M., Nepstad, D. C., Lefebvre, P., et al. (1999). Positive feedbacks in the fire dynamics of closed canopy tropical forests. *Science, 284*, 1832–1835.
Confalonieri, U., Menne, B., Akhtar, R., Ebi, K. L., Hauengue, M., Kovats, R. S., et al. (2007). Human health. In M. L. Parry, O. F. Canziani, J. P. Palutikof, P. J. van der Linden, & C. E. Hanson (Eds.), *Climate change 2007: Impacts, adaptation and vulnerability* (pp. 391–431). Contribution of Working Group II to the Fourth Assessment Report of the Intergovernmental Panel on Climate Change. Cambridge: Cambridge University Press.
Craig, J., & Haskey, J. (1978). The relationships between the population, area and density of urban areas. *Urban Studies, 15*(1), 101–107.
Dearing, J. A., Braimoh, A. K., Reenberg, A., Turner, B. L. II, & van der Leeuw, S. (2010). Complex land systems: The need for long time perspectives to assess their future. *Ecology and Society, 15*(4), 21 [online]. URL: http://www.ecologyandsociety.org/vol15/iss4/art21/
DeFries, R. S., Houghton, R. A., Hansen, M. C., Field, C. B., Skole, D., & Townshend, J. (2002). Carbon emissions from tropical deforestation and regrowth based on satellite observations for the 1980s and 1990s. *Proceedings of the National Academy of Sciences of the United States of America, 99*(22), 14256–14261.
Fissore, C., Baker, L. A., Hobbie, S. E., King, J. Y., Mcfadden, J. P., Nelson, K. C., et al. (2011). Carbon, nitrogen, and phosphorus fluxes in household ecosystems in the Minneapolis-Saint Paul, Minnesota, urban region. *Ecological Applications, 21*(3), 619–639.
Geist, H. J., & Lambin, E. F. (2004). Dynamic causal patterns of desertification. *BioScience, 54*, 817–829.
Geyer, H. S., & Kontuly, T. M. (1993). A theoretical foundation for the concept of differential urbanization. *International Regional Science Review, 15*(2), 157–177.
Gubler, D. (1998). Resurgent vector-borne diseases as a global health problem. *Emerging Infectious Diseases, 4*(3), 442–450.
Halstead, S. B. (2008). Dengue virus-mosquito interactions. *Annual Review of Entomology, 53*, 273–291.
Harpham, T. (1997). Urbanisation and health in transition. *The Lancet, 349*, 11–13.
Hasse, J. E., & Lathrop, R. G. (2003). Land resource impact indicators of urban sprawl. *Applied Geography, 23*(2–3), 159–175.
Hopp, M. J., & Foley, J. A. (2003). Worldwide fluctuations in dengue fever cases related to climate variability. *Climate Research, 25*(1), 85–94.

Jackson, L. E. (2003). The relationship of urban design to human health and condition. *Landscape and Urban Planning, 64*(4), 191–200.
Jantz, P., Goetz, S., & Jantz, C. (2005). Urbanization and the loss of resource lands in the Chesapeake Bay watershed. *Environmental Management, 36*(6), 808–825.
Kau, J. B., & Lee, T. F. (1976). Functional, density gradient and price elasticity of demand for housing. *Urban Studies, 13*(2), 193–200.
Kelly, P. F. (1998). The politics of urban–rural relations: Land use conversion in the Philippines. *Environment and Urbanization, 10*(1), 35–54.
Laurance, W. F., & Williamson, G. B. (2001). Positive feedbacks among forest fragmentation, drought and climate change in the Amazon. *Conservation Biology, 15*(6), 1529–1535.
Laurance, W. F., Cochrane, M. A., Bergen, S., Fearnside, P. A., Delamonica, P., Barber, C., et al. (2001). The future of the Brazilian Amazon. *Science, 291*, 438–439.
Lenton, T. M., Held, H., Kriegler, E., Hall, J. W., Lucht, W., Rahmstorf, S., et al. (2008). Tipping elements in the Earth's climate system. *Proceedings of the National Academy of Sciences of the United States of America, 105*(6), 1786–1793.
Medley, K. E., Pickett, S. T. A., & McDonnell, M. J. (1995). Forest-landscape structure along an urban-to-rural gradient. *The Professional Geographer, 47*(2), 159–168.
Melles, S., Glenn, S., & Martin, K. (2003). Urban bird diversity and landscape complexity: Species-environment associations along a multiscale habitat gradient. *Conservation Ecology, 7*(1), 22.
Mieszkowski, P., & Mills, E. (1993). The causes of metropolitan suburbanization. *Economic Perspectives, 7*(3), 135–147.
Millennium Ecosystem Assessment. (2005). URL: http://www.maweb.org/en/Index.aspx
Mills, E. S. (1970). Urban density functions. *Urban Studies, 7*(1), 5–20.
Molyneux, D. H. (2002). Vector borne infections and health related to landscape changes. In A. A. Aguirre, M. C. Pearl, R. S. Ostfeld, C. House, & G. M. Tabor (Eds.), *Conservation medicine: Ecological health in practice* (pp. 194–206). Oxford: Oxford University Press.
Moran, E. F. (2010). *Environmental social science: Human-environment interactions and sustainability*. Oxford, UK: Wiley-Blackwell.
Munroe, D. K., Croissant, C., & York, A. M. (2005). Land use policy and landscape fragmentation in an urbanizing region: Assessing the impact of zoning. *Applied Geography, 25*, 121–141.
Muth, R. F. (1969). *Cities and housing: The spatial pattern of urban residential land use*. Chicago: University of Chicago Press.
National Research Council. (2010). *Understanding the changing planet: Strategic directions for the geographical sciences*. Washington, DC: National Academies Press.
Ostrom, E. (2009). A general framework for analyzing sustainability of social-ecological systems. *Science, 325*(5939), 419–422.
Patz, J. A., Daszak, P., Tabor, G. M., Alonso Aguirre, A., Pearl, M., Epsein, J., et al. (2004). Unhealthy landscapes: Policy recommendations on land use change and infectious disease emergence. *Environmental Health Perspectives, 112*(10), 1092–1098.
Petschel-Held, G., Block, A., Cassel-Gintz, M., Kropp, J., Lüdeke, M. K. B., Moldenhauer, O., et al. (1999). Syndromes of global change: A qualitative modelling approach to assist global environmental management. *Environmental Modelling and Assessment, 4*, 295–314.
Robbins, P. (2007). *Lawn people: How grasses, weeds and chemicals make us who we are*. Philadelphia: Temple University Press.
Rockström, J., Steffen, W., Noone, K., Persson, Å., Chapin, F. S., III, Lambin, E. F., et al. (2009). A safe operating space for humanity. *Nature, 461*, 472–475.
Roy Chowdhury, R., Larson, K., Grove, J. M., Polsky, C., Cook, E., Onsted, J., et al. (2011). A multi-scalar approach to theorizing socio-ecological dynamics of urban residential landscapes. *Cities and the Environment, 4*(1), 6 [online]. URL: http://digitalcommons.lmu.edu/cate/vol4/iss1/6
Somboon, P., Aramrattana, A., Lines, J., & Webber, R. (1998). Entomological and epidemiological investigations of malaria transmission in relation to population movements in forest areas of north-west Thailand. *The Southeast Asian Journal of Tropical Medicine and Public Health, 29*(1), 3–9.

Steffen, W., Sanderson, A., Tyson, P. D., Jäger, J., Matson, P. A., Moore, B. III, et al. (2004). *Global Change and the Earth System: A Planet under Pressure*. IGBP Global Change Series. Berlin, Germany: Springer-Verlag.

Stein, S. M., McRoberts, R. E., Alig, R. J., Nelson, M. D., Theobald, D. M., Eley, et al. (2005). *Forests on the edge: housing development on America's private forests* (General Technical Report PNW-GTR-636). Portland: Pacific Northwest Research Station, USDA Forest Service.

Stohlgren, T. J., Baron, J. S., Chase, T. N., Pielke, R. A., Sr., & Kittel, T. G. F. (1998). Evidence that local land use practices influence regional climate, vegetation, and stream flow patterns in adjacent natural areas. *Global Change Biology, 4*(5), 495–504.

Suwonkerd, W., Overgaard, H. J., Tsuda, Y., Prajakwong, S., & Takagi, M. (2002). Malaria vector densities in transmission and non-transmission areas during 23 years and land use in Chiang Mai province, northern Thailand. *Basic and Applied Ecology, 3*(3), 197–207.

Van Benthem, B. H. B., Vanwambeke, S. O., Khantikul, N., Burghoorn-Maas, C., Panart, K., Oskam, L., et al. (2005). Spatial patterns of and risk factors for seropositivity for dengue infection. *The American Journal of Tropical Medicine and Hygiene, 72*(2), 201–208.

Vanwambeke, S. O., Lambin, E. F., Eichhorn, M. P., Flasse, S. P., Harbach, R. E., Oskam, L., et al. (2007). Impact of land-use change on dengue and malaria in northern Thailand. *EcoHealth, 4*, 37–51.

Index

Note: The terms "agriculture," "Amazon," "household," "institutions," and "land use" appear more than one hundred and fifty times in the text. For these terms, page numbers refer to the first instance of the term in each chapter.

A

Adaptation, 2, 7, 8, 10, 11, 13, 17, 18, 25–27, 97, 104, 109, 111, 275, 277, 293, 327–330, 345, 346, 361, 385, 395
Adaptive approach, 109
Africa, 96, 165–203, 343, 376, 380, 395
Agricultural adoption, 284
Agriculture, 9, 33, 58, 84, 103, 117, 142, 170, 253, 281, 297, 322, 344, 371–385, 393
Agro-extractive settlement projects (PAEs), 211, 213–231
Agroforestry, 20, 257, 268, 270, 319–338, 345, 357–359, 362–365, 376, 379, 380
Amazon, 19, 33, 55, 212, 215, 237, 251–270, 319, 321–338, 371–385
Amazon floodplains, 211, 213–231
Archeological approaches to research, 319–320

B

Batistela, M., 319, 323
Biodiversity, 12, 81, 86, 98, 104, 114, 115, 125, 136, 166, 169, 171, 178, 182, 192, 193, 254, 257, 292, 394, 396
Biomass, 120, 167–169, 183, 192, 201, 331, 333, 335, 347, 363
Birth control policy, 26, 59, 61, 63, 67–69, 73
Black pepper, 20, 258, 319, 322, 324, 325, 328–330
Bolfe, E.L., 319, 331, 334, 336
Boserup(ian) theory, 56
Botswana, 19, 96, 170, 181–193, 202
Brazil, 19, 20, 25, 29–49, 212, 215, 217, 229, 253–255, 258, 259, 261, 265–267, 269, 322, 323, 325, 330, 375–377, 380, 382, 383
Brondizio, E.S., 98, 322, 323, 328, 395
Bushmeat, 82, 86, 88

C

Carbon sequestration, 20, 153–155, 319
Case studies, 15, 18, 25, 26, 35, 39, 42, 79, 95, 98–100, 103–105, 109, 115, 116, 120–135, 166, 167, 169–202, 214, 251–270, 297, 303, 322, 375, 380
Castro, 16, 211, 214, 216, 224, 226, 227
Cattle raising, 33, 212, 227, 257–259, 267
CBNRM. *See* Community-based natural resource management (CBNRM)
China's birth control policy, 54, 57, 61, 62, 70
Chowdhury, R.R., 12, 396
Climatic changes, 84, 376–381, 385
Coffee, 34, 117, 120, 267, 268, 279, 281, 285–292, 348, 349, 351–366
Colombia, 19, 211, 236–239, 248, 378, 383
Colonial policies, 20, 48
Co-management, 16, 98, 99, 102, 104, 211, 214–218, 221, 222, 228–231
Common pool resources (CPR), 14, 15, 153–155, 211, 239
Community-based natural resource management (CBNRM), 169, 170

E.S. Brondízio and E.F. Moran (eds.), *Human-Environment Interactions: Current and Future Directions*, Human-Environment Interactions 1,
DOI 10.1007/978-94-007-4780-7, © Springer Science+Business Media Dordrecht 2013

Comparative research, 17
Complexity, 2, 5–9, 12, 14, 15, 20, 31, 48, 154, 168, 169, 202, 214, 252, 311, 320, 359–366, 381–383, 385, 391–393, 397
Complexity theory, 7
Cooperation, 144, 153, 211, 213–231, 237, 238, 248, 329, 333
Cooperative behavior, 214, 215, 238
CPR. *See* Common pool resources (CPR)
Customary institutions, 355

D
Demographic transition, 54, 73, 392
Demography, 30, 31, 34–36, 48, 49, 58, 385
Dental health, 31, 35, 38, 39
Determinism, 1–8, 20, 320
Diffusionism, 373, 374
Domestication, 20, 320, 373, 376–380, 383, 384
Dynamic stability, 291–292

E
Ebola, 18, 26, 82
Economic development, 33, 59–64, 105, 285, 307, 308, 393
Ecosystem services, 18–20, 81, 96, 135, 136, 144, 150, 152, 155, 181, 301, 319, 394
Ecotourism, 18, 26, 86–88, 257, 259–260, 393
Emerging infectious diseases, 18, 26, 27, 79–88
Epidemiology, 30, 39, 48
Ethnography, 12, 19, 382
Europe, 3, 57, 84, 109, 237, 374, 376, 380
Experimental economics, 17, 19, 211, 237
Exurban, 143, 296, 309
Exurbanization, 20, 212, 296–298, 302, 303, 310, 312, 313

F
Farming-language dispersal hypothesis (FLDH), 320, 374, 376
Farmland, 173, 296, 309, 312, 313, 325, 395
Fertility decline, 26, 38, 54, 55, 57, 63, 67, 73
Fertility transition, 57, 64–73
Fiorini, S., 95, 102
Fischer, B.C., 96, 142, 153, 154
FLDH. *See* Farming-language dispersal hypothesis (FLDH)
Forest change, 95, 113–137, 141, 174, 291–292
Forest conditions, 275, 278
Forest governance, 142, 155–157, 285–286
Forest management, 121, 124, 142, 143, 145, 148–150, 157, 158, 212, 257, 260–262, 275–278, 281–284, 289–292, 376
Forest ownership, 144, 153–157
Forest reserves (FR), 125, 127–131, 134, 148, 172, 177, 263, 267–269, 345
Fragmentation, 81, 83, 86, 99, 111, 114, 135, 142–144, 153, 171, 175, 180, 192, 193, 296, 297, 302–305, 309–313, 320, 373, 380, 381, 395, 396
Futemma, C., 212, 220, 224, 226, 256, 260, 262, 264, 268

G
Geographic information systems (GIS), 19, 99, 115, 122, 126, 129, 132, 137, 211, 332
Gibbes, C., 166, 197, 198
GIS. *See* Geographic information systems (GIS)
Globalization, 17, 25, 276, 367, 396, 397
Global markets, 306, 319, 357–359

H
HDI. *See* Human Development Index (HDI)
Honduras, 19, 212, 253, 275–293
Household, 12, 34, 54, 127, 152, 171, 224, 252, 278, 303, 322, 349, 393
Household survey, 59, 67, 278
Human Development Index (HDI), 279, 325
Human-wildlife interactions, 26, 88

I
IKONOS, 168, 195, 196, 198
Importance value, 363–365
India, 19, 95, 113–137, 237
Indigenous health, 34, 37
Infectious diseases, 18, 26, 27, 79–88
Institutional adaptation, 293
Institutional analysis, 14–20, 95, 211, 212, 270, 296, 319, 333
Institutional design, 228
Institutional evolution, 19, 212, 275–293
Institutions, 4, 31, 65, 99, 142, 214, 237, 266, 275, 295–314, 327, 345, 396

L
Labor arrangement, 322
Land cover change, 20, 26, 83–86, 88, 115–117, 124, 126, 127, 129, 130, 132–134, 137, 165–203, 263, 277, 291, 319, 322, 347

Land cover classification, 96, 167, 174–177, 180, 181, 193
Land-labor demand hypothesis, 55, 56
Landsat ETM+, 122, 126, 129
Landsat MSS, 122, 123, 132
Landsat TM, 122, 126, 129, 132, 174, 175, 195–197, 199–201, 256, 278, 324
Landscape approach, 95
Landscape gradients, 96, 165–203
Land scarcity, 55, 282, 285
Land-security hypothesis, 55
Land tenure, 6, 19, 20, 54, 55, 57, 59–65, 74, 217, 218, 262, 319, 320, 346, 394, 397
Land use, 19, 20, 83–86, 100, 114, 142, 166–167, 216, 253, 284, 295–314, 322, 343, 395
Land use and cover change (LUCC), 17, 18, 26, 96, 367
Land use change, 115, 135, 212, 286, 309, 312, 323, 396
Land use intensification, 20, 319
Land-use/land-cover change (LULCC), 83–86, 88, 115, 137, 166–169, 322
Land use policy, 20, 212, 296, 297, 304, 306–309
Language dispersal, 20, 320, 373–376
Language diversity, 20, 320, 380
Late Holocene, 376–380
Longitudinal research, 25, 212
Longitudinal studies, 31, 39
Lopez, M.C., 211, 212, 238, 239, 243, 244, 246
Lowland South America, 18, 45, 373–376, 380
LUCC. *See* Land use and cover change (LUCC)
LULCC. *See* Land-use/land-cover change (LULCC)
Lyme disease, 18, 26, 81, 83, 84

M
Malaria, 18, 26, 81, 83–85, 323, 395
Malthus(ian) theory, 7
Manioc, 20, 254, 257, 258, 262, 269, 327, 349, 378–380
Marriage systems, 66, 68
Meta analysis, 18, 20, 95, 115–121, 319
Migrant community, 259
Modernization, 58, 73, 284, 287
Monitoring change, 101, 104, 166
Moran, E.F., 2, 11, 13, 30, 252, 270, 319, 322, 323, 327, 334, 392, 395
Multi-scale analysis, 167
Munroe, D.K., 149, 212, 303, 304, 306, 309, 397

N
Nagendra, H., 16, 95, 98, 114, 115, 118–121, 125–129, 134, 135, 137, 141, 168
Namibia, 19, 96, 170, 193–202
Natural resource management, 19, 95, 97–111, 143, 156, 170, 211, 213, 236, 279, 293
NDVI. *See* Normalized Difference Vegetation Index (NDVI)
Neves, E.G., 320, 374, 375, 377, 379, 381
Nipah virus, 18, 26, 83
Non-industrial private forests (NIPF), 147–149, 151, 306
Non-timber forest products, 131, 171, 212, 261, 285, 356, 362
Normalized Difference Vegetation Index (NDVI), 167, 174, 175, 177–179, 196–202, 332, 337
Nutrition transition, 31, 41–47

O
Own-children method, 59

P
PA. *See* Protected areas (PA)
PAEs. *See* Agro-extractive settlement projects (PAEs)
Participatory GIS (PGIS), 17, 18, 97–111
Pasture, 20, 118, 136, 171, 174, 260, 264, 269, 270, 281, 285, 292, 319, 321–338, 345, 367
Pattern and process, 166, 168
People and parks, 114, 120, 122, 127, 128, 132, 134, 135
PGIS. *See* Participatory GIS (PGIS)
Polyandry, 65–68, 72, 73, 75
Population pressure, 19, 95, 121, 296, 309, 345
Pre-Colombian Amazon, 20
Primate, 18, 26, 27, 82, 84–88
Private forests, 96, 141–158, 289, 353
Privatization, 15, 26, 281, 286, 288
Property-level analysis, 319
Property rights, 19, 55, 96, 142, 150, 153–158, 214, 217, 236, 238, 239, 280, 286, 289, 298, 300
Protected areas (PA), 17–19, 86, 95, 96, 114–118, 120, 121, 125, 128, 129, 134–137, 169, 198, 199, 252–255, 257, 268, 270, 289, 325, 335, 394
Public goods, 142, 147, 153–155, 211, 238–240, 244, 245, 248, 298, 301
Public-private interaction, 141–158

R

Red deer, 18, 95, 100
Regeneration, 102, 105, 107, 117, 168, 169, 263, 276, 283, 359, 361, 365, 367
Regrowth, 115, 117–120, 127, 129, 130, 135, 283
Remote sensing, 17–20, 95, 96, 115, 137, 166, 167, 170, 181, 193, 194, 202, 211, 212, 319, 331, 334
Remote sensing classification, 167, 168, 170, 201, 202
Resilience, 277, 283, 292, 293
Robustness, 277
Rural populations, 19, 59, 211, 347
Rural settlements, 253, 255–256, 259, 267
Rural tourism, 212, 256, 257, 259–260, 267
Rural-urban fringe, 295–314
Ruseva, T.B., 96, 142, 153, 154

S

SARS. *See* Severe acute respiratory syndrome (SARS)
Savanna, 168–173, 177, 180–182, 186, 193–194, 196, 200, 201, 348, 378
Scotland, 6, 18, 95, 98–105, 109
Severe acute respiratory syndrome (SARS), 18, 26, 82, 83
Small farmers, 19
Smallholders, 20, 56, 74, 171, 267–269, 320, 322, 327, 343–346, 349–351, 354–359, 361–367
Socio-ecological systems, 16, 153, 158
Socioeconomic differentiation, 25, 31, 41–48
Spatial analysis, 180, 314
Stakeholder, 18, 95, 96, 101, 102, 110, 111, 211, 214–231, 238, 252, 253, 289, 313
Stressors, 277–278, 293, 396
Structural adjustment, 320, 359–365, 393
Subsistence, 5–10, 30, 44, 47, 54, 121, 129, 136, 174, 257, 264, 270, 279, 281–283, 286, 289, 290, 322, 323, 350, 357, 361–363
Survey, 17, 19, 26, 39, 44, 55, 58–60, 64, 67–69, 72, 74, 87, 100, 101, 105, 126, 127, 129, 132, 144, 152, 174, 192, 211, 255, 256, 259, 262, 278, 303, 310, 381
Sustainable private forests, 143, 144

T

Tibet, 18, 26, 53–75
Traditional institutions, 285, 361, 366
Tucker, A.R., 349
Tucker, C., 16
Tucker, C.J., 323
Tucker, C.M., 155, 156, 212, 276, 278, 280, 282, 284, 287, 291

U

Uganda, 19, 20, 87, 96, 170–181, 202, 319, 343–367
United Kingdom (UK), 222
United States, 19, 20, 40, 55, 79, 81, 83, 84, 96, 141–158, 212, 237, 296, 298, 300, 313, 314, 397

V

Várzea, 215, 216, 218, 220–223, 226, 229, 230
Vegetation indices, 167, 170, 186, 202, 332, 336
Vegetation inventory, 331
Velez, M.A., 238, 239, 248
Vogt, N.D., 319, 320, 347, 364, 366
Vulnerability, 27, 39, 277, 284, 287, 292, 293, 367, 396

W

WWII, 20, 319, 320, 344, 351–359, 366, 393

X

Xavante indians, 29–49

Y

York, A.M., 147, 149, 212, 299, 301, 304, 306, 310, 311

Z

Zoning, 149, 212, 216, 296–304, 306–314, 351–356, 396
Zoonotic pathogens, 80, 86

MIX
Papier aus verantwortungsvollen Quellen
Paper from responsible sources
FSC® C105338

If you have any concerns about our products,
you can contact us on
ProductSafety@springernature.com

In case Publisher is established outside the EU,
the EU authorized representative is:
Springer Nature Customer Service Center GmbH
Europaplatz 3, 69115 Heidelberg, Germany

Printed by Libri Plureos GmbH
in Hamburg, Germany